Thermodynamics
for
Engineers

Thermodynamics
for
Engineers

B. V. Karlekar

Professor of Mechanical Engineering
Head of Department of Mechanical Engineering
Rochester Institute of Technology
Rochester, NY

PRENTICE-HALL, INC., Englewood Cliffs, NJ 07632

Library of Congress Cataloging in Publication Data

KARLEKAR, BHALCHANDRA V. (date)
  Thermodynamics for engineers.

  Bibliography: p.
  Includes index.
  1. Thermodynamics. I. Title.
TJ265.K29   536'.7   82-456
ISBN 0-13-914986-4   AACR2

Editorial/production supervision by Karen Skrable
Manufacturing buyers: Joyce Levatino and Anthony Caruso
Cover design by 20/20 Services, Inc.
Cover photo courtesy of Ingersoll-Rand

Printed in the United States of America

10 9 8 7 6 5 4 3 2

ISBN 0-13-914986-4

Prentice-Hall International, Inc., *London*
Prentice-Hall of Australia Pty. Limited, *Sydney*
Prentice-Hall Canada Inc., *Toronto*
Prentice-Hall of India Private Limited, *New Delhi*
Prentice-Hall of Japan, Inc., *Tokyo*
Prentice-Hall of Southeast Asia Pte. Ltd., *Singapore*
Whitehall Books Limited, *Wellington, New Zealand*

In memory of my father,

Vasudeo Trimbak Karlekar

Contents

PREFACE   xiii

NOMENCLATURE   xvii

1 BASIC PROPERTIES AND CONCEPTS   1

    1.1 Energy and Thermodynamics   1
    1.2 Volume   5
    1.3 Pressure   6
    1.4 Temperature   16
    1.5 Ideal Gas Law   20
    1.6 Some Thermodynamic Definitions   27
    1.7 Thermodynamic Processes   29
    1.8 Summary   35
    Problems   35

2 WORK AND HEAT   39

    2.1 Work   39
    2.2 Thermodynamic Work   40
    2.3 Work in a Compressible System   48
    2.4 Useful and Nonuseful Work   57
    2.5 Heat   62
    2.6 Summary   65
    Problems   67

3   THE FIRST LAW OF THERMODYNAMICS—
       CLOSED SYSTEMS   72

3.1 Introduction   72
3.2 Energy of a System   73
3.3 The First Law of Thermodynamics   73
3.4 Internal Energy   81
3.5 Adiabatic Process   89
3.6 Cyclic Process   100
3.7 Summary   110
       Problems   112

4   THERMODYNAMIC PROPERTIES OF PURE SUBSTANCES
       AND EQUATIONS OF STATE   117

4.1 Introduction   117
4.2 Liquid and Vapor Phases   118
4.3 The First Law and Liquid and Vapor Phases   131
4.4 Solid and Liquid Phases   143
4.5 Coefficients of Isothermal Compressibility and Isobaric Expansion   149
4.6 The Compressibility Factor   151
4.7 Equations of State   156
4.8 Summary   160
       Problems   161

5   OPEN SYSTEMS AND THE FIRST LAW   164

5.1   Introduction   164
5.2   Control Volume   164
5.3   Conservation of Mass and Control Volume   168
5.4   Flow Work and Transfer of Energy Across Control Surface   171
5.5   Energy Equation for a Control Volume   174
5.6   Applications of the Steady-Flow Energy Equation   178
5.7   Throttle Valve and Joule-Thomson Coefficient   192
5.8   Further Applications of the Energy Equation   196
5.9   Energy Equation for Unsteady Flow Processes   207
5.10 Summary   209
       Problems   211

6 THE SECOND LAW OF THERMODYNAMICS
AND ENTROPY   218

6.1   Introduction   218
6.2   Reversible and Irreversible Processes   219
6.3   The Second Law of Thermodynamics   223
6.4   Clausius Inequality   230
6.5   Entropy   233
6.6   Entropy Change for Ideal Gas   249
6.7   Entropy Change for a Control Volume   257
6.8   Available Work   261
6.9   Availability   277
6.10 Summary   280
       Problems   281

7 ENERGY CONVERSION—GAS CYCLES   286

7.1   Introduction   286
7.2   Nomenclature for a Reciprocating Device   287
7.3   Reciprocating Compressor   288
7.4   Gas Power Cycles—Otto Cycle   297
7.5   The Diesel Cycle   309
7.6   The Dual Cycle   317
7.7   The Stirling Cycle   318
7.8   The Ericsson Cycle   322
7.9   The Atkinson Cycle   322
7.10 The Brayton Cycle   323
7.11 Summary   341
       Problems   341

8 ENERGY CONVERSION—VAPOR CYCLES   348

8.1   Introduction   348
8.2   Mollier Diagram   349
8.3   The Rankine Cycle   350
8.4   Refrigeration   389
8.5   Summary   409
       Problems   410

9   GENERAL THERMODYNAMIC RELATIONS   417

9.1 Introduction   417
9.2 Partial Derivatives   418
9.3 Maxwell Relations   421
9.4 The Clapeyron Equation   424
9.5 Thermodynamic Relations for Internal Energy, Enthalpy, and Entropy   427
9.6 Relations for c_p and c_v 430
9.7 Joule-Thomson Coefficient   434
9.8 Summary   435
       Problems   436

10 NONREACTING MIXTURES   438

10.1 Introduction   438
10.2 Mixtures of Ideal Gases   439
10.3 Properties of Moist Air   446
10.4 Air Conditioning   457
10.5 Chemical Potential and Gibbs Function   459
10.6 Solutions   461
10.7 Summary   465
        Problems   466

11 CHEMICAL REACTIONS AND EQUILIBRIUM   471

11.1 Introduction   471
11.2 Fuels   472
11.3 The Combustion Process   474
11.4 Chemical Reaction and the First Law   481
11.5 The Third Law of Thermodynamics and Absolute Entropy   495
11.6 Chemical Reactions and the Second Law   498
11.7 Equilibrium   502
11.8 Summary   513
        Problems   514

BIBLIOGRAPHY   518

ANSWERS TO SELECTED PROBLEMS   522

APPENDICES   525

APPENDIX A   PROPERTIES OF SOME IDEAL GASES AT 300 K   526
APPENDIX B1   PROPERTIES OF COMPRESSED LIQUID WATER   527
APPENDIX B2.1   PROPERTIES OF SATURATED STEAM
                    (TEMPERATURE TABLE)   529
APPENDIX B2.2   PROPERTIES OF SATURATED STEAM
                    (PRESSURE TABLE)   532
APPENDIX B3   PROPERTIES OF SUPERHEATED STEAM   535
APPENDIX B4   PROPERTIES OF WATER IN THE SATURATED
                    SOLID-VAPOR REGION   543
APPENDIX C1   PROPERTIES OF FREON-12
                    (DICHLORODIFLUOROMETHANE)   544
APPENDIX C2   PROPERTIES OF SUPERHEATED FREON-12   546
APPENDIX D1   PROPERTIES OF SATURATED AMMONIA   550
APPENDIX D2   PROPERTIES OF SUPERHEATED AMMONIA   552
APPENDIX E   PROPERTIES OF SATURATED MERCURY   554
APPENDIX F1   PROPERTIES OF SATURATED NITROGEN   556
APPENDIX F2   PROPERTIES OF SUPERHEATED NITROGEN   557
APPENDIX G1   PROPERTIES OF SATURATED OXYGEN   559
APPENDIX G2   PROPERTIES OF SUPERHEATED OXYGEN   560
APPENDIX H   CONSTANT-PRESSURE SPECIFIC HEATS OF
                    VARIOUS IDEAL GASES   562
APPENDIX I1   ENTHALPY OF FORMATION AT 25°C, IDEAL
                    GAS ENTHALPY AND ABSOLUTE ENTROPY
                    AT 100 kPa PRESSURE   563
APPENDIX I2   PROPERTIES OF AIR AT LOW PRESSURE   569
APPENDIX J1(a)   TEMPERATURE-ENTROPY DIAGRAM FOR H_2O 572
APPENDIX J1(b)   MOLLIER DIAGRAM FOR STEAM   573
APPENDIX J2   THE PRESSURE-ENTHALPY DIAGRAM FOR
                    FREON-12   574
APPENDIX K   PSYCHROMETRIC CHART   575
APPENDIX L   CRITICAL CONSTANTS OF SEVERAL SUBSTANCES   576
APPENDIX M   LOGARITHMS TO THE BASE e OF THE EQUILIBRIUM
                    CONSTANT K 577
APPENDIX N   CONVERSION FACTORS   578

INDEX   579

Preface

Thermodynamics is an important subject in engineering curricula and is generally a prerequisite for courses in fluid mechanics and heat transfer. This text on classical thermodynamics is a student-oriented text in which a rigorous and comprehensive treatment of the subject is presented. It is recognized that the subject matter has been well developed over the years and that there is relatively little new information that can be added in a fundamental course. The text is designed for a first-level under-graduate course in thermodynamics, and it can be used for a single course given in a quarter or a semester system or as a two-course sequence. Although the text is designed for an engineering student, a technology student with some calculus background should find the text material relevant.

It is the author's experience that a typical student is awed by the prospect of going through a course in thermodynamics. This is partly due to the horror stories told by contemporaries who have already taken the course and discovered that a multitude of thermodynamic variables are introduced at a rapid pace and that the applications are generally deferred until after the second law of thermodynamics is covered. Therefore, it has been one of the objectives in writing this text to introduce the thermodynamic variables at a carefully structured pace and to present applications generously throughout the text. Since the observations of phenomena have led to the thermodynamic concepts and laws, as far as possible the presentation of a concept is preceded by an everyday example.

In the first chapter, after giving a brief historical background of the subject of thermodynamics, volume, pressure, and temperature are introduced. These properties are capable of being measured with relative ease and are known to the student. After presenting the ideal gas law and working out a few examples, some thermodynamic definitions (such as equilibrium, quasistatic process, and cyclic process) are given. Chapter 2 is devoted to a discussion of work and heat. Also discussed in this chapter are useful work, nonuseful work, and lost work. The mathematical background on exact and inexact differentials and on path functions and point functions is presented as notes at the end of the chapter. In Chapter 3, the first law of thermodynamics is presented, and internal energy and enthalpy are introduced. Once again, the relevant mathematics is presented in the form of notes at the end of the chapter. After defining an adiabatic process and the thermal efficiency of a cyclic process, a first-law analysis of the Carnot and the Otto cycles is presented. Chapter 4 is devoted to the thermodynamic properties of pure substances and to the equations of state. To make the discussion of the p-v-T behavior of the solid, liquid, and vapor phases meaningful, water is used as a vehicle for the presentation. The equations of state and the generalized compressibility chart are presented at this point so that the student can see these in the overall perspective of the general behavior of a substance. The application of the first law to open systems forms the subject matter of Chapter 5. Control volume, conservation of mass, flow work, throttling process, and enthalpy are discussed in this chapter. Numerous examples of flow processes are also presented. The chapter is concluded with a discussion of a gas turbine power plant and of a heat pump. The second law of thermodynamics, entropy, reversibility, entropy change, available work, availability, and irreversibility are discussed in Chapter 6. Every attempt is made to maintain clarity and logic in explaining these topics with which a student is often uncomfortable.

The air standard cycles for energy conversion are discussed in Chapter 7. At the beginning of the chapter the nomenclature for a reciprocating device is introduced, and the reciprocating compressor is discussed. This is a relatively simple device operating on a two-stroke cycle and using a gas as a working medium. It is hoped that the student will be at ease in studying applications if he or she is first introduced to a relatively simple system such as a reciprocating compressor. Among the major cycles discussed in this chapter are the Otto, the Diesel, and the Brayton cycles, ideal and actual. The vapor cycles for power and refrigeration are presented in Chapter 8. In addition to covering the Rankine cycle with all the usual modifications and the vapor compression refrigeration system—including multistage vapor compression—liquefaction of gases and absorption refrigeration are also discussed. General thermodynamic relations form the subject matter of Chapter 9. Thermodynamics of nonreacting mixtures and solutions, and an introduction to air conditioning constitute the topics for Chapter 10. Finally, Chapter 11 includes a discussion of reacting mixtures and of related topics such as air-fuel ratio, enthalpy of formation, adiabatic flame temperature, explosion pressure, extent of reaction, chemical equilibrium, and equilibrium constant. Also presented in this chapter are the third law of thermodynamics and absolute entropy.

It is believed that a certain degree of repetition is desirable for effective teaching. Consequently, the student will find repetition of some concepts and topics. These include, among others, the Carnot cycle, the Otto cycle, the Brayton cycle, thermal efficiency, enthalpy, and the combined first and second laws.

All illustrative examples in the text have been worked out in great detail. The solution to an example is, typically, categorized into *given, objective, diagram, assumptions, relevant physics, analysis,* and *comments.* It is hoped that with this approach the student will gain some insight into solving a new problem.

An increasing number of schools and universities are using the SI (Système Internationale) units. Keeping this trend in mind, the entire text is written using SI units.

If the text is to be used for a one-term course, the author suggests the following material for the course:

Chapter 1:   All sections.
Chapter 2:   Sections 2-1, 2-2, 2-3, 2-5, and 2-6.
Chapter 3:   All sections.
Chapter 4:   Sections 4-1 to 4-4, 4-6, 4-7, and 4-8.
Chapter 5:   Sections 5-1 to 5-8, and 5-10.
Chapter 6:   Sections 6-1 to 6-6, and 6-10.
Chapter 7:   Sections 7-1 to 7-4, 7-10, and 7-11.
Chapter 8:   Sections 8-1 to 8-4, and 8-5.

The author wishes to thank Professor E.M. Sparrow of the Mechanical Engineering Department of the University of Minnesota for his meticulous review of the manuscript during its preparation. The exchange of ideas that the author has had with his friend, Professor Robert Ellson, was instrumental in arriving at the sequence of topics presented in this text. Thanks are also due to Professor Satish Kandlikar, Mr. Michael Lints, and Mr. Mark McVea for their numerous suggestions.

It is a pleasure to acknowledge Mrs. Margaret Urckfitz, Mrs. Marjorie Sawyer, Mr. Gary Compagna and Mr. Michael Lints of the Rochester Institute of Technology, and Mr. Anant P. Oka and Mr. Shirish P. Mulay for their efforts in assisting with the typing of the manuscript and the preparation of the figures.

Last, but not least, I appreciate the patience and understanding shown and the encouragement given by my wife, Lata, during the preparation of this text; and the patience and understanding shown by my daughter, Mohana, and son, Saagar, during the long evening hours of the writing that kept me away from them.

B. V. KARLEKAR

Nomenclature

A	area	c	specific heat
AF	air fuel ratio	c_p	specific heat, constant pressure
AW	available work	c_v	specific heat, constant volume
C	Celsius	$\bar{c}_p$	specific heat, constant pressure per mole
D	diameter	$\bar{c}_v$	specific heat, constant volume per mole
E	energy	d	diameter
G	Gibbs Function	e	specific energy
H	enthalpy	g	gravitational acceleration, specific Gibbs function (per unit mass)
I	Irreversibility		
K	Kelvin	$\bar{g}$	specific Gibbs function (per mole)
K_p	equilibrium constant	h	specific enthalpy (per unit mass)
L	length, stroke	$\bar{h}$	specific enthalpy (per mole)
M	molar mass (molecular weight)	$\bar{h}_f$	enthalpy of formation (per mole)
N	rpm	l	length
N_C	number of cylinders	m	mass
P	power	n	number of moles, number of thermodynamic cycles
Q	heat energy		
R	gas constant	p	pressure
$\bar{R}$	universal gas constant	q	heat energy
S	entropy	r	radius
T	temperature in Kelvin	s	specific entropy (per unit mass)
U	internal energy	$\bar{s}$	specific entropy (per mole)
V	volume	t	temperature in Celsius
$V\!\!\!/$	velocity	t_d	dew point temperature
W	work, humidity ratio	t^*	wet-bulb temperature
Z	compressibility factor, elevation	u	specific internal energy (per unit mass)
a	specific Helm-holtz function	$\bar{u}$	specific internal energy (per mole)
a	area	v	specific volume (per unit mass)

$\bar{v}$	specific volume (per mole)	κ	coefficient of isothermal compressibility
w	work per unit mass	μ	chemical potential, degree of saturation
w_s	shaft work	ν	stoichiometric coefficient, Joule Thomson
x	Cartesian coordinate, mole fraction		coefficient
y	Cartesian coordinate, mass fraction	τ	time
z	Cartesian coordinate	Φ	availability of a system
α	coefficient of linear expansion	φ	relative humidity
β	coefficient of isothermal expansion	λ	extent of a chemical reaction
γ	ratio of specific heats at constant pressure	σ	surface tension force
	and constant volume respectively	ψ	availability of a unit mass in an open
δ	a small quantity		system
η	efficiency	$\dot{\Psi}$	time rate of availability of an open system
Δ	change in a quantity		

Thermodynamics
for
Engineers

-1-

Basic Properties and Concepts

1.1 ENERGY AND THERMODYNAMICS

The word *energy* is based on a combination of two Greek words, which mean capacity and work. A dictionary often defines energy as capacity to do work. Thermodynamicists postulate that energy is a fundamental property of a substance. We learn from physics and chemistry that thermal, electrical, chemical, potential, kinetic, and nuclear energy are some of the forms of energy. The science of thermodynamics deals with the conversion of energy from one form into another. Thermodynamics is also concerned with various physical and chemical properties of substances and the changes of these properties when energy is transferred to or from the substance.

Historically, man has had to use his physical energy, a mechanical form of energy, to hunt for food and to build a shelter. When he discovered fire, a release of chemical energy, he used it to keep himself warm and to cook food. In the course of the development of civilization, when he needed more than his own physical energy, he used other men and animals to build structures and to plough fields (3000 B.C.). He invented the water wheel, which converts potential and kinetic energies into mechanical energy, around 100 B.C. He harnessed wind energy first for sailing and then for windmills by 900 A.D. The first steam engine, converting thermal energy into mechanical energy, was developed in the year 1711. The development of the steam engine marked the beginning of the industrial society in the Western world. Electric generators converting mechanical energy into electrical energy and electric motors converting electrical energy into mechanical energy were in practical use by 1831. Einstein proved that mass can be converted into energy ($e = mc^2$) in 1905, and the atomic bomb was exploded in 1945.

Currently, research work is being conducted on several energy conversion devices, such as, fuel cells that convert chemical energy directly into electrical energy, nuclear fusion, photo cells, etc. Also under investigation are several other processes that will efficiently utilize energy from tidal waves, wind, the sun, coal, and high-temperature, high-pressure water beneath the crust of the earth. The science of thermodynamics plays an important role in all such processes. It is interesting to note that although the steam engine—the first energy conversion device to have a profound impact on mankind—was well developed by the early part of the eighteenth century, the laws of thermodynamics were not clearly understood until about 100 years later.

It has been known for a long time that the temperature of a body could be changed by contact with a hotter or a colder body. Since the body's temperature changed, it was thought that something must have been gained or lost by the body. The concept of *caloric* was introduced around the 1750s to explain such a gain or loss. Caloric was considered to be a substance with neither mass nor volume that could flow between bodies with different temperatures. A body at a high temperature was thought to possess a high caloric content. In the late 1700s Count Rumford discovered that boring solid metal submerged in water caused the water to boil. Even though no substance of a higher caloric content was brought in contact with the water, the water temperature rose until it reached the boiling point. Rumford then concluded that there was no flow of any caloric into the water, but that the friction between the boring tool and the metal was responsible for the "heating" of the metal and of the water. Count Rumford's observation had a profound impact on the scientific thinking of the time. Shortly after Rumford's experiment, James Joule conclusively demonstrated that mechanical work can be used to produce a heating effect.

It was known from physics that potential energy and kinetic energy are mutually convertible. For example, when a pendulum is oscillating, the potential energy is highest and the kinetic energy is lowest when the bob of the pendulum is at its highest point; the converse is true when the bob is at its lowest point. It was also known that passing an electric current through a resistor caused a heating effect on the resistor. Finally, chemists had observed that certain chemical reactions resulted in a rise in the temperature of the chemicals involved in the reactions (exothermic reactions).

The concept of energy, along with its conservation and convertibility, evolved in an effort to present a unified approach for the analysis of these apparently unrelated phenomena. It is now known that in the absence of nuclear reactions, energy is conserved and that energy cannot be created or destroyed. This represents the first law of thermodynamics. Also, any form of energy can be completely converted into thermal energy or heat energy, but thermal energy cannot be completely converted into other forms of energy. This observation is the basis of the second law of thermodynamics. The second law determines the upper limit of the efficiency of energy conversion devices. The laws of thermodynamics play an important role in the design of a wide range of devices, such as automotive engines, air conditioners, rocket motors, high-power lasers, refineries, and windmills.

Thermodynamics is studied by physicists, chemists, and engineers. Physicists and chemists are concerned with basic laws, properties of substances, and changes in the properties caused by interaction of different forms of energy. Engineers are interested not only in all these aspects, but also in the application of thermodynamic principles to the design of machines that will convert energy from one form into another. Electrical engineers are often interested in the conversion of mechanical energy into electrical energy. Mechanical engineers are frequently concerned with the design of a system that will most efficiently convert thermal or heat energy into mechanical energy.

In the modern industrial society it is the mechanical form of energy that is ultimately responsible for performing a large number of household chores, as well as industrial tasks. Some examples of the former are washing clothes in a washing machine, vacuuming carpets with a vacuum cleaner, and beating eggs with an egg-beater. Industrial tasks are of a diverse nature; they include rolling a steel ingot into different shapes, such as flats, channels, or wheels for railroad cars, digging deep trenches for laying a pipeline, excavating for minerals, and propelling ships on waterways and automobiles on roadways. Often the mechanical energy is obtained from thermal energy at the site of use, as in the case of automobiles, tractors, and aircraft. Mechanical energy is produced on large scale in central thermal power plants. The mechanical energy obtained in such plants is further converted into electrical energy and the electrical energy is transmitted over long distances. Finally the energy is used in an electrical, a mechanical, or a thermal form.

The conversion of energy from one form into another is of importance in the modern life. As noted previously, energy conversion is the domain of thermodynamics. Basic to the study of thermodynamics are certain definitions and concepts, properties of substances and changes thereof due to energy transfer processes, the principle of conservation of energy, and the consequences of the observation that thermal energy can be transferred from a body at a high temperature to a body at a low temperature without the aid of any device. In this chapter we shall discuss units, basic thermo-dynamic properties and concepts, and the behavior of an ideal gas.

1.1.1 Units

There are many different systems of units in use. Traditionally physicists have used the CGS system, in which the units of length, mass, and time are centimeter, gram, and second, respectively. In the United States the engineering profession has used the English system of units. In this system the unit of length is the foot (ft), the unit of mass is the pound (lb), and the unit of time is the second (s). The pound is also the unit of force in the English system, the unit being defined as the gravitational force on 1 pound mass at sea level. To avoid confusion between pound mass and pound force, a unit of mass—the *slug*—is also used. One slug equals 32.174 pounds mass.

Most of the industrialized world has been using the metric system of units. In 1970 the Systéme International, usually abbreviated SI, was accepted as the system of units for scientific and engineering work. The SI units are gradually being introduced in U.S. industries, and it is expected that a changeover from English units to SI units will be completed in the near future. We shall use only SI units in this text. The units for length, time, mass, and force in the SI system are meter (m), second (s), kilogram (kg), and newton (N), respectively. We note that

$$1 \text{ newton} = (1 \text{ kilogram}) \cdot (1 \text{ meter/second}^2)$$

Weight is the gravitational force on a body and, therefore, has the same units as force. Since, at sea level, the gravitational acceleration is 9.81 m/s², the weight of a kilogram mass at sea level on earth is 9.81 newtons. The weight of the same mass on moon is (9.81/6) or 1.63 newtons since the gravitational acceleration on moon is one-sixth the value on earth. The units for pressure, energy, and power are pascal (Pa), joule (J), and watt (W), respectively.

$$1 \text{ pascal} = 1 \text{ newton/meter}^2$$

$$1 \text{ joule} = 1 \text{ newton-meter}$$

$$1 \text{ watt} = 1 \text{ joule/second}$$

In engineering applications, the joule and pascal often prove to be rather small, as we frequently encounter several thousand joules or pascals. In such cases—and particularly in the tables of properties—we shall use kilojoule, kilopascal, and megapascal as the additional units of energy and pressure. We note that

$$1 \text{ kilojoule} = 1000 \text{ joules} = 1 \text{ kJ}$$

$$1 \text{ kilopascal} = 1000 \text{ pascals} = 1 \text{ kPa}$$

$$1 \text{ megapascal} = 1,000,000 \text{ pascals} = 1 \text{ MPa}$$

1.1.2 Microscopic and Macroscopic Approaches

There are two viewpoints for dealing with the subject of thermodynamics: *macroscopic* and *microscopic*. In the first approach, also known as the *classical approach*, the facts that matter is made up of molecules and that the molecules have motions are completely ignored. When a body is subjected to transfer of energy or other thermodynamic processes, attention is focused on the behavior of the body as a whole. This approach is mathematically rather simple, and permits us to analyze different types of systems that are encountered in industrial processes in a straightforward manner. The microscopic approach inquires into the motion of molecules, assumes certain mathematical models for the molecular behavior, and draws conclusions regarding the behavior of a substance. This approach is mathematically complex. We shall use the classical approach in this text.

1.1.3 Thermodynamic Properties

A *thermodynamic property* of a substance is a characteristic or an attribute that can be used to describe the substance. When we describe a person to a friend who does not know the person, we talk of such features as the person's height, weight, complexion, hair, glasses, or dress. In so doing we hope to paint a vivid picture of the person so that our friend can identify this person if he or she happens to see the person. Thermodynamic properties serve exactly the same purpose. Thermodynamic properties are quantified and measured. For describing a substance, we naturally use those properties that are easy to measure. Also, we must use enough properties to describe a substance so that it is uniquely identified.

It is useful to define two classes of properties, *extensive properties* and *intensive properties*. An extensive property of a substance is proportional to the mass of the substance. Because the volume of a substance is directly proportional to the mass of the substance, volume is an extensive property. Likewise, the weight of a substance is an extensive property. An intensive property of a substance is independent of the mass of the substance. The pressure of a substance in a given situation is the same throughout the substance regardless of the quantity of the mass that we examine; therefore pressure is an intensive property. The temperature of a substance is also an intensive property, for a similar reason. Other intensive properties are density, elevation, and velocity. As various properties are introduced, we shall point out whether they are extensive or intensive.

1.2 VOLUME

There are three basic thermodynamic properties that are commonly used to describe a substance: volume, pressure, and temperature. The *volume*, V, is the physical space occupied by a body. The body itself can be in a solid, liquid, or gaseous state. The volume can be regular, such as that of a cube, cylinder, or sphere; or it can be irregular, such as the volume of a person or the volume of air in a balloon. The volume of a body is proportional to the mass of the body, and therefore volume is an extensive property. It is useful to define *specific volume*, v, which is the volume per unit mass. If a body of volume V has a mass m, then the specific volume is

$$v = \frac{V}{m} \tag{1.1}$$

and the unit is m^3/kg. Since the density, ρ, of a substance is defined as its mass per unit volume, kg/m^3, we conclude that density and specific volume are reciprocals of each other; that is,

$$v = \frac{1}{\rho} \tag{1.1a}$$

When we talk about volume or specific volume, we are usually dealing with dimensions that are of the order of several billion times the molecular size. In other words, even if we are dealing with a very small volume, it is large enough to accommodate a very large number of molecules. Under this condition the facts that molecules are in constant motion and that the intermolecular distances are large compared to the molecular dimensions do not obscure our measurement of volume.

1.3 PRESSURE

The concept of pressure is connected with the idea of force. You are already familiar with the interaction of forces with a body through your study of statics. In statics we encounter a force acting at a point or a force distributed over an area. An example of the first is a weight suspended by a rope [Figure 1.1(a)], while an example of the latter is a number of sand bags supported by a slab [Figure 1.1(b)]. When a force is uniformly distributed over an area, it is customary to define a load intensity equal to force per unit area (N/m²).

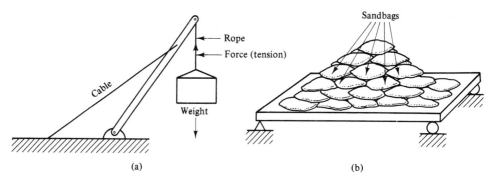

Figure 1.1
(a) Force acting at a point.
(b) Distributed force.

When we are dealing with a liquid or a gas, which is very often the case in engineering thermodynamics, the concept of force per unit area has a somewhat different meaning. Consider a fluid confined to a leakproof, frictionless piston-cylinder arrangement as shown in Figure 1.2(a). Let the area of the piston face be A. Let a force of magnitude $\mathfrak{F}$ act in the positive x-direction on the piston. A free-body diagram for a portion of the fluid adjacent to the piston is shown in Figure 1.2(b). Because there is no friction, the force exerted by the fluid in volume $BCCB$ on the fluid of length ℓ in a volume $BDDB$ must be numerically equal to $\mathfrak{F}$ and in the negative x-direction. Let the magnitude of the force exerted by the fluid be F. Contributions to F are made by all the fluid particles adjacent to the layer BB. If we divide F by the cross-sectional area of the fluid at BB (which is the same as the area A of the piston face), we have the quantity F/A, which has the units of load intensity. This quantity is called *fluid pressure* and is denoted by the symbol p. Pressure is the force exerted by a fluid on a unit area. The pressure of a fluid can be uniform or nonuniform. If it is nonuniform, then the pressure p is defined as the limit of the ratio, F/A, as A goes to a very small dimension; that is,

$$p = \lim_{A \to 0} \frac{F}{A} \tag{1.2}$$

We do not permit A to become zero since we know that matter is made of molecules, intermolecular spaces exist, and molecules are not stacked tightly.

6

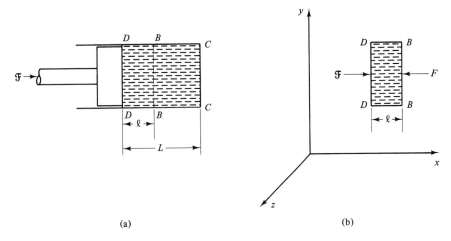

Figure 1.2
(a) Force acting on the piston of a piston-cylinder system.
(b) Free-body diagram for the fluid in (a).

A qualitative explanation of pressure can be given on a molecular basis. Molecules of a gas are in constant random motion. They have velocities of varying magnitudes. Also, the number of molecules in a unit volume is very large; under ordinary conditions in 1 mm³ there are more than a billion molecules. As a first approximation, the molecules are assumed to behave as perfect elastic spheres. With the molecules in constant random motion, there are collisions among the molecules themselves, and between the molecules and the container walls. The walls experience a continuous bombardment of molecules and a whole series of impulses due to the changes in the momenta of the molecules hitting the walls. An elemental area on one of the walls feels only an overall integrated or average effect of these impulses. We call this effect *pressure.*

Since the pressure of a substance does not depend on its mass, pressure is an intensive property. The unit of pressure in SI units is the pascal (Pa) or newtons per square meter (N/m²). We shall also use kilopascal (kPa) and megapascal (MPa) as units of pressure. Sometimes the unit bar is used for pressure. One bar equals 100 kPa.

Referring to Figure 1.2, we can draw a free-body diagram for a portion of the fluid whose length is greater than ℓ and then make a force analysis. We should then find that the fluid pressure has the same value as before. For a stationary fluid, the pressure has the same value at those points in the body of the fluid that have the same elevation. The fluid pressure p always acts normal to a surface. The pressure at a point in a fluid has the same value regardless of the orientations of all the areas that can be drawn through the point. This can be proved in the following manner.

Consider a small element of fluid consisting of one-half the volume of a rectangular parallelopiped of size $(dx \cdot dy \cdot 1)$, which is shown in Figure 1.3(a). If we ignore shear forces and consider the force equilibrium of this element, we can write

$$\Sigma F_x = 0$$

where the left side represents the sum of the forces in the x-direction. On the left face of the element, there is pressure p_x and a force of magnitude $(p_x \, dy \cdot 1)$ acting in positive x-direction. On the inclined face there is pressure p and a force of magnitude

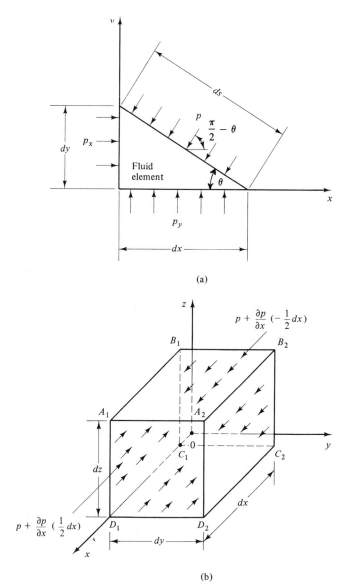

(a)

(b)

Figure 1.3
(a) Pressure distribution on a fluid element.
(b) Pressure and body forces on a fluid element.

$(p \, ds \cdot 1)$ in a direction normal to ds. The component of this force along the positive x-direction is $-(p \, ds \cdot 1) \sin \theta$. Thus we have

$$+p_x \, dy \cdot 1 - (p \, ds \cdot 1) \sin \theta = 0$$

Since $dy = ds \sin \theta$, the above becomes

$$p_x = p$$

Also, we can now consider the equilibrium of the y-components of forces. These are due to the fluid pressure and the weight of the fluid. Proceeding as before, we obtain

$$+p_y \, dx \cdot 1 - (p \, ds \cdot 1) \cos \theta - (\tfrac{1}{2}) \cdot dx \cdot dy \cdot 1 \cdot \rho g = 0$$

In the limit, the product $dx \, dy$ approaches zero faster than does dx or $ds \cos \theta$ (which equals dx). Thus we obtain

$$p_y = p$$

or

$$p_x = p_y = p$$

Since the orientation of the inclined face of the element was arbitrary, we have shown that the pressure of a fluid at a point is the same in all directions. This result is of great importance in analyzing fluid systems.

Now let us consider the equilibrium of a stationary fluid element $A_1B_1C_1D_1$-$A_2B_2C_2D_2$, shown in Figure 1.3(b), under the influence of body forces and pressure forces. In most engineering applications, the body force consists of the gravitational force, or the weight, of a body. Referring to Figure 1.3(b), this force acts in the negative z-direction. For the pictured fluid element $A_1B_1C_1D_1A_2B_2C_2D_2$, whose sides are dx, dy, and dz, the magnitude of the body force is $\rho g \, dx \, dy \, dz$.

The pressure p can change as we move away from the origin in the x-, y-, and z-directions. That is, p is a function of three variables x, y, and z. This is sometimes expressed by the phrase *pressure field* in a body of fluid. Let $p(x, y, z)$ be the value of pressure at the origin. As we move a distance $(-dx/2)$ and arrive at the center of face $B_1C_1C_2B_2$, the pressure at this point is given by

$$p + \frac{\partial p}{\partial x}\left(-\frac{1}{2} \, dx\right)$$

and we may assume that this is the average pressure on face $B_1C_1C_2B_2$. Likewise

$$p + \frac{\partial p}{\partial x}\left(\frac{1}{2} \, dx\right)$$

is the average pressure on face $A_1D_1D_2A_2$. Similar expressions can readily be developed for the average pressures on each of the remaining faces of the fluid element.

For the static equilibrium of the fluid element, the vector sum of the external forces on the element must be zero.

$$\sum \mathbf{F}_{\text{surface}} + \sum \mathbf{F}_{\text{body}} = 0$$

In the above equation, the forces are expressed in vector form. Noting that the surface forces for a stationary fluid are due to pressure alone and the body forces are due only to the gravity in the negative z-direction, we can write the equilibrium equation in a component form as follows:

$$\Sigma F_{x,\,pressure} = 0$$
$$\Sigma F_{y,\,pressure} = 0$$
$$\Sigma F_{z,\,pressure} + \Sigma F_{z,\,gravity} = 0$$

The force due to pressure in a given direction equals the product of the pressure in that direction and the area normal to the direction. For the x-direction, we can write

$$\Sigma F_{x,\,pressure} = \underbrace{\left(p - \frac{1}{2}\frac{\partial p}{\partial x}dx\right)dy\,dz}_{\text{face } B_1C_1C_2B_2} + \underbrace{\left[-\left(p + \frac{1}{2}\frac{\partial p}{\partial x}dx\right)dy\,dz\right]}_{\text{face } A_1D_1D_2A_2} = 0$$

or

$$\frac{\partial p}{\partial x} = 0$$

Similarly, for the y-direction,

$$\Sigma F_{y,\,pressure} = \underbrace{\left(p - \frac{1}{2}\frac{\partial p}{\partial y}dy\right)dz\,dx}_{\text{face } A_1B_1C_1D_1} + \underbrace{\left[-\left(p + \frac{1}{2}\frac{\partial p}{\partial y}dy\right)dz\,dx\right]}_{\text{face } A_2B_2C_2D_2} = 0$$

or

$$\frac{\partial p}{\partial y} = 0$$

For the z-component of forces, we have

$$\Sigma F_{z,\,pressure} + \Sigma F_{z,\,body\;force} =$$

$$\underbrace{\left(p - \frac{1}{2}\frac{\partial p}{\partial z}dz\right)dx\,dy}_{\text{face } C_1D_1D_2C_2} + \underbrace{\left[-\left(p + \frac{1}{2}\frac{\partial p}{\partial z}dz\right)dx\,dy\right]}_{\text{face } B_1A_1A_2B_2} + \underbrace{(-\rho g\,dx\,dy\,dz)}_{\text{body force}} = 0$$

or

$$\frac{\partial p}{\partial z} + \rho g = 0$$

The above results tell us that for a stationary fluid, the pressure is a function of that direction in which the gravity acts. With our nomenclature, this means that pressure is a function of the z-direction only, and it is independent of the x- and y-directions. Consequently we can drop the partial derivative sign and write

$$\frac{dp}{dz} + \rho g = 0 \tag{1.3}$$

Atmospheric Pressure.   The air around us can be treated as a homogeneous gas, that is, as if it were a single component; in reality it is a mixture of oxygen, nitrogen, and traces of other gases. The surface of the earth is covered by a layer of air, which we call the *atmosphere*. This layer is usually taken to be 160 km thick. The mass of this air exerts a force. The pressure due to this force is called *atmospheric* *pressure*. At sea level it has a value of 101,325 N/m², or 101.325 kPa. As we go up in elevation the atmospheric pressure decreases.

Evangelista Torricelli (1608–1647) discovered that atmospheric pressure can be measured by the height of a column of mercury in a tube, as shown in Figure 1.4. The tube is initially filled with mercury. The open end of the tube is covered, the tube is inverted in a trough of mercury, and the cover is removed. The level of mercury in the tube drops initially and then stabilizes. It is often assumed that there is no substance above point C, in Figure 1.4, and there is perfect vacuum, so that pressure at C is zero. Strictly speaking, this is not true because there will be some mercury vapor at a vapor pressure corresponding to the temperature (see Chapter 4).

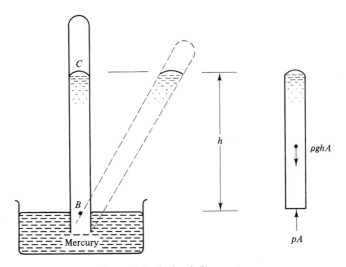

Figure 1.4 A simple barometer.

Consider the atmospheric pressure p_a on the free surface of the trough and the pressure at point B in the tube containing mercury. Since the pressure is independent of x- and y-coordinates and the elevation of the free surface (z-coordinate) and that of point B are identical, the pressure at B is exactly equal to that at the free surface. But the pressure at the free surface is the same as the atmospheric pressure, p_a. Therefore pressure at B equals p_a. Also, rewriting Equation 1.3, we have

$$dp = -\rho g\, dz$$

Integrating between points B and C, we obtain

$$\int_B^C dp = -\int_{z=0}^{z=h} \rho g\, dz$$

or

$$p_C - p_B = -\rho g(h - 0)$$

Noting that $p_C = 0$ and $p_B = p_a$, we have

$$p_a = \rho g h \tag{1.3a}$$

The above result can also be obtained by considering the free-body diagram in Figure 1.4. The vertical forces acting are the weight of the column of mercury and the pressure force. If p is the density of the mercury, g is the acceleration due to gravity, h is the height of the column of mercury above the free surface of mercury in the trough, and A is the cross-sectional area of the tube, then the weight of the mercury is $pghA$. The pressure force is p_A. Thus if p_a is the atmospheric pressure, a force balance for the y-direction gives

$$p_a A + (-pghA) = 0$$

or

$$p_a = pgh \tag{1.3a}$$

If the tube in Figure 1.4 is tilted, we find that the *vertical* height of the mercury is still h, consistent with Equation 1.3.

The device shown in Figure 1.4 is a simple barometer and it measures atmospheric pressure. A typical statement, such as the barometer is at 760 mm of mercury, means that the height h of the column of mercury in the tube is 760 mm. If the fluid were water, the length of the tube required would be about 10.3 m. Often pressure is expressed in millimeters of mercury or in atmospheres (atm), where 1 atm is 760 mm of mercury or 101,325 Pa.

Sometimes pressure is stated as a gauge pressure. The *gauge pressure* p_g is the excess of the absolute pressure p over the local atmospheric pressure p_a, or

$$p_g = p - p_a \tag{1.4}$$

A negative value for the gauge pressure is called a *vacuum reading* and is usually expressed in millimeters of mercury. For example, an absolute pressure of 21 kPa represents -80 kPa gauge pressure, or a vacuum of 608 mm of mercury. A pressure that is below atmospheric pressure is sometimes expressed in torr, after Torricelli, where 1 torr is equal to 1 mm of mercury. A pressure expressed in torr is absolute pressure. The basic instruments used for measuring fluid pressure are a manometer and a pressure gauge.

Manometer.   When the pressure of a fluid is to be measured in a laboratory, a manometer is often used. A typical U-tube manometer is shown in Figure 1.5. The left leg of the tube is connected to a pipe carrying a fluid. The pressure of the fluid at point E is to be measured. The lower portion of the U-tube contains a manometer fluid, such as mercury or colored water. The right leg of the manometer is open to atmosphere. We observe that since points E and D are at the same horizontal level, the pressures at these two points are identical. Therefore it will suffice to determine the pressure, p_D, at point D. Also, the pressure at C equals the pressure at B, since C and B are at the same horizontal level. The pressure at C, p_C, is due to the pressure p_D in the pipe and the weight of a fluid of height h_ℓ in the left leg of the manometer. If p_ℓ is the density of the pipe fluid that occupies the left leg, we can write from Equation 1.3a

$$p_C = p_D + p_\ell g h_\ell$$

where g is the gravitational acceleration.

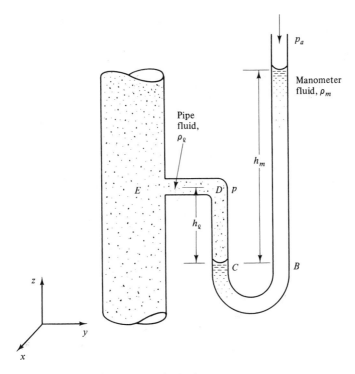

Figure 1.5 A U-tube manometer.

The pressure at B equals the sum of the atmospheric pressure p_a and the pressure due to the weight of the manometer fluid above point B in the right leg, or

$$p_B = p_a + \rho_m g h_m$$

where ρ_m and h_m are the density and the height of the manometer fluid in the right leg above point B, respectively.

We note that for equilibrium of forces,

$$p_C A = p_B A$$

where A is the cross-sectional area of the manometer tube. Substituting for p_C and p_B gives

$$p_D + \rho_\ell g h_\ell = p_a + \rho_m g h_m$$

or

$$p_D = p_a + (\rho_m g h_m - \rho_\ell g h_\ell)$$

The above equation can be used to determine the unknown pressure of the fluid in the pipe at point C. If it is known that the pressure p differs from p_a by only a small amount, then the manometer fluid is usually water. If the difference between p and p_a is large, then mercury is used as the manometer fluid.

It should be noted that Figure 1.5 implies that the pressure in the pipe is greater than the atmospheric pressure p_a, pushing the manometer fluid in the right leg of the manometer up. If the pipe pressure were less than p_a, the level of the manometer fluid in the left leg would be higher than that in the right leg. Manometer configurations vary significantly, depending on their applications. It is, therefore, advisable to develop a relation for the desired pressure from first principles.

   Pressure Gauge.   The primary element of a pressure gauge consists of a tube bent in the form of an arc. The cross section of the tube is usually elliptic. One end of the tube is closed and the other is connected to the fluid whose pressure is to be measured (Figure 1.6). As the pressurized fluid enters the bent tube, the resulting forces cause the closed end of the tube to move outward. The movement of the closed end of the tube is transmitted to a pointer through a suitable mechanism. Using known pressures, the sweep of the pointer is calibrated to give the fluid pressure being measured. The reading on the dial of such a pressure gauge indicates gauge pressure. Local atmospheric pressure must be added to this value to obtain the absolute pressure of the fluid.

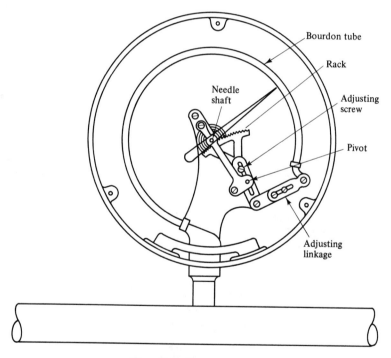

Figure 1.6   Pressure gauge.

Example 1.1

   An inclined tube manometer is used to measure gas pressure in a pipe, as shown in Figure 1.7. The manometer fluid is water. Determine the gas pressure in the pipe if the atmospheric pressure is 1.01×10^5 Pa.

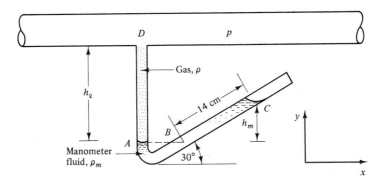

Figure 1.7

Solution:

Given: We know that the fluid in the pipe—and therefore the fluid in the upper left leg of the manometer—is gas. Also, the fluid in the manometer is liquid water, whose properties are available. Additionally, the atmospheric pressure p_a is given to be 1.01×10^5 Pa.

Objective: The gas pressure p at point D is to be calculated.

Assumptions: Densities of the gas and water are uniform. Since the ratio of the two densities is considerably less than unity, the ratio may be neglected to simplify the analysis and computations.

Relevant physics: Pressure is a function of the gravitational acceleration g and elevation z (Equation 1.3(a)).

Analysis: The pressure at A is due to the pressure p of the gas in the pipe and the pressure due to the weight of the gas of height h_ℓ in the left leg of the manometer. Denoting the density of the gas by ρ, we have

$$p_A = p + \rho g h_\ell$$

The pressure at B is due to the atmospheric pressure p_a and the pressure due to the weight of the manometer fluid of vertical height h_m. If ρ_m is the density of the manometer fluid,

$$p_B = p_a + \rho_m g h_m$$

Pressures at A and B are equal, because their z-coordinates are identical and the manometer fluid is stationary; we have

$$p + \rho g h_\ell = p_a + \rho_m g h_m$$

A rearrangement gives

$$p = p_a + \rho_m g \left(h_m - \left(\frac{\rho}{\rho_m} \right) h_\ell \right)$$

The ratio (ρ/ρ_m) is of the order of 0.001 for a gas in the pipe and water in the manometer. Therefore we can write the following equation for p, after neglecting the second term in brackets.

$$p = p_a + \rho_m g h_m$$

Observing that the vertical height $h_m = (\frac{14}{100}) \sin 30°$ and substituting for p, g, and p_a, we have

$$p = 1.01 \times 10^5 + 1000 \times 9.81 \times (\frac{14}{100}) \sin 30°$$
$$= 1.017 \times 10^5 \text{ Pa}$$

1.4 TEMPERATURE

A commonly encountered interpretation of temperature is that it is a measure of the degree of hotness or the degree of coolness. Words like red-hot, hot, lukewarm, and freezing cold serve to express the level of temperature in a qualitative way. Temperature and heat exchange are intimately related. It is a common experience that a hot plate, when turned off after cooking, becomes cool if sufficient time is allowed to pass. We say that heat energy has been transferred from the hot plate to the surrounding air. This transfer is possible because the degrees of hotness of the hot plate and the air are different.

In most cases when heat energy is supplied to a body, its level of hotness increases. Heat energy also causes a change in some of the characteristics of a body. For example, a long, slender metal rod when heated experiences a measurable change in its length. When an electric current is passed through a thin platinum wire, the wire gets warm and its electrical resistance changes. Mercury in a capillary tube undergoes a pronounced expansion when heated. A thermocouple junction is formed when two dissimilar metal wires are connected; when the junction is heated, an electromotive force develops. In practice, such changes in properties are used to construct temperature-measuring devices, such as a mercury-in-glass thermometer or iron-constantan thermocouple.

1.4.1 Zeroth Law of Thermodynamics

All temperature-measuring devices that use a direct contact with the body whose temperature is to be measured have one thing in common. It is assumed that the temperature-measuring device, when brought in contact with a body whose temperature is to be measured, will assume the temperature of the body after sufficient time has elapsed. Furthermore, if the device shows the same reading when brought in contact with another body, then the temperatures of the two bodies are considered to be identical. This is known as the *zeroth law of thermodynamics*. It can be restated as follows: If body A is in thermal contact with body B so that heat exchange can take place and thermal equilibrium eventually results and if body A is also in thermal equilibrium with body C, then we say that bodies B and C are in thermal equilibrium. When a body is in contact with another body through a common boundary for a sufficiently long time, no further change occurs after some initial change in the properties of the bodies. The bodies are then said to be in *thermal equilibrium*. If the two bodies are initially in thermal equilibrium, then no change will occur in their properties. At this point we note that thermal equilibrium between two bodies implies uniformity of temperature at the common boundary of the two bodies. The zeroth law forms the basis for temperature-measuring devices, some of which will now be discussed.

1.4.2 Mercury-in-Glass Thermometer

The most common type of temperature measuring device is the mercury-in-glass thermometer. In such a thermometer, a small glass bulb holds mercury and the glass bulb is connected to a closed capillary tube. The thermometer is placed in a reference state, such as the melting point of ice at sea level, and the position of mercury is marked on the capillary. Next, the thermometer is placed in another reference state, such as boiling water at sea level, and the new position of the mercury is marked. These two reference states are called the ice point and the steam point, respectively. Strictly speaking, the *ice point* is defined as the temperature of a mixture of ice and water that is in equilibrium with saturated air at a pressure of 1 atm. The *steam point* is likewise defined as the temperature of water and steam that are in equilibrium at a pressure of 1 atm.

   The distance between these two marks is the expansion experienced by mercury as the thermometer was moved from the ice point to the steam point. If this distance is divided into 100 equal parts, we have the *Celsius* system of temperature measurement, formerly called the centigrade scale. In 1954 the scale was renamed the Celsius scale after the astronomer who devised it.

   The method of creating a temperature scale in the manner described above depends on two reference points and also on the medium used, namely, mercury. The accuracy depends on such variables as the uniformity of the cross section of the capillary, properties of mercury and glass, and the surrounding temperature. The constant-volume gas thermometer, described next, requires only one reference state and does not depend upon the medium used.

1.4.3 Constant-Volume Gas Thermometer

The constant-volume gas thermometer consists of a glass bulb with a bent neck, a three-way valve, one bent glass tube, a rubber tube, a straight glass tube, and a scale. These components are assembled to form apparatus such as the one shown in Figure 1.8(a). One leg of the three-way valve is connected to the glass bulb, the second leg is connected to the bent tube, and the third leg is open to the atmosphere. First, the valve is turned in such a way that the atmospheric air can traverse path *C–V–B* and enter the bulb. Then the valve is turned so that air is trapped in the bulb and the air in the bent tube can escape. Mercury is then poured through the open end of the straight tube on the right side of the apparatus until the level of mercury reaches *AA*. The bent tube has a mark at *A*. At this point the valve is turned so that path *B–V–D* is open and paths *C–V–B* and *D–V–C* are closed.

   The bulb is then placed in a mixture of ice and water. The air in the bulb contracts and the level of mercury in the left leg rises. By moving the right leg up or down, the level of mercury in the left leg can be brought back to mark A so that the volume of the trapped air in the bulb is restored to its original value. The pressure of the air in the bulb at ice point is then recorded as p_i.

Next, the glass bulb is placed in boiling water. After sufficient time has elapsed, the mercury level in the left side falls below A as the warm air in the bulb pushes the mercury down. The tube on the right side is again moved until mercury rises to level A in the left tube. Once again we have the same volume of air in the bulb as was there before the bulb was placed in boiling water; hence the name constant-volume gas thermometer. Let the pressure under this condition be p_s, the pressure at boiling point or steam point. Note that both p_s and p_i are absolute pressures. We can now calculate the ratio p_s/p_i corresponding to the ice point pressure p_i.

Next, we proceed to determine a few more values of the ratio p_s/p_i. This is done as follows. By connecting leg C of the three-way valve to a vacuum pump and operating the pump for a short time, we can remove some of the air from the bulb. Then we can repeat the experiment to obtain a new set of values for p_s and p_i. Corresponding to the new value of p_i, we can calculate the ratio p_s/p_i. Typically this new ratio will be smaller than the previous one. A whole series of such points can be obtained and plotted, as shown in Figure 1.8(b).

If the experiment is repeated by replacing the air in the bulb by some other gas, another curve is obtained. Extrapolation of all such curves to $p_i = 0$ results in a unique value of the ordinate. The limiting value of the ordinate is found to be

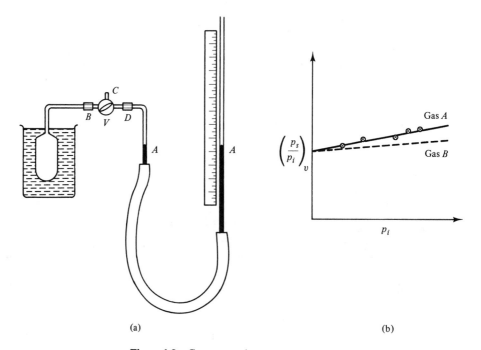

(a)   (b)

Figure 1.8   Constant volume gas thermometer.

$$\lim_{p_i \to 0} \left(\frac{p_s}{p_i}\right)_V = 1.36609$$

The subscript V in the above ratio $(p_s/p_i)_V$ denotes constant volume. we define a temperature scale such that

$$\frac{T_s}{T_i} = 1.36609$$

Also, a value of 100 is now assigned to the difference between T_s and T_i so that

$$T_s - T_i = 100$$

Solving these two equations, we obtain

$$T_s = 373.15 \quad \text{and} \quad T_i = 273.15$$

This temperature scale is called the *absolute temperature scale* and is referred to as the Kelvin scale. Temperature on the Kelvin scale is designated by K, without a degree symbol.

The temperature absolute zero provides an excellent reference level. When the Celsius scale is used, absolute zero has a value of $-273.15°C$, and we have

$$K = °C + 273.15 \tag{1.5}$$

In 1954, the Tenth General Conference of Weights and Measurement redefined the temperature scale in terms of a single fixed point instead of defining the difference between the steam point and ice point temperatures to be 100. The fixed point used is the triple point of water, which is the temperature at which ice, liquid water, and water vapor coexist in equilibrium. A value of $0.01°C$ is assigned to the temperature at triple point. The steam point on this scale is found to be $100.00°C$. A revised International Practical Temperature Scale was adopted by the International Committee on Weights and Measurements in 1968. This scale utilizes a number of primary fixed and readily reproducible points and specific formulas for interpolating temperatures between these fixed points. These fixed points are based on the properties of hydrogen, neon, oxygen, water, zinc, silver, and gold. The primary fixed points used in the International Practical Temperature Scale are listed below in degrees Kelvin.

1. Triple point of equilibrium—hydrogen   13.81
2. Boiling point of equilibrium—hydrogen at 33.33 kPa   17.042
3. Boiling point at 1 atm pressure of equilibrium —hydrogen   20.28
4. Boiling point of neon at 1 atm   27.102
5. Triple point of oxygen   54.361
6. Boiling point of oxygen at 1 atm   90.188
7. Triple point of water   273.16
8. Boiling point of water at 1 atm   373.15
9. Freezing point of zinc at 1 atm   692.73
10. Freezing point of silver at 1 atm   1235.08
11. Freezing point of gold at 1 atm   1377.58

Thus far we have examined three basic properties, volume, pressure, and temperature, in some detail. Next we consider a relationship among these for a gaseous substance.

1.5 IDEAL GAS LAW

In thermodynamics we are concerned with various substances and their properties. Substances can exist in three phases: solid, liquid, and gaseous. Solids and liquids usually exhibit very small changes in volume when they are subjected to large changes in pressure or temperature. Such is not the case with gases. The volume of a gas is sensitive to changes in pressure, as well as to changes in temperature. A complete discussion of the properties of solids, liquids, and gases is presented in Chapter 4. However, in this section, we shall consider the behavior only of gases, since we intend to use them as a vehicle to discuss the interaction of work and heat energy. Gaseous substances are used more often than other substances in the first few chapters of this text since they facilitate visualization of some of the processes in which energy interaction occurs. Also, large quantities of mechanical work can be obtained from a gaseous substance by subjecting it to suitable processes.

It is important to be able to distinguish between two samples of the same gas. This can be done if we know the specific volume, the pressure, and the temperature of the gas in each sample. If the respective values of these properties are identical for two samples, then we say that both the samples have the same *thermodynamic state* or condition. If we use the specific volume, which is the ratio of the volume to the mass of the gas in the sample, we have three parameters that can be used to identify the state of a gas: pressure, specific volume, and temperature. The parameters that are used to identify the thermodynamic state of a substance are called *thermodynamic*, or *state*, variables. For a gas, these are pressure, p, specific volume, v, and temperature, T. If we examine the behavior of a gas we find that once any two state variables (p and v, v and T, or T and p) are fixed, the remaining third variable (T, p, or v) is automatically fixed by nature. This is equivalent to saying that a functional relationship exists among the three variables p, v, and T; that is,

$$f(p, v, T) = 0$$

Such a relationship is called *equation of state*.

Many different forms have been proposed for this functional relationship. Any proposed relation must be verified experimentally. One of the simplest relations in use is

$$pv = RT \tag{1.6}$$

or

$$pV = mRT \tag{1.6a}$$

where R is a constant and the definition of specific volume v has been used. This equation gives reasonably good results when pressures are low and temperatures are high. Equation 1.6 is called the *ideal gas law*, or the *perfect gas equation*. The quantity R in Equation 1.6 is known as the *gas constant*, and it has different values for different gases. The units for R are joules per kilogram-degree Kelvin. Values of R for some gases are listed in Table 1-1. The table also gives value of R for air. Although air is a mixture of several gases and water vapor, it can be treated as a homogeneous gas as if it had only a single component or constituent.

TABLE 1.1 R-VALUES FOR GASES

Gas	Gas Constant, R J/kg K	Molar Mass, M kg/kg-mol
Carbon dioxide, CO_2	188.92	44.01
Carbon monoxide, CO	296.83	28.01
Helium, He	2007.03	4.0003
Hydrogen, H_2	4124.18	2.016
Methane, CH_4	518.35	16.04
Nitrogen, N_2	296.80	28.013
Oxygen, O_2	259.83	31.999
Air	287.00	28.97
Steam	461.52	18.015
Ammonia	488.20	17.03

We know from chemistry that the *atomic weight* of an element signifies the ratio of the weighted average mass (considering isotopes) of one atom of the element to one-twelfth the mass of carbon atom. The atomic weight is a number without dimensions. The *gram atomic weight* of an element is the weight in grams that contains the same number of atoms as twelve grams of carbon-12. The *molecular weight* of a compound expresses how heavy a molecule of the compound is compared to one-twelfth the weight of a carbon atom. A mole of a compound contains Avogadro's number of molecules, Avogadro's number being 6.023×10^{26} per kilogram mole. The mass of one mole of a compound is called *molar mass* and its units are kilograms per kilogram-mole (kg/kg-mol). Some authors call molar mass molecular weight and do not associate any units with the latter. We prefer to use the phrase molar mass to molecular weight because weight implies force whereas we are concerned with mass in the thermodynamic context.

Since the mass m of a substance equals the product of its molar mass (molecular weight) M and its number of moles n, we can write

$$m = nM \tag{1.7}$$

Since Equation 1-7 is applicable to any substance, we can substitute for m from Equation 1-7 in Equation 1-6(a) to obtain

$$pV = nMRT$$

It turns out that the quantity MR is constant regardless of the gas under consideration. It is therefore called the *universal gas constant* $\bar{R}$. Its value is

$$\bar{R} = 8314.3 \text{ J/kg-mol K} \tag{1.8}$$

Equations 1.6 and 1.6(a) can now be rewritten as

$$p\bar{v} = \bar{R}T \tag{1.9}$$

and

$$pV = n\bar{R}T \tag{1.9a}$$

where $\bar{v}$ is the volume per mole of the gas, or the *molal volume*.

Figure 1.9 shows a plot of the quantity pv/RT against pressure p, with tempera-

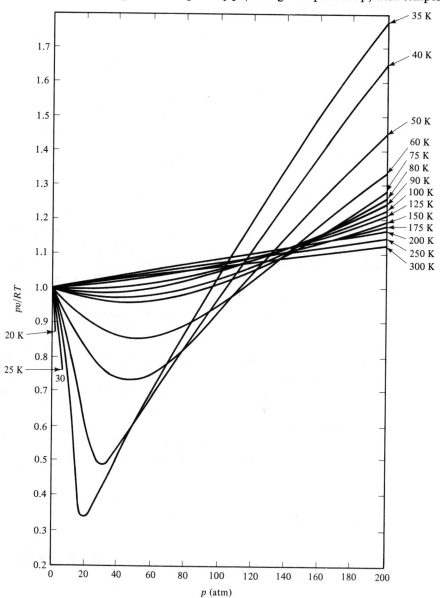

Figure 1.9 A plot of pv/RT against pressure for nitrogen. (A. S. Friedman, 1950. Used with permission.)

ture T as a parameter for nitrogen. The graph is typical of many gases. If Equation 1.6 were to hold true for all values of temperature and pressure, we would see a horizontal line with an ordinate value of unity and parallel to the p-axis. Instead, we see a number of curves of different shapes. However, all of these curves merge as pressure approaches zero. This means that for low pressures, the behavior of a gas can be approximated by Equation 1.6. Also, the figure shows that as the temperature goes up, the deviation from the ideal gas law is less pronounced, allowing us to represent the behavior of a gas by Equation 1.6.

The ideal gas law, Equation 1.6, thus provides us with a simple relation among the state variables of a gas that is satisfactory for low pressures and high temperatures. More complex relations for the behavior of gases, which have better accuracy than the ideal gas law, will be discussed in Chapter 4.

p-v-T Surface of an Ideal Gas.   We recall that the pressure, volume, and temperature of an ideal gas are related by the equation

$$pv = RT \tag{1.6}$$

If we choose p, v, and T as the coordinate axes, analogous to the x-, y-, and z-axes in analytical geometry, we find that Equation 1.6 represents a three-dimensional surface, which is shown in Figure 1.10. The p-v-T-surface in Figure 1.10 represents the equation $pv = RT$, just as a spherical surface represents the equation $x^2 + y^2 + z^2 = 1$. A point on the surface represents a possible state of the ideal gas.

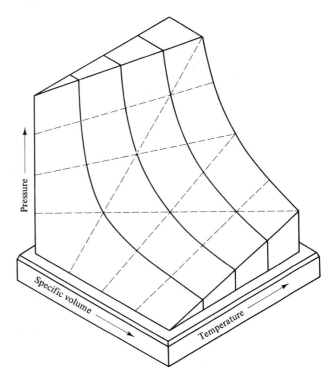

Figure 1.10   The p-v-T surface of an ideal gas.

Example 1.2

A 2.5-m³ tank contains nitrogen gas at a pressure of 100 kPa and a temperature of 30°C. What is the mass of the nitrogen in the tank?

Solution:

Given: The pressure, volume, and temperature of the nitrogen gas in a tank are specified. The pressure is relatively low.

Objective: The mass of the nitrogen is to be determined.

Diagram:

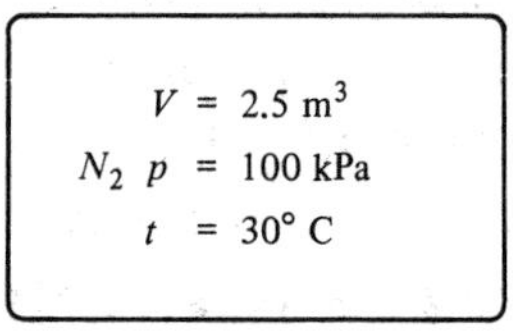

$$
\begin{array}{l}
V = 2.5 \text{ m}^3 \\
N_2\ p\ = 100 \text{ kPa} \\
t\ = 30° \text{ C}
\end{array}
$$

Figure 1.11

Assumptions: The nitrogen obeys the ideal gas law. We have to assume that the properties are uniform throughout the body of the gas.

Relevant physics: The application of the ideal gas equation, Equation 1.6(a), to a sample of gas requires that the properties entering the equation be uniform throughout the sample. Also, it requires a knowledge of the gas constant. Further, the temperature has to be expressed in absolute scale, that is, in degrees Kelvin.

Analysis: We use the ideal gas law to determine the mass of the gas since the pressure of the gas is low (see Figure 1.9). When Equation 1.6(a) is rearranged, we can write

$$ m = \frac{pV}{RT} $$

From Table 1.1,

$$ R = 296.80 \text{ J/kg K} \qquad (\text{for N}_2) $$

Also

$$ T = 273.15 + 30 = 303.15 \text{ K} $$

Substitution gives

$$ m = \frac{100 \times 10^3 \times 2.5}{(296.80 \times 303.15)} $$

$$ = 2.78 \text{ kg} $$

Example 1.3

Let the temperature of the air surrounding the tank in Example 1.1 be raised and maintained at 150°C. What is the new pressure in the tank?

Solution:

Given: The initial pressure, temperature, and volume of the nitrogen gas in a tank are specified. The initial pressure is relatively low. The tank is now in a high temperature environment.

Objective: We wish to calculate the final gas pressure.

Diagram:

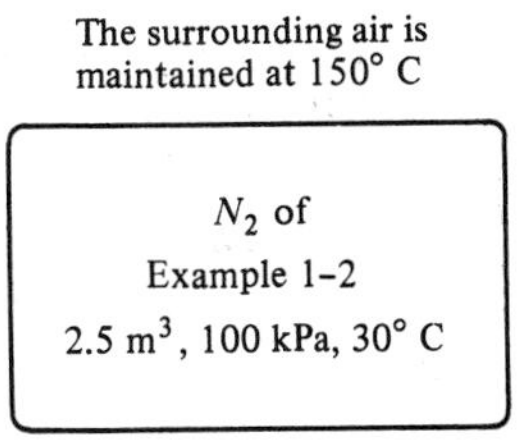

The surrounding air is
maintained at 150° C

N_2 of

Example 1-2

2.5 m³, 100 kPa, 30° C

Figure 1.12

Assumptions: The ideal gas law is applicable and the properties are uniform at the
initial and final states. The gas tank wall permits heat transfer between the
surrounding air and the gas in the tank.

Relevant physics: Since the walls are assumed to permit heat transfer, heat energy will
be transferred from the high-temperature surroundings to the low-temperature
gas in the tank. As a result of this heat transfer, the temperature of the gas will
rise until it equals the surroundings temperature, namely 150°C or 423.15 K.

Analysis: The solution involves identifying the initial and final thermodynamic states
of the gas. Since the gas does not leave the tank and the tank is rigid, both the
mass m and the volume V of the gas are constant and are the same as before.
We may write Equation 1.6(a) for the initial and final states.

$$\text{Initial state:} \quad p_1 V = mRT_1$$

$$\text{Final state:} \quad p_2 V = mRT_2$$

Dividing the second equation by the first, we obtain

$$\frac{p_2}{p_1} = \frac{T_2}{T_1}$$

or

$$p_2 = \frac{p_1 T_2}{T_1}$$

Substitution gives

$$p_2 = 100 \times 10^3 \times \frac{423.15}{303.15} = 139.6 \text{ kPa}$$

Comments: If the tank wall did not permit heat transfer, the temperature of the gas in
the tank would remain unchanged and so would the gas pressure.

Example 1.4

A 2-m³ tank contains oxygen at 50 kPa and at a temperature of 60°C. Another tank
of the same volume contains oxygen at 30 kPa and at a temperature of 25°C. The two
tanks are connected and pressures in the two tanks are allowed to equalize. After a
sufficiently long time, the temperature of oxygen in both the tanks is observed to be
20°C. What is the final pressure in the tanks?

Solution:

Given: Two tanks of identical volumes contain oxygen. The volumes, pressures, and
temperatures of both tanks are specified. The oxygen gas is initially in two
separate tanks. Then the tanks are connected so that the oxygen is effectively in

a single large tank whose volume equals the sum of the volumes of the two original tanks. The final temperature of this oxygen is prescribed.

Objective: The final pressure of the oxygen gas residing in two connecting tanks is to be determined.

Diagram:

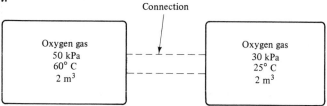

Connection

| Oxygen gas 50 kPa 60° C 2 m³ | Oxygen gas 30 kPa 25° C 2 m³ |

The final oxygen temperature after connecting the tanks is 20° C.

Figure 1.13

Assumptions: The ideal gas law (Equation 1.6(a)) holds since the pressures are low. The final properties are uniform throughout both tanks.

Relevant physics: Heat transfer from the tanks to the surroundings must occur, as the final temperature is less than the initial temperatures in the two tanks. The mass of oxygen is conserved; that is, it does not change. We shall need the gas constant for oxygen and we have to express the temperatures in degrees Kelvin.

Analysis: We observe that in the final state the oxygen in the two tanks is at a uniform pressure p_0 (unknown) and at a temperature 20°C. Further, the total volume occupied by the gas is $2 + 2 = 4$ m³. If we can calculate the total mass of the oxygen, then we are in a position to compute the unknown pressure by employing the ideal gas law.

The total mass of the gas must be equal to the sum of the masses m_1 and m_2 of the gas initially in tanks 1 and 2. Since the initial pressures, volumes, and temperatures of oxygen in the two tanks are known, we can calculate m_1 and m_2. Denoting the initial conditions in tanks 1 and 2 by, subscripts 1 and 2, we can write Equation 1.6(a) for the oxygen in the two tanks.

$$p_1 V_1 = m_1 R T_1 \quad \text{and} \quad p_2 V_2 = m_2 R T_2$$

or

$$m_1 = \frac{p_1 V_1}{R T_1} \quad \text{and} \quad m_2 = \frac{p_2 V_2}{R T_2}$$

We note from Table 1.1 that for oxygen

$$R = 259.83 \text{ J/kg K}$$

Also

$$p_1 = 50 \text{ kPa} \qquad\qquad p_2 = 30 \text{ kPa}$$

$$V_1 = 2 \text{ m}^3 \qquad\qquad V_2 = 2 \text{ m}^3$$

$$T_1 = (273.15 + 60) \text{ K} \qquad T_2 = (273.15 + 25) \text{ K}$$

Substitution gives

$$m_1 = \frac{50 \times 10^3 \times 2}{259.83 \times (273.15 + 60)} \quad \text{and} \quad m_2 = \frac{30 \times 10^3 \times 2}{259.83 \times (273.15 + 25)}$$

or
$$m_1 = 1.16 \text{ kg} \quad \text{and} \quad m_2 = 0.78 \text{ kg}$$

The final conditions are designated by subscript zero. In the final state, after the two tanks are connected and pressure and temperature are allowed to stabilize, we have
$$p_0 V_0 = m_0 R T_0$$

Also
$$V_0 = V_1 + V_2 \quad \text{and} \quad m_0 = m_1 + m_2$$

so that
$$p_0(V_1 + V_2) = (m_1 + m_2) R T_0$$

or
$$p_0 = \frac{m_1 + m_2}{V_1 + V_2} R T_0$$

Substitution gives
$$p_0 = \frac{1.16 + 0.78}{2 + 2} \times 259.83 \times (273.15 + 20) = 36.9 \text{ kPa}$$

1.6 SOME THERMODYNAMIC DEFINITIONS

In Example 1.4 we first analyzed the oxygen in tank 1 and then the oxygen in tank 2. Finally, we analyzed the combined masses. In thermodynamic analysis it is useful to identify that which we wish to analyze. The concept of *a system* is useful in this context. A system is anything that we wish to analyze thermodynamically. In Example 1.4, there were three systems: the oxygen in tank 1, the oxygen in tank 2, and the combined masses.

We distinguish the contents of a system from things outside the system by visualizing a *boundary* of the system that separates the two. The boundary may be real as the walls of the oxygen tank of Example 1.3, imaginary, or a combination of both. To illustrate this point we consider a fluid in a container placed on a hot plate and heated by electrical energy (Figure 1.14(a)). If we wish to analyze the fluid alone, then the fluid is our system and the system boundary is as shown in Figure 1.14(b). Notice that part of the boundary is real (walls of the container) and part is imaginary (separating the fluid from the atmosphere). If we wish to examine the fluid and the container, then these two become our system and the system boundary is as shown in Figure 1.14(c). Figure 1.14(d) shows that the fluid, the container, and the hot plate are all part of yet another system. Note that in this case the system boundary cuts the electric power cord, which is perfectly acceptable. When we proceed to analyze such a system we may want to know what comes through that cord at the point where is enters the system.

A system is considered to be a *closed system* if the mass within the system remains constant; that is, there is no mass crossing the system boundary. The system of the water in the container is a closed system so long as no water vapor escapes from the system boundary. In this and the next three chapters, we shall deal only with closed

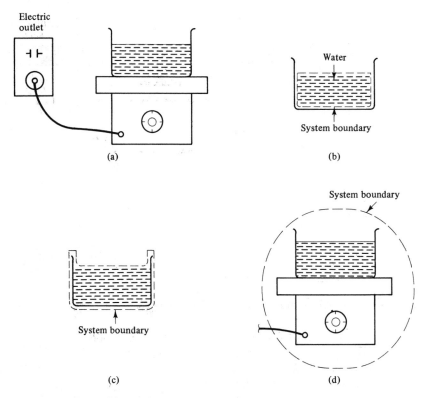

Figure 1.14 System and system boundary.

systems. If the mass within a system does not remain constant or if there is some mass crossing the system boundary, then the system is called an *open system*. One example of an open system is a bath tub being filled or emptied. Another example is an automobile engine which takes in fuel and air and discharges exhaust gases. A number of open systems will be analyzed in Chapter 5.

Whatever is outside the boundary of a system is called *surroundings*. When we talk of the surroundings of a system, we refer to the surroundings that are in the immediate vicinity of the system and have a capacity to interact with the system. The objects that are far away from the system and unable to interact with the system are not of concern. The phrase *local surroundings*, is sometimes used to emphasize the restricted nature of the surroundings of a system. Thus for the system in Figure 1.14(b), the pot, the hot plate, and the source of electric power are part of the surroundings; for the system in Figure 1.14(c), the hot plate and the source for electric power are part of the surroundings.

The combination of a system and its surroundings is called the *universe*. The universe as defined here is of limited scope, because it includes only the system and its local surroundings. Again referring to Figure 1.14, the fluid, the pot, the hot plate, and the power source are all part of the universe.

If we deal with an elastic body as a system, we consider such relevant properties as its Young's modulus, Poisson ratio, physical dimensions, and temperature distribution. On the other hand, if we examine an electric battery as our system, we need to know something about the total electric charge in the battery, the ion concentration in the sulphuric acid of the battery, the e.m.f. that it can generate, and its internal resistance. To look into the power that can be generated by letting water from an elevated lake drive a hydraulic turbine, we should consider the pressure and velocity of the water at the turbine inlet and outlet and the quantity of water that can flow through the turbine on a unit time basis. In considering the air-conditioning requirement for a room, a large number of parameters enter the picture: temperature and humidity of the air inside and outside the system (the room), the number of people in the system and the nature of their activities, the number and type of electric and gas applicances in the room, and requirements for ventilation are a few of these parameters. It is hoped that enumeration of these examples helps you appreciate the need to know about all pertinent details of the system to be analyzed.

When we specify pressure p and temperature T of a gaseous system, it is implied that no matter at what point in the system the pressure is measured, we shall find the same values of pressure and temperature. If this implication is valid, then we have *thermodynamic equilibrium*. Thermodynamic equilibrium requires that all properties be uniform and homogeneous throughout the system, with no variations or gradients present. This requires that the system not be disturbed on any account. Thermodynamic equilibrium requires that there be no unbalanced forces between a system and its surroundings and no spontaneous changes in internal structure or chemical reactions should occur. The air in a room heated by a baseboard heater is not in a thermodynamic equilibrium because the temperature of the air changes as we move from floor to ceiling. Air blown into a small balloon and examined after some time will essentially be in thermodynamic equilibrium if the thermodynamic conditions of the surroundings do not change with time. As another example, imagine a leakproof piston cylinder arrangement wherein a gas is enclosed and the force on the piston is balanced by the force due to the pressure of the gas. After sufficient time has elapsed, this gas is in thermodynamic equilibrium. If the force on the piston is abruptly removed, we find that at any given moment the pressure and temperature have non-uniform values within the gaseous system and that these values change with time for some period of time. The gas in the cylinder is no longer in thermodynamic equilibrium; that is, the gas is in a state of nonequilibrium. The concept of thermodynamic equilibrium is important. In most of the systems that we shall study, it will be assumed that there is thermodynamic equilibrium. Also, a specification of a set of properties for a system has meaning only if the system is in thermodynamic equilibrium.

1.7 THERMODYNAMIC PROCESSES

In Examples 1.2 through 1.4, we determined the final state of a gas given initial values of the three properties pressure, volume, and temperature and final values of any two of the three properties. It was implied there that the gaseous system under consideration was in thermodynamic equilibrium at both the initial and the final states. Such states of equilibrium can be represented on a p-V, or pressure-volume, diagram. A p-V diagram for a system graphically shows the pressure-volume history of the system.

Let us assume that we have a gaseous system and it goes from an initial equilibrium state i to a final equilibrium state f (Figure 1.15). This assumption on our part appears to contradict the requirements of thermodynamic equilibrium as discussed in the preceding section. We circumvent this contradiction by saying that the system is disturbed infinitesimally and the resulting change in the properties is infinitesimal. Under these conditions we do not have a genuine thermodynamic equilibrium, but we have something very close to it. We call such a state *quasi-equilibrium*, distinct from states of equilibrium or nonequilibrium. Reverting to our system we find that it goes through a whole series of intermediate states of quasiequilibrium as it passes from state i to state f. These states are shown on a p-V diagram in Figure 1.15(a). We note that the change in pressure and volume as the system goes from one state to the next is extremely small, in fact infinitesimal. Under these circumstances we can draw a smooth curve connecting points i and f on the diagram and passing through all the intermediate points. This is shown in Figure 1.15(b). We now have a *quasi-equilibrium*, or *quasistatic thermodynamic process*, taking the gaseous system from state i to state f. In defining a quasistatic process, we are not concerned with what causes the change in the properties of a system or how long it takes to bring about this change. The time interval between the initial and the final states may be a second, an hour, or a year. It is irrelevant to the analysis of the system. Quasistatic process is an idealization and no real process follows the definition of a quasistatic process. This idealization in thermodynamics is analogous to the concept of a point mass or a frictionless drive in mechanics.

The following thought experiment illustrates how a quasistatic process might be accomplished. Consider a gas (p_1, V_1, T_1) confined in a piston-cylinder arrange-

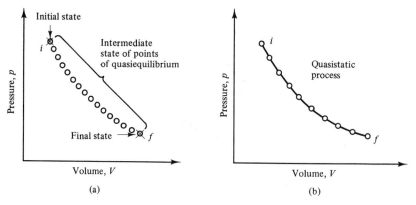

Figure 1.15
(a) Thermodynamic states on a p-V diagram.
(b) Thermodynamic process on a p-V diagram.

ment, where the piston is frictionless and weightless (Figure 1.16). The force due to the pressure of the gas on the piston is balanced by suitable weights on the piston. Let these weights be in the form of a large number of thin disks, each of a very small mass. If we remove one disk from these weights, we disturb the force equilibrium. There results a net upward force, and an upward motion of the remaining disks ensues. As soon as this happens, the volume of the gas effectively increases and its pressure and temperature are perturbed. If nothing else is done, some heat transfer will take place between the surroundings and the gas, due to the temperature difference created. Eventually the temperature of the gas equals that of the surroundings. Also, a new value of gas pressure is established.

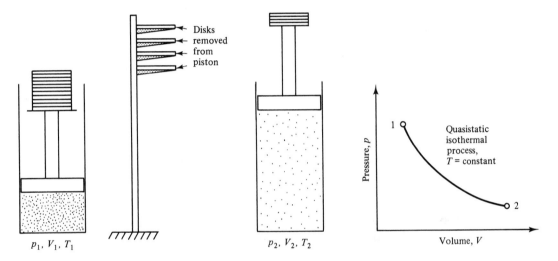

Figure 1.16   A gas undergoing a quasistatic isothermal process.

If the mass of the disk removed is extremely small, then the changes in the volume and pressure of the gas are of infinitesimal order and we can assume that the gas is essentially in thermodynamic equilibrium, or *quasi-equilibrium* and that the process is *quasi-equilibrium*, or *quasistatic process*. The new state can be represented by a point just to the right of and below point 1 on the *p-V* diagram.

If the above process is repeated and a number of disks are removed one at a time, then we bring about a finite change in pressure and volume; the final values are p_2 and V_2. No change in temperature occurs since we permit heat transfer between the gas and the surroundings. All the intermediate state points can then be plotted on a *p-V* diagram. We can draw a curve on the *p-V* diagram that passes through all the intermediate state points and connects points 1 and 2. We can say with some confidence that every single point on the curve between points 1 and 2 represents a state experienced by the gas during expansion. What the gas experienced in the expansion is called a quasi-static *constant temperature process* which can be represented by a curve on a *p-V* diagram. Most of the processes that we shall study and use will be assumed to be quasistatic processes, although not necessarily at constant temperature. We show a quasistatic process on a p-V diagram by a solid line.

Let us now suppose that, instead of removing one small disk at a time from the piston, we removed all the disks at once. The piston will literally fly away. Eventually the gas will come to state 2. This state 2 will be the same as state 2 at the end of the quasistatic constant temperature process discussed above if the number of disks removed is the same in both the processes and if a thermal contact exists between the system and the surroundings. However, we have a more or less chaotic state of affairs between points 1 and 2 when the disks are removed abruptly. There is no way in which the history between points 1 and 2 can be recorded on a p-V diagram. In such transition from point 1 to point 2, the gaseous system is in nonequilibrium. We show a nonequilibrium process on a p-V diagram by a dashed line.

The equation for a quasistatic constant-temperature process is

$$T = \text{constant}$$

If we have an ideai gas undergoing a quasistatic constant-temperature process, we can write

$$pV = mRT = \text{constant}$$

or

$$pV = C \tag{1.10a}$$

where C is a constant. Alternately, if subscripts 1 and 2 represent the initial and final states in a constant-temperature process, we have

$$p_1 V_1 = p_2 V_2 \tag{1.10b}$$

Equations 1.10(a) and 1.10(b) represent *Boyle's law*, first encountered in physics.

The quasistatic constant-temperature process was achieved by removing the disks from the top of the piston one at a time. The process involved transfer of heat from the surroundings to the gaseous system due to an infinitesimal temperature difference, dT. The surroundings functioned as an infinite source of energy, or a *reservoir* at temperature T. In thermodynamic analysis, it is useful to visualize reservoirs of energy at various temperatures. Depending on the nature of the desired process, these reservoirs can be made to function as sources or sinks of energy. A *source* is a reservoir of energy that supplies energy to a system, while a *sink* is a reservoir of energy that receives energy from a system.

There are two other processes that can conveniently be discussed at this point. They are the *quasistatic constant pressure process* and the *quasistatic constant volume process*. In the constant pressure process, as the name implies, the system pressure is maintained as a constant. For the gas in a piston–cylinder system, this means that we leave the weights on the top of the piston as they are. The temperature of the gas is gradually changed by bringing the gas into thermal contact with a number of reservoirs at temperatures $T_1 + \Delta T, T_1 + 2\Delta T, T_1 + 3\Delta T, \ldots, T_2 - 2\Delta T, T_2 - \Delta T,$ and T_2, successively. Each reservoir then imparts a small quantity of energy to the gas due to a heat-transfer process, raising the gas temperature by a small amount at a time. This causes a corresponding series of small increases in the volume. The final state then becomes p_2, V_2, T_2, with $p_2 = p_1$. The constant pressure process appears as a horizontal line on a p-V diagram. A practical way of accomplishing a constant pressure process is described below.

Imagine a gaseous system confined in a piston-cylinder arrangement with suitable weights balancing the gas pressure. Further, imagine the cylinder to be surrounded by an electrical heating coil. Let a very small current pass continuously through the coil. Its effect will be to raise the surroundings temperature, and, therefore, cause a transfer of heat to the gas. The gas temperature will rise by a small amount. Since the weights are still on the piston, the gas pressure remains constant. So, with the rise in the temperature of the gas, an increase in the volume of the gas will occur. This results in an upward movement of the weights. If the current through the heating coil is now increased slightly, the temperature and volume of the gas will increase by small amounts. Repetition of the process will lead to a quasistatic constant pressure process. The equation for a quasistatic constant pressure process is

$$p = \text{constant} \tag{1.11}$$

For an ideal gas

$$pV = mRT$$

For a constant pressure process, dividing by pT gives

$$\frac{V}{T} = \frac{mR}{p} = \text{constant} \tag{1.11a}$$

and

$$\frac{V_1}{T_1} = \frac{V_2}{T_2} \tag{1.11b}$$

This result is known as *Charles' law*.

We can visualize a quasistatic constant volume process to proceed in much the same way as the quasistatic constant pressure process. This time, however, we confine the gas in a rigid container, so that the gas volume cannot change. Progressively changing temperatures will cause corresponding changes in the gas pressure, while the volume is maintained constant. The equation for a quasistatic constant volume process is

$$V = \text{constant} \tag{1.12}$$

For an ideal gas

$$pV = mRT$$

For a constant volume process, dividing by VT yields

$$\frac{p}{T} = \frac{mR}{V} = \text{constant} \tag{1.12a}$$

and

$$\frac{p_1}{T_1} = \frac{p_2}{T_2} \tag{1.12b}$$

This result is *Gay Lussac's law*.

We have examined three types of quasistatic processes. In the future, unless otherwise stated, it will be implied that a process is quasistatic. These three processes are shown in Figure 1.17(a). On a p-V diagram, a constant temperature process appears as a curve like the one connecting 1 and 2, curve 1–2; for an ideal gas, the curve has the shape of a hyperbola ($pV = $ constant). A constant pressure process appears as a horizontal line on a p-V diagram, as shown by the line 1–3 in Figure 1.17(a). This is true for all substances undergoing a constant pressure process. A constant volume process appears as a vertical line on a p-V diagram, as shown by line 1–4 in Figure 1.17(a). This is also true for all substances. In Figure 1.17(a) we observe that the initial state for all three processes is the same; but the terminal states, points 2, 3, and 4 on the p-V diagram, are different.

Let us now consider the same three processes, but this time we take the initial state for one process as the final state for another process. Thus in Figure 1.17(b), path 1–2 is a constant pressure process, path 2–3 is a constant volume process, and path 3–1 is a constant temperature process. What we now have in Figure 1.17(b) is not separate segments or curves, but rather a closed curve, 1–2–3–1. A combination of processes taking a system through a succession of states and returning the system to its initial state is called a *cyclic process*, and it appears as a closed curve on a p-V diagram. When the nature of the processes making up a cyclic process is not specified, it is customary to show the cyclic process by a closed curve, as shown in Figure 1.17(c). A cyclic process may be traversed clockwise, as in Figure 1.17(b), or counterclockwise, as in Figure 1.17(c). A cyclic process always appears as a closed curve in a graphical representation regardless of the thermodynamic properties chosen for the abscissa and the oridinate of the graphical plot. Cyclic processes have important applications in engines, compressors, refrigerators, and the like.

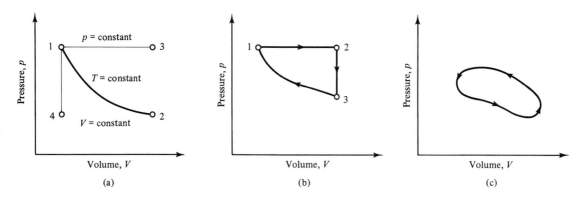

Figure 1.17
(a) p-V diagrams for different quasistatic processes.
(b) A thermodynamic cycle comprising constant pressure, constant volume, and constant temperature processes.
(c) A thermodynamic cycle comprising arbitrary processes.

1.8 SUMMARY

We can summarize the concepts studied in this chapter in the following manner:

1. There are three basic properties of substances; namely, specific volume, pressure, and temperature. Specification of these three properties for a gas uniquely defines the state of the gas.

2. A mathematical relationship among the three properties—pressure, volume, and temperature—of a gas is called the equation of state. If the equation of state for a gas is known, specification of only two properties is needed, since the third property is fixed by the equation of state.

3. The ideal gas is the simplest model of behavior of a gas.

4. System, surroundings, closed and open systems, thermodynamic equilibrium, quasi-equilibrium, quasistatic process, and cyclical process are some of the basic concepts of thermodynamics.

5. In analyzing a closed system, the time required for the process is of no consideration.

PROBLEMS

1.1. Express a pressure of 4 atmospheres in millimeters of a column of liquid, where the liquid is (a) water, (b) ethyl alcohol, and (c) liquid sodium. The densities are 1000 kg/m^3, 789 kg/m^3, and 860 kg/m^3, respectively.

1.2. Determine the pressure at the bottom of a reservoir of water if the level of the water is 100 m above the bottom of the reservoir. The atmospheric pressure is 1.013×10^5 Pa and the density of water is 980 kg/m^3.

1.3. A gauge on a steam generator shows a pressure of 700 kPa. The barometer is at 755 mm of Hg. What is the absolute pressure of the steam in the steam generator? The specific gravity of mercury is 13.6.

1.4. A manometer is used to measure the pressure drop experienced by air as it flows in a pipe, past an obstruction in its path. The manometer is a U-tube type and its two legs are connected to the pipe fluid on either side of the obstruction. The manometer fluid is water. The difference in the heights of the manometer fluid in its two legs is 12 mm. Determine the drop in the air pressure in pascals and millimeters of mercury.

1.5. A manometer is connected to a pipe carrying a liquid, as shown in Figure 1.5. The heights h_ℓ and h_m are 15 cm and 54 cm, respectively. The atmospheric pressure is 1.01×10^5 Pa. The pipe fluid has a density of 900 kg/m^3, while the manometer fluid is water with a density of 1005 kg/m^3. Determine the pressure of the pipe fluid at E (Figure 1.5) and express it in millimeters of mercury. The density of mercury is $13,600 \text{ kg/m}^3$.

1.6. An inclined manometer of the type shown in Figure 1.7 is used to measure the pressure of water in a pipe. The manometer fluid is mercury. The inclined leg makes an angle of 15° with the horizontal. The total length of the mercury in the inclined leg is 77 mm, while that in the vertical leg is 8 mm. The height of the water above the mercury in the vertical column is 60 cm. What is the pressure of the water in the pipe? The atmospheric pressure is 763 mm of Hg. The specific gravity of mercury is 13.6.

1.7. One may assume that the pressure p and the density ρ of the atmospheric air are related by $p = $ (constant) $\rho^{1.4}$. The pressure and the density of air at sea level are 1.013×10^5 Pa and 1.177 kg/m³, respectively. If the pressure on the top of a mountain is 20,000 Pa, determine the height of the mountain. Assume that the gravitational acceleration is constant at 9.81 m/sec².

1.8. A person took barometer readings at the bottom and at the top of a building and found them to be 766 mm of Hg and 727 mm of Hg, respectively. If the average density of the air is 1.2 kg/m³, estimate the height of the building.

1.9. The vacuum reading on a condensor is 750 mm of Hg. The barometer reads 761 mm of Hg. Express the pressure in the condensor in pascals and Torrs.

1.10. The barometric pressure at a certain mountain location is 742 mm of Hg. The vacuum gauge connected to the intake manifold of an automobile engine reads 510 mm of Hg vacuum. What is the absolute pressure in the manifold in millimeters of mercury and in pascals?

1.11. Instead of defining temperature t as a linear function of some thermodynamic property Z, a student has defined a temperature t^* as a logarithmic function

$$t^* = a \ln Z + b$$

The student has found that the values of Z are 6 cm and 36 cm for $t_i^* = 0°$ and $t_s^* = 100°$, respectively. Find the values of Z for $t^* = 10°$ and $t^* = 90°$.

1.12. The nitrogen in a constant volume gas thermometer has a specific volume of 1.0 m³/kg. The height of mercury above AA (see Figure 1.8) is -5 cm and the density of the mercury is 13,600 kg/m³. What is the temperature indicated by the thermometer? The atmospheric pressure is 765 mm of Hg.

1.13. **(a)** A new temperature scale is defined on which the ice point and steam points are $100°$ N and $400°$ N, respectively. Correlate these temperatures with temperatures on the Celsius scale.

(b) If oxygen is at $600°$ N, what is the temperature in degrees Kelvin?

1.14. The resistance of a platinum wire is found to be $10.000 \ \Omega$ at the ice point, $14.247 \ \Omega$ at the steam point, and $27.887 \ \Omega$ at the sulphur point ($446°C$). Find the constants A and B in the equation $R = R_0(1 + AT + BT^2)$, where T is in degrees Celsius.

1.15. When the ice point and the steam point are chosen as fixed points with $100°C$ between them, the ideal gas temperature of the ice point may be written as $\theta_i = 100/(r_s - 1)$, where

$$r_s = \lim_{p_i \to 0} \left(\frac{p_s}{p_i} \right)$$

at constant volume.

(a) Show that the fractional error in θ_i produced by an error in r_s is nearly 3.73 times the fractional error in r_s, or $d\theta_i/\theta_i = -3.73 \ dr_s/r_s$.

(b) Show that the fractional error in the temperature θ is

$$\frac{d\theta}{\theta} = \frac{dr}{r} - 3.73 \frac{dr_s}{r_s}$$

1.16. A constant volume gas thermometer is at an unknown temperature T^*. A series of values of pressure p^* corresponding to T^* are obtained by varying the quantity of the gas in the thermometer. Values of ice point pressure p_i for the various quantities of gas in the thermometer are also obtained. From the data given below, determine the limiting value of the ratio p^*/p_i and find the unknown temperature T^* if the temperature corresponding to p_i is 50 units.

p^* (mm of Hg)	128.0	256.4	385.9	515.9
p_i (mm of Hg)	200	400	600	800

1.17. A thermocouple is a device for measuring temperature. It consists of two wires of dissimilar metals connected at each end. One junction is kept at a reference temperature t_{ref}, and the other junction is brought in contact with the body whose temperature t is to be measured. The difference in the temperatures of the two junctions produces an electrical potential difference e. The relation between e and t is often given by

$$e = c_1(t - t_{ref}) + c_2(t - t_{ref})^2$$

For a certain thermocouple, $c_1 = 0.24$ and $c_2 = -5.6 \times 10^{-4}$, with e in millivolts and t in degrees Celsius. Determine the values of e for values of t ranging from $-100°C$ to $500°C$ in increments of $100°C$. Plot a graph of e versus t. Assume $t_{ref} = 20°C$.

1.18. If for the thermocouple in Problem 1.17 the voltage is 20.1 mV, what is the temperature of the hot junction?

1.19. A pump discharges water at a rate of 4000 cm³/s. If the density of water is 975 kg/m³, express the discharge rate in kilogram per second. How long will it take the pump to fill up a tank 4 m in diameter and 40 m in height? What is the minimums pressure in pascals that the pump must develop?

1.20. A rigid tank holds oxygen gas at 100 atmospheres at a temperature of 30°C. The volume of the tank is 0.6 m³. Determine the quantity of the oxygen gas in kilograms and kilogram-moles. Use the ideal gas law.

1.21. Some of the oxygen gas in Problem 1.20 is used up and the values of pressure and temperature are found to be 7080 kPa and $-20°C$. Determine the quantity of the gas used.

1.22. A figure similar to Figure 1.9 gives a value of pv/RT for oxygen as 1.06 for $p = 100$ atmospheres and $T = 30°C$. Determine the specific volume under these conditions. Using these values, rework Problem 1.20 and compare your result with the one obtained by using the ideal gas law.

1.23. A tank of 0.25 m³ holds hydrogen gas at 22°C and 750 kPa. Another tank of 0.2 m³ holds hydrogen gas at 18°C and 120 kPa. Both the tanks are connected to each other by a valve. The surroundings temperature is 10°C. If the valve is opened and sufficient time is allowed to pass, what is the final pressure of the hydrogen gas?

1.24. A tank of volume 176 m³ holds carbon dioxide gas at a pressure of 750 kPa and 25°C. A leak has developed in the tank and is not detected until the pressure in the tank has dropped to 640 kPa. When the leak is detected, the temperature is observed to be 15°C. Determine the mass of carbon dioxide that has leaked out.

1.25. Rework Problem 1.24 if the gas in the tank is helium.

1.26. Hydrogen gas was used in balloons for transatlantic flights before the advent of air-planes. The payload of one such balloon was 40 metric tons and the balloon was intended to cruise at an altitude of 3000 m, where the atmospheric pressure is 72 kPa. Estimate the quantity of the hydrogen in the balloon. Assume the temperature to be 8°C.

1.27. Two tanks of equal volume containing ammonia at 500 kPa and 30°C and at 400 kPa and 10°C are connected by a valve. One of these tanks is connected by a valve to a third evacuated tank of twice the volume. If both valves are opened, what is the final pressure of ammonia? Assume the surroundings to be at 20°C.

1.28. Air, which may be treated as an ideal gas, is compressed from 100 kPa to 1000 kPa in a quasistatic isothermal manner. Its initial temperature is 18°C and its mass is 40 g. Plot the process on a p-V diagram by calculating a minimum of five points.

1.29. If the air in Problem 1.28 is to expand in a quasistatic constant pressure process after its isothermal compression to 1000 kPa, describe how it can be done. If the atmospheric temperature is 18°C, what conclusion can you draw?.

1.30. State with your reasons whether the following systems are closed systems or open systems. Draw a system boundary in each case and show if mass crosses the boundary.
(a) Air in a bicycle tube.
(b) Water in a whistling tea kettle.
(c) Air freshener in an aerosol can.
(d) Water in a drain pipe.
(e) Hot air in heating ducts.
(f) The refrigerent in a household refrigerator.
(g) The lubricant oil in an automobile engine.

-2-

Work and Heat

2.1 WORK

Ordinarily we think of work as something that requires a physical or mental effort and expenditure of energy. In thermodynamics, we are interested in physical work, whether it is performed by a man or a machine. When primitive man discovered a way to produce fire by rubbing two sticks of dry wood, he did work to produce a spark. When a person climbs up a hill, he or she does work. In the process, the person expends energy from his or her body, and thus feels a bit tired at the end of the climb. After a person cuts a lawn using a push-type lawnmower, he or she feels that some work has been done. When a swimmer swims some distance, he or she does work in pushing the water in the pool. When a sump pump in the basement of a home is turned on, it lifts the water in the sump through some height, and in the process the pump does work. The sump pump by itself is not capable of doing work unless it is supplied with rotational energy. This rotational energy is supplied to it by an electric motor. The electric current going through the motor pushes electrons over a potential difference, and the result is work. The work done by the current is called *electrical work* since it is done electrically and it results in the mechanical rotation of the shaft of the motor. When the springs in a chest exerciser are stretched during an exercise, work is required to stretch the springs; this work comes from the person who is exercising. The blower in a forced-air home-heating system does work in circulating the hot air around the home.

In all of the foregoing examples, we find that there are two requirements for work to be performed. They are the force (push, pull, pressure, exertion) and displacement (movement, motion, lifting, lowering). We know that force and displacement are vector quantities, and a vector quantity has a magnitude, a direction, and a point of application. According to mechanics, when a force **F**, acts on a body and produces a movement **ds** of the body, work is said to be done by the force on the body (Figure 2.1). The magnitude of the mechanical work done, W, is the product of the magnitude of the force and the component of the movement or displacement of the body in the direction of the applied force. Mathematically we have

$$W = \mathbf{F} \cdot \mathbf{ds} = F \, ds \cos \theta \tag{2.1}$$

It is important to realize that if a force acts on a body with no resulting movement of the body, then there is no work. Suppose you are waiting for a bus at a bus stop with an armful of bags. You may get tired, but you are not doing any work because you are in one spot and not moving. Only when you step up into the bus with your load of bags do you work. Likewise, compressed air in a tank does not do any work, although it causes a reasonable level of stress in the tank material. When compressed air coming through an air gun helps tighten the nuts on an automobile wheel, the air does work. Motion is essential to work.

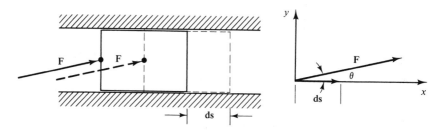

Figure 2.1   Work done by a force acting on a body.

When we consider the force involved in doing mechanical work, only that force necessary to overcome the opposing or restraining force is used in the computation of work. We know that a strong person can easily lift a mass of 25 kg. If a person lifts a briefcase of 2 kg mass, then he or she does work equal to the product of 19.62 N (9.81 N is the force needed to lift 1 kg mass on the earth's surface) and the height through which the person lifts the mass. A person's ability to exert a force of 245 N to lift a mass of 25 kg is irrelevant. If a 40-kW motor drives an air compressor and if the system is idling, then only a small fraction of the 40-kW capacity is used in doing work against the friction in the system. Similarly, a 7200-cm³ engine under the hood of an automobile is able to deliver some 225 kW of power; yet when the car is cruising, only a small portion of that capacity—something like 18 kW—is utilized.

2.2 THERMODYNAMIC WORK

Any definition of work in thermodynamics ought to relate to the concepts of system and surroundings and also be consistent with the definition of work in mechanics. Therefore we define work as follows: Work is said to be done when a system interacts with its surroundings, and the sole effect of this interaction, external to the boundaries of the system, could have been raising or lowering of a weight. Work is an energy-transfer process. The definition does not call for actual raising or lowering of a weight but rather the possibility of raising or lowering of a weight. The effect of raising or lowering of a weight entails displacement of a force. Also, the work effect must be external to a system. Consider a person and a suitcase in an elevator; with the elevator and its contents as a system, no work is done if the person lifts the suitcase, since the raising of weight (the suitcase) is not external to the system (the elevator).

Consider a paddle wheel immersed in a fluid, where the paddle wheel is driven by an electric motor (Figure 2.2a). Let the motor be turned on for a short period of

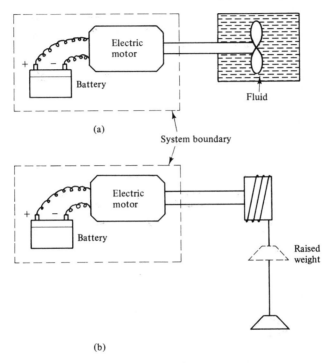

Figure 2.2
(a) Electric motor driving a paddle wheel immersed in fluid.
(b) Electric motor doing work raising weights.

time, resulting in churning of the fluid. Is work involved? The question can be answered if we first define our system. Let the electric motor be the system, with the shaft cutting the boundary of the system. We note that this is a closed system. Let the paddle wheel and the fluid be replaced by a rope and weight arrangement, as shown in Figure 2.2b. If the electric motor is now turned on for the same duration of time as before, drawing the same amount of current and with the same voltage impressed on the motor as before, we would find that a weight is raised by the motor through some height. Therefore work is done and energy is transferred. Also implied is the fact that an electrical current due to a potential difference passing through a system does work. This can be verified by drawing a system boundry around the battery in Figure 2.2. The current may be used to drive a motor and to raise a weight or the current may be used to produce a heating effect. In both instances work is done.

When there is a work interaction between a system and its surroundings, the effect is *equivalent* to a raising of a weight. In determining the work done by a body, we must define a closed system and its boundaries, identify the change in the thermodynamic state variables of the system, and ask this question: Can there be a raising of a weight with the *same* change in the thermodynamic state variables? For the production of a work effect, there must be an interaction between a system and its surroundings.

In thermodynamics, work done *by* a system is considered *positive*. When a system does work, energy leaves the system, and the surroundings receive energy; work is positive from the system point of view. If work is done on a system, the effect is equivalent to lowering a weight. Work done *on* a system is considered *negative*. Occasionally we may use symbols like W_{in} or W_{out} to denote work input or work output of a system; when we do use them, W_{in} and W_{out} will represent negative and positive quantities, respectively.

When we analyze a simple physical system, we shall represent the small quantity of work done during a small change in the state of the system by $F\,dX$, where F and X are two different properties of the system; the property F will represent generalized force and the property X will represent generalized displacement. We shall represent this small quantity of work by the symbol δW instead of dW to emphasize the fact that we are not dealing with an exact differential. The distinction between exact and inexact differentials is discussed in the notes at the end of this chapter. We call attention to the presence of an inexact differential by the symbol δ placed in front of the function.* Thus, for example, the quantity $F\,dx$, representing a small quantity of work, is an inexact differential and we represent it by the symbol δW. In general, the amount of work depends on the manner in which the change in the thermodynamic coordinates or state variables of a system is brought about. Furthermore, work is not a property and it is meaningless to talk about a value of the function W corresponding to a given state of a system. Thus when we have to refer to a small quantity of work due to a small change in the state of a system, we use the symbol δW. The quantity W represents the cumulative work done by a system when it undergoes a change of state due to a work interaction with its surroundings. Thus

$$W = \int_{path} \delta W \tag{2.2}$$

The quantity W does not represent an integral of a differential quantity.

We shall now develop expressions for work in some simple systems.

Gravitational field.   When an object of mass m is under the influence of gravity it experiences a gravitational force. If g is the acceleration due to gravity,

---

*This convention is not usually followed in the textbooks on calculus, but is standard in thermodynamic texts. For further details, see *Notes* at the end of this chapter.

then the gravitational force is mg and it acts toward the center of the earth. Let our closed system consist of an object of mass m at an elevation z relative to the surface of the earth. Let the mass be displaced upward so that the change in its elevation is dZ (see Figure 2.3).

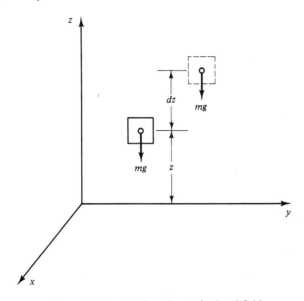

Figure 2.3   Work done in gravitational field.

We are not concerned with the horizontal displacement of the object, since there is no horizontal force acting on the object. In changing the elevation of the object by dZ (a positive quantity) the gravitational force mg is overcome, and work is expended by an external agent as the object is raised. From the thermodynamic viewpoint, work is done on the system (negative work), so that

$$\delta W = -mg\,dZ \tag{2.3}$$

and

$$W = -\int_{Z_1}^{Z_2} mg\,dZ = mg(Z_1 - Z_2) \tag{2.3a}$$

An elastic bar under tension.   Consider a system consisting of an elastic bar, as shown in Figure 2.4(a). When a tensile force of magnitude P_0 is gradually applied to the elastic bar it stretches by $\delta\ell$. Gradual application of the force means starting with an unloaded bar and then placing small weights, one at a time, in the pan until the desired magnitude of force P_0 due to the weights is reached. This assures a quasistatic process. In the process of loading, the bar is stretched by $\delta\ell$. The stretching phenomenon is sketched on a plot of force magnitude versus elongation in Figure 2.4(b). The graph shows, for example, that when the applied force is $P_0/3$, the elongation is $\delta\ell/3$. At successive loading points on the graph, the internal force in the elastic bar is exactly matched by the applied force. The magnitude of the

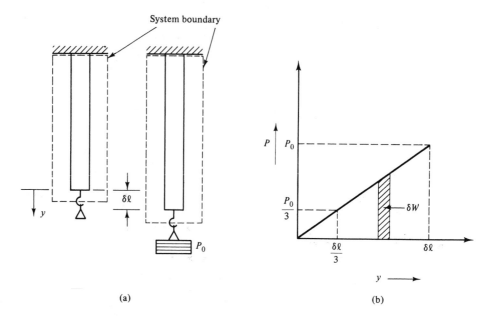

(a) (b)

Figure 2.4 Work done in an elastic system.

average force applied in the process of stretching is $\frac{1}{2}(0 + P_0)$ or $P_0/2$. The displacement is $\delta\ell$, and the magnitude of the work done on the system is $\frac{1}{2}(P\delta\ell)$. This result can be obtained in a rigorous manner as follows.

Let the magnitude of the force on the bar be P, corresponding to an elongation y of the bar. Let the magnitude be increased by δP, causing an additional elongation, dy. The magnitude of the average force acting on the bar during the displacement dy is $\frac{1}{2}[P + (P + \delta P)]$. The elemental work done in stretching the bar by dy is given by

$$\delta W = -\frac{1}{2}[P + (P + \delta P)]\, dy$$
$$= -P\, dy - \frac{1}{2}\delta P\, dy$$

Neglecting the higher-order term, $\frac{1}{2}\delta P\, dy$, we obtain

$$\delta W = -P\, dy$$

The negative sign tells us that work is done on the system. The total work done by the system of the elastic bar is

$$W_{1\text{-}2} = \int \delta W = -\int P\, dy$$
$$= -\int_0^{\delta\ell} y\frac{P_0}{\delta\ell}\, dy$$
$$= -\frac{P_0}{\delta\ell}\left[\frac{y^2}{2}\right]_0^{\delta\ell}$$
$$= -\frac{1}{2}P_0\, \delta\ell \tag{2.4}$$

44

Note that in evaluating the integral we substituted a relationship between P and y, the variables inside the integral sign. We know that the relationship between P and y for elastic bodies is linear.

A spring. Figure 2.5 shows a coiled spring. Also shown in the figure is a force of magnitude P_0 applied to the tip T of the spring and acting tangentially to the spring. The radius of the spring at T is r. The force P_0 is applied gradually to the spring to ensure a quasistatic process. The force P_0 causes an angular displacement θ_0. This, in turn, causes the applied force to experience a displacement of $r\theta_0$. We assume that throughout this displacement, the applied force acts tangentially to the spring at T.

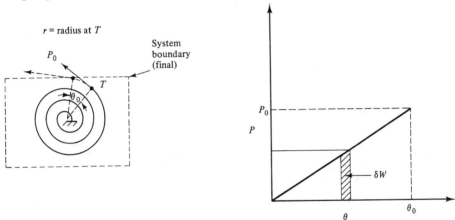

Figure 2.5 Work done by a coiled spring.

When the applied force is increased slightly, the spring tip experiences a small displacement, given by $r\,d\theta$. By a reasoning similar to the one used for the elastic bar, the elemental work done by the spring is given by

$$\delta W = -Pr\,d\theta$$

and

$$W = -\int_0^{\theta_0} Pr\,d\theta = -\tfrac{1}{2}P_0r\theta_0 \qquad (2.5)$$

Example 2.1

A helical spring has its load-deflection characteristic given by the equation

$$P = 4640y + 240{,}000y^2$$

where P is the magnitude of the applied load in newtons and y is the deflection in meters. Determine the work done when a load P_0 of 70 N is applied to the spring.

Solution:

Given: The relationship between the force on a helical spring and the deflection the spring experiences due to the force is prescribed. A force of 70 N is applied to the tip of the spring.

Objective: We wish to calculate the work done by the spring.
Diagram:

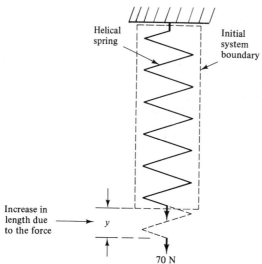

Increase in length due to the force

y

70 N

Figure 2.6

Assumptions: We assume that the force is gradually applied to the spring and that the process is quasistatic.

Relevant physics: The force-displacement relation for the spring is nonlinear. From this relation, the magnitude of the force necessary to produce a given deflection can be calculated. For known values of the initial and final deflections, the work done by the spring is obtained by integrating $-P\,dy$ between appropriate limits.

Analysis: The work done by the spring is given by

$$W = -\int_0^{y_0} P\,dy$$

The upper limit of the integral represents the deflection of the spring corresponding to the load P_0, or 70 N. It can be determined from the load-deflection characteristic of the spring. Thus we have

$$70 = 4640y_0 + 240{,}000y_0^2$$

or

$$240{,}000y_0^2 + 4640y_0 - 70 = 0$$

This is a quadratic in y_0, giving

$$y_0 = \frac{-4640 \pm \sqrt{4640^2 - 4(240{,}000)(-70)}}{2 \times 240{,}000}$$

Discarding the negative value of y_0, we have $y_0 = 0.01$ m. The work is given by

$$W = -\int_0^{0.01} (4640y + 240{,}000y^2)\,dy$$

Integration gives

$$W = -\left[4640\frac{y^2}{2} + 240{,}000\frac{y^3}{3}\right]_0^{0.01}$$

$$= -[2320(0.01)^2 + 80{,}000(0.01)^3]$$

$$= -0.312 \text{ J}$$

Comments: The negative sign indicates that work is done on the spring.

Surface tension.   Consider a system consisting of a C-shaped wire *ABCD* with a slider *GH* (Figure 2.7) that can move in the x-direction. Immerse the system in a soap-water solution, and take it out. A thin film of the solution will be formed over *BGCH*. The surface tension characteristics of the soap film and the wire material permit the film to oppose an externally applied force up to a certain limit. If this limit is exceeded, the film will rupture. If the slider *GH* is moved slowly to the right by gradually applying a force of magnitude *F*, the film stretches in the quasistatic process and the area of the film increases by *dA*. The elemental work done by the film in stretching is

$$\delta W = -F\,dA$$

The work done by the film in a finite displacement is obtained by integrating the above equation, which gives

$$W_{1\text{-}2} = -\int_1^2 F\,dA \tag{2.6}$$

The surface tension force *F* is usually expressed in newtons per meter length of the perimeter of area *A*.

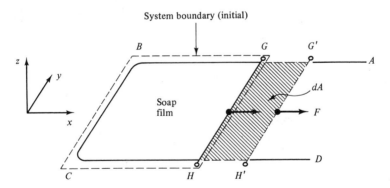

Figure 2.7   Work done against surface tension force.

An electrical system.   We saw earlier in this chapter that when electrons move through a potential difference *E*, work is done. This is due to the fact that electrons are negatively charged particles and it takes work to move them in an electrical field. One electron volt (eV) is a unit of energy, and it is the amount of work necessary to move an electron through a potential difference of 1 V. Also, if *N* is the number of electrons and τ is time, the current *i*, which is the rate of electrons moving per unit time, is given by

$$i = \frac{dN}{d\tau}$$

Let us now consider a system carrying a current i with a potential difference E impressed on the system. Work done in moving dN electrons across the potential E is

$$\delta W = -E\, dN$$

But the number of electrons, dN, equals $i\, d\tau$, since $i = dN/d\tau$. Therefore

$$\delta W = -Ei\, d\tau$$

Integration gives

$$W_{1\text{-}2} = -\int_{1}^{2} Ei\, d\tau \tag{2.7}$$

The negative sign in Equation 2.7 tells us that work is done on the system and electrical energy flows into the system. The quantity Ei in Equation 2.7 is the time rate of work done and is usually referred to as *power*.

2.3 WORK IN A COMPRESSIBLE SYSTEM

Consider a gas in thermal equilibrium in a piston-cyclinder arrangement (see Figure 2.8(a)). The piston is assumed to be frictionless. Let the gas be at pressure p, temperature T, and occupy a volume V. The gas pressure causes a force F on the piston that is equal to the product of the pressure p of the gas and the area A of the piston face. Under static equilibrium (the piston at rest), the weights on the top of the piston exert a downward force that is equal and opposite to the force F due to the gas pressure. Now suppose one small weight is removed from the top of the piston, as shown in Figure 2.8(b). The gas experiences a small reduction in pressure and it

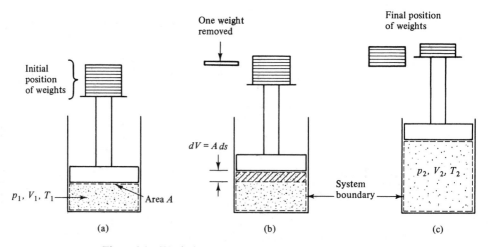

Figure 2.8 Work done by a gas in a piston-cylinder system.

undergoes a slight expansion. The new gas pressure is now able to support the reduced weights on the piston. The weights have been raised through a small height, ds, and therefore work is done. It is

$$\delta W = (pA)\,ds = p\,dV \tag{2.8}$$

Equation 2.8 implies that the gas experienced a quasistatic process; that is, the small disturbance caused by the removal of a small weight from the piston altered the state of the gas only infinitesimally, and therefore the gas was essentially at equilibrium during and after the removal of the small weight.

Now, let us assume that a finite number of weights are removed from the piston while ensuring a quasistatic process. The resulting change in the volume of the gas is finite—from V_1 to V_2. The new state, state 2, is shown in Figure 2.8(c). The work done by the gas in this quasistatic process is given by

$$W_{1\text{-}2} = \int_1^2 p\,dV \tag{2.8a}$$

Work in a compressible system is often stated as specific work. Specific work is work per unit mass of a system and is denoted by the symbol w. Noting that $V = mv$ and $W = mw$, we have

$$\delta w = p\,dv \tag{2.9}$$

and

$$w_{1\text{-}2} = \int_1^2 p\,dv \tag{2.9a}$$

Equations 2.8 and 2.9 can be applied to any body of any arbitrary shape undergoing a quasistatic change of volume. Figure 2.9 shows a body of arbitrary shape undergoing a change in volume dV. The quantity $p\,dV$ in Equation 2.8 represents an elemental work done by a body experiencing a change in its volume. If the body experiences an expansion or an increase in volume, dV is positive and work is done by the body. If the body experiences a compression or a decrease in volume, dV is negative and work is done on the body.

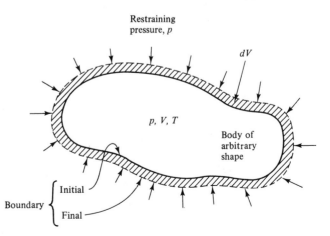

Figure 2.9 Work done by a compressible system.

In Chapter 1 we saw that the history of a quasistatic process for a gaseous system can be plotted on a graph with suitable thermodynamic coordinates as the abscissa and ordinate. Figure 2.10 shows such a plot on a p-V diagram. The quantity $p\, dV$ represents the area of a small vertical strip and the quantity

$$\int_1^2 p\, dV$$

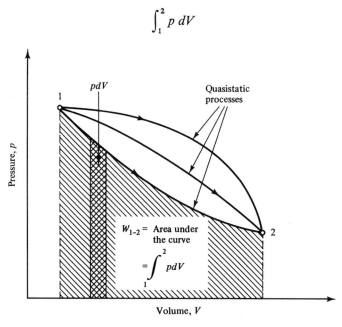

Figure 2.10 Representation of work done by a compressible system on a p-V diagram.

represents the area under the curve on the p-V diagram for the quasistatic process 1-2. There is a variety of quasistatic processes taking a system from state 1 to state 2, and for each of these processes there is a unique curve connecting points 1 and 2 on a p-V diagram. The areas under such curves are different and so is the work for each of these processes. An area under a curve on a p-V diagram represents work only if the curve represents a quasistatic process. We can determine the nature of the curve if we know the equation of state of the compressible system. For an ideal gas, the task becomes particularly simple. Its equation of state is

$$pV = mRT$$

Constant Temperature Process.   For a constant temperature process, the right side of the equation represents a single constant. Therefore the curve for expansion is a hyperbola on a p-V plot. From Equation 2.8(a).

$$W_{1\text{-}2} = \int_1^2 p\, dV$$

Substituting for p from the equation of state, we have

$$\underset{\substack{\text{isothermal}\\\text{ideal gas}}}{W_{1\text{-}2}} = \int_1^2 mRT\,\frac{dV}{V} = mRT\ln\left(\frac{V_2}{V_1}\right) \tag{2.10}$$

Constant Pressure Process. A quasistatic constant pressure process is shown in Figure 2.11. For this process, the evaluation of the integral in Equation 2.8(a) does not require a knowledge of the equation of state. Since the pressure is constant, we treat p in Equation 2.8(a) as a constant and write

$$\underset{\substack{\text{constant}\\\text{pressure}}}{W_{1\text{-}2}} = p\int_1^2 dV = p(V_2 - V_1) \tag{2.11}$$

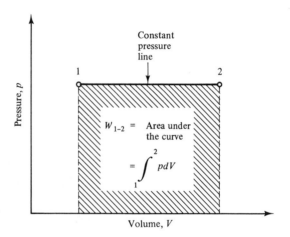

Figure 2.11 Work done in a constant pressure process.

Constant Volume Process. For a constant volume process there is no work done. Work requires displacement and displacement is absent in a constant volume process.

Example 2.2

An ideal gas expands in a quasistatic manner from 0.25 m³ to 0.75 m³. The initial pressure of the gas is 800 kPa. Determine the work done in the expansion process if the temperature is maintained constant at 300 K.

Solution:

Given: For convenience, we assume the gas to be in a piston-cylinder system. The initial pressure, volume, and temperature of the ideal gas are given. This gas expands in a constant temperature process.

Objective: The work done by the gas in the expansion process is to be determined.

Diagram:

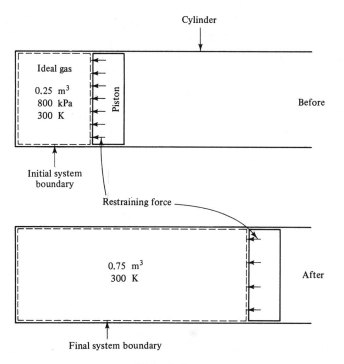

Figure 2.12

Assumptions: The properties are uniform at every stage of the expansion; that is, the expansion is quasistatic.

Relevant physics: The restraining force exerted by the piston on the gas is precisely equal to the force exerted by the gas on the piston during the entire expansion process. Sufficient heat exchange occurs between the system and the surroundings during the expansion process so that the temperature of the gas remains constant.

Analysis: The quasistatic work done is given by

$$W_{1\text{-}2} = \int_1^2 p \, dV$$

The relation between p and V in this particular problem is such that the temperature remains constant (isothermal process). Since the gas is known to be ideal, we have

$$pV = mRT$$

or

$$pV = C$$

where C is a constant with a unique value for this problem. Thus

$$p = \frac{C}{V}$$

and

$$W_{1\text{-}2} = \int_1^2 \frac{C}{V}\,dV = C\ln\frac{V_2}{V_1}$$

We can substitute for C to yield

$$W_{1\text{-}2} = (p_1 V_1)\ln\frac{V_2}{V_1}$$

Figure 2.13 shows the quasistatic isothermal expansion and the work done in the process. Inserting the values of p_1, V_1, and V_2, we obtain

$$W_{1\text{-}2} = (800 \times 10^3 \times 0.25)\ln\frac{0.75}{0.25}\ \text{J} = 220\ \text{kJ}$$

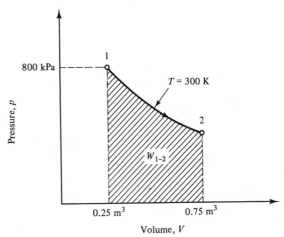

Figure 2.13 Work done in an isothermal expansion.

Comments: Since neither the nature of the ideal gas nor its R value was given, we could not use Equation 2.10. Also, we did not need the given value of the temperature to calculate the work done; however, we did use the information that the temperature is maintained constant.

Example 2.3

Determine the work per unit mass for the ideal gas in Example 2.2 if $R = 287$ J/kg K.

Solution:

Given: The gas constant for the gas in Example 2.2 is given.

Objective: The objective is to determine the work per unit mass of the gas. The *diagram, assumptions,* and the *relevant physics* for this example are the same as those for Example 2.2.

Analysis: In order to calculate the work per unit mass, we need to know the total mass in the system. It is given by

$$m = \frac{pV}{RT} \qquad\qquad \text{(ideal gas)}$$

$$= \frac{800 \times 10^3 \times 0.25}{287 \times 300} = 2.323\ \text{kg}$$

Therefore the work per unit-mass, or the specific work output, is

$$w_{1\text{-}2} = \frac{W_{1\text{-}2}}{m} = \frac{220}{2.323} = 94.7 \text{ kJ/kg}$$

Polytropic Process.   The relation $pV^n = C$ is often used to represent the actual expansion or compression of a gas, and the process is known as a *polytropic process*. The use of the relation does not mean in practice that every single state point during an actual expansion lies on the curve defined by the equation $pV^n = C$ on a p-V plot. It is only an approximation of an actual process to facilitate analysis. We now develop a general expression for work in a polytropic process (see Figure 2.14).

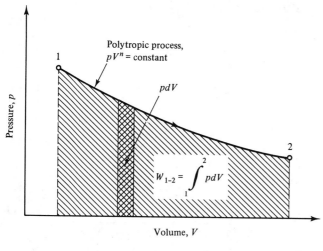

Figure 2.14   Work done in a polytropic process.

In order to evaluate $\int p \, dV$ we must know the specific relationship between p and V, express p in terms of V using the relationship, and then carry out the integration. Using the equation for the polytropic process we can write

$$p = CV^{-n}$$

and

$$W_{1\text{-}2} = \int_1^2 p \, dV = \int_1^2 CV^{-n} \, dV$$

$$= \left(\frac{CV^{-n+1}}{-n+1} \right)_1^2$$

$$= \frac{1}{1-n}(CV_2^{-n+1} - CV_1^{-n+1})$$

We observe that

$$C = p_1 V_1^n = p_2 V_2^n$$

Using the two expressions for C, we obtain

$$W_{1\text{-}2} = \frac{1}{1-n}(p_2 V_2^n V_2^{-n+1} - p_1 V_1^n V_1^{-n+1})$$

$$= \frac{1}{1-n}(p_2 V_2 - p_1 V_1) \qquad (2.12)$$

The above expression for the work done in a polytropic process involves the exponent n and the values of p and V at the initial and final states. It is left as an exercise to show that the foregoing expression can be recast as

$$W_{1\text{-}2} = \frac{p_1 V_1}{1-n}\left[\left(\frac{V_1}{V_2}\right)^{n-1} - 1\right] \tag{2.12a}$$

or

$$W_{1\text{-}2} = \frac{p_1 V_1}{1-n}\left[\left(\frac{P_2}{P_1}\right)^{\frac{n-1}{n}} - 1\right] \tag{2.12b}$$

Example 2.4

Air, which may be treated as an ideal gas, is compressed from 100 kPa to 1500 kPa in a quasistatic manner so that it obeys the relation

$$pV^{1.3} = \text{constant}$$

Determine the amount of work necessary to compress 100 kg of air. The initial temperautre of the air is 300 K and $R = 0.287$ kJ/kg K.

Solution:

Given: Air is compressed in a polytropic process. The initial pressure and temperature and the final pressure are prescribed.

Objective: We have to calculate the amount of work to be done on 100 kg of air to compress it from 100 kPa to 1500 kPa.

Diagram: For convenience, we can assume the air to be in a large piston-cylinder system.

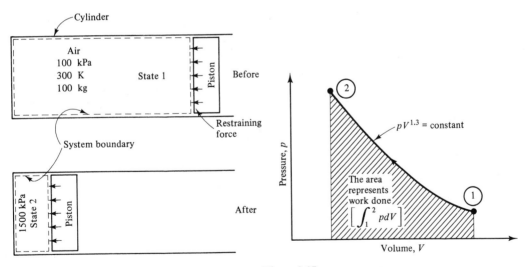

Figure 2.15

Assumptions: The compression process is quasistatic.

Relevant physics: There is a force equilibrium at the piston and the air interface at every stage during the compression. During the compression process, the pressure increases and the volume decreases. Also, the temperature increases causing heat transfer across the system boundary.

Analysis: We can list the given data as follows:

$$p_1 = 100 \text{ kPa} \qquad R = 287 \text{ J/kg K}$$

$$p_2 = 1500 \text{ kPa} \qquad m = 100 \text{ kg}$$

$$T_1 = 300 \text{ K} \qquad n = 1.3$$

The initial volume can be found from

$$V_1 = \frac{mRT_1}{p_1} = \frac{100 \times 287 \times 300}{100 \times 10^3}$$

$$= 86.1 \text{ m}^3$$

and

$$W_{1\text{-}2} = \frac{100 \times 10^3 \times 86.1}{1 - 1.3}\left[\left(\frac{1500 \times 10^3}{100 \times 10^3}\right)^{(1.3-1)/1.3} - 1\right]$$

$$= -2.49 \times 10^7 \text{ J}$$

$$= -24{,}900 \text{ kJ}$$

Comments: The negative sign implies that the work is done on the air. Thus 2.49×10^7 J of work must be done to compress 100 kg of air.

Several Simultaneous Modes of Work. Thus far we have examined work done by a gravitational system, an elastic bar, a spring, a soap film, an electrical system, and a compressible system. The expression for elemental work done in each case is a product of two terms, one term representing force—mechanical, electrical, surface tension, or pressure—and the other term representing an infinitesimal displacement. This is true of all systems. For example, in magnetic cooling—which requires work—the basic properties are the intensity of magnetic field H and magnetization M. It can be shown that in a quasistatic process the work done is

$$\delta W = -\mu H d(V\text{M}) \tag{2.13}$$

where μ is the permeability of free space and V is the volume.

If we use the symbol F_k to represent the magnitude of a generalized force, with a subscript k on F to denote a specific mode of work, and the symbol X_k to denote a generalized displacement in such a mode, we can write the following equation for total work when several modes of work simultaneously exist.

$$\delta W_{\text{total}} = \sum F_k dX_k$$

$$= -mg\, dZ - P\, dy - Pr\, d\theta - F\, dA - E\, dN$$

$$+ p\, dV - \mu H\, d(V\text{M}) + \cdots$$

Table 2.1 shows the generalized force and displacement for different modes of work.

TABLE 2.1 EXPRESSIONS FOR WORK

Mode of work	Generalized force F_k	Generalized displacement X_k	Expression for work
Gravitational	mg	Z	$-mg\, dZ$
Elastic	P	y	$-P\, dy$
Surface	F	A	$-F\, dA$
Electrical	E	N	$-E\, dN$
Compressible	p	V	$p\, dV$
Magnetic	H	M	$-\mu H\, d(V M)$

2.4 USEFUL AND NONUSEFUL WORK*

In discussing work we relied on the definition of work from mechanics. We required that the sole effect of the work interaction be equivalent to raising or lowering a weight. We shall now examine useful and nonuseful work, concepts usually not discussed in mechanics. Work for quasistatic processes is defined by the various equations for the systems discussed in this chapter. It can be stated to be equal to the sum of useful and nonuseful work, or

$$W = W_{useful} + W_{nonuseful} \qquad (2.14)$$

Work from a system is *useful* if that work is available and is used to fulfill a definite objective; otherwise it is *nonuseful*. The objective can be one of many possible diverse types: raising a weight, stressing and straining an elastic body, winding up a spring, charging a battery, supplying water to an overhead tank, circulating air in a room on a hot summer afternoon, heating a room by passing electric current through a resistor, or toasting a slice of bread in a toaster. As long as part or all of the work delivered by a system meets a certain objective, we say that useful work is done. In a given situation W_{useful} can be anywhere from zero to 100 percent of W. Naturally, we would like to maximize W_{useful} and minimize $W_{nonuseful}$. The conditions under which W_{useful} is a maximum involve the second law of thermodynamics and entropy, which are discussed in Chapter 6. At this point, we note that W_{useful} is a maximum if a given process is quasistatic. However, this does not mean that the nonuseful work in a quasistatic process is always zero. It may or may not be zero.

To illustrate this point, let us reconsider the process of expansion of a gas in a piston-cylinder system, shown in Figure 2.8. As depicted there, the process is quasistatic and the expansion of the gaseous system results in the weights being raised. It also results in the atmosphere being pushed. The former produces useful work. The latter is nonuseful work, since it is not generally an objective to push the atmosphere. The amount of this nonuseful work equals the product of the atmospheric pressure

*Optional material.

and the volume by which the atmosphere is displaced, that is, the change in the volume of the gas in the cylinder. Work, given by Equation 2.8(a), is the sum of the useful and nonuseful work. Thus we find that we have a quasistatic process and also a nonzero amount of nonuseful work.

When an electric current drives an electrical motor, which in turn drives a pump to lift water through a height, work is done by the electrical battery. Only under ideal conditions, such as 100 percent efficiency of the motor and the pump and zero electrical resistance of the connecting wires, is the entire work output from the battery useful work. In reality, the efficiencies of the motor and the pump are less than 100 percent and the connecting wires have a finite resistance. Consequently, there is some nonuseful work along with the useful work.

Now consider a battery that supplies current to a resistor. We know from experience that there is a heating effect and the resistor becomes warm. We also know that the battery does work. Is it useful or nonuseful? The answer depends on the function of the resistor. If the resistor is part of the heating element of a toaster or a space heater, then the energy dissipated in the resistor is used for toasting a slice of bread or for heating a room, and the work is useful. On the other hand, if the resistor is due to the finite electrical conductivity of, for example, an amplifier circuit, the work energy is dissipated as undesired heating and is considered as nonuseful work. These examples demonstrate that in a quasistatic work interaction, the entire work output is not necessarily useful work.

Let us examine a nonquasistatic process. Referring again to Figure 2.8, consider a process where the weights are removed all at once from the top of the piston. Such a process is called *free expansion*. When the gas does not experience a restraining force, it wants to go in all directions. As long as it is within the confines of the cylinder, its primary direction of movement is upward. When it streams out of the cylinder, it goes in all possible directions and eventually becomes part of the atmosphere. The process is not quasistatic, and thermal equilibrium does not exist during such a free expansion process. At no point is there an external restraining force balancing the internal forces. We can argue that the expanding gas has pushed the atmosphere away and therefore has done some work in the process of pushing. But that work is not useful work. Nothing is to be gained by pushing around the atmosphere. The only effect of this process on the surroundings is an infinitesimal increase in the temperature of the surroundings. Whatever work is done is all nonuseful work, and the *potential* to do some useful work is lost forever due to the nonquasistatic nature of the process. However, this does not mean that there is never any useful work in a nonquasistatic process, as the following discussion will show.

We may now inquire what happens when only a few of the weights on the piston are removed. This is more representative of a rapid, but controlled, expansion, such as would be encountered in actual devices like engines and compressors. This results in a nonquasistatic process. As a few weights are removed from the piston, the restraining force on the piston decreases by a finite amount and the piston moves up with a finite acceleration. This causes finite pressure gradients within the gas. Eventually, the gas pressure adjusts to the new restrictive force. The useful work done by the gas in this nonquasistatic process equals the product of the *new* restrictive force and the change in the volume of the gas, and it is less than the useful work that would be obtained had the process been quasistatic. The nonuseful work equals the product of the atmospheric pressure and the volume by which the atmosphere is displaced; that is, the change in the volume of the gas. Nonuseful work results in a dissipation of energy and an eventual infinitesimal change in the temperature of the surroundings. All real processes involving work interactions are nonquasistatic and, generally speaking, they will have useful and nonuseful work. It should be noted that, for the same end states of a system, the useful work in a nonquasistatic process is less than the useful work in a quasistatic process. Unless otherwise stated, the term *work* will refer to quasistatic work. It will include useful and nonuseful work.

2.4.1 Lost Work

In general, for a given change of state of a system, a quasistatic process results in a larger work output than a nonquasistatic process. Due to the nonquasistatic nature of the process, some of the potential to do work is irretrievably lost. The difference between the quasistatic and nonquasistatic work outputs for the same end states of a system is called *lost work*, W_{lost}.

$$W_{lost} = W_{quasistatic} - W_{nonquasistatic} \qquad (2.15)$$

The lost work is dissipated as heat and results in a small increase in the temperature of the surroundings. There is no other effect. The free expansion of the gas in a piston-cylinder system and the rapid but controlled expansion of the gas, both discussed in the preceding section, are nonquasistatic processes. In both these examples, the piston accelerates and acquires a finite velocity due to a finite change in the restraining force. The piston gains in kinetic energy, but this gain is eventually dissipated—that is, converted into heat—since the piston finally comes to rest due to friction of the air molecules. Also, the sudden removal of weights creates pressure waves within the gas, and the energy of these waves is eventually dissipated. Such factors contribute to lost work. Some other examples of processes in which there is lost work are a free-falling body and a gas expanding in a piston-cylinder arrangement with friction present.

Lost work can be due to dry friction, viscosity, finite electrical resistance, hysteresis effects, and finite changes in restraining forces. All real processes involving work interactions are nonquasistatic and, in general, they will have lost work. Lost work will be discussed further in Chapter 6.

It should be reiterated that work can exist only during a process. To compute work, we need to know about the process used in producing the work and the initial and final state points. Its effect might be changes in the kinetic energy (accelerating a mass), potential energy (lifting of weights), or strain energy (elastic solid). Work itself is not stored in a body. Work is energy in transit, and its end result can be a change in the energy level of a body. This point will be discussed more fully in the section on internal energy.

Example 2.5

A gas, at an initial pressure 10^6 Pa, is confined in a frictionless piston-cylinder system and occupies a volume of 0.05 m³. The area of the piston face is 100 cm². The piston is held in position by suitable weights placed on the piston. When 31.27% of the weights supported by the gas are abruptly removed from the top of the piston, the volume of the gas increses by 39%. Determine the useful work done, the work done in pushing the atmosphere, and the total work done. If a quasistatic process were to take the system from the same initial state to the same final state, determine the quasistatic work. Also, determine the lost work.

Solution:

Given: The initial pressure and volume of a gas in a piston-cylinder system are prescribed. The gas is restrained by suitable weights on the piston top. Some of the weights are removed, allowing the gas to expand.

Objective: The useful work, the work on the atmosphere, the total work and the lost work are to be evaluated.

Diagram:

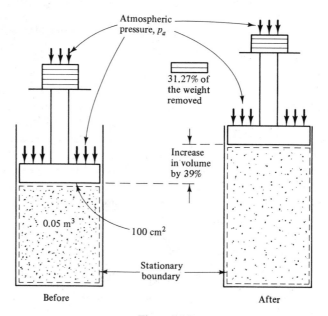

Figure 2.16

Assumption: The properties of the gas are uniform when sufficient time has elapsed after some of the weights are removed.

Relevant physics: This is not a quasistatic process and the process cannot be depicted by a curve on a p-V diagram. We can show only the end states on a p-V diagram. Therefore we cannot evaluate the integral $\int p \, dV$. Instead, we look at the equivalent effect of raising weights. We shall first determine the downward force due to the weights remaining on the piston top and the height through which these

weights are lifted. Also, there is a certain amount of work that is irretrievably
lost due to the abrupt removal of 31.27% of the weights on the piston. For the
same terminal states of a system, the lost work, W_{lost}, equals the difference
between the quasistatic work and the nonquasistatic work.

Analysis: The initial absolute pressure of the gas is 10^6 Pa. The atmospheric pressure
is taken as 1.013×10^5 Pa. Thus the gauge pressure is $(10^6 - 1.013 \times 10^5)$ Pa,
or 8.987×10^5 Pa. Since the area of the piston face is 100 cm², or 100×10^{-4} m², the initial downward force due to the weights on the piston top is
$(8.987 \times 10^5) \cdot (100 \times 10^{-4})$, or 8987 N. The downward force on the piston
after 31.27% of the weights are removed is

$$F = (1.0 - 0.3127) \times (8987) = 6177 \text{ N}$$

and the final pressure is

$$p_2 = \frac{6177}{(100 \times 10^{-4})} + 1.013 \times 10^5 = 7.19 \times 10^5 \text{ Pa}$$

The cylindrical volume of the gas is known to increase by 39%. Since the
initial volume of the gas is 0.05 m³, the final volume is

$$V_2 = 1.39 V_1 = 1.39 \times 0.05 = 0.0695 \text{ m}^3$$

The change in the volume of the gas is then $0.0695 - 0.05$, or 0.0195 m³.
Since the area of the piston face is 100 cm², the piston displacement is

$$s = \frac{0.0195}{(100 \times 10^{-4})} = 1.95 \text{ m}$$

The useful work done by the gas is then given by

$$W_{useful} = Fs = 6177 \times 1.95 = 12{,}045 \text{ J} \simeq 12 \text{ kJ}$$

The work done in pushing the atmosphere equals the product of the atmospheric pressure p_a and the volume by which the atmosphere is displaced, or

$$W_{nonuseful} = p_a(V_2 - V_1) = 1.013 \times 10^5 \times 0.0195 = 1{,}975 \text{ J} \simeq 2 \text{ kJ}$$

Hence the total work done in this nonquasistatic process is

$$\begin{aligned} W_{\substack{1\text{-}2 \\ nonquasistatic}} &= W_{useful} + W_{nonuseful} \\ &= 12{,}045 + 1975 \\ &= 14{,}020 \text{ J} \\ &\simeq 14 \text{ kJ} \end{aligned}$$

We shall now calculate the lost work for the present problem. We observe that

$$p_1 = 10^6 \text{ Pa} \qquad p_2 = 7.19 \times 10^5 \text{ Pa}$$
$$V_1 = 0.05 \text{ m}^3 \qquad V_2 = 0.05 \times 1.30 = 0.0695 \text{ m}^3$$

and assuming an ideal gas behavior,

$$\frac{T_2}{T_1} = \frac{p_2 V_2}{p_1 V_1} = \frac{7.19 \times 10^5 \times 0.0695}{10^6 \times 0.05} = 1$$

We therefore consider a quasistatic isothermal process between points 1 and 2.
If we were to find $p_1 V_1 \neq p_2 V_2$, we should use a quasistatic polytropic process.

For a quasistatic isothermal process, from Equation 2.10

$$W_{1-2} = p_1 V_1 \ln \frac{V_2}{V_1}$$

For our problem, then,

$$\underset{\text{quasistatic}}{W_{1-2}} = 10^6 \times 0.05 \ln \frac{0.0695}{0.05} = 16{,}465 \text{ J} \simeq 16.4 \text{ kJ}$$

We observe that of the 16,465 J of quasistatic work, the nonuseful work would still be 1975 J, since the volume displacement is the same as before. The lost work is then given by Equation 2.15.

$$W_{\text{lost }1-2} = \underset{\text{quasistatic}}{W_{1-2}} - \underset{\text{nonquasistatic}}{W_{1-2}}$$

$$= 16{,}465 - 14{,}020$$

$$= 2445 \text{ J} \simeq 2.4 \text{ kJ}$$

2.5 HEAT

When we sit in front of a fireplace, we soon feel warm. This is due to a transfer of heat energy from the high-temperature hearth to our low-temperature body. When we work on our stalled car on a winter day, we feel cold. There is a transfer of heat energy from our body to the surrounding air. When water is poured over ice cubes in a glass, the ice cubes soon melt and the water becomes cold. Once again, heat energy is transferred from the moderate-temperature water to the low-temperature ice cubes. Temperature difference is essential for heat transfer. When there is a temperature difference between a system and its surroundings and the system and the surroundings are able to interact, a transfer of heat energy results. Heat energy exists only *during* an interaction. If work interaction is absent, then the effect of a heat transfer process is to equalize the temperature of a system and its surroundings. If the boundaries of a system are insulated, then heat transfer to or from the system is not possible even though there may be a temperature difference between the system and the surroundings.

The thermodynamic definition of heat energy is different from the ordinary day-to-day usage of the term heat. We often hear phrases like *heat stored in a body* or *heat content of a substance.* Such phrases have no place in the study of thermodynamics. We recognize *heat as energy in transit,* or energy that is transferred from one body to another by virtue of temperature difference. It is acceptable to speak of the energy level of a body, that is, the internal energy of the body. The concept of internal energy will be discussed in Chapter 3.

The unit of heat is the same as that of work, namely, a joule. Consider 1 g mass of air at 100 K. The quantity of heat energy necessary to raise the temperature of this air by 1 K, under a constant pressure condition is 1 J. Heat energy supplied to a system of mass m is denoted by the symbol Q while that for a unit mass by the symbol q.

As pointed out at the beginning of Chapter 1, in most practical situations it is the mechanical work that is ultimately required at the point of use. Such mechanical work is usually obtained by converting heat energy into mechanical work to produce electricity and by reconverting the electrical energy into mechanical work on site. In certain applications the intermediary, the electrical energy, is absent. Also, historically, people have been interested in extracting mechanical work by supplying heat energy to a system. Therefore at a time when the nature of heat and the laws of thermodynamics were not well understood, it seemed reasonable to assign a positive sign to heat transferred to a system. This sign convention is generally followed even today and we shall adopt the same sign convention. This sign convention for heat transferred to a system is opposite to that for work. Although heat and work are both different forms of energy in transit, heat energy transferred to a system is considered *positive*, while work energy transferred to a system is considered *negative*. Heat energy supplied to a system is treated as a positive quantity, while heat energy rejected by a system is considered a negative quantity. Thus Q can be positive or negative. Sometimes we may use subscripts such as in and out on Q to denote heat supplied to a system and heat removed from a system. Thus Q_{in} will denote a positive quantity and Q_{out} will represent a negative quantity. If Q is zero we have no heat transfer to or from a system. Such a process is called an *adiabatic process*. An adiabatic process may involve work interactions. If a system has neither heat nor work interactions with the surroundings, then the system is called an *isolated system*.

In our discussion of work, we required that a process be quasistatic, so that a system was essentially in equilibrium in the work process and the process could be analyzed and graphically represented. This also resulted in a system delivering maximum useful work. In thermodynamic analysis of a system undergoing a process with heat and work interactions, we require that the system be in equilibrium or quasiequilibrium (Section 1.5) during such interactions. It was stated earlier that we shall be considering mainly quasistatic work processes. Quasistatic work processes need quasistatic heat interactions. Consequently, we require that all heat transfer processes be quasistatic for a thermodynamic analysis. A quasistatic process requires that there be only an infinitesimal variation of thermodynamic properties within a system at any given moment during the process. This applies to changes in temperature as well. To ensure a quasistatic heat transfer process resulting in heating or cooling of a system, we bring one or more reservoirs of energy in thermal contact with the system. A *reservoir of energy* is a body whose temperature does not change regardless of the quantity of heat transferred to or from it. The temperatures of the reservoirs are such that during a thermal contact with the system, the quasistatic heat transfer is due to an infinitesimal temperature difference dT between a reservoir and the system.

It is known from experience that the rate of heat transfer between two bodies is directly proportional to the temperature difference between the two bodies. It follows, therefore, that the rate of heat transfer in a quasistatic process is necessarily infinitesimal. In practical situations we cannot afford to have infinitesimal rates of heat transfer for real processes must occur in finite periods of time. Finite rates of heat transfer mean nonuniform properties in a system resulting in nonequilibrium processes.

The quantity of heat energy transferred to a system depends on the process that the system experiences; in other words, heat energy is a function of path. This is similar to work energy. Heat energy is not a function of the state of a body. Therefore we denote small quantities of heat energy by the symbol δQ and the amount of heat energy transferred to a body resulting in a change of its state from state 1 to state 2 by $Q_{1\text{-}2}$, which equals $\int_1^2 \delta Q$. In order to evaluate the integral of δQ, we must work with specific expressions for δQ for specific processes and integrate such expressions.

Example 2.6

A certain gas is heated in a quasistatic process, and δQ for the process is given by

$$\delta Q = 6(2.47 - 0.025\theta^{0.75})\, d\theta \text{ kJ}$$

where $\theta = T/100$ and T is the temperature of the gas in degrees Kelvin. Determine the amount of heat that must be transferred to the gas to raise its temperature from 300 K to 400 K.

Solution:

Given: The relationship between the amount of heat supplied to the gas and its temperature change is given in a differential form. The initial and final temperatures of the gas are given.

Objective: We wish to determine the quantity of heat supplied to the gas.

Diagram:

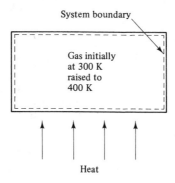

Figure 2.17

Assumption: The heating process is quasistatic.

Relevant physics: In addition to the temperature change, there may be volume and pressure changes for the gas. However, these do not concern us. The quantity of heat supplied can be obtained by integrating the given relationship between δQ and θ.

Analysis: The amount of heat supplied is given by

$$Q_{1\text{-}2} = \int_1^2 \delta Q$$

$$= \int_{\theta_1}^{\theta_2} 6(2.47 - 0.025\theta^{0.75})\, d\theta$$

With $\theta_1 = (300/100) = 3$ and $\theta_2 = (400/100) = 4$, we obtain

$$Q_{1\text{-}2} = \int_3^4 6(2.47 - 0.025\theta^{0.75})\, d\theta$$

$$= \left[6\left(2.47\theta - 0.025 \cdot \frac{\theta^{1.75}}{1.75} \right) \right]_3^4$$

$$= 6\left[2.47\,(4 - 3) - \frac{0.025}{1.75}(4^{1.75} - 3^{1.75}) \right]$$

$$= 15.20 \text{ kJ}$$

2.6 SUMMARY

It is convenient to summarize the major concepts studied in this chapter in the following manner:

1. Work is a form of energy and it exists only in transition. It does not exist in a body nor can it be stored in a body. Work can result only if there is a suitable interaction between a system and its surroundings.
2. Force and displacement are essential to the process of work. For a work interaction to occur, there must be an imbalance between the external force exerted by the surroundings and the internal force exerted by the system. If this imbalance is infinitesimal, the ensuing process is quasistatic and maximum useful work is obtained. If this imbalance is finite, then the resulting process is nonquasistatic and the useful work is reduced, resulting in lost work.
3. Computation of work requires the knowledge of variation of force with respect to the displacement it produces. Work depends upon path.
4. Heat is a form of energy and it exists only during the heat transfer process. The quantity of heat energy involved in a system-surroundings interaction depends upon the nature of the process that the system experiences.
5. Temperature difference is essential to the process of heat transfer. Assumption of an infinitesimal temperature difference dT permits us to have a quasistatic heating or cooling process. A finite temperature difference always leads to a nonequilibrium process.

Notes

Exact and Inexact Differentials.   Consider the following relations

$$df = (x^2 + \sin x)\, dx$$

$$df = \left[e^{4y} + \left(\frac{1}{y} \right) \right] dy$$

$$df = z\, dz$$

In each of the above equations there is only one variable on the right side; that is, we have functions of a single variable. We can readily integrate each one of these equations to obtain a unique functional relationship between f and the single independent variable to yield

$$f(x) = \frac{x^3}{3} - \cos x$$

$$f(y) = \frac{1}{4}e^{4y} + \ln y$$

$$f(z) = \frac{z^2}{2}$$

We therefore call df in the foregoing equations an *exact differential*.

Now consider the following examples.

$$dg = (x + y) \, dx + (x^2 - y^2) \, dy$$
$$dg = y \, dx$$
$$dg = 2xy \, dx + x^2 \, dy$$

We observe that there are two variables, x and y, on the right side of each of these equations. To facilitate a formal discussion, we can express these equations by

$$dg \, (x, y) = M(x, y) \, dx + N(x, y) \, dy \qquad (2.16)$$

In order to integrate this type of equation, we must know either that (1) the condition

$$\frac{\partial M(x, y)}{\partial y} = \frac{\partial N(x, y)}{\partial x} \qquad (2.17)$$

is fulfilled by the functions $M(x, y)$ and $N(x, y)$ or that (2) a relationship exists between x and y and is valid for the interval of integration. This situation does not arise in dealing with a function of a single variable.

If the condition in (2.17) is fulfilled, the right side of Equation 2.16 is an *exact differential* of the function $g(x, y)$; otherwise, the right side of Equation 2.16 is called an *inexact differential*. If $dg \, (x, y)$ is an exact differential, it follows that

$$dg \, (x, y) = \frac{\partial g(x, y)}{\partial x} \, dx + \frac{\partial g(x, y)}{\partial y} \, dy$$

so that

$$M(x, y) = \frac{\partial g(x, y)}{\partial x} \quad \text{and} \quad N(x, y) = \frac{\partial g(x, y)}{\partial y}$$

After verifying that Condition 2.17 is fulfilled, we can obtain the integral of Equation 2.16 from

$$g(x, y) = \int M(x, y) \, dx + G(y)$$

where y in $M(x, y)$ is held constant for the integration, and

$$g(x, y) = \int N(x, y) \, dy + F(x)$$

where x in $N(x, y)$ is held constant for the integration. The unknown functions $F(x)$ and $G(y)$ can be determined by a comparison of the results from the two integrations.

If we now examine the three examples presented earlier, we find that only

$$dg = 2xy \, dx + x^2 \, dy$$

is an exact differential since

$$\frac{\partial}{\partial y}(2xy) = 2x = \frac{\partial}{\partial x}(x^2)$$

Then

$$g(x, y) = \int 2xy \, dx + G(y) = x^2 y + G(y)$$

Also

$$g(x, y) = \int x^2 \, dy + F(x) = x^2 \, y + F(y)$$

On comparing the two expressions for $g(x, y)$, we conclude that

$$G(y) = F(x) = \text{Constant} = C$$

and

$$g(x, y) = x^2 y + C$$

It should be noted that there is a class of functions that appear to be inexact differentials, but these functions can be rendered exact differentials by suitable algebraic manipulations involving integrating factors. (See any textbook on calculus.)

If we do encounter an inexact differential, then we must know the functional relationship between x and y in order to integrate Equation 2.16. Using this relationship, we can solve for y in terms of x, eliminate y from $M(x, y)$, and generate a new function $\bar{M}(x)$. In a like manner, using the given functional relationship, we can solve for x in terms of y, eliminate x from $N(x, y)$, and generate a new function $\bar{N}(y)$. This procedure will result in

$$dg \, (x, y) = \bar{M}(x) \, dx + \bar{N}(y) \, dy$$

a form that can readily be integrated. This procedure is called *contour integration*. When a function can be integrated only in this manner, the function is called a *path function*. The value of a path function depends on the relationship between the variables of integration—that is, the path taken by the variables of integration; hence the name path function.

PROBLEMS

2.1. A system consisting of 5 kg of a substance is stirred with a torque of 0.6 N·m at a speed of 500 rpm for 24 h. The system meanwhile expands from 1.5 m³ to 2.0 m³ against a constant pressure of 100 kPa. What will be the net work done in kilojoules?

2.2. A force of 28 N is exerted by a tool bit on a work piece on a lathe. The diameter of the work piece is 2 cm and it is rotating at 600 rpm. Determine the rate of work done by the work piece.

2.3. How much work energy should be supplied to accelerate a 1000-kg mass from (a) 0 to 90 km/s, and (b) 45 to 90 km/s?

2.4. A satellite of mass 200 kg is to be boosted to 400 km above the surface of the earth. The gravitational acceleration g varies with the elevation Z according to the relation

$$g = 9.81 - 3.32 \times 10^{-6}Z$$

Determine the work required to boost the satelite.

2.5. Helium is compressed in a quasistatic process according to the relation $pV^{1.2} = C$, where C is a constant. The initial conditions are 98 kPa and 20°C, and the final pressure is 1000 kPa. Assuming an ideal gas behavior, determine the work required to compress 100 kg of oxygen. Compare this work with the work of an isothermal compression.

2.6. Figure 2.18 shows a channel of width b normal to the plane of the paper, height h, and length L. The pressure on the free surface of the fluid in the channel is p_0.
   (a) Show that the force exerted by the water on the piston is $F = (p_0 + \rho gh/2)hb$, where ρ is the density of the water.
   (b) Calculate the work done by water on the piston and by the atmosphere on water when the length is slowly increased from L_1 to L_2. Express your answer in terms of p_0, b, h, L_1, L_2, ρ, and g.

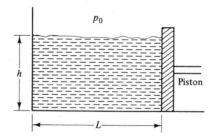

Figure 2-18

2.7. The equation of state of an ideal elastic substance is given by

$$F = KT\left[\frac{L_0}{L} - \frac{L_0^2}{L^2}\right]$$

where K is a constant and L_0 (the value of L at zero tension) is a function of the temperature alone. Calculate the work necessary to compress the substance from $L = L_0$ to $L = L_0/2$ quasistatically and isothermally.

2.8. The load-displacement characteristic of a spring is given by $P = 93y - 3y^2$ where P is the force in newtons and y is the displacement in centimeter. If the spring is stretched from its unloaded condition by 1.5 cm, determine the necessary work.

2.9. Determine the work necessary to stretch the spring in Problem 2.8 from 0.5 cm to 1.5 cm.

2.10. The tension in a wire is increased quasistatically and isothermally from F_i to F_f. If the length L, cross-sectional area A, and isothermal Young's modulus Y remain constant, show that the work done is $W = -L(F_f^2 - F_i^2)/2AY$. The Young's modulus is $Y = L/A(\partial F/\partial L)_T$.

2.11. The tensile force F in a spring is given by $F = Cy^n$ where y is the elongation of the spring, and C and n are constants. Derive an expression for the work done by the spring when it is stretched from y_1 to y_2.

2.12. A soap film is held in a rectangular area of 15 mm by 25 mm by a surface tension force of 17.2 N·m/m. Determine the work done by the film if the long side of the rectangle is moved by 7 mm to stretch the film.

2.13. A soap film is stretched across a loop with a diameter of 4 cm. A soap bubble of radius 6 cm is formed by blowing air through the loop. The surface tension of the film is 40×10^{-5} N·m^{-1}. Assuming that all soap film goes into making the bubble, determine the total work required in making the bubble.

2.14. The alternator in an automobile supplies 8.5 A of current at 12.8 V to the battery. The battery is rated at 12 V. Determine the amount of work done by the battery in 20 min.

2.15. A substance of mass 5 kg undergoes the following processes in a cylinder-piston arrangement, starting with $V_i = 0.4$ m^3 and $p_i = 500$ kPa. The final volume is $V_f = 0.2$ m^3. Calculate the work done in the following processes.
(a) p = constant.
(b) pV = constant.

2.16. Determine the work done by the substance in Problem 2.15 in a process where (a) the pressure is proportional to the volume and (b) the pressure is proportional to the square of the volume. Show the work done on a p-V diagram.

2.17. Nitrogen gas expands from 1000 kPa to 200 kPa in a quasistatic isothermal process. The initial temperature is 290 K. Determine the amount of work done per kilogram of gas in the expansion process. What would be the useful work if the gas were simply let into the atmosphere? Show both processes on a p-V plot.

2.18. When heated, nitrogen gas expands from 0.2 m^3 to 0.85 m^3 in a quasistatic process. The pressure is maintained at 1000 kPa. Determine the amount of work done by the gas and the final temperature. Assume a mass of 1 kg.

2.19. Suppose that the gas in Problem 2.18 is confined in a piston-cylinder system and that we wish to restore the gas to the original volume by pushing the piston inward. If the process is carried out by holding the temperature constant, determine the amount of work done in the process and the pressure at the end of the compression. Show the state points on a p-V diagram.

2.20. What process would restore the gas in Problem 2.19 after its isothermal compression, back to its original state of 0.2 m^3 volume and 1000 kPa pressure? What is the work done in this process?

2.21. Air at 2500 kPa and 300 K is confined in a piston-cylinder system. The piston is held in place by a spring that has a stiffness of 1000 N/mm (Figure 2.19). At this point the

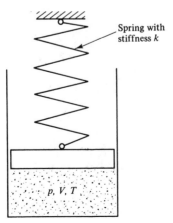

Figure 2.19

spring is in a compressed condition. The diameter of the piston face is 25 cm. Determine the amount of work done by the gas on the spring when the gas is heated in a quasistatic process until its volume is doubled. Assume the gas volume to be very small when the spring is not stressed.

2.22. A piston-cylinder system holds gas which undergoes a quasistatic process defined by $p = C/V^2$, where C is a constant. Derive an expression for the work done when the gas expands from volume V_1 to V_2. Assume ideal gas behavior.

2.23. The p-v-T relationship for gas is given by

$$\left(p + \frac{a}{v^2}\right)(v - b) = RT$$

where a and b are constants and R is the gas constant. Derive an expression for the work done by the gas in a constant temperature process.

2.24. The diameter of a spherical balloon increases from 0.2 m to 0.3 m due to heating. The initial pressure in the balloon is 200 kPa, and the pressure in the balloon is known to be proportional to its diameter. Assuming that the shape of the balloon remains spherical, calculate the work done in this process. What is the nonuseful work in this process if the atmospheric pressure is 100 kPa?

2.25. Oxygen is compressed in a quasistatic process according to the relation $pV^{1.2} = C$, where C is a constant. The initial conditions are 98 kPa and 20°C, and the final pressure is 1000 kPa. Assuming an ideal gas behavior, determine the work required to compress 100 kg of oxygen. Compare this work with the work of an isothermal compression.

2.26. The relation $M = C/T$, where C is a constant, is known as *Curie's law for paramagnetic substances*. A specified paramagnetic substance undergoes a quasistatic isothermal change of state. Show that the work done in changing the magnetization of a unit volume is

$$w = \mu \frac{T}{2C}(M_i^2 - M_f^2) = \mu \frac{C}{2T}(H_i^2 - H_f^2)$$

2.27. A sealed cylinder of diameter D and height H floats in water with one of its circular faces in contact with the water. Derive an expression for the work required to push the cylinder into the water by L meters. L is less than the initial height of the portion of the cylinder that was above the water.

2.28. The gas in a certain piston-cylinder system (see Figure 2.8) is at 500 kPa, 60°C, and it occupies 0.02 m³. The diameter of the piston is 80 cm. There are three identical weights on the piston counteracting the gas pressure. One of the weights is removed from the piston top and, after a sufficient amount of time has elapsed, the temperature is found to be 25°C. What is the nature of the process? Determine the final volume and calculate the useful and nonuseful work. The atmospheric pressure is known to be 100 kPa. Assuming that the initial and final states for a quasistatic process can be related by $p_1 V_1^n = p_2 V_2^n$, determine the exponent n, quasistatic work, and the lost work in the actual process ($R = 188.92$ J/kg K).

2.29. An insulated chamber is partitioned into two parts. One part of the chamber contains a gas at pressure p and temperature T. The other part of the chamber is evacuated. The partition is punctured at a certain moment. Determine the amount of work done by the expanding gas.

2.30. The heat supplied to a gas is given by the following relation

$$\delta q = [-0.0849 + 0.6938 \, (T/100)^{0.5} - 0.09326 \, (T/100)] \, dT$$

where q is in kJ, v in m³/kg, and p in Pa. Initially, the volume and temperature of the gas are 0.2 m³/kg and $-100°$C. Also, the R-value for the gas is 188.92 J/kg K. Determine the amount of heat that must be supplied in a quasistatic constant pressure process to raise the temperature of 1 kg of gas to $+100°$C.

2.31. An electric water heater has a resistance of 40 Ω. If it is connected across a voltage supply of 240 V for a period of 1 h, determine the nature and the magnitude of the energy interaction.

2.32. Figure 2.20 shows a battery supplying current to a resistance heater coil wrapped around a container that contains a gas. Four possible systems are shown in the figure. Determine the nature of the interaction at the system boundaries in each case—that is, if the interaction involves work, heat, both, or none.

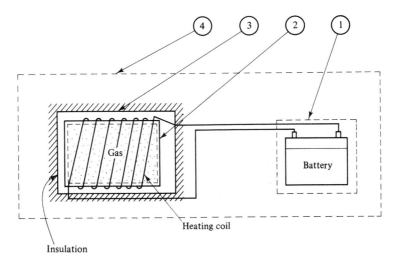

Figure 2.20

3

The First Law
of Thermodynamics
—Closed Systems

3.1 INTRODUCTION

In the preceding chapter we examined work and heat energy and some of the processes involving work for a closed system. We recall that the mass of a closed system is fixed and does not change during a process. If an elastic bar is stretched by a tensile force, work is done on the bar and strain energy is stored in the bar. When an object is raised in the gravitational field, work is done on the object and the potential energy of the object increases. When a coiled spring is wound, as in a clock, work is done on the spring, energy is stored in the spring of the clock, and this energy is available to operate the clock. In these examples we find that a supply of work results in an increase in the energy level of a system. Now let us consider some other examples. In passing an electric current through a heating coil, work is done as the electrons move against a potential difference. This work produces a heating effect on the coil. A metal rod held in a flame receives heat energy from the flame, and a heating effect is produced on the rod. In these two examples, both a supply of work energy and a supply of heat energy led to a heating effect and to an increase in the temperature. Now consider a gas in a piston-cylinder system. Let the system receive some heat energy. If we allow the gas to expand, we find that it does some work. We may also find a change in the temperature of the gas.

These interactions among work, heat, and energy can be explained only with the help of the first law of thermodynamics. Before we begin a study of the first law, it is useful to introduce the concept of the energy of a system.

3.2 ENERGY OF A SYSTEM

The energy of a system depends on its physical and chemical states, for instance, on its temperature, pressure, velocity, and elevation. The energy E of a system consists of internal energy, strain energy, electrostatic energy, magnetic energy, surface energy, potential energy, and kinetic energy. The internal energy U of a system is the total of the microscopic contributions of the energies due to the motions of the molecules and atoms that form the body. This will be discussed further in Section 3.4. The strain energy of a body is due to the change in the volume of a body when forces are applied to it. The electrostatic energy is associated with an electrically charged body moving in an electric field. Similarly, the magnetic energy is due to the movement of a polarized body in a magnetic field. You are already familiar with potential energy, PE, and kinetic energy, KE. Potential and kinetic energies are related to the coordinate system used to describe a system. The potential energy of a body of mass m under the influence of gravitational acceleration g equals mgZ, where Z is the Z-coordinate of the mass, measured in a direction opposite to that of the gravitational acceleration. Note that the Z-value is with reference to a fixed corrdinate system. The kinetic energy of a body of mass m and possessing velocity V, measured with reference to a fixed coordinate system, equals $(mV^2/2)$. The sum of all the foregoing forms of energy is the energy of a system.

For the systems that we shall consider in this text we will be chiefly concerned with internal energy U, potential energy PE, and kinetic energy KE. We therefore write

$$E = U + \text{PE} + \text{KE} \tag{3.1}$$

where

$$\text{PE} = mgZ \tag{3.1a}$$

and

$$\text{KE} = \frac{mV^2}{2} \tag{3.1b}$$

The quantity E is an extensive property. We use the symbol e (J/kg) to denote specific energy, or energy per unit mass. It is not possible to determine the absolute energy E of a system. However, changes in the energy level of a system can be determined.

For a stationary system of fixed mass, there are only two forms of energy that can be transferred to or from the system: work and heat. Whenever there is a work or heat interaction with a system, there can be an effect on the energy level of the system. The first law of thermodynamics correlates the work and heat interaction with the change in the energy level of a system.

3.3 THE FIRST LAW OF THERMODYNAMICS

It was known for many years before James Prescott Joule (1818–89) that the different forms of mechanical energy (kinetic energy, potential energy, mechanical work, and elastic energy) are mutually convertible. Thermal energy was considered rather unique. It was Joule who conclusively demonstrated that mechanical work can be converted into heat energy. His and many other experiments have established that energy cannot be created or destroyed; it can only change forms. This is known as the *first law of thermodynamics*, or the *principle of conservation of energy*. The law cannot be proved mathematically. It is based on experimental observations and nothing has yet been found in nature that violates this law. A nuclear reaction in which mass is converted into energy is not considered a violation of the law; rather mass is treated as yet another form of energy.

The law can be written for a closed system in an equation form in the following manner.

$$\begin{pmatrix}\text{Energy supplied}\\\text{to a system}\end{pmatrix}-\begin{pmatrix}\text{energy removed}\\\text{from the system}\end{pmatrix}=\begin{pmatrix}\text{increase in the energy}\\\text{level of the system}\end{pmatrix} \quad (3.2)$$

For a constant mass system, the only forms of energy that can be supplied or removed from a system are heat energy and work energy. Our adopted sign convention is such that the *heat energy entering* a system is *positive* and the work energy *leaving a system* is also *positive*. We can now write the first law for a constant mass system as

$$Q - W = \Delta E = E_2 - E_1 \quad (3.3)$$

On a unit-mass basis the above becomes

$$q - w = \Delta e = e_2 - e_1 \quad (3.3a)$$

Equation 3.3 is illustrated in Figure 3.1. A system is initially shown to have an energy level E_1. Over a period of time—the actual duration is irrelevent—a quantity of heat energy Q is supplied to the system and the system delivers a quantity of work W. After this has happened, the system has a new energy level E_2. The increase in the energy level of the system—that is, the final value of the energy of the system minus the initial value of the energy of the system—equals the quantity $Q - W$.

If very small quantities of energy are involved, we write the first law as

$$\delta Q - \delta W = dE \quad (3.3b)$$

or on a unit-mass basis we write

$$\delta q - \delta w = de \quad (3.3c)$$

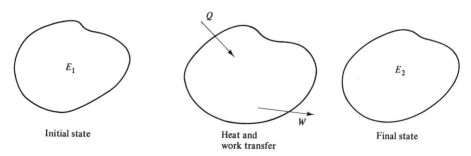

$$Q - W = E_2 - E_1 = \Delta E$$

E_1 Q E_2

W

Initial state   Heat and work transfer   Final state

Figure 3.1   The first law of thermodynamics applied to a system.

Note that on the left side of Equations 3.3(b) and 3.3(c) we have *small quantities* of heat energy and work energy, signified by the symbol δ. On the right side of the equations we have infinitesimal changes in E or e, signified by the differentials dE or de. In the discussion of the energy E of a system, it was stated that E is a property of the system. This means that a change in E can be calculated if we know a functional relationship between E and the state variables of the system. This point will be further explored in Section 3.4.

To facilitate understanding of the first law, consider the following analogy. Imagine an empty bath tub equipped with faucets and with a drain plug. We assume that the water entering the tub from the faucets is analogous to heat input, water leaving the drain is analogous to work output, and the quantity of water in the tub is analogous to the energy E of a system. Let the faucets be opened for a period of time and then closed. Initially the tub was empty; now there is some water in it. This is analogous to supplying energy to a body in the form of heat and observing an increase in its energy. Now let the drain plug be removed for some time. We then observe the amount of water in the tub has decreased, because some amount of water has left the tub. This is analogous to the decrease in the energy of a system that has provided a work output. Now let the faucets be opened and the drain plug removed at the same time, permitting a quantity of water to flow into and a quantity of water to flow out of the tub. Depending on which quantity is greater, the amount of the water in the tub will rise or fall. If the two quantities are equal, then there will not be any change in the amount of water in the tub. We can express this experience in the following statement: In a given time interval, the quantity of water that enters the tub minus the quantity of water that leaves the tub equals the change in the amount of water in the tub. This statement is analogous to Equation 3.2.

Returning to Equation 3.3 and recalling our sign convention, we note that the quantity $Q - W$ represents the excess of heat energy supplied to a system over the work done by the system. This excess of energy manifests itself as an increase in the energy E of the system. If the amount of work done by a system is greater than the amount of heat energy it receives then the extra work is done at the expense of the energy of the system. The system then experiences a decrease in its energy and ΔE is a negative quantity. There is a third possibility, namely, that the quantity of heat energy supplied is exactly equal to the quantity of work delivered by the system. In such case the quantity $Q - W$ is zero, and so is ΔE. The energy level of the system experiences no change.

Let us now apply the first law to some of the examples that were discussed earlier in a qualitative way. In heating a pot of water, the heat energy transferred to the water causes a large change in the temperature of the water and a very small change in the volume of the water. If we assume that the water is at atmospheric pressure, then the small expansion of water results in pushing the atmosphere, and a small quantity of work is done. The net effect of heating the water is a small work output from the system and an increase in the internal energy of the system. None of the terms in Equation 3.3 vanish for this case.

In lifting a weight, there is no heat interaction. The work done in lifting the weight causes an increase in the energy of the weight, and we have a positive value of ΔE. Thus

$$\overset{0}{\cancel{Q}} - W = \Delta E$$

for this system. Note that when the negative value of W in this example is subtracted, the result is a positive value for ΔE.

In the application of the first law to a system, you should identify the quantity of heat energy that is transferred across the boundaries of the system, determine if the heat energy is entering or leaving the system, and assign a sign to Q accordingly. In a similar manner the quantity of work energy that is transferred across the system boundaries should be identified. If information on the initial and final levels of the energy of a system is available, it should be accounted for in accordance with Equation 3.3. All of this information should be properly recorded on a system diagram.

Example 3.1

An elastic bar of diameter 40 mm is stretched 0.1 mm by a tensile force of 500,000 N. Determine the change in the energy, ΔE, of the rod.

Solution:

Given: The elongation of a bar of diameter 40 mm is specified as 0.1 mm. This elongation is caused by a force of 500,000 N.

Objective: We are to calculate the change in the energy of the rod.

Diagram:

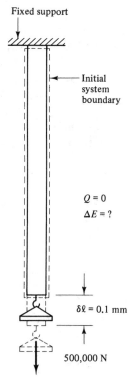

Fixed support

Initial system boundary

$Q = 0$
$\Delta E = ?$

$\delta \ell = 0.1$ mm

500,000 N

Figure 3.2

Assumptions: The force is gradually applied to the bar. Also, there are no other interactions between the system consisting of the bar and the surroundings.

Relevant physics: There is a work interaction between the system and the surroundings.
This causes a change in the energy level of the system.

Analysis: The work done by the rod is given by Equation 2.4,

$$W = -\tfrac{1}{2}P\delta\ell$$

where P is the magnitude of the applied force and $\delta\ell$ is the change in the length
of the rod. Substitution gives

$$W = -\tfrac{1}{2}(500{,}000)(0.1 \times 10^{-3})$$
$$= -25 \text{ J}$$

Since there are no other interactions between the system and the surroundings,
the heat supplied to the bar is zero. The system with its Q and W is shown in
Figure 3.2. Application of the first law gives

$$\overset{0}{\cancel{Q}} - W = \Delta E$$

or
$$0 - (-25) = \Delta E$$

or
$$\Delta E = +25 \text{ J}$$

The energy of the rod has increased by 25 J.

Example 3.2

An experimental vehicle uses a massive flywheel as an energy storage device. The
flywheel is brought up to its design speed by suitably connecting it to a stationary
energy source. The system, consisting of the vehicle and the flywheel, has an initial
energy level of 2×10^6 J. Its mass is 1500 kg. The coefficient of friction μ between
the wheels of the vehicle and road surface is 0.01. What is the final energy level of the
system if it travels a distance of 10 km on a level road?

Solution:

Given: A vehicular system of mass 1500 kg has an energy storage device, a flywheel.
The device is charged and the vehicle travels a distance of 10 km on a level road.
There is a frictional force between the wheel surface and the road. The coefficient
of friction is given to be 0.01

Objective: We are to calculate the final energy level of the system.

Diagram:

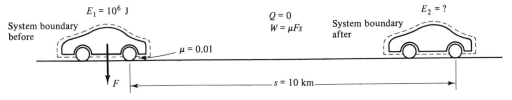

Figure 3.3

Assumptions: The only interaction between the system and the surroundings is the
work interaction. Also, the kinetic and potential energies of the vehicle at any
point are included in the energy level of the system, and so is the energy of the
flywheel.

Relevant physics: The system is required to do work in overcoming the frictional force between the road surface and the wheel surface. It is equal to the product of the coefficient of friction μ, the weight of the vehicle F, and the distance traveled s: $W = \mu F s$. The first law can then be used to determine the final energy level.

Analysis: We are given that $\mu = 0.01$, $F = 1500 \times 9.81$ N, and $s = 10$ km. Substitution gives

$$W = (0.01) \times (1500 \times 9.81) \times (10 \times 10^3)$$
$$= 1.47 \times 10^6 \text{ J}$$

These quantities are indicated in Figure 3.3. Initially, the energy level of the system is 2×10^6 J, so

$$E_1 = 2 \times 10^6 \text{ J}$$

The first law requires

$$Q - W = E_2 - E_1$$

or

$$0 - 1.47 \times 10^6 = E_2 - 2 \times 10^6$$

or

$$E_2 = 0.53 \times 10^6 \text{ J}$$

The final energy level of the system is 0.53×10^6 J.

Example 3.3

A heat engine produces power at a rate of 100 horsepower (hp). It consumes fuel at a rate of 8 g/s. The fuel when burned releases 42×10^6 J of energy per kilogram of fuel. Apply the first law and discuss the results.

Solution:

Given: The fuel consumption rate of a heat engine is given as 8 g/s. The quantity of heat energy released on combustion of 1 kg of fuel is specified as 42×10^6 J. The power output of the engine is given to be 100 hp. The horsepower is a unit of power in the English system of units. One horsepower equals 746 W.

Objective: We are to make a first-law analysis of the engine.

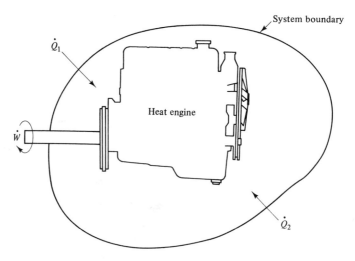

Figure 3.4

Assumptions: The only interactions between the system and the surroundings are work transfer and heat transfer. We further assume that the quantities of these interactions per unit time do not change with time.

Relevant physics: This problem involves *time rates* of supply and energy and of work output. Typically, once an engine is "warmed up," it is able to supply power (energy per unit time) uniformly. After the warming-up period, the engine itself does not experience any change in its energy, and ΔE in such a case is identically zero. The application of the first law then leads to the following equation:

$$\begin{pmatrix} \text{Energy entering the} \\ \text{heat engine per unit} \\ \text{time} \end{pmatrix} - \begin{pmatrix} \text{energy leaving the heat} \\ \text{engine per unit time} \end{pmatrix} = 0$$

The fuel that is supplied to the engine has a certain amount of chemical energy. This energy becomes available upon combustion of the fuel inside the engine. Thus a quantity of heat energy Q_1, due to the chemical energy in the fuel, enters the engine. The rate of supply of heat energy $\dot{Q}_1$ to the engine is the product of the mass flow rate of the fuel in kilogram per second, $\dot{m}$, and the amount of heat energy released by combustion of a unit mass of fuel in J/kg, HV. Thus

$$\dot{Q}_1 = \dot{m} HV$$

In Figure 3.4 we have shown $\dot{Q}_1$, as well as another rate of heat supply, $\dot{Q}_2$. It is a common experience that an engine is warm when running. Since heat energy is transferred from a warm body to a cool body, we expect heat transfer from the engine to the surroundings. This is represented by $\dot{Q}_2$. To be consistent with our sign convention, we assume an unknown entity to be positive unless proven otherwise. Consequently we have shown $\dot{Q}_2$ in Figure 3.4 by an arrow pointing into the system.

Analysis: For the system

$$\dot{Q}_1 = 8 \times 10^{-3} \times 42 \times 10^6 = 336 \times 10^3 \text{ J/s, or W}$$

The rate at which work is done is given to be 100 hp, or work done in 1 s is

$$\dot{W} = 100 \text{ hp}$$
$$= 74.6 \times 10^3 \text{ J/s, or W}$$

Application of the first law gives

$$(\dot{Q}_1 + \dot{Q}_2) - \dot{W} = 0$$

or

$$\dot{Q}_2 = \dot{W} - \dot{Q}_1 = 74.6 \times 10^3 - 336 \times 10^3$$
$$= -261 \times 10^3 \text{ J/s, or W}$$

Comments: The negative sign tells us that the engine actually rejects heat energy. We do know from experience that a heat engine rejects heat through its exhaust and via cooling water.

Note that we assumed that a quantity of heat, $\dot{Q}_2$, enters the engine per unit time that is distinct from $\dot{Q}_1$. We did not assume that a quantity of work energy distinct from $\dot{W}$ enters the engine per unit time. Natural processes are such that there is a spontaneous transfer of heat and not of work.

3.3.1 The First Law for a Cyclic Process

The cyclic process was defined in Chapter 1. To recapitulate, a cyclic process is one in which a system undergoes a series of thermodynamic processes and finally returns to its original state. A system undergoing a cyclic process has the same thermodynamic state at the end of the cycle that it had at the beginning of the cycle. To a casual observer of the system who is away during the cycle, the state of the system appears exactly the same as that before the cycle.

The thermodynamic state of a system before the beginning of the cyclic process is exactly the same as the state after the cyclic process. This means that the energy level E of the system remains unchanged as a result of the cyclic process. If it were not so, the system would register a change in its energy. This change must manifest itself as a change in some of the properties of the system, contradicting the cyclic nature of the process. Thus, in view of the first law, we conclude that in a cyclic process the net heat energy supplied to a system is equal to the net work supplied by the system. The net quantity of heat energy supplied to a system during a cyclic process is the *algebraic* sum of the quantities of heat energy supplied during all of the individual processes comprising the cyclic process. The word *algebraic* serves to emphasize that a quantity may be positive, negative, or zero. Mathematically, such an algebraic sum for a cyclic process is conveniently represented by

$$\oint \delta Q$$

The circle on the integral sign represents a cyclic process such as the one shown in Figure 3.5. In a like manner,

$$\oint \delta W$$

represents the algebraic sum of the work output from a system undergoing a cyclic process. The first law then takes the following form for a cyclic process:

$$\oint \delta Q = \oint \delta W \tag{3.4}$$

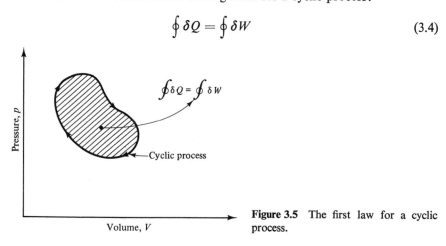

Figure 3.5   The first law for a cyclic process.

80

3.4 INTERNAL ENERGY

If we deal with a system in which the system velocity is small and changes in the elevation are negligible, then we can ignore the changes in the kinetic and potential energies of the system. Let us further assume an absence of electromagnetic and surface forces. In such a case, the energy E of a system is its internal energy U. The internal energy is due to such factors as electron spin and vibrations, molecular motion, and chemical bond. When we are dealing with a thermodynamic system on a macroscopic level, it is not necessary to inquire into the details of various microscopic contributions to the energy of a system. It is, however, instructive to examine briefly those forms of molecular energy that produce pronounced macroscopic effects. In the case of a gas, all the molecules possess varying amounts of kinetic energy because they are in constant motion. The sum total of these kinetic energies is part of the internal energy of a gas.

The molecular structure of a diatomic gas is often represented by a dumbbell. The molecule can translate in three principal directions and it can rotate about three principal axes. Such translations and rotations produce six energy contributions to the internal energy of the gas. However, no significant energy contribution comes from the rotation about the axis joining the two atoms. The diatomic molecule is therefore said to have five degrees of freedom. If we imagine that the two atoms are connected by a spring, then the atoms are capable of vibrating along the axis joining the two atoms. Thus the vibrational energy of molecules also contributes to the internal energy of a gas. Also, there are coulombic and gravitational forces that hold molecules together, and there is a potential energy due to such forces of attraction between molecules. This potential energy is relatively quite large for gases. Furthermore, on a submolecular basis, there are orbiting electrons, electrical forces, and nuclear spin, all of which contribute to the internal energy of a substance. All such contributions, as well as other microscopic contributions, are collectively called the *internal energy* U of a substance.

It can be proved that the internal energy of a system or a body is a property of the body and that it is a function of its state. This can be done by starting with the first law for a cyclic process, Equation 3.4. The concepts of point function and path function are presented in the notes at the end of this chapter. Consider a stationary system with negligible changes in its potential energy. We represent its initial state by a point X on a p-V diagram, as shown in Figure 3.6. Let the system experience a quasistatic process taking it from point X to point Y along path X–A–Y. In the process, there will be heat and work interactions, Q_A and W_A, between the system and its surroundings. These are given by

$$Q_A = \int_{XAY} \delta Q \quad \text{and} \quad W_A = \int_{XAY} \delta W$$

Let the system now return to point X from Y along a different path, path YBX, There will again be some interactions of heat and work energy between the system and the surroundings. We can identify these as

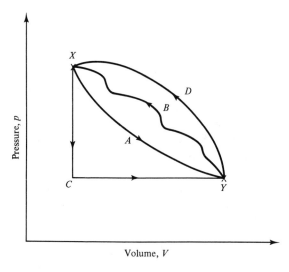

Figure 3.6 Path function and point function.

$$Q_B = \int_{YBX} \delta Q \quad \text{and} \quad W_B = \int_{YBX} \delta W$$

If we combine the two operations—paths X–A–Y and Y–B–X—we have a complete cyclic process. We have the same thermodynamic state at the end of the cyclic process as we started out with. During this cyclic operation, there is a net quantity of heat transferred to the system, which is given by $Q_A + Q_B$, and a net quantity of work delivered by the system, namely, the quantity $W_A + W_B$. According to the first law, Equation 3.4, these two quantities must be equal since there is no change in the state of the system; that is

$$\oint \delta Q = \oint \delta W \tag{3.4}$$

For the cyclic path X–A–Y–B–X the above becomes

$$Q_A + Q_B = W_A + W_B$$

Rearrangement gives

$$Q_A - W_A = W_B - Q_B$$

Substituting the integral expressions for the Ws and Qs, we obtain

$$\int_{XAY} \delta Q - \int_{XAY} \delta W = \int_{YBX} \delta W - \int_{YBX} \delta Q$$

Since the same path is used for the two integrals on the left side of the equation, we can combine the two integrals. With a similar reasoning for the right side of the equation, the last equation becomes

$$\int_{XAY} (\delta Q - \delta W) = \int_{YBX} (\delta W - \delta Q)$$

or

$$\int_{XAY} (\delta Q - \delta W) = \int_{XBY} (\delta Q - \delta W)$$

The last step is justified since we have reversed the path of integration on the right side of the equation and at the same time changed the sign of the integrand.

We know that paths $X–A–Y$ and $X–B–Y$ are completely arbitrary paths. We could very well choose paths $X–C–Y$ and $X–D–Y$ and arrive at the conclusion that no matter which path we use to take a system from point X to point Y, the integral

$$\int_{XY} (\delta Q - \delta W)$$

has the same value. We therefore conclude that the integrand $\delta Q - \delta W$ must be an exact differential, which we call dU. The integral of dU, which is the same as $\int_{XY} (\delta Q - \delta W)$, has a unique value if points X and Y are specified. Thus we write

$$\int(\delta Q - \delta W) = \int dU$$

or, in a differential form,

$$\delta Q - \delta W = dU \tag{3.5}$$

For finite changes

$$Q - W = \Delta U = U_2 - U_1 \tag{3.5a}$$

where the subscripts 1 and 2 represent initial and final states, respectively.

We have shown that the change in the internal energy of a system depends only on the initial and the final states of the system. In other words, internal energy of a system or a body is a function of its state and is therefore a property.

It is interesting to observe that δQ and δW represent small amounts of heat energy and work output, but their difference when related to a system equals an infinitesimal change in the internal energy of the system. Experience has shown that for most substances with no phase change involved, internal energy strongly depends on temperature. Its dependence on pressure and volume is relatively small. Although it is not possible to calculate the absolute value of the internal energy of a body, we can calculate changes in the internal energy. In most thermodynamic calculations it is sufficient to know about the changes in the internal energy of a body. Internal energy is an extensive property. We use the symbol u (J/kg) to denote the specific internal energy of a body.

Example 3.4

A system receives 10^7 J in the form of heat energy in a specified process, and it produces work of 4×10^6 J. The system velocity changes from 10 km/s to 25 km/s and its elevation changes from 20 m to 12 m. The mass of the system is 50 kg. Determine the change in the internal energy of the system.

Solution:

Given: The quantities of heat and work interactions of a system of 50 kg mass with its surroundings are given. Also given are the initial and final velocities and elevations of the system, $Q = +10^7$ J, $W = +4 \times 10^6$ J, $V_1 = 10$ km/s, $V_2 = 25$ km/s, $Z_1 = 20$ m, $Z_2 = 12$ m, and $m = 50$ kg.

Objective: We are to determine the change in the internal energy of the system.

Diagram:

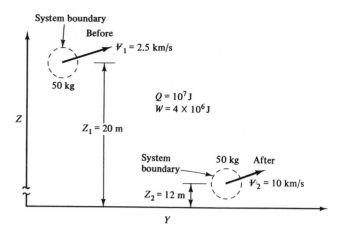

Figure 3.7

Assumptions: The mass of the system is fixed. The work and heat interactions are completed as the system goes from its initial state to the final state.

Relevant physics: This example involves a direct application of the first law to a system of fixed mass.

Analysis:

$$\Delta KE = \tfrac{1}{2}m\,(V_2^2 - V_1^2)$$
$$= \tfrac{1}{2} \times 50 \times (25^2 - 10^2)$$
$$= 13{,}125\ \text{J} \simeq 13\ \text{kJ}$$
$$\Delta PE = mg(Z_2 - Z_1)$$
$$= 50 \times 9.81 \times (12 - 20)$$
$$= -3924\ \text{J} \simeq -4\ \text{kJ}$$

According to the first law, Equation 3.3,

$$Q - W = \Delta U + \Delta PE + \Delta KE$$

Substituting, we have

$$10^7 - 4 \times 10^6 = \Delta U + 13{,}125 - 3924$$

or

$$\Delta U = 5.99 \times 10^6\ \text{J} \simeq 6000\ \text{kJ}$$

Comments: The internal energy of the system increases by almost 6×10^6 J as a result of the work and heat interactions and the changes in the velocity and the elevation of the system.

3.4.1 Enthalpy

Enthalpy is a thermodynamic property of a substance and is defined as the sum of its internal energy and the product of its pressure and volume. That is, enthaply is given by

$$H = U + pV \qquad\qquad (3.6)$$

Enthalpy is an extensive property. Specific enthalpy h, J/kg, is given by

$$h = u + pv \qquad (3.6a)$$

For a body undergoing a constant pressure process, its work is given by

$$W = \int_1^2 p \, dV = p(V_2 - V_1)$$

Application of the first law gives

$$Q - p(V_2 - V_1) = U_2 - U_1$$

or

$$\begin{aligned}
Q &= U_2 - U_1 + p(V_2 - V_1)\\
&= (U_2 + pV_2) - (U_1 + pV_1)\\
&= H_2 - H_1\\
&= \Delta H
\end{aligned}$$

Thus for a constant pressure process, the amount of heat supplied to a body equals the change in its enthalpy.

Values of enthalpy are usually listed in tables of properties (see Chapter 4). Enthalpy is particularly useful in dealing with problems involving flows. We have introduced enthalpy at this point to facilitate discussion of specific heats.

3.4.2 Specific Heats at Constant Volume and at Constant Pressure

The *specific heat* of a substance at *constant volume*, c_v, is the rate of change of specific internal energy of the substance with respect to a change in the temperature of the substance while maintaining a constant volume. It is given by

$$c_v = \left(\frac{\partial u}{\partial T}\right)_v \qquad (3.7)$$

The specific heat of a substance at *constant pressure*, c_p, is the rate of change of specific enthalpy of the substance with respect to a change in the temperature of the substance while maintaining a constant pressure. It is given by

$$c_p = \left(\frac{\partial h}{\partial T}\right)_p \qquad (3.8)$$

Specific heat is a property of a substance. Sometimes the term *heat capacity* is used. Heat capacity is the product of the mass of a substance and its specific heat.

For a substance of unit mass undergoing a constant volume process, the first law takes the form

$$\delta q - \overset{0}{\cancel{\delta w}} = du$$

With Equation 3.7, the above becomes

$$\delta q_v = c_v \, dT \qquad (3.9)$$

where the subscript v on δq is a reminder that the equation is valid only for a constant volume process.

For a substance of unit mass undergoing a constant pressure process, its enthalpy change dh is given by

$$dh = d(u + pv) = du + p\,dv + \overset{0}{\cancel{v\,dp}}$$

The last term on the right side drops out since in a constant pressure process, dp is identically zero. Now if the work done by the substance is due only to a quasistatic change in its volume, then $p\,dv = \delta w$, and $dh = du + \delta w$. Also, if the changes in the potential and kinetic energies of the substance are negligible, we can use the first law in the form of equation 3.5 and write

$$dh = du + \delta w = \delta q_p$$

The subscript p on δq is a reminder that the equation is valid only for a constant pressure process. Employing Equation 3.8, we have

$$\delta q_p = c_p\,dT \tag{3.10}$$

Equations 3.9 and 3.10 enable us to calculate the quantities of heat transferred in a constant volume process and in a constant pressure process. At one time these equations were considered to be the basis for the definition of the specific heats, resulting in the word *heat* in the property specific heat. As can be seen from the following example, the word *heat* in specific heat is a misnomer.

Imagine two identical systems with fixed volumes. Let one of them be supplied with 1000 J of heat energy produced by combustion of fuel. Let the other system be supplied with 1000 J of work in the form of electrical energy. In both cases the work done by the systems is zero since we have assumed fixed volumes. The change produced in the internal energy of both the systems is identical. In one case it is produced by a supply of heat energy, while no such supply of heat is involved in the second system. Yet the same average value of specific heat at constant volume is found (Equation 3.7) in both cases.

The variation in c_p and c_v for copper with temperature is shown in Figure 3.8. The curves are for a constant pressure of 1 atm. For low temperatures the values of c_p and c_v are nearly equal, and rapidly approach zero near a temperature of absolute zero. This behavior is typical of most solid substances. As the temperature increases, c_p continues to rise; however, c_v approaches a fixed value. It was discovered by Dulong and Petit that many solids approach a value of about 25×10^3 J/mol K (on a mole basis). It is interesting to note that the value of the universal gas constant $\bar{R}$ (8.31 J/mol K) is one-third of this limiting value of c_v. It is possible to show that the value of c_v for solids at high temperatures approaches $3\bar{R}$.

The effect of pressure on c_p and c_v is shown for mercury in Figure 3.9. The curves are typical of a number of substances.

Figure 3.10 shows the effect of temperature on the constant pressure specific heat for a very low pressure—in effect, for pressure tending to zero—for several gases. Such specific heat values are referred to as constant pressure specific heats at

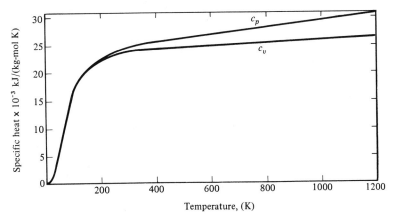

Figure 3.8 The variation in c_p and c_v with temperature for copper. (F. W. Sears, *An Introduction to Thermodynamics; The Kinetic Theory of Gasses and Statistical Mechanics*, 2/E. Reading, Mass.: Addison-Wesley © 1953. Reprinted with permission.)

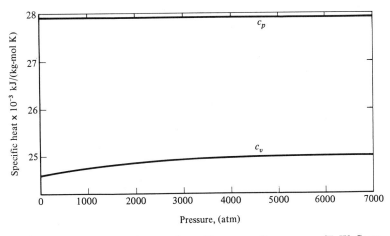

Figure 3.9 The variation in c_p and c_v with pressure for mercury. (F. W. Sears, *An Introduction to Thermodynamics; The Kinetic Theory of Gases and Statistical Mechanics*, 2/E. Reading, Mass.: Addison-Wesley © 1953. Reprinted with permission.)

zero pressure. The figure shows that as the number of atoms in the molecules of a gas increases, the curves exhibit increasing slopes. Recall from the discussion of internal energy that there are several modes of motion, such as translational, rotational, and vibrational, contributing to the total kinetic energy of a molecule. As the temperature increases, more and more of these modes make increasingly significant contributions to the internal energy. This explains the trend of the curves in Figure

3.10. In general, the effect of temperature on the specific heats is much stronger than is the effect of pressure.

It is possible to show that the internal energy of an ideal gas depends only on temperature (see Chapter 9). Joule demonstrated this fact in 1843. Thus for an ideal gas,

$$u = u(T)$$

This means that the specific heat of an ideal gas is determined solely by its temperature.

Since $h = u + pv$ and $pv = RT$ for an ideal gas, it follows that enthalpy for an ideal gas is

$$h = u(T) + RT = h(T)$$

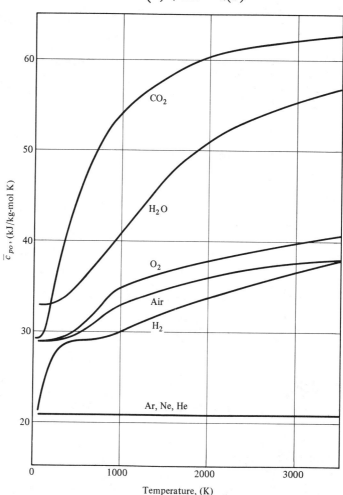

Figure 3.10 The variation in $\bar{c}_{po}$ with temperature for several gases. (G. J. Van Wylen and R. E. Sonntag, *Fundamentals of Classical Thermodynamics.* New York: John Wiley & Sons, Inc. © 1976. Used with permission.)

That is the enthalpy of an ideal gas is a function only of its temperature. Differentiating the above equation, we get

$$dh = du + R\,dT$$

For an ideal gas, from Equations 3.7 and 3.8,

$$dh = c_p\,dT \quad \text{and} \quad du = c_v\,dT$$

Substitution gives

$$c_p\,dT = c_v\,dT + R\,dT$$

or

$$c_p - c_v = R \tag{3.11}$$

Also, it follows from the above equations that, for an ideal gas, c_p and c_v are functions solely of temperature.

Since all gases behave like an ideal gas at very low pressures, the ideal gas specific heat of a substance is often called zero-pressure specific heat. Symbols c_{p0} and c_{v0} are used to denote such specific heats at constant pressure and constant volume, respectively, while symbols $\bar{c}_{p0}$ and $\bar{c}_{v0}$ denote molal specific heats. Values of c_{p0} for a number of substances are listed in Appendix A. Values of c_{v0} can be determined using Equation 3.11.

3.5 ADIABATIC PROCESS

In an adiabatic process, heat energy is neither supplied to a system nor rejected by the system. Under this condition the first law becomes

$$\overset{0}{\cancel{\delta Q}} - \delta W = dU$$

The above equation means that work is done by a system solely at the expense of its internal energy and, conversely, adiabatic work done on a system results only in an increase in the internal energy of the system. Rapid strokes of a bicycle pump result in essentially an adiabatic compression of air. An adiabatic process is an idealization, since there is a heat transfer with the slightest temperature difference between a system and its surroundings. However, such idealization is of immense use in the analyses of theoretical cycles for power generation and refrigeration.

3.5.1 Adiabatic Process for an Ideal Gas

We now develop an equation that governs the quasistatic adiabatic process for an ideal gas. An application of the first law to an adiabatic process, on a unit-mass basis, gives

$$\overset{0}{\cancel{\delta q}} - \delta w = du$$

From Equation 3.7, for an ideal gas

$$du = c_v\,dT$$

Also, if the work done by a system is due solely to a quasistatic change in its volume, then

$$\delta w = p\, dv$$

Since these equations hold for an ideal gas, we can substitute these two equations in the first-law equation to obtain

$$-p\, dv = c_v\, dT$$

Expressing p in terms of v and T using the ideal gas equation, we obtain

$$RT\frac{dv}{v} = -c_v\, dT$$

or

$$\frac{dv}{v} = -\frac{c_v}{R}\frac{dT}{T}$$

We know that R is constant; if we assume that c_v is constant, we can integrate the last equation to yield

$$\frac{R}{c_v}\ln v = -\ln T + \ln C$$

where $\ln C$ is a constant of integration. Simplification gives

$$v^{R/c_v}T = C \tag{3.12}$$

The above equation interrelates the temperature and specific volume of an ideal gas undergoing a quasistatic adiabatic process. The equation can be further simplified by the use of Equation 3.11.

$$c_p - c_v = R \tag{3.11}$$

Dividing by c_v, we get

$$\frac{c_p}{c_v} - \frac{c_v}{c_v} = \frac{R}{c_v} \quad \text{or} \quad \frac{c_p}{c_v} - 1 = \frac{R}{c_v}$$

Introducing

$$\gamma = \frac{c_p}{c_v} \tag{3.13}$$

we obtain

$$\gamma - 1 = \frac{R}{c_v} \tag{3.13a}$$

After substitution in the relation for an adiabatic process, Equation 3.12, we have

$$Tv^{\gamma-1} = C \quad \text{or} \quad TV^{\gamma-1} = C \tag{3.14}$$

where C is a constant. We can eliminate T in favor of p, the pressure, by using the ideal gas equation

$$T = \frac{pv}{R}$$

Substitution in Equation 3.14 gives

$$\frac{pv}{R} \cdot v^{\gamma-1} = \text{constant}$$

or

$$pv^\gamma = C \quad \text{or} \quad pV^\gamma = C \tag{3.14a}$$

where C is a constant. It should be emphasized that Equations 3.14 and 3.14a are valid only for an ideal gas with constant specific heat. The commonly used γ-value of 1.4 for air is for 300 K.

Equation 3.14 tells us that if the volume of a gas increases in a quasistatic adiabatic process, its temperature decreases. This is shown for air ($\gamma = 1.4$) in Figure 3.11(a). The figure shows, for instance, that to bring about a 50 percent decrease in temperature, the volume has to increase by a factor of 5.65. Also shown in the figure is a constant temperature line.

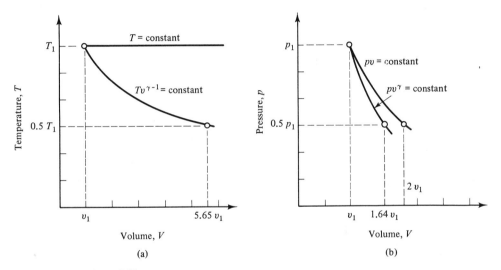

Figure 3.11
(a) T-v diagrams for air undergoing isothermal and adiabatic processes.
(b) p-v diagrams for air undergoing isothermal and adiabatic processes.

Equation 3.14(a) implies that as the volume increases in a quasistatic adiabatic process, the pressure falls. Figure 3.11(b) shows this behavior for air. It can be seen from the figure that for a 50 percent decrease in pressure, the volume increase is only 64 percent. A constant temperature curve is also shown in the figure. For this curve, a 50 percent decrease in pressure requires a 100 percent increase in the volume.

3.5.2 Determination of γ

It is possible to devise a simple experiment to determine the value of the ratio c_p/c_v $= \gamma$. We shall utilize the relationship developed for an adiabatic process.

Imagine a large glass flask fitted with a cover through which a thermometer or a thermocouple can be inserted. Let there be a pressure gauge mounted on the cover to sense the pressure inside the flask (Figure 3.12). Let some air be pumped into the flask using a hand pump. Now we have a system containing air at pressure p_1 and tempera-

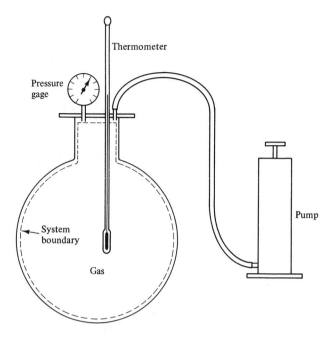

Figure 3.12 Schematic of an experiment to determine γ for air.

ture T_1. Next, slide the cover so that some air from the flask escapes and then slide the cover back in place. Air in the flask is now at pressure p_2 and temperature T_2. We can treat this process essentially as an adiabatic process if we allow very little time for any heat transfer to take place. This would, of course, require a temperature-measuring device with a short response time. We assume that the air in the flask behaves like an ideal gas.

Using Equation 3.14, we can write for the adiabatic process

$$T_1 v_1^{\gamma-1} = T_2 v_2^{\gamma-1}$$

Also, from the ideal gas equation,

$$\frac{p_1 v_1}{T_1} = \frac{p_2 v_2}{T_2}$$

Combining these two equations, we can write

$$\frac{p_1}{p_2} \cdot \frac{T_2}{T_1} = \frac{v_2}{v_2} = \left(\frac{T_1}{T_2}\right)^{1/(\gamma-1)}$$

Further simplification yields

$$\frac{p_1}{p_2} = \left(\frac{T_1}{T_2}\right)^{1/(\gamma-1)+1} = \left(\frac{T_1}{T_2}\right)^{\gamma/(\gamma-1)}$$

Taking the log of both sides, we obtain

$$\frac{\gamma}{\gamma-1} = \frac{\log(p_1/p_2)}{\log(T_1/T_2)}$$

Solving for γ gives

$$\gamma = \frac{\log(p_1/p_2)}{[\log(p_1/p_2) - \log(T_1/T_2)]} \qquad (3.14\text{b})$$

Since the values of p_1, p_2, T_1, and T_2 are known, we can determine the value of γ. For air, the value of γ determined in this manner is usually found to be close to 1.4. Values of γ for a number of gases are listed in Appendix A.

It is interesting to derive an expression for work done by a gaseous system in a quasistatic adiabatic process. We begin with the equation for work done by a gaseous system.

$$W_{1-2} = \int_1^2 p \, dV$$

We can use Equation 3.14(a),

$$pV^\gamma = C$$

to eliminate p, since the process is quasistatic adiabatic. Thus

$$W_{1-2} = \int_1^2 CV^{-\gamma} \, dV$$

Integration gives

$$W_{1-2} = \frac{CV_2^{-\gamma+1} - CV_1^{-\gamma+1}}{-\gamma + 1}$$

Substituting pV^γ for C, we obtain

$$W_{1-2} = \frac{p_2 V_2^\gamma V_2^{-\gamma+1} - p_1 V_1^\gamma V_1^{-\gamma+1}}{-\gamma + 1}$$

or

$$W_{1-2} = \frac{p_2 V_2 - p_1 V_1}{-\gamma + 1} \qquad (3.14\text{c})$$

We have so far examined the concepts of work (W), heat (Q), energy of a system (E), and internal energy of a system (U). In order to get more experience in dealing with these quantities, we now consider a few additional examples.

Example 3.5

Air at 600 K and 1200 kPa (state 1) undergoes an adiabatic expansion. The new value of pressure is 600 kPa (state 2). The mass of the air, which may be treated as an ideal gas, is 0.12 kg. Determine the work done by the air and the change in the internal energy of the air.

Solution:

Given: Air, which may be treated as an ideal gas, expands adiabatically. The initial state and the final pressure are prescribed. We are given that $T_1 = 600$ K, $p_1 = 1200$ kPa, $m = 0.12$ kg, and $p_2 = 600$ kPa.

Objective: We are to determine the work done by the air and its change of internal energy.

Diagram: For convenience, we assume that the air is confined in a piston-cylinder system.

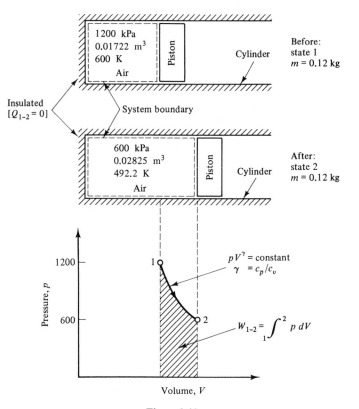

Figure 3.13

Assumptions: The adiabatic process is quasistatic. The specific heats c_p and c_v are constant: $c_v = 0.7165 \times 10^3$ kJ/kg K.

Relevant physics: The initial volume of the gas can be calculated from the ideal gas equation, $pV = mRT$. The final pressure of the gas may be worked out from the equation for a quasistatic adiabatic process, namely, $pV^\gamma =$ constant. The final temperature can be calculated from the final pressure and volume. For computation of the work done in the quasistatic adiabatic process, we need to evaluate $\int p\, dV$, bearing in mind that p must be expressed in terms of V before integrating.

Analysis: We can calculate the initial volume, V_1, from

$$V_1 = \frac{mRT_1}{p_1}$$

or

$$V_1 = \frac{0.12 \times 0.287 \times 10^3 \times 600}{1200 \times 10^3} = 0.01722 \text{ m}^3$$

The governing equation for a quasistatic adiabatic process is

$$pV^\gamma = \text{constant} = C$$

or

$$p_1 V_1^\gamma = p_2 V_2^\gamma \qquad\qquad (3.14a)$$

94

Noting that $\gamma = (c_p/c_v) = 1.4$ for air, we obtain

$$\left(\frac{V_2}{V_1}\right)^{1.4} = \frac{P_1}{P_2} = \frac{1200 \times 10^3}{600 \times 10^3} = 2$$

or

$$\frac{V_2}{V_1} = (2)^{1/1.4} = 1.64$$

and

$$V_2 = 1.64 \times 0.01722 = 0.02825 \text{ m}^3$$

Since the mass of the system remains unchanged, we can obtain the new value of the temperature T_2 from

$$T_2 = \frac{p_2 V_2}{mR} = \frac{600 \times 10^3 \times 0.02825}{0.12 \times 0.287 \times 10^3} = 492.2 \text{ K}$$

We can compile the property values in a tabular form as follows:

Property	State 1	State 2
m (kg)	0.12	0.12
p (kPa)	1200	600
V (m³)	0.01722	0.02825
T (K)	600	492.2

The expansion process is shown in Figure 3.13.
    The work done is given by

$$W_{1-2} = \int_1^2 p \, dV$$

The integral was evaluated in Equation 3.14(c) for a quasistatic adiabatic process, so that the work done is

$$W_{1-2} = \frac{p_2 V_2 - p_1 V_1}{-\gamma + 1} \tag{3.14c}$$

Inserting values of p_1, V_1, p_2, V_2, and γ, we obtain

$$W_{1-2} = \frac{600 \times 10^3 \times 0.02825 - 1200 \times 10^3 \times 0.01722}{(-1.4 + 1)} \text{ J}$$

$$= 9.28 \text{ kJ}$$

The sign of W_{1-2} is positive, meaning that work is done by the air in the expansion process.
    Irrespective of the process, the change in the internal energy of an ideal gas is given by

$$\Delta U_{1-2} = \int_1^2 mc_v \, dT$$

Since we have assumed that c_v is constant over the temperature range of interest, we can write

$$\Delta U_{1-2} = mc_v(T_2 - T_1)$$

The value of c_v for air is 0.7165 kJ/kg K, so that

$$\Delta U_{1-2} = 0.12 \times 0.7165 \times 10^3 \times (492.2 - 600)$$

$$= -9.28 \text{ kJ}$$

Comments: We could have obtained the same result by noting that the process is quasistatic adiabatic, and therefore

$$Q = 0$$

Application of the first law then gives

$$\cancel{\overset{0}{Q}} \; - W = \Delta U$$

or

$$\Delta U_{1-2} = -W = -9.28 \text{ kJ}$$

Work done by a system in an adiabatic process is at the expense of the system's internal energy.

Example 3.6

Let the system in Example 3.5 be compressed from its state 2 to state 3, to a volume of 0.01722 m³, while maintaining the pressure constant. Determine the work done, the heat supplied, and the change in the internal energy of the system.

Solution:

Given: The air in Example 3.5, after undergoing a quasistatic adibatic expansion, $Q = 0$, experiences a constant pressure compression to state 3. The new volume, V_3, equals 0.01722 m³.

Objective: We have to determine the heat and work interactions and the resulting change of internal energy for the constant pressure process.

Diagram:

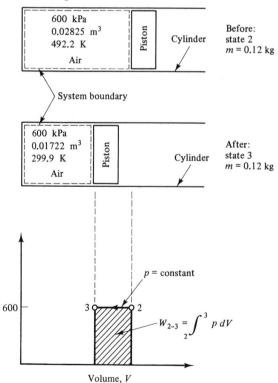

Figure 3.14

Assumptions: The compression process is quasistatic and the specific heats, c_p and c_v, are constant, with $c_v = 0.7165 \times 10^3$ kJ/kg K.

Relevant physics: In the compression process, volume decreases and work is done on the system. The temperature also decreases, so there is a decrease in the internal energy of the system. The heat interaction can be determined by applying the first law of thermodynamics.

Analysis: The system pressure, volume, and mass at state 3 are known. The temperature, T_3, can be calculated from the ideal gas law with $p =$ constant.

$$\frac{T_3}{T_2} = \frac{V_3}{V_2}$$

or

$$T_3 = T_2 \frac{V_3}{V_2} = 492.2 \times \frac{0.01722}{0.02825} = 299.9 \text{ K}$$

Properties at states 2 and 3 are tabulated below.

Property	State 2	State 3
m (kg)	0.12	0.12
p (kPa)	600	600
V (m³)	0.02825	0.01722
T (K)	492.2	299.9

The work done is shown graphically in Figure 3.14. It is

$$W_{2-3} = \int_2^3 p \, dV = p \int_2^3 dV$$

$$= p_2(V_3 - V_2)$$

$$= 600 \times 10^3(0.01722 - 0.02825) \text{ J}$$

$$= -6.62 \text{ kJ}$$

The negative sign indicates that work input is needed to compress the system. Also,

$$\Delta U_{2-3} = \int_2^3 mc_v \, dT = mc_v \, (T_3 - T_2)$$

or

$$\Delta U_{2-3} = 0.12 \times 0.7165 \times 10^3 \times (299.9 - 492.2) \text{ J}$$

$$= -16.52 \text{ kJ}$$

Application of the first law gives

$$Q_{2-3} = W_{2-3} + \Delta U$$

$$= -6.62 \times 10^3 + (-16.52 \times 10^3) \text{ J}$$

$$= -23.14 \text{ kJ}$$

That is, 23.14 kJ of heat energy is removed from the system.

Example 3.7

The system of Examples 3.5 and 3.6 is now restored to state 1 by a constant volume process. Determine the work done, the heat input, and the change in the internal energy in the cyclic process 1–2–3–1.

Solution:

Given: The air that experienced adabatic process in Example 3.5 and then a constant pressure process in Example 3.6 is now returned to its original state by a constant volume process. This means that the air has experienced a cyclic process.

Objective: We are to determine the heat and work interactions and the change of internal energy for the cyclic process.

Diagram:

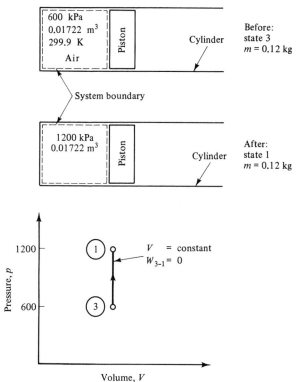

Figure 3.15

Assumptions: The constant volume process is quasistatic. The specific heats, c_p and c_v, are constant, with $c_v = 07165 \times 10^3$ kJ/kg K.

Relevant physics: The heat and work interactions for the cyclic process are the sums of the respective interactions for the three individual processes making up the cyclic process. This means that the interactions for the constant volume process will have to be calculated. Also, for the cyclic process, the two interactions are equal and the change of internal energy for the system is zero.

Analysis: The $p\,dV$ work done in a constant volume process is zero; that is, $W_{3-1} = 0$.

The work done in the cyclic process, W_{cycle}, is given by

$$W_{cycle} = W_{1-2} + W_{2-3} + W_{3-1}$$

Noting that $W_{1-2} = 9.28$ kJ and $W_{2-3} = -6.62$ kJ, we obtain

$$W_{cycle} = 9.28 \times 10^3 + (-6.62) \times 10^3 + 0 = 2.66 \text{ kJ}$$

The quantity of heat supplied in the constant volume process 3–1 can be obtained in the following manner. Application of the first law to process 3–1 gives

$$Q_{3-1} - \cancel{W_{3-1}}^{0} = \Delta U_{3-1}$$

$$Q_{3-1} = U_1 - U_3 = \int_3^1 mc_v \, dT = mc_v(T_1 - T_3)$$

$$= 0.12 \times 0.7165 \times 10^3 \times (600 - 299.9) \text{ J}$$

$$= 25.80 \text{ kJ}$$

The total quantity of heat supplied during the cyclic process is

$$Q_{cycle} = Q_{1-2} + Q_{2-3} + Q_{3-1}$$

$$= 0 + (-23.14 \times 10^3) + (25.8 \times 10^3) \text{ J}$$

$$= 2.66 \text{ kJ}$$

The above results are summarized in a tabular form:

Property	State 1	State 2	State 3
m (kg)	0.12	0.12	0.12
p (kPa)	1200	600	600
V (m³)	0.01722	0.02825	0.01722
T (K)	600	492.2	299.9

Process		W (kJ)	Q (kJ)	ΔU (kJ)
Adiabatic	1–2	9.28	0	−9.28
Constant pressure	2–3	−6.62	−23.14	−16.52
Constant volume	3–1	0	25.80	25.80
Cycle 1–2–3–1		+2.66	+2.66	0

The cyclic process is shown in Figure 3.15.

Since the air is returned to its original state in the cyclical process, the change in the internal energy of the air is zero.

Comments: The specific heat at constant pressure, c_p, does not enter the calculations for process 3–1 since it is at constant volume.

3.6 CYCLIC PROCESS

Example 3.7 illustrates a cyclic process in which three different processes took a gaseous system through a thermodynamic cycle (Figure 3.15). A thermodynamic cycle can be constructed by using two or more processes. We are already familiar with four different processes, namely, constant pressure, constant volume, constant temperature, and adiabatic. We can visualize a number of cyclic processes by combining some of these four, or any other processes, in a suitable manner. A thermodynamic cycle involves a supply of heat energy, Q_{in}, a removal of heat energy, Q_{out}, a supply of work, W_{in}, and an extraction of work, W_{out}. A cyclic process obeys the first law so that

$$\oint \delta Q = \oint \delta W$$

For N individual processes making a cyclic process, we write

$$\sum_{i=1}^{N} Q_i = \sum_{i=1}^{N} W_i$$

or

$$Q_{in} - Q_{out} = W_{in} - W_{out}$$

The quantities on the left side of the last three equations represent the algebraic sum of the heat supplied to a system during a cyclic process; this sum is referred to as Q_{cycle}, or Q_{net}, and it may be positive or negative. Likewise, the quantities on the right side of these equations represent the algebraic sum of the work done by a system during a cyclic process; this sum is referred to as W_{cycle}, or W_{net}, and it may be positive or negative. The net work of a quasistatic cycle is graphically represented by the area enclosed by the cycle on a p-V diagram. If the cycle is traversed clockwise on a p-V diagram, W_{cycle} is positive, and the cycle is said to function as a *power cycle*. If the cycle is traversed counterclockwose, W_{cycle} is negative, and the cycle is said to function as a *refrigerator*. Practical applications of counterclockwise cycles are found in a refrigerator, a compressor, and a heat pump.

People have had to devise machines to supply work. These machines, known as *heat engines*, employ thermodynamic cycles (power cycles) and the objective of such machines is to deliver work. They fulfill this objective by receiving heat energy from a high-temperature source, converting part of this energy into work, and rejecting the rest of it to a low-temperature sink. The high-tempreature source can be hot gases of combustion, heat from a nuclear reaction, or solar energy, while the low temperature sink is usually the atmosphere or large bodies of water, such as a river or a lake.

The performance of a thermodynamic cycle, a dimensionless quantity, can be considered to be given by

$$\text{Performance} = \frac{\text{useful output}}{\text{price paid to obtain the useful output}} \quad (3.15)$$

For a class of thermodynamic cycles whose objective is to produce mechanical work, the numerator in Equation 3.15 represents the *net* work output from a cycle. The price paid is the quantity of heat energy *supplied* to the cycle. Also, instead of using the word *performance*, we use the phrase *thermal efficiency* of a cycle, and denote it by η_t. Thus the thermal efficiency of a cycle is defined as

$$\eta_t = \frac{W_{\text{cycle}}}{Q_{\text{in}}} \tag{3.15a}$$

The denominator in Equation 3.15(a) represents only that quantity of heat energy that is supplied to a system during a cyclic process. Any heat energy that is removed from a system during a cycle does not enter the denominator in Equation 3.15(a). We note that the thermal efficiency of a power cycle is always less than one.

It may seem baffling that the *gross* amount of heat supplied to a cycle is used in the efficiency calculation, while in the same calculation the *net* work output of the cycle is used. The underlying logic becomes clear if we consider the objective of such cyclic processes. The principal objective of a power cycle is to extract mechanical work by supplying heat energy. What counts is the net work available at the shaft of an engine. This net work is compared with the price we have to pay to produce that work. Until recently, the heat rejected by a power cycle was not used in any manner and it was, therefore, treated as "waste" heat. Under this condition, the price to be paid is for the gross amount of heat energy entering the cycle.

In recent times, waste heat has been recovered, that is, the waste heat has been put to use for space heating, operating absorption air conditioners, and supplying domestic hot water. Whenever the waste heat from a power cycle is recovered, the amount of heat recovered is added to W_{cycle} in Equation 3.15(a), and the resulting efficiency is called *combined efficiency* of the power cycle and the recovery system. Evidently recovery of waste heat improves performance. The typical efficiency of a power plant producing electricity by burning coal, oil, or gas is about 35 percent. If such a plant is located in an area of predominantly cold climate, the waste heat can be recovered economically for space heating and the combined performance can be as high as 65 percent.

The concept of performance as defined by Equation 3.15 is applicable not only to a thermodynamic cycle, but to many other situations as well, such as a thermoelectric generator, a solar cell, a water pump, and an electric motor. The performances of these devices range from about 5 percent to 99 percent.

In Example 3.7,

$$W_{\text{cycle}} = 2.66 \text{ kJ} \quad \text{and} \quad Q_{\text{in}} = 25.8 \text{ kJ}$$

so that

$$\eta_t = \frac{2.66 \times 10^3}{25.8 \times 10^3} = 0.103, \text{ or } 10.3\%$$

We now consider two cycles, the Carnot cycle and the Otto cycle. The former is of profound importance in the theory of thermodynamics, and the latter has been adapted in the applications of thermodynamics, namely, in the automobile engine.

3.6.1 The Carnot Cycle

The Carnot cycle consists of two quasistatic adiabatic processes and two quasistatic isothermal processes. The cycle is shown in Figure 3.16(a) on a p-V plot and in Figure 3.16(b) on a T-V plot. For convenience, consider the working medium to be an ideal gas that is enclosed in a piston-cylinder system. We now procede to make first-law analyses of the four processes that comprise a Carnot cycle. We shall obtain expressions for the work done, the heat supplied, and the change in the internal energy for each of the four processes.

Let us begin by tracing the cycle starting at point 1 in Figure 3.16. Let the system be brought in thermal contact with a reservoir of energy whose temperature is greater

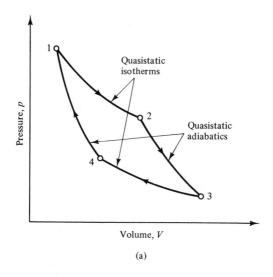

(a)

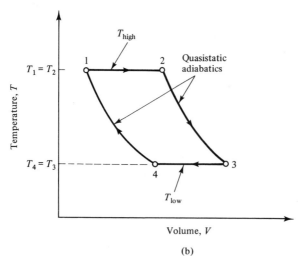

(b)

Figure 3.16 The Carnot cycle.

than the system temperature (T_1) by an infinitesimal amount. Such contact causes a quasistatic heat transfer to the system and the gas experiences an isothermal expansion, as shown by curve 1–2 in the figure. The new state of the gas at the end of this quasistatic isothermal expansion is represented by point 2. For this process,

$$\Delta U_{1-2} = \int_1^2 mc_v \, dT = 0$$

$$W_{1-2} = \int_1^2 p \, dV = mRT_1 \int_1^2 \frac{dV}{V}$$

$$= mRT_1 \ln \frac{V_2}{V_1}$$

Here Q_{1-2} can be determined from the first law.

$$Q_{1-2} - W_{1-2} = \Delta U_{1-2}^{\nearrow 0}$$

or

$$Q_{1-2} = W_{1-2} = mRT_1 \ln \frac{V_2}{V_1} \tag{3.16}$$

After the gas has reached state 2, we remove the reservoir and insulate the system so that no heat transfer takes place between the system and its surroundings. By suitably manipulating the restraining force on the system, we allow the gas to expand to state 3 in a quasistatic adiabatic fashion. For this process

$$Q_{2-3} = 0$$

and

$$W_{2-3} = \int_2^3 p \, dV = \int_2^3 CV^{-\gamma} \, dV$$

$$= \frac{p_3 V_3 - p_2 V_2}{-\gamma + 1} \qquad \text{(See Equation 3.14c.)}$$

Also, for constant specific heats,

$$\Delta U_{2-3} = \int_2^3 mc_v \, dT = mc_v(T_3 - T_2)$$

We note that in view of the first law,

$$\cancel{Q_{2-3}}^{\,0} - W_{2-3} = \Delta U_{2-3}$$

Next, we remove the insulation from the system and bring it into thermal contact with a reservoir of energy whose temperature is infinitesimally smaller than T_3, the temperature of the gaseous system at this point. The ensuing heat transfer from the system to the reservoir results in a quasistatic isothermal compression of the gas to point 4, and we have

$$\Delta U_{3-4} = 0$$

$$W_{3-4} = mRT_3 \ln \frac{V_4}{V_3}$$

and

$$Q_{3-4} = mRT_3 \ln \frac{V_4}{V_3}$$

Finally, we bring the system back to its initial state, point 1 on the diagrams, by a quasistatic adiabatic compression. For this process, we have

$$Q_{4-1} = 0$$

$$W_{4-1} = \frac{p_1 V_1 - p_4 V_4}{-\gamma + 1}$$

$$\Delta U_{4-1} = mc_v(T_1 - T_4) = -W_{4-1} = \frac{p_1 V_1 - p_4 V_4}{1 - \gamma}$$

We can now write an expression for the net work done in the Carnot cycle.

$$W_{cycle} = W_{1-2} + W_{2-3} + W_{3-4} + W_{4-1}$$

or

$$W_{cycle} = mRT_1 \ln \frac{V_2}{V_1} + \frac{p_3 V_3 - p_2 V_2}{-\gamma + 1} + mRT_3 \ln \frac{V_4}{V_3} + \frac{p_1 V_1 - p_4 V_4}{-\gamma + 1} \quad (3.17)$$

We can simplify the right side of Equation 3.17 if some relationship is established for the volumes at the beginning of each process.

We observe that processes 2–3 and 4–1 are quasistatic adiabatic, so that from Equation 3.14

$$T_2 V_2^{\gamma-1} = T_3 V_3^{\gamma-1} \quad \text{and} \quad T_4 V_4^{\gamma-1} = T_1 V_1^{\gamma-1}$$

or

$$\frac{V_2}{V_3} = \left(\frac{T_3}{T_2}\right)^{1/(\gamma-1)} \quad \text{and} \quad \frac{V_1}{V_4} = \left(\frac{T_4}{T_1}\right)^{1/(\gamma-1)}$$

Since 1–2 and 3–4 are isothermal processes,

$$T_1 = T_2 \quad \text{and} \quad T_3 = T_4$$

and the volume relations become

$$\frac{V_2}{V_3} = \left(\frac{T_3}{T_1}\right)^{1/(\gamma-1)} = \frac{V_1}{V_4}$$

Rearrangement gives

$$\frac{V_2}{V_1} = \frac{V_3}{V_4}$$

Equation 3.17 for the work done in the Carnot cycle can now be recast. To this end we use the above relationship among the four volumes and the equations for the two isothermal processes

$$p_1 V_1 = p_2 V_2 \quad \text{and} \quad p_3 V_3 = p_4 V_4$$

to yield

$$W_{cycle} = mRT_1 \ln \frac{V_2}{V_1} + \frac{p_3 V_3 - p_1 V_1}{-\gamma + 1} + mRT_3 \ln \frac{V_1}{V_2} + \frac{p_1 V_1 - p_3 V_3}{-\gamma + 1}$$

The second and the fourth terms on the right side cancel each other. The remaining terms can be combined using the property of logarithms that $\ln(a/b) = -\ln(b/a)$. Thus

$$W_{cycle} = mR(T_1 - T_3) \ln \frac{V_2}{V_1} \qquad (3.17a)$$

An expression for the thermal efficiency of the Carnot cycle is now given by

$$\eta_t = \frac{W_{cycle}}{Q_{in}} = \frac{W_{cycle}}{Q_{1-2}} \qquad (3.15a)$$

Employing Equations 3.16 and 3.17(a) for the denominator and the numerator respectively, in the above equations, we obtain

$$\eta_t = \frac{mR(T_1 - T_3) \ln (V_2/V_1)}{mRT_1 \ln (V_2/V_1)} = \frac{T_1 - T_3}{T_1}$$

or

$$\eta_t = \frac{T_1 - T_3}{T_1} = 1 - \frac{T_3}{T_1} \qquad (3.18)$$

The efficiency of a Carnot cycle is given in more descriptive terms by

$$\eta_{Carnot} = 1 - \frac{T_{low}}{T_{high}} \qquad (3.18a)$$

where T_{high} and T_{low} are the high and the low temperatures, respectively, of the Carnot cycle. For convenience, the subscript t on η in Equation (3.18) is omitted in Equation (3.18a).

Thus the thermal efficiency of a Carnot cycle depends only on the two temperatures at which the working medium receives and rejects heat energy. Equation 3.18(a) was obtained by considering an ideal gas as the working medium. It will be shown in Chapter 6 that the result is valid regardless of the type of the medium used for the Carnot cycle. It will also be demonstrated there that with T_1 and T_3 fixed as the temperatures for the heat source and the heat sink, no thermodynamic cycle can have its thermal efficiency greater than that of the Carnot cycle operating between the same temperature limits.

Example 3.8

Determine the thermal efficiency of a Carnot cycle that operates between 600°C and 300°C.

Solution:

Given: The upper and lower temperature limits of a Carnot cycle are given in degrees Celsius, $t_{high} = 600°C$ and $t_{low} = 300°C$.

Objective: We are to calculate the thermal efficiency of the Carnot cycle.

Diagram:

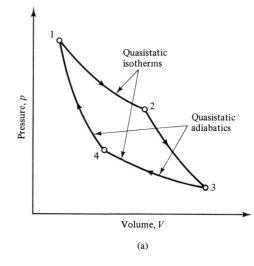

(a)

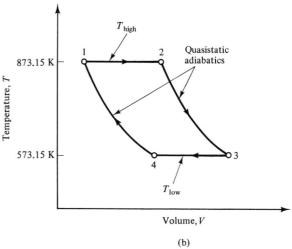

(b)

Figure 3.17

Assumptions: None

Relevant physics: By definition, all processes in a Carnot cycle are quasistatic. The cycle efficiency, Equation 3.18(a), is a function of the temperature limits within which the cycle operates. The cycle is shown in Figure 3.17.

Analysis: The thermal efficiency of a Carnot cycle is given by Equation 3.18(a) and it requires that the temperatures be expressed on an absolute scale.

$$\eta_{\text{Carnot}} = 1 - \frac{T_{\text{low}}}{T_{\text{high}}}$$
$$= 1 - \frac{273.15 + 300}{273.15 + 600}$$
$$= 0.344, \text{ or } 34.4 \text{ percent}$$

Comments: There is no way to calculate the work output of the cycle using the given data. To do so, we would need additional information such as the volume ratios in the isothermal expansion or the quantity of heat supplied.

3.6.2 The Otto Cycle

The Otto cycle is an idealized cycle consisting of two quasistatic adiabatic processes and two quasistatic constant volume processes. The Otto cycle is the basis of the reciprocating engine that uses gasoline as fuel. This engine is used extensively in automobiles. It is to be noted that the actual thermodynamic cycle used in an automobile engine deviates somewhat from the idealized Otto cycle. It is not the intent here to discuss the automobile engine and its performance, but rather to develop an expression for the thermal efficiency of the Otto cycle.

The cycle is sketched on a *p-V* diagram in Figure 3.18. For convenience, we consider an ideal gas enclosed in a piston-cylinder system as the working medium. We now proceed to analyze each of the four processes that comprise the Otto cycle by assuming constant specific heats.

Point 1 on the *p-V* diagram in Figure 3.18 represents the high-pressure, high-temperature, and low-volume point of the cycle. Let the gas expand in a quasistatic adiabatic manner to point 2, experiencing an increase in its volume but a drop in its temperature and pressure. The work done in such an expansion process is given by Equation 3.14(c).

$$W_{1-2} = \int_1^2 p\, dV = \int_1^2 CV^{-\gamma}\, dV$$
$$= \frac{p_2 V_2 - p_1 V_1}{-\gamma + 1}$$

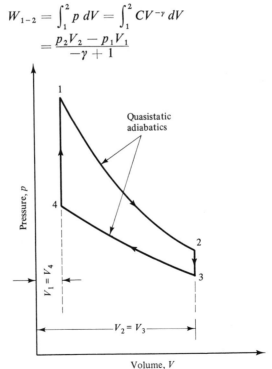

Figure 3.18   The Otto cycle.

Also,

$$Q_{1-2} = 0$$

and, in view of the first law applied to this process,

$$-W_{1-2} = \Delta U_{1-2} = mc_v(T_2 - T_1)$$

Now let the working medium be cooled under a constant volume condition. This can be accomplished in a quasistatic way if several heat sinks of progressively lower temperatures are brought in contact with the system, starting with a heat sink of temperature $T_2 - dT$. This would result in a quasistatic heat transfer from the system to the sinks caused by a temperature differential of dT. For this quasistatic constant volume process,

$$W_{2-3} = 0$$

and an application of the first law gives

$$Q_{2-3} = \Delta U_{2-3} = mc_v(T_3 - T_2)$$

We now subject the system to another quasistatic adiabatic compression process, process 3–4. For this adiabatic process, we have

$$W_{3-4} = \int_3^4 p \, dV = \int_3^4 CV^{-\gamma} \, dV$$

$$= \frac{p_4 V_4 - p_3 V_3}{-\gamma + 1}$$

$$Q_{3-4} = 0$$

and

$$-W_{3-4} = \Delta U_{3-4} = mc_v(T_4 - T_3)$$

Finally, we take the system back to its original state by employing a constant volume process, process 4–1. A supply of heat to the system is required in this process. We can write

$$W_{4-1} = 0$$

Application of the first law yields

$$Q_{4-1} = \Delta U_{4-1} = mc_v(T_1 - T_4)$$

The net work for the Otto cycle is given by

$$W_{\text{cycle}} = W_{1-2} + W_{2-3} + W_{3-4} + W_{4-1}$$

$$= \frac{p_2 V_2 - p_1 V_1}{-\gamma + 1} + 0 + \frac{p_4 V_4 - p_3 V_3}{-\gamma + 1} + 0$$

$$= \frac{mRT_2 - mRT_1}{-\gamma + 1} + \frac{mRT_4 - mRT_3}{-\gamma + 1}$$

where the ideal gas equation, $pV = mRT$, has been used. Simplification yields

$$W_{\text{cycle}} = \frac{mR}{-\gamma + 1}(T_2 - T_1 + T_4 - T_3)$$

It can be verified that the above expression also represents the net quantity of heat supplied to the cycle. The gross amount of heat supplied is in the constant volume process, namely, Q_{4-1}. We can develop now an expression for the thermal efficiency of the Otto cycle as follows.

$$\eta_t = \frac{W_{\text{cycle}}}{Q_{\text{in}}} = \frac{W_{\text{cycle}}}{Q_{4-1}}$$

Substitution of the expressions for W_{cycle} and Q_{4-1} gives

$$\eta_t = \frac{[mR/(-\gamma+1)](T_2 - T_1 + T_4 - T_3)}{mc_v(T_1 - T_4)}$$

From Equation 3.13(a),

$$c_v = \frac{R}{\gamma - 1}$$

so the expression for η_t becomes

$$\eta_t = \frac{T_1 - T_4 - T_2 + T_3}{T_1 - T_4} = 1 - \frac{T_2 - T_3}{T_1 - T_4}$$

This expression contains four different temperatures in the cycle. It is possible to simplify the expression further if we introduce the *compression ratio* r of the Otto cycle. It is the ratio of the largest and the smallest volumes of the system in the cyclic operation. (The compression ratio as understood in the automobile industry has the same meaning.) It is

$$r = \frac{\text{largest volume in the cycle}}{\text{smallest volume in the cycle}} = \frac{V_2}{V_1} = \frac{V_3}{V_4} \tag{3.19}$$

Since 1–2 and 3–4 are quasistatic adiabatic processes, from Equation 3.14

$$T_1 V_1^{\gamma-1} = T_2 V_2^{\gamma-1} \quad \text{and} \quad T_3 V_3^{\gamma-1} = T_4 V_4^{\gamma-1}$$

or

$$\frac{T_1}{T_2} = \left(\frac{V_2}{V_1}\right)^{\gamma-1} = r^{\gamma-1} \quad \text{and} \quad \frac{T_4}{T_3} = \left(\frac{V_3}{V_4}\right)^{\gamma-1} = r^{\gamma-1}$$

Therefore we conclude that

$$\frac{T_1}{T_2} = \frac{T_4}{T_3} = r^{\gamma-1} \quad \text{or} \quad \frac{T_3}{T_2} = \frac{T_4}{T_1}$$

Rewriting the expression for η_t, we obtain

$$\eta_t = 1 - \frac{T_2[1 - (T_3/T_2)]}{T_1[1 - (T_4/T_1)]}$$

$$= 1 - \frac{T_2}{T_1} \tag{3.20}$$

Substituting for the ratio (T_2/T_1) gives the following expression for the efficiency of an Otto cycle, where the subscript t on η is omitted.

$$\eta_{\text{Otto}} = 1 - \frac{1}{r^{\gamma-1}} \tag{3.21}$$

The above expression tells us that the efficiency of an Otto cycle increases as its compression ratio increases. The p-V diagram for the actual Otto cycle looks much different from that shown in Figure 3.18. There are several reasons for the deviation, which will be discussed in Chapter 7.

It is interesting to note that the expressions for the thermal efficiencies of the Carnot cycle and the Otto cycle given by Equations 3.18 and 3.20, respectively, contain ratios of two tempreatures. In Equation 3.18, the two temperatures are the two extreme temperatures in the cycle (see Figure 3.16(b)). In Equation 3.20, the two temperatures are the temperatures at the beginning and at the end of the quasistatic adiabatic expansion (see Figure 3.18). These two temperatures in the Otto cycle are fixed by the pressure ratio, and therefore it is customary to express the thermal efficiency of an Otto cycle in terms of its compression ratio r, as is done in Equation 3.21. In a Carnot cycle, the two extreme temperatures in the cycle are fixed by the temperatures of the two reservoirs exchanging heat with the cycle so these temperatures appear as such in equation 3.18 or 3.18(a) for the thermal efficiency.

3.7 SUMMARY

The basic ideas developed in this chapter can be summarized as follows.

1. Energy is conserved. If there are an inflow and an outflow of energy—in the form of heat and work—for a closed system, the system will experience a change in its energy. The change may be positive, zero, or negative.
2. Internal energy is a property of a body that depends on the physical state of the body. Only changes in the internal energy of a body can be calculated using the first law.
3. An adiabatic process is one in which no heat transfer takes place between a system and its surroundings. Since heat transfer in practice requires some time, a very rapid process may be treated as adiabatic, although it would not be quasistatic. Insulation placed on a system will also result in a reduction in heat transfer between the system and the surroundings, thus approximating the adiabatic boundary.
4. In a system undergoing a cyclic process, the final internal energy is the same as the initial internal energy. Also, for a cyclic process the net quantity of heat energy supplied to a system equals the net work output from the system.
5. The thermal efficiency of a thermodynamic cycle for power is the ratio of the net work output (useful output) to the gross amount of heat energy supplied (price paid).
6. A Carnot cycle consists of two quasistatic isothermals and two quasistatic adiabatics. Its thermal efficiency is given by $\eta_t = 1 - (T_{\text{low}}/T_{\text{high}})$, where the two temperatures are the temperatures of the two reservoirs exchanging heat with the cycle. The Carnot cycle is of profound importance in thermodynamics.

7. An Otto cycle consists of two quasistatic adiabatics and two quasistatic constant volume processes. Its thermal efficiency is given by $\eta_t = 1 - (1/r^{\gamma-1})$, where r is the compression ratio. The Otto cycle forms the basis for the thermodynamic cycle of a reciprocating automobile engine using gasoline for fuel.

Notes

Point and Path Functions. Consider two points $A(x_1, y_1)$ and $B(x_2, y_2)$ on a curve $y = f(x)$, as shown in Figure 3.19. The distance OA between the origin and point A on the curve is given by

$$OA = (x_1^2 + y_1^2)^{1/2}$$

If we arbitrarily select a point P on the curve, then its distance OP from the origin is given by

$$OP = (x^2 + y^2)^{1/2}$$

The distance of point P from the origin is a function of its coordinates (or state); in other words the distance is a *point function*.

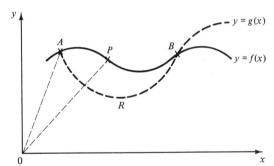

Figure 3.19   0

Now suppose that a traveler moves along the curve from point A toward point P and that the traveler is interested in the distance traveled along the journey. We cannot express this distance as $[(x_1 - x)^2 + (y_1 - y)^2]^{1/2}$. This expression represents a straight line distance between points P and A. The distance traveled by our traveler must be expressed as

$$\int_A^P ds$$

where

$$ds^2 = dx^2 + dy^2$$

We note that dx and dy appearing in ds are not arbitrary, but are regulated by the function $f(x)$ that represents the curve. That is,

$$y = f(x) \quad \text{and} \quad dy = f'(x)dx$$

where the prime denotes the first derivative of $f(x)$ with respect to x. Then the distance traveled by our traveler is given by

$$\int_{A(x_1, y_1)}^{P(x, y)} \{1 + [f'(x)]^2\}^{1/2} \, dx$$

Thus the distance traveled by the traveler is a function of path, whereas the distance separating the person from the origin is a point function, or a function of the traveler's coordinates or state.

As the traveler leaves point A for point B, there is a choice of two paths: path APB and path ARB. The traveler may cover more or less distance, depending on the path chosen. On the other hand, the distance separating the traveler from the origin depends not on the path chosen, but on position or coordinates. Work and heat depend on path, but internal energy is a point function.

PROBLEMS

3.1. A typical garage door has springs that release energy while opening the door and store energy while closing the door. One door weighs 1000 N and must be raised through 2.2 m. If there are two springs in the mechanism to open the door and if these springs can stretch no more than 0.2 m each, determine the force in the springs. Neglect friction.

3.2. The stiffness of a spring is defined as the force needed to produce a unit deflection. A helical spring has a constant stiffness value of 100 N/mm. How much force should be gradually applied so that the energy of the spring increases by 2.5 J?

3.3. How far will the experimental vehicle of Example 3.2 go on a level road before coming to a dead stop? If the road had a grade 1 in 100, how far up the road will the vehicle go?

3.4. An inventor claims to have developed a machine that will collect solar energy over an area of 10 m² and will produce power at a rate of 5 hp. Determine the rate of collection of solar energy (W/m²) that must be maintained. Comment if this rate is a realistic one.

3.5. A certain heat engine burns fuel at a rate of 3 gm/s and produces 40 kW of mechanical output. The heating value of the fuel is 40,000 J/gm. It is known that 20 percent of the energy supplied to the engine as fuel leaves the engine in its exhaust gases. What is the rate of energy given out to the cooling water of the engine? Express your answer in watts.

3.6. In modeling the variation of atmospheric pressure p with the elevation z above the surface of the earth, it is sometimes assumed that the temperature of the atmosphere remains constant. This results in the following equation for the atmospheric pressure:

$$p = p_0 \exp \frac{-\rho_0 g z}{p_0}$$

where the subscript 0 denotes values at $z = 0$. (a) Derive this equation by assuming an ideal gas law. (b) A balloon of initial volume V_1 slowly rises from $z = z_1$ to $z = z_2$ in the atmosphere. As the balloon rises, its volume increases so that $pV = $ constant. Assuming the foregoing pressure distribution, derive expressions for the work done by the balloon on the atmosphere and the change in the energy of the balloon.

3.7. A cube of side L and mass m is held in water by a rope. The cube is slowly raised through a height z_0 by pulling the rope. Determine the change in the energy level of the cube.

3.8. A fluid is held in a rigid container. It can receive energy in four different ways, as depicted in Figure 3.20. Determine Q, W, and ΔU for each of the possibilities.

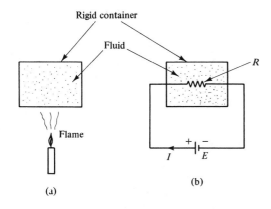

(ɹ)

(b)

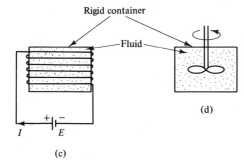

(c)

(d)

Figure 3.20

3.9. A spherical balloon contains an ideal gas. When the gas in the balloon is heated, its pressure increases in proportion to the diameter of the balloon. Derive an expression to give the amount of heat that must be supplied to the balloon to double its diameter.

3.10. Hydrogen gas is initially at 1600 kPa and 400 K. It expands to 400 kPa under quasi-static isothermal condition. Determine the work done, the heat supplied, and the change in the internal energy of the hydrogen gas. Assume the mass of the gas to be 1 kg. Draw a suitable p-V diagram and tabulate your results.

3.11. One kilogram of hydrogen gas is compressed from 400 kPa and 400 K to 1600 kPa using a quasistatic constant volume process. Determine the work done, the heat supplied, and the change in the internal energy of the gas. Show the process on a p-V diagram and tabulate your results.

3.12. Do Problem 3.10 if the process is quasistatic adiabatic.

3.13. Sketch the processes of Problems 3.10 and 3.12 on a graph paper by calculating at least four intermediate state points. What do you conclude about the relative slopes of the two curves?

3.14. Prove that the magnitude of the slope of an adiabatic curve is always greater than that for an isothermal curve if c_p is greater than c_v.

3.15. The specific heat of water vapor at constant pressure is given in (kJ/kg-mol K) by $\bar{c}_p = 143.05 - 183.54(T/100)^{0.25} + 82.751(T/100)^{0.5} - 3.6989(T/100)$. Treating water vapor as an ideal gas, determine the change in its specific enthalpy if its temperature were to change from 100°C to 200°C.

3.16. In an experiment to determine the value of γ similar to the one in Section 3.5.1, the following values were obtained: $p_1 = 2$ atm, $p_2 = 1.8$ atm, $T_1 = 27°C$, $T_2 = 23.5°C$. Determine the value of γ. By examining Table 2.1, what do you suspect to be the nature of the gas under investigation?

3.17. The work and heat transfer per degree change of temperature for a system are given by $\delta W/dT = +90$ J/K and $\delta Q/dT = +30$ J/K, respectively. Determine the change of internal energy as its temperature is changed from 160°C to 120°C.

3.18. The internal energy of a system is given by $U = 45 + 0.275T$ (kJ), where T is in degrees Kelvin. The work done on the system can be expressed by $\delta W/dT = 200$ J/K. Calculate the heat transferred if the temperature rises from 0°C to 80°C.

3.19. The internal energy (in joules) of a closed system can be expressed as $U = 100 + 50T + 0.04T^2$, and the heat absorbed (in joules) as $Q = 4000 + 16T$, where T is in degrees Kelvin. If the system changes its state from 500 K to 1000 K by a nonquasistatic process, what will be the work done in kilojoules?

3.20. A gaseous system executes a quasistatic process from an initial state 1 to a final state 2, rejecting 80 kJ of heat and expanding from 2 m³ to 2.25 m³ against a constant pressure of 400 kPa. The system is brought back to its initial state by a nonquasistatic process, absorbing 100 kJ of heat. What is the work done in the second process?

3.21. A rigid container holds 3 kg of air. The air is stirred so that its pressure changes from 500 kPa to 2000 kPa. The initial temperature is held at 50°C. The heat transferred is 200 kJ. Assume the air to behave like an ideal gas with $c_v = 0.714$ kJ/kg K. Find the final temperature, the change in internal energy, and the work done.

3.22. Nitrogen at a pressure of 200 kPa and 30°C undergoes a constant pressure process with a heat addition of 1000 kJ. The mass of the air is 4 kg. Compute (a) final temperature, (b) change in enthalpy, (c) change in internal energy, and (d) work done. Assume an ideal gas behavior.

3.23. A perfectly insulated piston-cylinder system contains 0.01 m³ of hydrogen at 25°C and 600 kPa. It is stirred at constant pressure until the temperature reaches 50°C. Determine (a) the amount of heat transferred, (b) the change in the internal energy, (c) the stirrer work, and (d) the net work. Treat hydrogen as an ideal gas.

3.24. An insulated piston-cylinder system contains helium gas. The initial state is 800 kPa, 190°C, and 0.06 m³. The gas undergoes a quasistatic process governed by the equation $pv^{1.8} = $ constant. If the final pressure of the gas is 250 kPa, determine the work done and the change in the internal energy.

3.25. What is the enthalpy change of the gas in Problem 3.24?

3.26. A piston-cylinder system, oriented vertically, contains 1 kg of a gas. The gas pressure is maintained at 600 kPa by having suitable weights on the top of the piston. In a quasistatic process, the volume changes from 0.025 m³ to 0.1 m³, and the internal energy increases by 100 kJ/kg. Determine the work done by the gas, the heat supplied to the gas, and the change in enthalpy of the gas.

3.27. Using the equations for c_p in Appendix H, determine the change in the internal energy of 5 kg of nitrogen gas when its temperature is changed from 10°C to 100°C at a constant pressure of 1 atm.

3.28. Electrical energy of 0.02 kWh is dissipated through an electric resistor placed in carbon dioxide gas in a piston-cylinder system. This causes the volume of the gas to change from 0.09 m³ to 0.21 m³ at constant pressure. The initial temperature is 300 K. Determine the magnitude and direction of heat transfer, if any. Also, determine the changes in the internal energy and the enthalpy of the gas.

3.29. An insulated piston-cylinder system contains 0.4 kg of carbon monoxide at 300° K and 500 kPa. A current of 10 A flows through a resistor within the system for 1 min. The gas expands quasistatically until its volume increases by 50%. What is the resistance of the resistor?

3.30. The internal energy of a certain material is given by $u = a + bpv$, where a and b are constants. Show that when the material undergoes a quasistatic adiabatic process, the p-v relation can be expressed by $pv^n = $ constant, where $n = (b + 1)/b$.

3.31. A parachutist jumps from an elevation of 2500 m and lands on ground. If the mass of the parachutist is 90 kg and if the heat transferred to the surroundings during the descent is 4 kJ, determine the change in the internal energy of the parachutist.

3.32. One kilogram of hydrogen gas undergoes a constant pressure process. Its initial pressure is 1600 kPa and temperature is 1600 K. Its final temperature is 400 K. Determine the work done, the change in the internal energy of the hydrogen gas and the heat supplied to it. Sketch the process on a p-V plot and tabulate your results.

3.33. Consider Problems 3.10, 3.11, and 3.32 together. You will find that the hydrogen gas has gone through a cyclic process. Determine $Q_{supplied}$, $Q_{rejected}$, and W_{cycle}. Tabulate all results and sketch the cycle on a p-V plot.

3.34. Air is used as the working medium for a Carnot cycle. The two isotherms are at 600°C and 60°C. Determine the thermal efficiency of the cycle. If the net work output of the cycle is 170 J determine the quantity of heat supplied.

3.35. One kilogram of air at 20°C and 1 atm undergoes an Otto cycle. It is compressed in a quasistatic adiabatic fashion to a pressure of 12 atm. At this point the air receives 400 kJ of heat energy under a constant-volume process. It then follows the Otto cycle. Determine all the state points and the compression ratio r. Apply the first law to determine Q, W, and ΔU for each leg of the cycle. What is the net work output of the cycle and its thermal efficiency? Tabulate your results and draw an appropriate p-V diagram for $c_p = 1.0035$ kJ/kg K and $c_v = 0.7165$ kJ/kg K.

3.36. One kilogram of air at 20°C and 1 atm (state 1) is compressed adiabatically to 24 atm (state 2). It then receives 400 kJ of heat energy at constant pressure, reaching state 3. Next it expands adiabatically to state 4, and finally a constant-volume process takes it from state 4 to state 1. Determine all the state points in the cycle and sketch the cycle on a p-V diagram. Apply the first law to determine Q, W, and ΔU for each leg of the cycle. Determine the net work output and the thermal efficiency of the cycle.

3.37. Figure 3.21 shows a quasistatic cycle executed by 0.4 kg of helium. Determine the work done, the heat supplied, and the change of internal energy for processes 1–2, 2–3, and 3–1 and for the entire cycle.

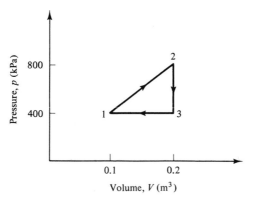

Figure 3.21

3.38. Rework Problem 3.37 if state 3 is 800 kPa and 0.1 m³.

3.39. Write a computer program that will compute thermal efficiency of an Otto cycle for different values of the compression ratio for different gases (different values of γ).

3.40. Write a short paper, based on library research, on the theoretical Otto cycle and its adaption to the automobile engine.

-4

Thermodynamic Properties of Pure Substances and Equations of State

4.1 INTRODUCTION

There is large variety of substances in nature, such as water, petroleum, air, gold, and wood. We learn from chemistry that some substances are composed of single-atom molecules. Most metals and such gases as neon, argon, and helium fall in this category. Other substances are formed of molecules made up of two or more atoms. The atoms in a molecule may belong to the same species, as in the case of such basic elements as hydrogen and oxygen, or they may belong to different species, examples of which are water (H_2O), carbon dioxide (CO_2), and methane (CH_4). These are called *chemical compounds*. Thermodynamic systems employing such chemical compounds or elements are called *single-component* systems. There are yet other types of substances that are made up of mixtures of different chemical elements and compounds. They are often called *multicomponent* systems. Air is a mixture of oxygen, nitrogen, and traces of other gases. Commercial gasoline is a blend of different hydrocarbons. All of these substances possess a variety of properties. These properties could be structural (Young's modulus, Poisson's ratio), electrical (dielectric constant, resistance), physical (volume, refractive index), or thermodynamic (specific heat, internal energy). In this chapter, we shall be concerned with the thermodynamic properties of pure substances; that is, we shall consider single component systems containing a single chemical species and their properties.

We live in an environment where temperature can change from $-50°C$ to $+45°C$ while pressure remains relatively constant. Under these conditions we observe that a piece of copper remains in a solid state, mercury remains in a liquid state, and nitrogen gas in air remains in a gaseous state. Water, however, is found in a solid, liquid, and vapor—or gaseous—state, depending on temperature. In thermodynamics the word *phase* is used to identify a system enclosed in a boundary that has the same chemical species, physical structure, and properties throughout. In general, there are three principal phases of any substance—solid, liquid, and gaseous. It is possible, however, for a solid or a liquid to exist in two or more phases. Liquid helium has two phases, helium I and helium II. Ice has seven different crystalline forms, or phases.

Water changes from a solid state to a liquid state whenever the environmental temperature changes from negative Celsius values to positive Celsius values. It is further observed that water boils at temperatures lower than 100°C on mountain tops. This phenomenon is usually explained by the statement that a drop in pressure causes a drop in the boiling point of water and the pressure on mountain tops is less than the pressure at sea level. These experiences tell us that variations in pressure and temperature influence the phases of water. In fact, changes in pressure and temperature can cause changes in the phases of all substances.

Molecular bonds are strong in solids, but weak in gases. Consequently, solids occupy a fixed volume at a given temperature and pressure, while gas molecules tend to move in random directions, so they require a closed container to hold them. When heat energy is supplied to a solid substance at constant pressure, the temperature of the solid substance increases until its melting point is reached. A further supply of heat to the substance at this point weakens the intermolecular bonds. Weakening of these bonds causes a substance to change from a solid state to a liquid state and then, eventually, from a liquid state to a gaseous state. Typically, the addition of heat energy causes an increase in the volume of a substance. An exception is a solid substance, such as ice, that contracts on melting. When work energy is supplied to a substance at constant temperature, the substance's pressure increases and its volume contracts. This contraction in volume eventually causes a substance to pass from a gaseous state to a liquid state and then to a solid state. Supply of heat energy at constant pressure causes a liquid to vaporize, while an isothermal supply of work energy to a liquid causes an increase in its pressure and a solidification of the liquid.

We shall use water to describe phase changes, although the definitions and concepts introduced will be applicable to all other substances as well. We shall use the first law to enhance the understanding of the properties of the three phases and then discuss the properties of some substances of engineering interest. Finally, real gases—which behave nonideally, (Section 1.5)—will be examined to complete our study of the three phases of substances.

4.2 LIQUID AND VAPOR PHASES

Consider a container holding water in the liquid state at room temperature and atmospheric pressure. Let a weightless, frictionless piston be placed over the water, as shown in Figure 4.1. Note the external pressure p on the piston. The thermodynamic, or state, variables for the water are pressure p, specific volume v, and temperature T. Let heat energy be supplied to this body of water in a quasistatic manner.

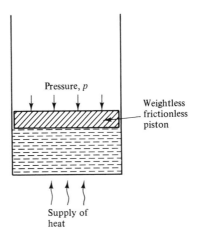

Pressure, *p*

Weightless
frictionless
piston

Figure 4.1 Constant pressure heat
supply to liquid water.

Supply of
heat

This can be accomplished by successively bringing a series of heat reservoirs at temperatures $T + \Delta T, T + 2\Delta T, \ldots$ in contact with the water. Since a frictionless,
weightless piston is placed on the top of the water, the pressure of the water equals
the atmospheric pressure *p* experienced by the piston. As the constant pressure heating
process progresses, the temperature and the volume of the water will increase. The
process is shown in the *T-v* and *p-v* diagrams in Figure 4.2. There are a number of
curves in these diagrams, but for now, let us look at the curve in each graph that
is labeled by the unsubscripted letters *a*, *b*, *c*, *d*, *e*, *f*. This curve shows the change in
the temperature and specific volume of water as the water is heated at constant pressure *p*.

  After the temperature has reached a sufficiently high level, point *c* in the figure,
the water will start to boil. In the boiling process, the liquid changes into vapor.
The temperature at which boiling occurs is called the boiling point, or the *saturation
temperature*. The state at the start of boiling is known as the *saturated liquid state*.
This is shown by point *c* in Figure 4.2. Once a liquid reaches its saturation temperature, a further supply of heat energy, however small, causes evaporation. On the
other hand, when a liquid is at its saturation temperature, removal of a small quantity
of heat energy causes the temperature of the liquid to decrease. The state of a liquid
below its saturation point, point *b* in Figure 4.2, is called a *subcooled liquid state*,
or a *compressed liquid state*. All the points on the curve *abc* except point *c* represent
subcooled liquid states.

  As water continues to be heated, more and more liquid is converted into vapor.
We find that during evaporation, the volume of the water shows a large increase as the
boiling occurs, and that the temperature remains constant. This is shown by the
line *cde* in Figure 4.2. The figure also shows that the specific volume of water vapor
is much larger than the specific volume of liquid water at atmospheric pressure. The
process along *cde* is a *constant temperature, constant pressure process of evaporation*.
At any point along the line *cde* we have a *mixture of liquid water and water vapor*,
and the total volume of the water is the sum of the volumes occupied by the liquid
and the vapor.

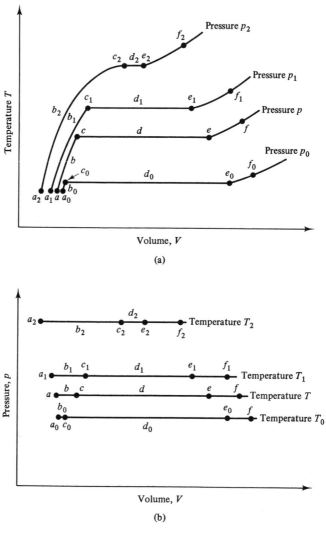

Figure 4.2 Constant pressure lines for liquid water undergoing a change of phase.
(a) on a T-v diagram.
(b) on a p-v diagram.

As heating is continued, a point is eventually reached when all the liquid has undergone a phase change and the enclosure is filled with water vapor. This state is represented by point e in Figure 4.2. Removal of a small quantity heat energy from the vapor causes condensation of a small quantity of vapor. On the other hand, further addition of heat causes a change (an increase) in the temperature of the water vapor. The water at point e is called *saturated vapor*. Addition of heat energy causes a change of phase at point c, while it causes a change of temperature at point e.

120

If the water vapor continues to receive heat energy, its temperature and volume will increase. This is a constant pressure heating process, represented by the curve *ef* in Figure 4.2. The state of the vapor at any point to the right of point *e* on curve *ef* is known as a *superheated vapor state*. We observe that *ef* in the *T-v* diagram of Figure 4.2(a) is not a straight line. Were the vapor to behave like an ideal gas, the constant pressure process *ef* would appear as a straight line. All state points on the curve *ef* except point *e* are called superheated vapor states.

Thus we see that a supply of heat energy at a constant pressure to liquid water at state *a* takes it through the various subcooled states. During this process the volume and the temperature change until the saturated liquid state is reached. Further addition of heat converts the liquid into vapor. During this phase change, both the pressure and the temperature remain constant until the saturated vapor state is reached. After the saturated vapor state is reached, no liquid is left and a further supply of heat causes an increase in the volume, as well as in the temperature.

Now let us suppose that the liquid water in the container, initially at atmospheric pressure, is pressurized to pressure p_1 while maintaining a constant temperature. The pressure of the water is now above atmospheric pressure and, referring to Figure 4.1, the water-pressure would be balanced by suitable weights on the top of the piston. Raising the pressure of water requires work input. This work input causes a slight decrease in the volume of the water. This gives us a state point, a_1, which is to the left of point *a* in Figure 4.2.

Let us now supply heat energy to the liquid water at constant pressure p_1, so that the temperature and the volume of the water increase. The constant pressure process will proceed along path $a_1 b_1 c_1$, as shown in Figure 4.2, until a new boiling point, point c_1, is reached. Point c_1 represents the saturated liquid state corresponding to pressure p_1. For water, c_1 is to the right of and above point *c*, because pressure at c_1 is greater than the pressure at *c*. As we continue heating the water beyond point c_1, more and more liquid is converted into vapor, and the process proceeds in a constant temperature, constant pressure manner along path $c_1 d_1 e_1$. The evaporation of the liquid is completed at point e_1. For water, point e_1 lies to the left of and above point *e*. Point e_1 represents the saturated vapor state corresponding to pressure p_1. The behavior of water at points *e* and e_1 is similar in that an addition of heat to the water causes an increase in the temperature and a removal of heat causes condensation of the vapor. Further addition of heat takes the water along path $e_1 f_1$. All state points on the curve $e_1 f_1$ except point e_1 are superheated vapor states.

If we were to begin with liquid water at pressure p_2 much higher than pressure p_1 and carry out a constant-pressure heating operation, the states of the water would pass along curve $a_2 b_2 c_2$, $c_2 d_2 e_2$, and $e_2 f_2$, as shown in Figure 4.2. These three curves represent the constant pressure heating of a liquid (the subcooled region), the constant pressure–constant temperature change of phase from a liquid to a vapor state (vaporization), and the constant pressure heating of a superheated vapor (superheating), respectively. The saturated liquid state, point c_2, lies to the right of and above points c_1 and *c*. The saturated vapor state, point e_2, lies to the left of and above points e_1 and *e*. This is shown in the *T-v* and *p-v* diagrams of Figure 4.2. If we were to carry out a constant pressure heating of water at a pressure below atmospheric pressure, we would find that the water passes through various states along the curve $a_0 b_0 c_0 d_0 e_0 f_0$, as shown in Figure 4.2(a). This curve also exhibits three distinct regions, namely, subcooled liquid, evaporation, and superheated vapor.

We can draw a curve through the points c_0, c, c_1, and c_2, as shown in Figure 4.3. Such a curve is called a *saturated liquid line*. A point on this curve represents a boiling point or a saturated liquid state. There is one-to-one correspondence between the temperature and the pressure at a saturation point. A constant pressure addition of heat to the liquid on the saturated liquid curve brings about a change of phase. A point to the left of this curve represents a subcooled liquid state, that is, a liquid whose temperature is below the boiling point, or the saturation temperature, corresponding to the pressure of the liquid. This point also represents a compressed liquid state, since at any such point the pressure is higher than the saturation pressure corresponding to the temperature at the point in question. On a *T-v* diagram, if we move to the left of the saturated liquid line along a constant temperature line, we shall encounter lines of successively higher pressures.

We can also draw a curve through the points e_0, e, e_1, and e_2. Such a curve, shown in Figure 4.3, is called a *saturated vapor line*. A point on this curve represents a saturated vapor state, or a condensation point. Removing heat energy at constant pressure from the vapor on a saturated vapor line causes condensation. Adding heat, on the other hand, raises the temperature of the vapor. A point to the right of the saturated vapor line represents a superheated vapor state, that is, a vapor whose temperature is above the boiling point, the saturation temperature, corresponding

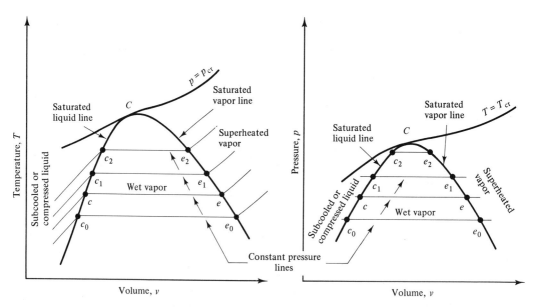

Figure 4.3 Saturated liquid and saturated vapor lines and subcooled, liquid-vapor, and superheated vapor regions on *T-v* and *p-v* diagrams.

to the pressure of the vapor. The saturated liquid line and the saturated vapor line are important because they delineate the liquid phase from the liquid-vapor mixture and the mixture from the vapor phase, respectively. It would be difficult to have a real process proceed along either of these two lines.

The distance ce represents the change in the specific volume of the water as it is taken from the saturated liquid state to the saturated vapor state. It is interesting to note that as the pressure is raised, the change in the volume of water due to evaporation becomes smaller and smaller. This is evident when we compare segments $c_0 e_0$, ce, $c_1 e_1$, and $c_2 e_2$. A question may then be asked: Is there a pressure at which the segment reduces to a point, that is, where there is no change in the volume of the water as it changes phase? The answer is yes. At a pressure called the *critical pressure*, there is no change in the volume as the liquid phase is transformed into the vapor phase. The saturated liquid line and the saturated vapor line meet at point C when the critical pressure is reached. The values of the pressure, the temperature, and the volume at point C are known as the *critical values*. Critical values for water and some other substances are listed in Table 4.1.

TABLE 4.1 CRITICAL POINT DATA FOR SELECTED SUBSTANCES

	Temperature (°C)	Pressure (MPa)	Volume (m³/kg)
Ammonia	132.35	11.28	0.004251
Carbon dioxide	31.05	7.39	0.002143
Carbon monoxide	−140.15	3.5	0.003320
Helium	−267.85	0.229	0.014439
Hydrogen	−239.85	1.3	0.032192
Oxygen	−118.35	5.08	0.002438
Nitrogen	−146.95	3.39	0.003209
Water	374.14	22.09	0.003155

At the critical point, a liquid does not exhibit a meniscus, and there is no surface separating the liquid and the vapor. It is impossible to indentify visually the difference between the liquid and the vapor, since the density of the substance changes continuously in the region near the critical point. For this reason, the substance in the critical point region is often referred to as a *fluid*, and not a liquid or vapor. The critical point provides an important reference point for the properties of a substance. The constant pressure line passing through the critical point is considered as the line separating the vapor region from the gaseous region, with the region below the line and to the right of point C being vapor and that above the line being gaseous. When the pressure of a gas is greater than its critical pressure, the gas is said to be in a *supercritical state*. The critical pressures for most gases are greater than 10 atm. Critical temperatures, however, range from 5 K to 647 K. In the selection of fluids for power plants or refrigeration plants, values of critical pressures and temperatures play a key role, since for ordinary applications it is desirable that the critical temperature and pressure of the working fluid for a power plant or refrigeration plant be well above the operating temperature and pressure ranges. It should be noted that the supercritical steam generating plants operate at pressures well above the critical pressure for water.

Referring to the process that took us from subcooled liquid states to the superheated vapor states, we note that the paths a–b–c, c–d–e, and e–f in Figures 4.2 and 4.3 represent the constant pressure processes. If we assume a quasistatic process and a stationary closed system, application of the first law on a unit mass basis gives

$$\delta q - \delta w = du$$

A rearrangement gives

$$\delta q = du + p\, dv$$

We can add the term $v\, dp$ to the right-hand side without affecting the equation, since $v\, dp$ is identically zero for a constant-pressure process. Thus

$$\delta q_p = du + p\, dv + v\, dp$$

where the subscript p on δq reminds us that we are dealing with a constant-pressure process. The two rightmost terms in this equation represent an exact differential since

$$d(pv) = p\, dv + v\, dp$$

The first law applied to a constant pressure heating process then takes the form

$$\delta q_p = du + d(pv) = d(u + pv) = dh$$

We recognize that the combination $u + pv$ is defined to be the specific enthalpy h (Section 3.4.1), which always appears in a constant pressure heating or cooling process of a substance. Ideally, in industrial applications involving devices such as a steam generator or a condenser, the heating or cooling process takes place at a constant pressure. The same is true of heat exchangers. It is therefore natural that the combination $u + pv$, or enthalpy h, should appear in thermodynamic analyses of such devices. The advantage of employing the enthalpy is that when a computation for an energy balance is involved, we have to deal with a single variable, h, and not with three variables, u, p, and v. Enthalpy plays the same role in a constant pressure process as internal energy does in a constant volume process.

4.2.1 Quality of a Liquid-Vapor Mixture

The saturated liquid line and the saturated vapor line for a substance form a bell-shaped curve that passes through the critical point, as seen on the T-v and p-v diagrams in Figure 4.3. This bell-shaped curve is sometimes referred to as the *saturation dome* for a substance. A state point inside the dome represents a liquid-vapor mixture in thermodynamic equilibrium. A liquid-vapor mixture in thermodynamic equilibrium has the same temperature and pressure throughout the mixture. It will be assumed throughout this text that a given mixture is in thermodynamic equilibrium unless otherwise stated. When we speak of a liquid-vapor mixture, we may have a container with a liquid at the bottom and some vapor at the top, or we may have vapor bubbles dispersed throughout a liquid. The mixture may also consist of a vapor cloud with fine liquid droplets dispersed throughout it.

In order to compute an extensive property of a mixture, such as volume or internal energy, we must take into account the relative proportions of the liquid and the vapor components in the mixture and their contributions to the extensive property in question. The quality x of a liquid-vapor mixture expresses the proportion of the vapor component in a liquid-vapor mixture. When steam is produced in small boilers, the resulting steam is not in a saturated vapor state, but it contains a certain amount of moisture, that is, saturated liquid. This means that the end product lacks in quality; the drier the steam, the higher the quality. The word *quality*, in reference to steam, is indicative of the extent of the dryness of the steam: 100 percent quality means completely dry steam. Quality, x, is defined to be the ratio of the mass of the vapor, m_g, to the mass of the misture, m. Or

$$x = \frac{m_g}{m} = \frac{m_g}{m_f + m_g} \tag{4.1}$$

where m_f is the mass of the liquid in the mixture and m_g is the mass of the vapor in the mixture. The subscripts f and g are generally used for saturated liquid and saturated vapor, respectively, in the literature in discussing properties of substances. The subscripts apparently have their origin in the German words *flussigkeit* for a liquid and *gas* for vapor. It is also customary to refer to x as the *quality* of a vapor. If the quality of a vapor is less than unity—that is, if it contains some moisture—the vapor is called *wet vapor*. The term *moisture* is used to represent the fraction of the liquid in a mixture. It is equal to $1 - x$.

An extensive property of a liquid-vapor mixture can then be calculated from the following.

$$
\begin{array}{l}
\text{Extensive property} \\
\text{of a liquid-vapor} \\
\text{mixture}
\end{array}
=
\begin{array}{l}
\text{(mass of the liquid component)} \cdot \text{(intensive} \\
\text{property of the liquid component)} \\
+ \text{(mass of the vapor component)} \cdot \text{(intensive} \\
\text{property of the vapor component)}
\end{array}
\tag{4.2}
$$

If we consider a unit mass of the mixture, then the above becomes

$$
\begin{array}{l}
\text{Specific intensive} \\
\text{property of the} \\
\text{mixture}
\end{array}
=
\begin{array}{l}
(1 - x) \text{ (specific intensive property of the} \\
\text{liquid component)} \\
+ x(\text{specific intensive property of the vapor} \\
\text{component)}
\end{array}
\tag{4.2a}
$$

where x is the quality of the mixture. Throttling and separating calorimeters are used for experimental determination of quality.

If we use the subscript f to denote the liquid component and the subscript g to denote the vapor component, then a property y of the mixture can be expressed as

$$y = (1 - x)y_f + xy_g \tag{4.2b}$$

where y_f is the property of the liquid component and y_g is the property of the vapor component.

4.2.2 Tables of Properties for the Liquid and the Vapor Phases

We saw in Chapter 3 that if we have an ideal gas and if we know the values of two of the three variables p, v, T describing the state, we can calculate the value of the third variable. The changes in the internal energy and enthalpy of the ideal gas can also be calculated if we know the initial and the final state points. When we are dealing with a subcooled or compressed liquid, a liquid-vapor mixture, or a superheated vapor, evidently we cannot use the ideal gas law. We must depend on the experimentally determined values of the properties of the phase being considered. In certain instances, it might be possible to fit an equation to the experimental values; this equation would then indeed be a convenient tool. Very often, though, the equation is too complicated to use for hand calculations, and we rely on tabulated values of properties. Tabulated values are obtained by evaluating that equation which accurately fits the experimental data. We shall now discuss the tables of properties—in particular, those for water.

When evaporation of a liquid or condensation of a vapor is considered, the liquid and vapor saturation points for a given pressure form excellent reference points. The saturated liquid point is designated by the subscript f (for fluid) and the saturated vapor state is denoted by the subscript g (for gas), as noted in the preceding section. The subscript fg denotes a complete change from the liquid to the vapor state. Thus we have the following definitions for specific volumes:

v_f Specific volume of saturated liquid (state point on the left leg of the saturation dome, Figure 4.3).

v_g Specific volume of saturated vapor (state point on the right leg of the saturation dome).

v_{fg} Change in the specific volume between the saturated liquid state and the saturated vapor state. It equals $v_g - v_f$.

We can define $u_f, u_g, u_{fg}, h_f, h_g$, and h_{fg} in a similar manner. The quantity h_{fg} is commonly known as the *latent heat of vaporization*, or simply as *heat of evaporation*.

Appendix Bl gives values of specific volume, specific internal energy, and specific enthalpy for compressed liquid water at pressures ranging from 5000 kPa to 50,000 kPa and temperatures ranging from 0°C to 380°C. Appendix B2.1 lists these properties for saturated liquid water and saturated water vapor. The key to the use of this appendix is the temperature of water (0.01°C to 374.14°C); that is, values of temperature appear in the leftmost column and a uniform increment for the temperature is used virtually throughout the temperature range for which properties are listed. Appendix B2.2 lists the same properties, but employs pressure (0.6113 kPa to 22,090 kPa) as the key to the table. Appendix B3 lists the properties of superheated steam. The tabulation of properties in Appendixes B2.1, B2.2, and B3 are collectively called *steam tables*.* In these tables the specific internal energy of saturated liquid water at 0.01°C is assigned a value of zero as a convenient reference state.

*An additional property called entropy (s) also appears in these tables. Its significance is discussed in Chapter 6.

126

Excerpts from the above Appendixes are reproduced in Table 4.2. Part A relates to compressed liquid, Parts B and C relate to saturated liquid water, and Part D relates to superheated steam. Let us suppose that we need the value of the enthalpy of water at 5000 kPa and 100°C. We know from the values in Appendix B2.2 that the saturation temperature for 5000 kPa is 263.99°C. This temperature is greater than the given temperature of 100°C. Therefore we surmise that we have a compressed or subcooled liquid on hand and we read the h-value for 100°C from Part A in the table, namely, 422.72 kJ/kg. The first column in Part B, Table 4.2 (and Appendix B2.1) lists temperatures, while that in Part C, Table 4.2 (and Appendix B2.2) lists pressures. Both parts list values of saturation pressure and saturation temperature. Values of specific volume, specific internal energy, and specific enthalpy for saturated liquid and saturated vapor, as well as changes in internal energy and enthalpy due to evaporation, are listed in Parts B and C and in Appendixes B2.1 and B2.2. For example, if we need the value of the specific internal energy of saturated water vapor at 105°C, we read 2512.4 kJ/kg as the u_g value in the line for 105°C from Part B. If we need the latent heat of vaporization at 200 kPa, we read 2201.9 kJ/kg as the value of h_{fg} from the line for 200 kPa from Part C. The superheated vapor table, part D (and Appendix B3), is divided into a number of sections. Each section relates to only one value of pressure, which appears at the top of the section. The number in the parentheses represents the saturation temperature. Under each section, values of specific volume, internal energy, and enthalpy are listed for a series of values of temperature.

Besides water, which is widely used in power cycles, there are other substances that are of significant importance in engineering applications. Some of these substances are Freon-12, ammonia, and mercury. Properties of these substances in saturated and superheated states are tabulated in Appendixes C, D, and E. Other substances of engineering interest are oxygen and nitrogen. The properties of these gases in superheated state are listed in Appendixes F and G. For Freon-12 and ammonia the specific enthalpy of saturated liquid at −40°C is assigned a value of zero as a convenient reference state. For other substances the reference states for zero enthalpy may be different.

When a substance in a liquid or a vapor state is to be described in engineering applications, pressure and temperature are often specified. It then becomes necessary to ascertain the phase of the substance. To determine the phase, look up the saturation temperature, t_{sat}, from Appendix B2.2 for the given pressure and compare t_{sat} with the given temperature t.

1. If $t < t_{sat}$, the given substance is in a subcooled or compressed liquid state.
2. If $t > t_{sat}$, the given substance is in a superheated vapor state.
3. If $t = t_{sat}$, the given substance can be in (a) a saturated liquid state, (b) a saturated vapor state, or (c) a liquid-vapor mixture under thermodynamic equilibrium. Some additional information is necessary to pinpoint the state.

Sometimes a system is described as a saturated liquid or a saturated vapor, and a value of either the temperature or the pressure is specified. In such cases, we know immediately that the state is represented by a point somewhere on the saturation dome.

TABLE 4.2 EXCERPTS FROM STEAM TABLES

A. Properties of Compressed Liquid Water

T	$p = 5000$ kPa (263.99)				$p = 10,000$ kPa (311.06)				$p = 15,000$ kPa (342.24)			
	v	u	h	s	v	u	h	s	v	u	h	s
80	0.0010268	333.72	338.85	1.0720	0.0010245	332.59	342.83	1.0688	0.0010222	331.48	346.81	1.0656
100	0.0010410	417.52	422.72	1.3030	0.0010385	416.12	426.50	1.2992	0.0010361	414.74	430.28	1.2955
120	0.0010576	501.80	507.09	1.5233	0.0010549	500.08	510.64	1.5189	0.0010522	498.40	514.19	1.5145
140	0.0010768	586.76	592.15	1.7343	0.0010737	584.68	595.42	1.7292	0.0010707	582.66	598.72	1.7242

B. Saturated Steam Temperature Table

Temp °C T	Press. MPa p	Specific volume m³/kg		Internal energy kJ/kg			Enthalpy kJ/kg		
		Sat. Liquid v_f	Sat. Vapor v_g	Sat. Liquid u_f	Evap. u_{fg}	Sat. Vapor u_g	Sat. Liquid h_f	Evap. h_{fg}	Sat. Vapor h_g
100	0.10135	0.001044	1.6729	418.94	2087.6	2506.5	419.04	2257.0	2676.1
105	0.12082	0.001048	1.4194	440.02	2072.3	2512.4	440.15	2243.7	2683.8
110	0.14327	0.001052	1.2102	461.14	2057.0	2518.1	461.30	2230.2	2691.5
115	0.16906	0.001056	1.0366	482.30	2041.4	2523.7	482.48	2216.5	2699.0

C. Saturated Steam Pressure Table

Press. kPa p	Temp. °C T	Specific volume m³/kg		Internal energy kJ/kg			Enthalpy kJ/kg		
		Sat. Liquid v_f	Sat. Vapor v_g	Sat. Liquid u_f	Evap. u_{fg}	Sat. Vapor u_g	Sat. Liquid h_f	Evap. h_{fg}	Sat. Vapor h_g
50	81.33	0.001030	3.240	340.44	2143.4	2483.9	340.49	2305.4	2645.9
75	91.78	0.001037	2.217	384.31	2112.4	2496.7	384.39	2278.6	2663.0
MPa									
0.100	99.63	0.001043	1.6940	417.36	2088.7	2506.1	417.46	2258.0	2675.5
0.125	105.99	0.001048	1.3749	444.19	2069.3	2513.5	444.32	2241.0	2685.4
0.150	111.37	0.001053	1.1593	466.94	2052.7	2519.7	467.11	2226.5	2693.6
0.175	116.06	0.001057	1.0036	486.80	2038.1	2524.9	486.99	2213.6	2700.6
0.200	120.23	0.001061	0.8857	504.49	2025.0	2529.5	504.70	2201.9	2706.7
0.225	124.00	0.001064	0.7933	520.47	2013.1	2533.6	520.72	2191.3	2712.1

D. Superheated Vapor

T	$p = 0.010$ MPa (45.81)			$p = 0.050$ MPa (81.33)		
	v	u	h	v	u	h
Sat.	14.674	2437.9	2584.7	3.240	2483.9	2645.9
50	14.869	2443.9	2592.6			
100	17.196	2515.5	2687.5	3.418	2511.6	2682.5
150	19.512	2587.9	2783.0	3.889	2585.6	2780.1
200	21.825	2661.3	2879.5	4.356	2659.9	2877.7
250	24.136	2736.0	2977.3	4.820	2735.0	2976.0

If a system is a mixture of liquid and vapor phases, then one more piece of information must be specified besides pressure and temperature. This information could be about the quality x, the volume v, the internal energy u, or the enthalpy h of the mixture. The values of x, v, u, or h can be used together with the following equations to determine the rest of the properties.

$$v = (1 - x)v_f + xv_g = v_f + xv_{fg} \tag{4.3}$$

$$u = (1 - x)u_f + xu_g = u_f + xu_{fg} \tag{4.3a}$$

$$h = (1 - x)h_f + xh_g = h_f + xh_{fg} \tag{4.3b}$$

Example 4.1

Determine the specific enthalpy of water at (a) 40°C and 25 kPa, (b) saturated at 120°C, (c) 300°C and 500 kPa, and (d) 150°C and a specific volume of 0.27 m³/kg.

Solution:

Given: Four different states of water are given.

Objective: The specific enthalpy for each state is to be determined.

Assumptions: None.

Relevant physics: For a given state there are five possible categories: subscooled or compressed liquid, saturated liquid, saturated liquid–saturated vapor mixture, saturated vapor, and superheated vapor. Once we determine the category to which a given state belongs, we can refer to the appropriate table in Appendix B.

Analysis: (a) On checking the saturated steam tables, Appendix B2.2, we find that for $p = 25$ kPa, $t_{sat} = 64.97°C > 40°C$. Therefore the given water is in a subcooled liquid state. To determine its specific enthalpy, we look up the table for compressed liquid water in Appendix Bl and discover that the smallest value of pressure in the table is 5 Mpa, or 5000 kPa. We can, however, approximately determine the specific enthalpy by using the saturated steam table. It will be demonstrated later in the text that the change in the value of h (and u) is much more sensitive to a change in temperature than it is to a change in pressure. Therefore we can use the value of h_f for saturated liquid at 40°C. Appendix B2.1 gives $h_f = 167.57$ kJ/kg for $t = 40°C$. Thus

$$h \text{ at 25 kPa and } 40°C \simeq h_f \text{ at } 40°C = 167.57 \text{ kJ/kg}$$

(b) Since it is not stated whether we have a saturated liquid or a saturated vapor, we shall determine the specific enthaplies for both the states. From the table of properties of saturated steam in Appendix B2.1

h at 120°C for saturated liquid $= h_f = 503.71$ kJ/kg

h at 120°C for saturated vapor $= h_g = 2706.3$ kJ/kg

and

h for saturated water at 120°C is in the range 503.71 kJ/kg and 2706.3 kJ/kg

(c) From Appendix B2.2, the saturation temperature for water corresponding to 500 kPa is 151.86°C. The temperature specified is 300°C. Thus we have superheated steam. From the table of properties of superheated steam, Appendix B3, we find that

$$h = 3064.2 \text{ kJ/kg}$$

(d) For $t = 150°C$, Appendix B2.1 gives

$$v_f = 0.001091 \text{ m}^3/\text{kg} \quad \text{and} \quad v_g = 0.3928 \text{ m}^3/\text{kg}$$

The volume given is $0.27 \text{ m}^3/\text{kg}$, which is greater than v_f but less than v_g. We conclude therefore that we have a mixture of a liquid and a vapor. We first determine the quality x from Equation 4.3:

$$v = (1 - x)v_f + xv_g$$

Solving for x, we obtain

$$x = \frac{v - v_f}{v_g - v_f}$$

$$= \frac{0.27 - 0.001091}{0.3928 - 0.001091}$$

or

$$x = 0.687$$

From Equation 4.3(b),

$$h = (1 - x)h_f + xh_g = 0.313(632.2) + 0.687(2746.5)$$
$$= 2083.7 \text{ kJ/kg}$$

4.3 THE FIRST LAW AND LIQUID AND VAPOR PHASES

Liquid Phase.   The application of the first law involves computation of the work done, heat energy supplied, and the change in the internal energy. When we want to calculate the quasistatic work done in compressing or expanding a liquid, we must evaluate the integral $\int p \, dV$. This evaluation requires a knowledge of the relationship between pressure, p, and volume, V, for the process. Since such a relationship is not always available, it is difficult to evaluate the integral. There are, however, two cases in which the integral can be evaluated. The two cases are (1) an ideal, or incompressible, liquid, and (2) a liquid undergoing a constant pressure process.

An *ideal, or incompressible, liquid* is one that experiences no change in its volume regardless of the changes in its pressure. This means that dV is zero for any process undergone by an ideal liquid. Consequently, the work of expansion or compression for an incompressible liquid is zero and $\int p \, dV = 0$ for an ideal liquid. On the other hand, when a real liquid undergoes a constant pressure process, we have

$$W_{1-2} = \int_1^2 p \, dV = p(V_2 - V_1)$$

In many practical situations, a liquid undergoes a constant pressure heating or cooling process. Since the change in the volume for a liquid is usually small, the amount of work involved in such a process is also small.

The change in the internal energy of the liquid can be calculated from $dU = mc_v \, dT$ if a relationship between c_v and T is known. Alternatively, with known end states, the u–values in the property tables can be used to determine the change in the internal energy. Finally, an application of the first law, Equation 2.3, gives the amount of heat supplied to the liquid.

Example 4.2

Fifteen kilograms of liquid water are heated at constant pressure from 20°C to 200°C. Determine the work done and the heat supplied if the pressure is (a) 5000 kPa, and (b) 50,000 kPa.

Solution:

Given: A mass of 15 kg of liquid water undergoes a constant pressure process.

$$\text{Initial state:} \quad t_1 = 20°C$$

$$\text{Final state:} \quad t_2 = 200°C$$

Objective: The work done and heat supplied in the constant pressure process are to be determined for two different values of pressures, 5000 kPa and 50,000 kPa.

Diagram:

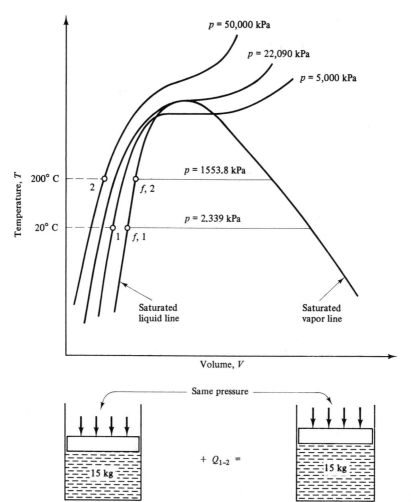

Figure 4.4

Assumptions: None.

Relevant physics: Since the process occurs at constant pressure

$$W_{1-2} = \int_1^2 p \, dV = p(V_2 - V_1) = mp(v_2 - v_1)$$

and from the first law

$$Q_{1-2} = \Delta U + W_{1-2}$$
$$= (U_2 - U_1) + p(V_2 - V_1)$$
$$= m(u_2 - u_1) + mp(v_2 - v_1)$$

Analysis: (a) We know $p = 5000$ kPa. We can prepare the following table using the property values from Appendix B1.

State	p(kPa)	t(°C)	v(m³/kg)	u(kJ/kg)
1	5000	20	0.0009995	83.65
2	5000	200	0.001153	848.1

Substitution of property values in the equation

$$W_{1-2} = p(V_2 - V_1) = mp(v_2 - v_1)$$

gives

$$W_{1-2} = 15 \times 5 \times 10^6(0.0011530 - 0.0009995) \text{ J}$$

or

$$W_{1-2} = 11.51 \text{ kJ}$$

Then

$$Q_{1-2} = m(u_2 - u_1) + W_{1-2}$$
$$= 15(848.1 - 83.65) + 11.51 \text{ kJ}$$

or

$$Q_{1-2} = 11.48 \times 10^3 \text{ kJ}$$

(b) Here $p = 50,000$ kPa. The relevant properties from Appendix B1 are tabulated below.

State	p(kPa)	t(°C)	v(m³/kg)	u(kJ/kg)
1	50,000	20	0.0009804	81
2	50,000	200	0.0011146	819.7

For the constant pressure process

$$W_{1-2} = mp(v_2 - v_1)$$

Substitution gives

$$W_{1-2} = 15 \times 50 \times 10^6(0.0011146 - 0.0009804)$$
$$= 1.0065 \times 10^5 \text{ J}$$

or

$$W_{1-2} = 100.65 \text{ kJ}$$

Also

$$Q_{1-2} = m(u_2 - u_1) + W_{1-2}$$
$$= 15(819.7 - 81) + 100.65$$
$$= 11.18 \times 10^3 \text{ kJ}$$

Comments: We observe that some work is done during the evaporation process because water expands during the process; this is nonuseful work as atmosphere is being pushed during the expansion process. This process is also shown in Figure 4.4.
    This example demonstrates that the work done by liquid water is very small when compared with the change in the internal energy. At the lower pressure, the ratio of the work done to the change in the internal energy is about 10^{-3} while at the higher pressure it is about 10^{-2}.

Alternative procedures: (a) Using the equation in Section 3.4.1, we can verify that the heat supplied during the constant pressure process can be calculated from

$$Q_{1-2} = m(h_2 - h_1)$$

At 5000 kPa, using the values of h_1 and h_2 from Appendix B1, we obtain

$$Q_{1-2} = 15(853.9 - 88.65)$$
$$= 1.1478 \times 10^4 \text{ kJ}$$

which is exactly the same result as before. At 50,000 kPa,

$$Q_{1-2} = 15(875.5 - 130.02)$$
$$= 1.118 \times 10^4 \text{ kJ}$$

which again is the same result as obtained earlier.
    (b) Observe that the two quantities of heat at the pressures of 5000 kPa and 50,000 kPa differ by only about 2.5%, although the two pressures differ by a factor of 10. This is typical of most liquids. Therefore it is a common practice to use the properties of the saturated liquid to calculate the heat supplied when the properties of the compressed liquid are not available. If we follow such an approximate method (Figure 4.4), we find from the tables of properties of saturated liquid water (Appendix B2.1) that the saturation pressure at $t = 20°C$ is 2.339 kPa and $h_{f,1} = 83.96 \text{ kJ/kg}$. Also, the saturation pressure at $t = 200°C$ is 1,553.8 kPa and $h_{f,2} = 852.45 \text{ kJ/kg}$. We then have

$$Q_{1-2} \simeq m(h_{f,2} - h_{f,1})$$
$$= 15(852.45 - 83.96)$$
$$= 1.153 \times 10^4 \text{ kJ}$$

This approximate value is greater than the value of Q_{1-2} obtained earlier for 50,000 kPa by 3.2%. At the lower pressure of 5000 kPa, the approximate method gives a value of Q_{1-2} that is greater than the exact value by 0.5%.

Liquid-Vapor Mixture.   If heat is supplied to a saturated liquid at a constant pressure, evaporation occurs. As long as the evaporation process is not completed, we have a mixture of saturated liquid and saturated vapor. Throughout the process of evaporation, the liquid and the vapor components are at identical pressure and temperature. The temperature will begin to increase only after the last drop of the liquid has been converted into vapor.

Example 4.3

Six kilograms of water at 20°C and atmospheric pressure are heated in a constant pressure process. Determine the quantity of heat energy that must be supplied to vaporize all the water.

Solution:

Given: A mass of 6 kg of water is completely vaporized in a constant pressure process.

Initial state: Subcooled liquid, $t_1 = 20°C$, $p_1 = 101.3$ kPa

Final state: saturated vapor, $t_2 = 100°C$, $p_2 = 101.3$ kPa

Objective: The quantity of heat supplied in the process is to be determined.

Diagram:

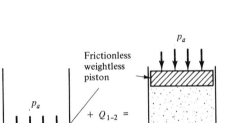 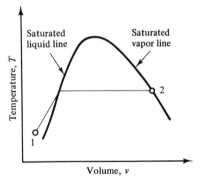

Figure 4.5

Relevant physics: The first law gives

$$Q_{1-2} - W_{1-2} = U_2 - U_1$$

or

$$Q_{1-2} = m(u_2 - u_1) + mp_1(v_2 - v_1) = m(h_2 - h_1)$$

We observe that some work is done during the evaporation process as water expands during the process, and that this is nonuseful work because atmosphere is being pushed during the expansion process. Nevertheless we must supply heat energy to produce this nonuseful work.

Analysis: Using Appendix B2.1 we find that for water vapor at 100°C and at atmospheric pressure,

$$h_{g,2} = h_2 = 2676.1 \text{ kJ/kg}$$

We approximate h_1 at 20°C and atmospheric pressure by h_f at 20°C. From Appendix B2.1, at 20°C

$$h_f = 83.96 \text{ kJ/kg}$$

and

$$h_1 \simeq h_{f,1} = 83.96 \text{ kJ/kg}$$

Substitution in the equation for Q_{1-2} gives

$$Q_{1-2} = 6(2676.1 \times 10^3 - 83.96 \times 10^3)$$
$$= 15.55 \times 10^6 \text{ J} = 15.55 \times 10^3 \text{ kJ}$$

Example 4.4

If the mass of the water in Example 4.3 was 10 kg at the beginning of the heating process, how much heat energy will be needed to evaporate 6 kg of the water?

Solution:

Given:

$$\text{Initial state: } m = 10 \text{ kg}, \qquad t_1 = 20°\text{C}, \qquad p_1 = 1.013 \text{ kPa}$$
$$\text{Final state: } \quad \text{Liquid mass} = 4 \text{ kg}, \qquad \text{vapor mass} = 6 \text{ kg},$$
$$t_2 = 100°\text{C}, \qquad p_2 = 1.013 \text{ kPa}$$

Objective: The quantity of heat energy that must be supplied to bring about vaporization of 6 kg of water out of a total mass of 10 kg is to be determined.

Diagram:

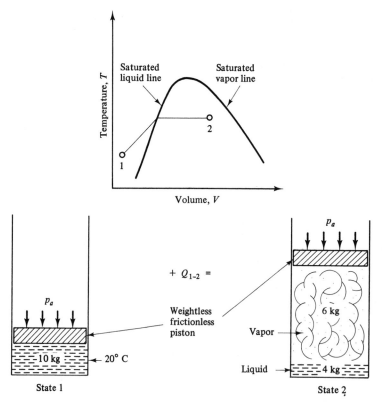

Figure 4.6

Assumptions: In the final state, the liquid and the vapor portions are in thermal equilibrium.

Relevant physics: In this problem we have a liquid-vapor mixture in the final state, so it is necessary to determine the quality of this mixture.

Analysis: The value of h_1 remains the same as before. The value of h_2 is different now, since we have a liquid-vapor mixture (Figure 4.6). The quality of this mixture is given by

$$x_2 = \frac{m_g}{m} = \frac{6}{10} = 0.6$$

Also, from Equation 4.3(b)

$$h_2 = h_{f,2} + xh_{fg,2} \qquad\qquad (h_f \text{ and } h_{fg} \text{ values at } 100°C)$$
$$= 419.04 \times 10^3 + 0.6 \times 2257 \times 10^3$$
$$= 1773.24 \times 10^3 \text{ J} = 1773.24 \text{ kJ}$$

Hence

$$Q_{1-2} = m(h_2 - h_1)$$
$$= 10(1773.24 \times 10^3 - 83.96 \times 10^3)$$
$$= 16.89 \times 10^6 \text{ J} = 16.89 \times 10^3 \text{ kJ}$$

Alternative procedure: The above result can also be obtained by considering two systems, one of 6 kg of water and another of 4 kg of water. The first system is taken from 20°C to the saturated vapor state, and the necessary heat input as found in Example 4.2 is 15.55×10^6 J. The second system is taken from 20°C to the saturated liquid state. The heat energy needed for the second system is

$$m(h_f \text{ at } 100°C - h_f \text{ at } 20°C) = 4(419.04 - 83.96) \text{ kJ}$$
$$= 1.34 \times 10^6 \text{ J} = 1.34 \times 10^3 \text{ kJ}$$

The sum of the two quantities of heat is 16.89×10^6 J. This is the same result we obtained earlier.

Example 4.5

A rigid, uninsulated container has a volume of 0.4 m³. It contains water vapor at 200°C at a pressure of 100 kPa. After some time, it is observed that the pressure has dropped to 25.03 kPa. Determine the quality of the water vapor in the container and the quantity of heat energy lost to the surroundings.

Solution:

Given: A rigid, uninsulated container contains water vapor that undergoes a constant volume process.

$$\text{Initial state:} \quad V_1 = 0.4 \text{ m}^3, \qquad t_1 = 200°C, \qquad p_1 = 100 \text{ kPa}$$
$$\text{Final state:} \quad V_2 = 0.4 \text{ m}^3, \qquad p_2 = 25.03 \text{ kPa}$$

The mass of the water is constant throughout the process.

Objective: We are to determine the final quality and the quantity of heat transferred to the surroundings in the constant volume process.

Diagram:

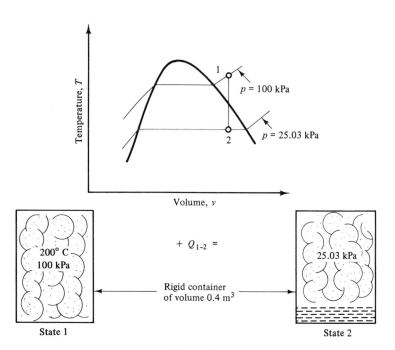

Figure 4.7

Assumptions: The surroundings are at room temperature.

Relevant physics: Since the tank is uninsulated and the contents of the tank are at a high temperature of 200°C, there is a heat transfer from the contents of the tank to the surroundings. Also, since the tank is rigid, the heat transfer occurs in a constant volume process. We may note that this is not a quasistatic process. For this constant volume process, work done is zero and

$$Q_{1-2} = U_2 - U_1$$

Analysis: From Appendix B2.2 we find that the saturation temperature for 100 kPa is 99.63°C, which is less than the initial temperature of the vapor in the container. Therefore we have a superheated vapor with the following properties.

$$v_1 = 2.172 \text{ m}^3/\text{kg}, \qquad u_1 = 2658.1 \text{ kJ/kg}$$

The mass of the water m is then given by

$$m = \frac{V_1}{v_1} = \frac{0.4}{2.172} = 0.1842 \text{ kg}$$

Corresponding to the new pressure $p_2 = 25.03$ kPa, we find $v_{f,2} = 0.001020$ m³/kg, $v_{g,2} = 6.197$ m³/kg, $u_{f,2} = 272.02$ kJ/kg, and $u_{g,2} = 2463.1$ kJ/kg. We observe that since the container is rigid and no mass leaves the container, the specific volume remains unchanged. Also,

$$v_2 = v_1 = 2.172 \text{ m}^3/\text{kg}$$

The constant volume process appears as a vertical line on the T-v diagram. We have

$$v_2 = (1 - x_2)v_{f,2} + x_2 v_{g,2}$$

Substitution gives

$$2.172 = (1 - x_2)0.00102 + 6.179x_2$$

or

$$x_2 = 0.35$$

Since

$$U_1 = mu_1 = 0.1842 \times 2658.1 = 489.6 \text{ kJ}$$

and

$$\begin{aligned} U_2 = mu_2 &= m[(1 - x_2)u_{f,2} + x_2 u_{g,2}] \\ &= 0.1842[(1 - 0.35)272.02 + 0.35 \times 2463.1] \\ &= 191.4 \text{ kJ} \end{aligned}$$

we obtain

$$Q_{1-2} = U_2 - U_1 = 191.4 - 489.6 = -298.2 \text{ kJ}$$

The negative sign indicates a heat loss from the system. Thus the quantity of heat lost to the surroundings is 298.2 kJ.

Superheated Vapor.   As in the case of a liquid, the work done and heat supplied to a superheated vapor can be calculated in a straightforward manner when the process involved is one of constant pressure. This is done by calculating the values of $p\Delta V$ and ΔH, respectively. When a constant volume process is considered, the work done is zero and ΔU and Q can be calculated using the values of the internal energy from the tables of properties.

A constant temperature process is seldom encountered in the applications of a superheated vapor. It would be a somewhat difficult task to handle such a situation by using the property tables alone. We can, however, draw some qualitative conclusions by examining the T-v and p-v diagrams in Figure 4.8. Consider a liquid at a moderately high pressure as a system. Let the temperature of the liquid be less than the critical temperature. A constant temperature heating process (see Figure 4.8(b)) will cause an expansion of the system. The volume will increase and the pressure will fall. As the expansion proceeds, the system will pass through a liquid zone, a liquid-vapor zone, and then a superheated vapor zone. Both the temperature and the pressure will remain constant in the liquid-vapor zone. If the initial temperature is greater than the critical temperature, we have the system in a fluid state as discussed earlier. As the fluid is taken through a constant temperature heating process, it expands to a gaseous state.

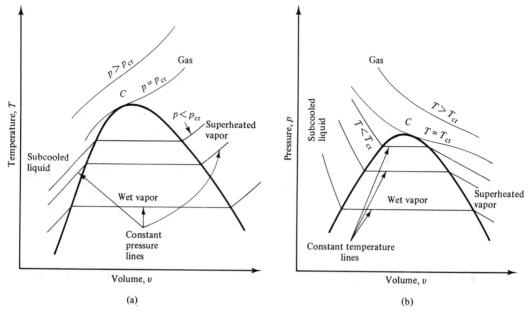

Figure 4.8
(a) Constant pressure lines ($p > p_{cr}, p = p_{cr}, p < p_{cr}$) in the liquid and vapor regions.
(b) Constant temperature lines ($T > T_{cr}, T = T_{cr}, T < T_{cr}$) in the liquid vapor regions.

The quasistatic adiabatic process is another process for a vapor that is a little difficult to handle using the tables of properties. In the absence of a relation between T and v or p and v, it cannot be presented graphically on a T-v or p-v diagram. Yet this is the process that is frequently employed in ideal power cycles. The analysis of a power cycle employing a quasistatic adiabatic process is greatly facilitated by the use of a thermodynamic property called *entropy*. This will be discussed in Chapter 6.

Although we have used water as the vehicle to discuss the liquid and the vapor phases, all the definitions and concepts introduced thus far are equally applicable to other substances, such as mercury, sodium, ammonia, and Freon. These substances are used in power and refrigeration plants. Property values for saturated and super-heated states for mercury, ammonia, and Freon-12 are given in the appendix.

Example 4.6

Determine the decrease in the enthalpy of 4 kg of Freon-12 when it is cooled from 70°C and 400 kPa to 10°C and a quality of 30%.

Solution:

Given: A mass of 4 kg of Freon-12 is cooled.

Initial state: $t_1 = 70°C$, $p_1 = 400 \text{ kPa}$

Final state: $t_2 = 10°C$, $x_2 = 0.30$

Objective: We are to determine the decrease in the enthalpy of the Freon.

140

Diagram:

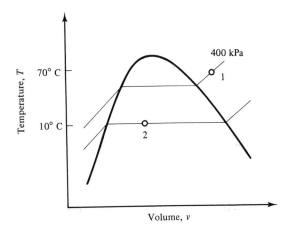

Figure 4.9

Relevant physics: As in the case of water in Example 4.5, we have to determine the initial phase of the Freon, which could be subcooled liquid, saturated liquid, saturated vapor, liquid-vapor mixture, or superheated vapor. Such a determination can be made by comparing the saturation temperature corresponding to the given initial pressure with the given temperature.

Analysis: From Appendix C1 we find that the saturation temperature of Freon-12 at 400 kPa is about 8°C, which is less than the initial temperature of 70°C. Therefore we have a superheated Freon-12 vapor, for which the specific enthalpy is

$$h_1 = 232.23 \times 10^3 \text{ J/kg}$$

Since a quality of 30% at the final state is prescribed, the final state is a mixture of liquid and vapor. For $t_2 = 10°C$

$$h_{f,2} = 45.337 \times 10^3 \text{ J/kg}$$

and

$$h_{g,2} = 191.602 \times 10^3 \text{ J/kg}$$

We can calculate h_2 from

$$h_2 = (1 - x_2)h_{f,2} + x_2 h_{g,2}$$

Substitution gives

$$h_2 = (1 - 0.30)45.337 \times 10^3 + 0.30 \times 191.602 \times 10^3$$

$$= 89.21 \times 10^3 \text{ J/kg} = 89.21 \text{ kJ/kg}$$

The change in the enthalpy of the Freon is

$$\Delta H = m(h_2 - h_1)$$

$$= 4(89.21 \times 10^3 - 232.23 \times 10^3)$$

$$= -572.08 \times 10^3 \text{ J} = -572.1 \text{ kJ}$$

The negative sign signifies a decrease in the enthalpy due to the cooling process.

Comments: The actual process is irrelevant here since we are interested in the change of enthalpy, which depends only on the end states.

Example 4.7

Saturated liquid ammonia at 50°C is expanded in such a way that its enthalpy remains constant. Its final pressure is 290.85 kPa. Determine the final state of the ammonia.

Solution:

Given: Ammonia expands in a constant enthalpy process.

$$\text{Initial state:} \quad \text{saturated liquid,} \quad t_1 = 50°C$$
$$\text{Final state:} \quad p_2 = 290.85 \text{ kPa}$$

Objective: We are to determine the final temperature and the phase of the ammonia.

Diagram:

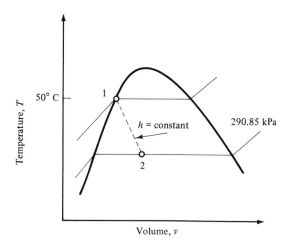

Figure 4.10

Assumptions: None.

Relevant physics: The enthalpy h_1 at the initial state can be determined from the given data. Also for the given process, $h_2 = h_1$. The value of h_2 should be compared with values of h_f and h_g corresponding to pressure p_2 to facilitate a determination of the final state.

Analysis: From the tables of properties of ammonia in Appendix D1, we find that for $t_1 = t_{sat} = 50°C$,

$$p_{sat} = 2032.62 \text{ kPa} \quad \text{and} \quad h_1 = h_f = 421.7 \times 10^3 \text{ J/kg}$$

For $p_2 = p_{sat} = 290.85$ kPa, we find

$$t_{sat} = -10°C, \quad h_{g,2} = 1432 \times 10^3 \text{ J/kg} \quad \text{and} \quad h_{f,2} = 135.2 \times 10^3 \text{ J/kg}$$

The enthalpy at state 2 equals, h_1, which is 421.7×10^3 J/kg. This value is less than $h_{g,2}$, but greater than $h_{f,2}$. Therefore we have a liquid-vapor mixture at state 2. The quality x_2 is related to the enthalpies h_2, $h_{f,2}$, and $h_{g,2}$ by

$$h_2 = (1 - x_2)h_{f,2} + x_2 h_{g,2}$$

Solving for x_2 yields

$$x_2 = \frac{h_2 - h_{f,2}}{h_{g,2} - h_{f,2}}$$

Substitution gives

$$x_2 = \frac{421.7 \times 10^3 - 135.2 \times 10^3}{1432 \times 10^3 - 135.2 \times 10^3} = 0.22$$

Thus the final state is a liquid-vapor mixture at $-10°C$ and 0.22 quality.

4.4 SOLID AND LIQUID PHASES

In engineering applications such as production of dry ice, refrigeration units, and power plants, we have to deal with solids, liquids, vapors, and gases. Thus far we have examined only the liquid and vapor phases; we have yet to discuss the solid and the gas phases. In a later section we shall present a number of equations of state and a chart method for predicting the behavior of gases. We shall now discuss the transition from the solid phase to the liquid or the vapor phase.

We begin with liquid water at atmospheric pressure and room temperature, point 1 in the T-v diagram of Figure 4.11(a). If we cool this water in a quasi-static constant pressure process, both the temperature and the volume decrease. Eventually, the water reaches a temperature of $0°C$ and any further removal of heat energy results in ice formation. This condition is shown as point 2 in Figure 4.11(a). At this point the ice and the liquid water are in thermal equilibrium and the ice and the liquid water are said to be in a *saturated state*. As the heat removal is continued, the temperature remains constant and the volume increases until all the liquid is converted into ice, a behavior unique to water. This is shown by line 2–3 in the figure. The state at point 3 is saturated solid state. Further heat removal causes both the temperature and the volume to decrease, as depicted by line 3–4 in the figure.

The amount of energy that must be removed from a unit mass of liquid at its saturation point (such as point 2 in Figure 4.11(a)) to convert it into a saturated solid is known as the *latent heat of fusion*. It is equal to the difference in the specific enthalpies of the saturated liquid and the saturated solid.

Now consider a substance like alcohol that contracts on solidification. Figure 4.11(b) shows a T-v diagram for such a substance. If we cool the substance from its liquid state (point 1 in the figure) in a constant pressure process, we find that the substance experiences a decrease in temperature and volume until the saturated liquid state, point 2, is reached. Then a change of phase from the liquid state to the solid state occurs (line 2–3), during which the temperature remains constant and the volume continues to decrease. Further cooling beyond point 3 in the figure results in a decrease in the temperature as well as in the volume.

If similar experiments are performed at different pressures, we obtain a series of points representing saturated liquid states and saturated solid states. A line drawn through the various saturated solid states corresponding to different pressures gives us the *saturated solid line*. Similarly, a line drawn through the various saturated liquid states gives us the *saturated liquid line* (incipient freezing). These lines are shown in Figure 4.11(a) and (b). So far there is no experimental evidence to suggest that these two lines meet in another critical point.

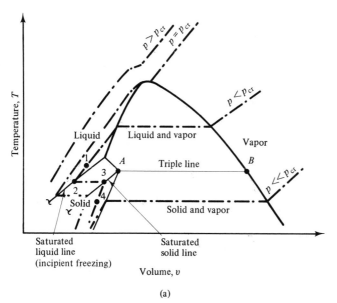

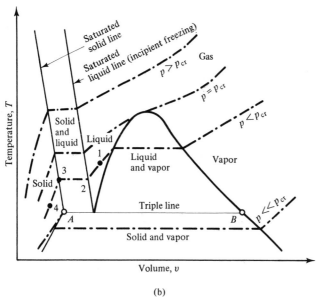

Figure 4.11
(a) Solid, liquid, and vapor phases on a T-v diagram for a substance that expands on freezing.
(b) Solid, liquid, and vapor phases on a T-v diagram for a substance that contracts on freezing.

Let us now consider heating of ice at constant pressure. Do we always expect first the formation of a liquid by melting of the solid and then vaporization of the liquid so formed? If we recall our observations of vapors coming off an ice cube freshly removed from a freezer, then we have to admit the possibility of a transformation directly from the solid phase to the vapor phase. There is a thin layer of vapor adjacent to the surface of an ice cube taken out of a freezer. The cube is at atmospheric pressure, which is the sum of the air pressure and the water vapor pressure. The latter is the partial pressure of the water vapor in a mixture of air and water vapor, existing in the proximity of the cube. It is at low values of the partial pressure that a transformation from a solid state directly into a vapor state is possible. Such a direct transformation is called *sublimation*. The dot-dashed lines labeled $p \ll p_{cr}$ in the T-v diagram of Figure 4.11(a) and (b) show the sublimation process. The solid and the vapor are in thermodynamic equilibrium until the sublimation process is complete. A finite quantity of energy is needed to bring about complete sublimation of a solid into a vapor, both being at the same temperature and pressure. This quantity of energy, on a unit-mass basis, is called *heat of sublimation* and is designated by the symbol h_{ig}. The subscript ig denotes a transition from solid (ice—i) to vapor (gas—g). The heat of sublimation is equal to the difference between the enthalpies of the saturated vapor and the saturated solid, both at the same temperature and pressure.

Four lines of constant pressure are shown in Figure 4.11(a) and (b), one for a pressure greater than the critical pressure, one for the critical pressure, one for a pressure less than the critical pressure, and one for a pressure much less than the critical pressure. The first line ($p > p_{cr}$) passes through solid, liquid, and gaseous regions. The second line ($p = p_{cr}$) does the same. The third line ($p < p_{cr}$) passes through solid, solid-liquid, liquid, liquid-vapor, and vapor regions. The fourth line ($p \ll p_{cr}$) passes only through solid, solid-vapor, and vapor regions.

As noted in the preceding paragraphs, a change from solid to vapor can occur in two regions. In the first, the solid melts into a liquid and the liquid formed subsequently evaporates. In the second region, the solid goes directly into a vapor state. These two regions are shown in Figure 4.11(a) and (b), Part (a) being for a substance that expands on freezing and Part (b) for a substance that contracts on freezing. These figures show that the two regions share a common boundary, line AB. This means that the solid, the liquid, and the vapor phases coexist at the common boundary, line AB. This common boundary is referred to as the *triple line*. Points on the triple line of a substance have the same pressure and temperature, but have different volumes due to different phases. It should be noted that it would be difficult to devise a process that would change the state of a substance along the triple line.

If we plot p-T diagrams for a substance that expands on freezing and for a substance that contracts on freezing, we obtain Figure 4.12(a) and Figure 4.12(b), respectively. The p-T diagram of a substance is called a *phase diagram*. These figures show a sublimation line separating the solid and the vapor domains, a fusion line separating the solid and the liquid domains, and an evaporation line separating the liquid and the vapor domains. These three lines meet at the *triple point*. The triple line on a T-v diagram becomes a triple point on a p-T diagram. Table 4.3 gives triple-point data for some substances. It can be seen from the table that the triple-point temperature varies greatly from one substance to another.

Figure 4.13(a) and (b) shows p-v diagrams for a substance that expands on freezing and for a substance that contracts on freezing, respectively. As do the T-v diagrams of Figure 4.11(a) and (b), the p-v diagrams show regions for the solid, liquid,

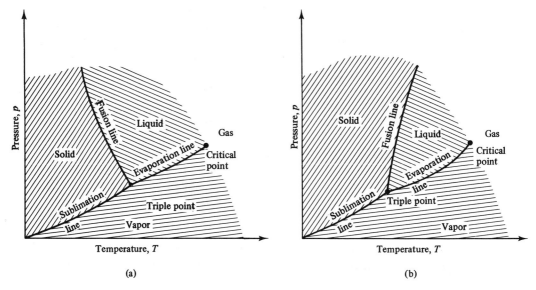

Figure 4.12 Solid, liquid, and vapor regions on a *p-T* diagram (a) for a substance that expands on freezing, and (b) for a substance that contracts on freezing.

TABLE 4.3 TRIPLE-POINT DATA FOR SELECTED SUBSTANCES

	Temperature (°C)	Pressure (kPa)
Copper	1083	79×10^{-6}
Hydrogen	−259	7.194
Mercury	−39	0.13×10^{-6}
Nitrogen	−210	12.53
Oxygen	−219	0.15
Silver	961	0.01
Zinc	419	5.066

vapor, and gaseous phases, regions of solid-liquid, solid-vapor, and liquid-vapor equilibrium, and the triple line. Four isotherms, for $T > T_{cr}$, $T = T_{cr}$, $T < T_{cr}$, and $T \ll T_{cr}$, are also shown in these figures. The isotherm for $T > T_{cr}$ passes through gaseous and vapor regions only. The critical isotherm delineates the liquid and the gaseous regions at supercritical pressures, and it passes through the vapor region at subcritical pressures. For a temperature moderately less than the critical temperature, an isotherm passes through solid, liquid, and vapor phases. Finally, for a temperature much less than the critical temperature, an isotherm passes through only solid and vapor regions.

It is worthwhile to recapitulate our experience with the three phases and their

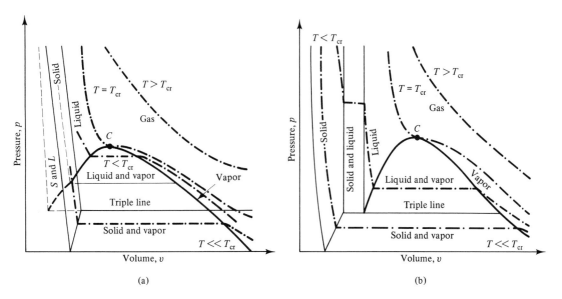

Figure 4.13 p-v diagram showing solid, liquid, and vapor phases.
(a) For a substance that expands on freezing and
(b) For a substance that contracts on freezing.

properties. Figure 4.14(a) and (b) shows p-v-T surfaces for a substance that expands on freezing and for a substance that contracts on freezing, respectively. If we start with a substance at a small specific volume and at high pressure, then we have the substance in the solid phase, provided that the temperature is low. If the temperature of the substance is high, the substance is in a liquid state. If the temperature is near the critical value, then the substance is in fluid state, a state in which it is difficult to distinguish between a liquid and a gas. When the temperature exceeds the critical temperature, by definition the substance is in a gaseous state.

As the substance is heated at a constant moderate pressure that is less than the critical pressure, it passes through solid, liquid, and vapor phases, as shown by line $p < p_{cr}$ in Figure 4.14. If the pressure during the heating process is maintained at a level much below the critical pressure, the substance will go from a solid phase directly into a vapor phase. This is shown by line $p \ll p_{cr}$ in Figure 4.14. If the pressure during heating is greater than the critical pressure then the substance will experience solid, liquid, and gaseous phases, as shown by line $p > p_{cr}$ in Figure 4.14.

Three isotherms, for $T = T_{cr}$, for $T < T_{cr}$, and for $T \ll T_{cr}$, are also shown on the p-v-T surfaces of Figure 4.14(a) and (b). The critical isotherm distinguishes between the liquid and the gaseous phases and then passes through the vapor region. The subcritical isotherm ($T < T_{cr}$) passes through the evaporation domain for temperatures above the triple point. For temperatures below the triple point, the isotherm ($T \ll T_{cr}$) passes through the sublimation region.

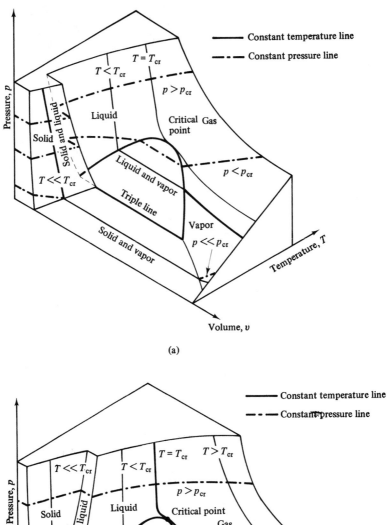

Constant temperature line
Constant pressure line

Pressure, p

$T = T_{cr}$
$T < T_{cr}$
$p > p_{cr}$

Liquid

Critical Gas
point

Solid

Solid and liquid

$T \ll T_{cr}$

Liquid and vapor

$p < p_{cr}$

Triple line

Solid and vapor

Vapor
$p \ll p_{cr}$

Temperature, T

Volume, v

(a)

Constant temperature line
Constant pressure line

Pressure, p

$T \ll T_{cr}$
$T < T_{cr}$
$T = T_{cr}$
$T > T_{cr}$

$p > p_{cr}$

Solid

Solid and liquid

Liquid

Critical point
Gas

Liquid and
vapor

$p < p_{cr}$

Triple line

Vapor
$p \ll p_{cr}$

Solid and vapor

Temperature, T

Volume, v

(b)

Figure 4.14
(a) p-v-T surface for a substance that expands on freezing.
(b) p-v-T surface for a substance that contracts on freezing.

4.4.1 Superheated and Supercooled Liquid

In the discussion of the saturated liquid line for boiling, we stated that a change of phase is imminent for a liquid on the saturation line. This is true only if there are sites within the fluid body where vapor bubbles can form and grow. Such locations are called *nuclei*, or *nucleation sites*. Solid particles, rough spots on a surface, or a dissolved gas can constitute nuclei. In the absence of such nuclei a liquid will not boil at the saturation temperature. As the heating is continued, the temperature of the liquid will rise beyond the saturation temperature. When this happens the liquid is said to be in a *superheated liquid* state. This phenomenon can be observed if a clean container holding pure liquid water is heated slowly under atmospheric pressure. The temperature of the water will rise several degrees beyond 100°C, the saturation temperature corresponding to 1 atm pressure. If a rod is inserted into the liquid water in the superheated state, vapor bubbles will suddenly form. The thermodynamic state of a superheated liquid is called a *metastable state*. A metastable state returns to a stable state if it experiences a finite disturbance. The saturated liquid state is a stable state.

It is possible to cool a liquid and have it remain in a liquid state at a temperature below its freezing point. Such a state of the liquid is called a *supercooled liquid state*. If we have supercooled liquid water in a beaker, a gentle tap on the beaker will initiate crystallization of the water into ice. The supercooled liquid state is also a metastable state. The property values of·liquids in the tables are only for the stable states of a liquids.

4.5 COEFFICIENTS OF ISOTHERMAL COMPRESSIBILITY AND ISOBARIC EXPANSION

In our discussion of the three principal phases of a substance, we used T-v and p-v diagrams without inquiring about the methods for constructing such diagrams. We also commented that it is difficult to represent the behavior of the gaseous phase (except for the ideal gas law) or the liquid phase by a simple equation. We do know, however, that there is a unique relationship among the three variables p, v, and T that represent the state of a substance. We can formally represent this relationship as

$$f(p, v, T) = 0 \qquad (4.4)$$

From the point of view of analytic geometry the above equation represents a three-dimensional surface, the dimensions being p, v, and T (instead of the usual x, y, and z in analytic geometry). The equation can be formally recast to read

$$v = v(p, T) \qquad (4.4a)$$

This equation can be interpreted as the one that predicts a value of v when values of p and T are prescribed. The equation also implies that a change in p or T will cause a change in v. Mathematically, we express this as

$$dv = \left(\frac{\partial v}{\partial p}\right)_T dp + \left(\frac{\partial v}{\partial T}\right)_p dT$$

On dividing the above equation by v, we obtain

$$\frac{dv}{v} = \frac{1}{v}\left(\frac{\partial v}{\partial p}\right)_T dp + \frac{1}{v}\left(\frac{\partial v}{\partial T}\right)_p dT \qquad (4.5)$$

The coefficient of dp in the above equation is the fractional change in the specific volume when only pressure is changed, holding the temperature constant. Experience tells us that the volume of a substance decreases as the pressure is increased, while maintaining the temperature constant. In other words, the coefficient of dp has a negative value. We define this coefficient, called the *coefficient of isothermal compressibility*, κ, by

$$\kappa = -\frac{1}{v}\left(\frac{\partial v}{\partial p}\right)_T \qquad (4.5a)$$

The coefficient of dT in Equation 4.5 is the fractional change in the specific volume as the temperature is changed, while holding the pressure constant. It is called the *coefficient of isobaric expansion*, β. We define

$$\beta = \frac{1}{v}\left(\frac{\partial v}{\partial T}\right)_p \qquad (4.5b)$$

Equation 4.5 can now be rewritten as

$$dv = v(\beta \, dT - \kappa \, dp) \qquad (4.5c)$$

The coefficients of isothermal compressibility and isobaric expansion can be determined experimentally and, consequently, changes in the volume of a substance can be determined using Equation 4.5(c). This information can then be used to construct the p-v-T surface of a substance. For most liquids and solids, the values of κ and β are relatively small, although they do show changes when temperature and pressure are changed. The value of κ for water at 1 atm changes from 51.1×10^{-6} atm^{-1} to 46.3×10^{-6} atm^{-1} when the temperature changes from 0°C to 80°C. It drops from 46.4×10^{-6} atm^{-1} to 39.5×10^{-6} atm^{-1} if the pressure is increased from 1 atm to 500 atm and temperature maintained at 20°C. For a temperature increase from 0°C to 80°C, the value of β for water at 1 atm increases from -67×10^{-6} K^{-1} to 643×10^{-6} K^{-1}.

For an ideal gas we have

$$v = RT/p$$

Partial differentiation gives

$$\left(\frac{\partial v}{\partial p}\right)_T = -\frac{RT}{p^2} = -\frac{v}{p}$$

and

$$\left(\frac{\partial v}{\partial T}\right)_p = \frac{R}{p} = \frac{v}{T}$$

Thus for an ideal gas

$$\kappa = -\frac{1}{v}\left(\frac{\partial v}{\partial p}\right)_T = \frac{1}{p} \qquad (4.6)$$

and

$$\beta = \frac{1}{v}\left(\frac{\partial v}{\partial T}\right)_p = \frac{1}{T} \qquad (4.6a)$$

Equations 4.6 and 4.6(a) indicate that for an ideal gas the value of κ is inversely proportional to the pressure and the value of β is inversely proportional to the temperature. For superheated vapors, we can calculate values of κ and β using the tables of properties.

4.6 THE COMPRESSIBILITY FACTOR

The ideal gas equation is a good approximation of the behavior of real gases at low pressures and high temperatures. In other words, the relation pv/RT has a value close to unity at low pressures and high temperatures. Any deviation of this ratio from unity means that the ideal gas equation is not suitable for the gas in question. This ratio is called the *compressibility factor*, Z, where

$$Z = \frac{pv}{RT} \qquad (4.7)$$

Since the quantity RT/p for any gas gives its volume v_{ideal} based on the ideal gas equation, we can also write

$$Z = \frac{v}{v_{\text{ideal}}} \qquad (4.8)$$

A graph of Z plotted against one of the thermodynamic variables for a gas shows the extent to which the gas deviates from the ideal gas. Figure 1.9 in Chapter 1 shows a plot of pv/RT against p for hydrogen. The critical temperature of hydrogen is $-259°C$, or 14.15 K. The figure shows that for temperatures close to the critical temperature, the behavior of hydrogen gas deviates considerably from the ideal gas law; that is, the Z-value is far from unity. On the other hand, at temperatures greater than three times the critical temperature, and particularly at pressures below 100 atm, the Z-value for hydrogen is close to unity and the gas behavior can be approximated by the ideal gas law.

It has been found to be convenient to use *reduced pressure* p_r, *reduced temperature* T_r, and *reduced volume* v_r for constructing a compressibility chart. These reduced properties are defined as follows.

$$p_r = \frac{\text{gas pressure}}{\text{critical pressure of the gas}} = \frac{p}{p_{\text{cr}}} \qquad (4.9)$$

$$T_r = \frac{\text{gas temperature}}{\text{critical temperature of the gas}} = \frac{T}{T_{\text{cr}}} \qquad (4.9a)$$

and

$$v_r = \frac{\text{specific volume of the gas}}{\text{critical specific volume of the gas}} = \frac{v}{v_{\text{cr}}} \qquad (4.9b)$$

A chart with Z as the ordinate, p_r as the abscissa, and T_r as the parameter is called a *generalized compressibility* chart. Figure 4.15 shows data points for ten different gases plotted in the form of a generalized compressibility chart. It is interesting to observe that the data for the ten different gases falls on a single set of curves. The curves are the best fits to the experimental data and the deviation is less than 5 percent.

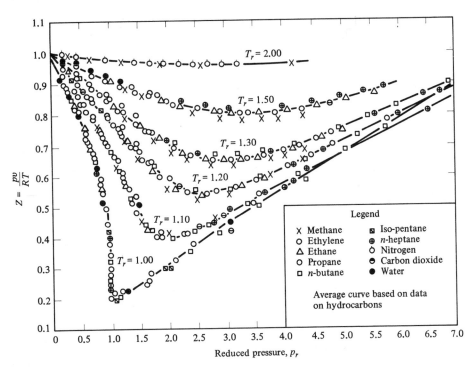

Figure 4.15 Curves of compressibility factor against reduced pressure for a number of substances. [Gour-Jen Su Modified Law of Corresponding States, *Ind. Eng. Chem.* (Intern. Edition), 38: 803 (1946). Reprinted with permission.]

If two gases have the same values of p_r and T_r, then the gases are said to be at *corresponding states*. The *law of corresponding states* postulates that for given values of p_r and T_r, the compressibility factor for all gases is approximately the same. We have seen that the law holds for the ten gases of Figure 4.15. The law of corresponding states is not a law in the sense that it is upheld in every situation, as is the first law of thermodynamics. The law of corresponding states, however, does provide a useful tool for engineering purposes in that—in the absence of any experimental data for a gas—we can use the generalized compressibility chart to correlate the pressure, volume, and temperature of the gas.

The generalized compressibility chart of Figure 4.16 has a p_r range of zero to 10 and a T_r range of 1.0 to 15.0. Also shown on this chart are dotted lines of constant

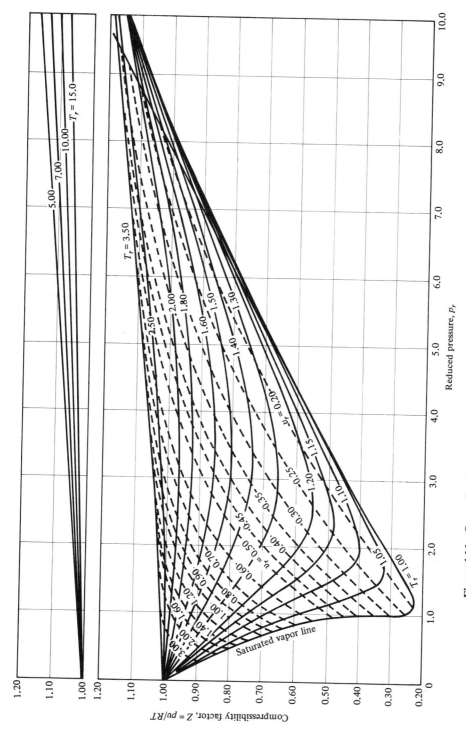

Figure 4.16 Generalized compressibility chart. (K. Wark, *Thermodynamics*, 3/E. New York: McGraw-Hill, © 1977. Used with permission.)

v_r. Figure 4.17 is a similar chart for a high-pressure range, with p_r ranging from 10 to 40. The left leg ($T_r = 1$) of the solid curve in Figure 4.16 is the saturated vapor line. The solid curves in Figures 4.16 and 4.17 give the Z-value of a gas for known values of p_r and T_r. These curves exhibit a minimum value of Z for low values of T_r. As the temperature increases, the solid curves flatten out. At low temperatures the attraction forces between molecules cause a gas to occupy a volume that is less than the ideal gas volume, which results in a value of Z less than one. At high temperatures the density of a gas decreases, causing an increase in the repulsion force between molecules. This results in a gas occupying a volume that is greater than the ideal gas volume.

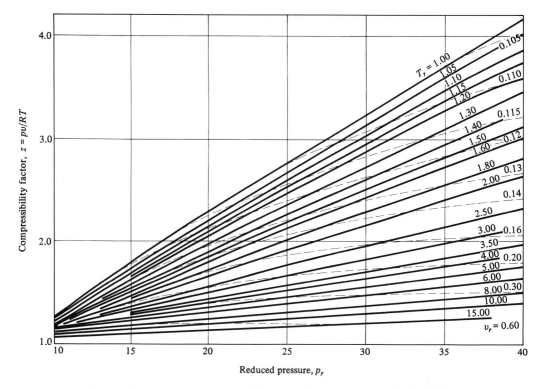

Figure 4.17 Generalized compressibility chart for high pressures. (K. Wark, *Thermodynamics*, 3/E. New York: McGraw-Hill, © 1977. Used with permission.)

The generalized compressibility chart should not be regarded as a substitute for accurate experimental data for a gas. The chart is a useful tool for determining whether a given gas can be approximated by the ideal gas equation for given values of p_r and T_r.

Consider water vapor in the range of 100°C to 600°C and 1 atm to 200 atm. This is the range of operation for a typical steam power plant. With $T_{cr} = 647$ K for water, the reduced temperature T_r ranges from $(100 + 273.15)/647$ to $(600 +$

273.15)/647, or 0.57 to 1.35, respectively. Also, with $p_{cr} = 22{,}090$ kPa the reduced pressure p_r ranges from $1/220.9$ to $220/220.9$, or 0 to 0.9, respectively. Figure 4.16 tells us that for these ranges of p_r and T_r, the behavior of water vapor deviates significantly from that of the ideal gas. We therefore use the tables of properties for water vapor.

Some conclusions can be drawn from a careful examination of the compressibility charts in Figures 4.16 and 4.17. The deviation from the ideal gas behavior is significant for reduced temperatures less than 2.5. The actual volume of a gas is greater than the volume predicted by the ideal gas law for reduced temperatures greater than 2.5. It is interesting to observe from the chart that for $8 < p_r < 10$, the deviation from the ideal gas behavior is independent of the nature of the gas. Also, the value of the compressibility factor approaches unity as the reduced pressure approaches zero.

Example 4.8

Determine the specific volume of ammonia at 2000 kPa and 355 K using the compressibility chart.

Solution:

Given: For ammonia, $p = 2000$ kPa and $T = 355$ K.

Objective: Specific volume, v, is to be determined by using the compressibility chart of Figure 4.16.

Assumptions: None.

Relevant physics: Ammonia vapor does not obey the ideal gas law at the given conditions. The compressibility chart can be used to determine the specific volume more accurately. Use of such a chart requires a calculation of the reduced temperature and pressure.

Analysis: From Table 4.1 we find that for ammonia, the critical values are

$$T_{cr} = 405.5 \text{ K} \quad \text{and} \quad p_{cr} = 11{,}280 \text{ kPa}$$

The reduced temperature and pressure are

$$T_r = \frac{T}{T_{cr}} = \frac{355}{405.5} = 0.88$$

and

$$p_r = \frac{p}{p_{cr}} = \frac{2000 \times 10^3}{11.28 \times 10^6} = 0.177$$

From Figure 4.16, we find

$$Z = 0.90$$

for the above T_{cr}, and p_{cr}. Also, the gas constant R for ammonia (Table 1.1) has a value of 488.2 J/kg. The specific volume is then given by Equation 4.7(a).

$$v = Z v_{ideal}$$

$$= Z \frac{RT}{p}$$

$$= 0.9 \times \frac{488.2 \times 355}{2000 \times 10^3}$$

$$= 0.078 \text{ m}^3/\text{kg}$$

Comments: Appendix D2 gives a value of 0.076 for ammonia at 2000 kPa and 355 K. Thus the compressibility chart gives a *v*-value that is larger than the exact value by about 2.6%.

4.7 EQUATIONS OF STATE

Carefully measured property values of a gaseous substance are the most reliable source of data for any application. Yet there is often a need for an analytical expression correlating pressure, volume, and temperature of a gas. The simplest expression is the ideal gas law. A variety of equations have been proposed to predict the *p-v-T* behavior of gases. These equations are called the *equations of state for gases.* Some of these equations are based on models that use kinetic theory of gases. In this section we shall examine some of the frequently encountered equations of state.

van der Waals Equation of State. The ideal gas equation ignores the molecular collisions, the volume occupied by the molecules, and the intermolecular forces. As the pressure increases, these factors become significant. To correct these shortcomings of the ideal gas equation, van der Waals proposed the following equation:

$$\left(p + \frac{a}{v^2}\right)(v - b) = RT \tag{4.10}$$

The term a/v^2 accounts for the increase in pressure due to intermolecular forces, and b represents the volume occupied by the molecules. The values of the constants a and b are given in Table 4.4. These are found by fitting the van der Waals equation

TABLE 4.4 CONSTANTS IN THE VAN DER WAALS EQUATION OF STATE

	a (kPa [m^3/kg-mol]^2)	b (m^3/kg-mol)
Ammonia	423.3	0.0373
Carbon dioxide	364.3	0.0427
Carbon monoxide	146.3	0.0394
Helium	3.41	0.0234
Hydrogen	24.7	0.0265
Nitrogen	136.1	0.0385
Oxygen	136.9	0.0315
Water	550.7	0.0304

of state to experimental data. The constants a and b can also be determined by observing that the isotherm for $T = T_{cr}$ on a *p-v* diagram for a gas exhibits a point of inflection. At the point of inflection

$$\left(\frac{\partial p}{\partial v}\right)_{T_{cr}} = 0 \quad \text{and} \quad \left(\frac{\partial^2 p}{\partial v^2}\right)_{T_{cr}} = 0$$

Differentiating Equation 4.10, we obtain

$$\left(\frac{\partial p}{\partial v}\right)_{T_{cr}} = \frac{-RT_{cr}}{(v_{cr} - b)^2} + \frac{2a}{v_{cr}^3} = 0$$

and

$$\left(\frac{\partial^2 p}{\partial v^2}\right)_{T_{cr}} = \frac{2RT_{cr}}{(v_{cr} - b)^3} - \frac{6a}{v_{cr}^4} = 0$$

The solution to the above equations in the unknowns a and b is

$$a = \frac{27R^2 T_{cr}^2}{64 p_{cr}} \quad \text{and} \quad b = \frac{RT_{cr}}{8 p_{cr}}$$

When the above expressions for a and b are inserted in the van der Waals equation, we find

$$\frac{p_{cr} v_{cr}}{RT_{cr}} = \frac{3}{8}$$

Experimental data indicates that the value of $p_{cr} v_{cr}/RT_{cr}$ is between 0.2 and 0.3 for most gases. Also, the "constants" a and b vary with temperature. Consequently, the van der Waals equation of state is of limited accuracy. It is, however, one of the first early efforts to predict real gas behavior.

Berthelot Equation of State.   The Berthelot equation of state takes into account the fact that the constant a in the van der Waals equation is influenced by temperature. This is done by replacing a in the van der Waals equation by a/T. The Berthelot equation is

$$\left(p + \frac{a}{Tv^2}\right)(v - b) = RT \tag{4.11}$$

Dieterici Equation of State.   The Dieterici equation is another modification of the van der Waals equation. It gives good agreement with the value of the quantity $p_{cr} v_{cr}/RT_{cr}$ obtained from experiments. It is

$$pe^{a/RTv}(v - b) = RT \tag{4.12}$$

Redlich-Kwong Equation of State.   One equation of state that has considerable accuracy over a wide range of p-v-T conditions is due to Redlich and Kwong:

$$\left(p + \frac{a}{\sqrt{T}\, v(v + b)}\right)(v - b) = \bar{R}T \tag{4.13}$$

The equation is empirical and contains only two constants, a and b. From the critical data we find that

$$a = \frac{0.4275 \bar{R}^2 T_{cr}^{2.5}}{p_{cr}}$$

and

$$b = \frac{0.0867 \bar{R} T_{cr}}{p_{cr}}$$

The equation gives good results at high pressures and at temperatures above the critical value. For temperatures less than the critical temperature, the difference between the values predicted by the equation and those obtained from experiments becomes significant.

Beattie-Bridgeman Equation of State. The equations of state discussed so far contain no more than two constants. Because of the inherent complexity of a p-v-T surface for a gas, it becomes necessary to have an equation containing several constants for an accurate representation of the p-v-T relationship for a gas over a wide range. One of the well-known equations of this class is the Beattie-Bridgeman equation. It is

$$p = \frac{\bar{R}T(1 - \epsilon)}{\bar{v}^2}(\bar{v} + B) - \frac{A}{\bar{v}^2} \tag{4.14}$$

where

$$A = A_0\left(1 - \frac{a}{\bar{v}}\right)$$

$$B = B_0\left(1 - \frac{b}{\bar{v}}\right)$$

$$\epsilon = \frac{c}{\bar{v}T^3}$$

$$\bar{v} = \text{molal specific volume}$$

$$\bar{R} = \text{universal gas constant}$$

and A_0, a, B, b, and c are constants for different gases. Table 4.5 lists the values of these constants for selected gases.

TABLE 4.5 CONSTANTS IN THE BEATTIE-BRIDGEMAN EQUATION OF STATE*

Gas	A_0	a	B_0	b	$c \times 10^{-4}$
Air	131.8441	0.01931	0.04611	−0.001101	4.34
Argon	130.7802	0.02328	0.03931	0.0	5.99
Carbon dioxide	507.2836	0.07132	0.10476	0.07235	66.00
Helium	2.1886	0.05984	0.01400	0.0	0.0040
Hydrogen	20.0117	−0.00506	0.02096	−0.04359	0.0504
Nitrogen	136.2315	0.02617	0.05046	0.00691	4.20
Oxygen	151.0857	0.02562	0.04624	0.004208	4.80

*p: kPa; $\bar{v}$: m³/kg-mol; T: K; $\bar{R} = 8.31434$ kJ/kg-mol K.

The Virial Equations of State. If values of pv are plotted against p (see Figure 4.18), a number of curves are obtained. This suggests a functional relationship between pv and p. The relationship may be expressed in the form of an infinite series.

$$pv = a_0 + a_1p + a_2p^2 + \cdots \tag{4.15}$$

Such an equation is called a *virial equation of state* and the coefficients $a_0, a_1, \ldots$

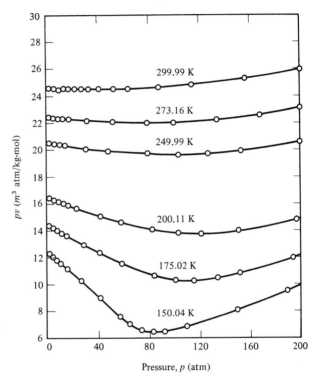

Figure 4.18   Curves of the product *pv* against pressure for nitrogen. (A. S. Friedman, 1950. Used with permission.)

are called *virial coefficients*. The virial coefficients are functions of temperature alone. The temperature at which the second virial coefficient, a_1, is zero is called *Boyle temperature*.

If we were to regard *pv* as a function of volume, we could write the following infinite series:

$$pv = v_0 + \frac{b_1}{v} + \frac{b_2}{v^2} + \cdots \qquad (4.15a)$$

This representation is also called a *virial equation of state*. The virial coefficients $b_0, b_1, b_2, \ldots$ are functions of temperature alone. It is possible to give physical significance to the virial coefficients on a molecular basis and to determine their values from statistical mechanics. The virial equation of state is useful in evaluating properties of matter.

4.8 SUMMARY

The major concepts studied in this chapter can be summarized in the following manner.

1. There are three principal phases of a substance—solid, liquid, and vapor. The state at which a liquid will begin to vaporize with addition of heat energy is called a *saturated liquid state*. During an evaporation process, the temperature and pressure remain constant. When the vaporization process is just completed, the resulting vapor is called a *saturated vapor*.

2. A mixture of a liquid and a vapor, both of which are at saturated states, is called *wet vapor*. The quality of a mixture is the mass fraction of the vapor portion in a liquid-vapor mixture.

3. The saturated liquid line and the saturated vapor line form the saturation dome. The apex of the dome is called the critical point. At the critical point of a liquid, its density changes continuously and it is difficult to differentiate between a liquid and a vapor. The division between a vapor and a gas is somewhat arbitrary. The critical isotherm $(T = T_{cr})$ is accepted as the line delineating the gas and the vapor of a substance.

4. Properties of saturated liquids and vapors and of superheated vapors are usually available in a tabular form. Note that the ideal gas law can not be used for determining such properties.

5. A change of phase from a solid state to a liquid state also occurs at a constant temperature and pressure. The volume may increase or decrease during melting, depending upon the substance. The saturated solid line and the saturated liquid line have not been known to converge to a point. If the pressure of a solid is sufficiently low, the solid can transfrom directly into a vapor phase as it is heated. This process is called *sublimation*.

6. There is a state, called the *triple point*, for every substance where the solid, the liquid, and the vapor phases of the substance coexist.

7. The coefficient of isothermal compressibility κ and the coefficient of isobaric expansion β are useful in predicting the change in the volume of a substance. These coefficients generally vary with temperature and pressure.

8. The law of corresponding states and the generalized compressibility chart are useful in determining whether the behavior of a given gas can be approximated by the ideal gas equation. The chart may be used for approximate calculations only.

9. A large number of equations exist to predict the behavior of real gases. The relatively simple-looking equations, such as the van der Waals equation and the Berthelot equation, contain only two constants. They offer good results over a limited range. Equations such as the Beattie-Bridgeman or the virial equation contain several constants, and they give accurate results over a wider range. They are complicated for routine use, however.

PROBLEMS

4.1. A rigid container contains water at 80°C. Determine the state of the contents if the pressure is (a) 10 kPa, (b) 47.39 kPa, and (c) 100 kPa.

4.2. The pressure of the water in a tank is 500 kPa. Determine the state of the water if the temperature is (a) 120°C, (b) 151.86°C, (c) 164.97°C, and (d) 200°C.

4.3. Determine the specific volume of the water in Problem 4.1.

4.4. Determine the specific internal energy and the specific enthalpy of the water in Problem 4.2.

4.5. Plot the saturation dome for water on a $T\text{-}v$ diagram and a $p\text{-}v$ diagram using the property values in the appendix.

4.6. The quality of a certain liquid-vapor mixture of water is 0.64. The pressure is known to be 1400 kPa. Determine the specific volume and the specific enthalpy of the mixture.

4.7. The temperature of a liquid-vapor mixture is 60°C and the quality is 0.3. Determine the specific internal energy of the mixture.

4.8. Determine the state of a closed system consisting of 1 kg of water under the following conditions. Also, determine the values of h, v, and x.
(a) Saturated liquid at 500 kPa.
(b) Saturated vapor at 1 MPa.
(c) Wet steam of quality 0.8 at 300 kPa.
(d) 500 kPa and 200°C.
(e) 10 MPa and 200°C.

4.9. The states of 1 kg of water substance are given below. Determine the nature of the phase of the water: namely, wet, dry saturated, superheated steam, or unsaturated liquid.
(a) $p = 100$ kPa, $T = 150$°C   (b) $p = 200$ kPa, $T = 200$°C
(c) $p = 300$ kPa, $V = 0.1$ m³   (d) $H = 2700$ kJ, $p = 50$ kPa
(e) $H = 2100$ kJ, $T = 50$°C   (f) $T = 100$°C, $x = 0.8$

4.10. Determine the changes in the enthalpy, specific volume, quality, and temperature for the following processes. The initial pressure, p_i, is 500 kPa. The final state is designated by subscript f.
(a) Constant volume: $v_i = 0.3$ m³/kg, $p_f = 350$ kPa.
(b) Constant volume: $h_i = 2500$ kJ/kg, $p_f = 190$ kPa.
(c) Constant volume: $T_i = 200$°C, $p_f = 240$ kPa.

4.11. Various states of oxygen are given. Determine whether the oxygen is a compressed liquid, a superheated vapor, or an equilibrium mixture of liquid and vapor.
(a) $T = 100$ K, $p = 150$ kPa
(b) $T = 125$ K, $u = -85$ kJ/kg
(c) $p = 1$ MPa, $h = 152.269$ kJ/kg

4.12. Rework Problem 4.11 if the given substance is nitrogen.

4.13. Saturated vapor of ammonia at -10°C is compressed to a pressure of 1033.97 kPa in accordance with the relation $pv^{1.3} = $ constant. Determine the work done in the process and estimate the final temperature of the vapor. (*Hint:* Estimate the average value of c_v for the temperature range.)

4.14. A rigid vessel contains a mixture of saturated liquid and saturated vapor of water in equilibrium at 400 kPa. What should be the proportions of the liquid and the vapor portions so that upon sufficient heating the system will pass through the critical state.

4.15. A well-insulated tank of volume 0.2 m³ holds 0.1 kg of steam at 110 kPa. Another tank holds saturated steam at 800 kPa. How much steam from the second tank should be added to the first so as to raise its pressure to 800 kPa? Also, determine the final temperature of the steam in the first tank.

4.16. A mass of 1.8 kg of saturated liquid water at 200°C is mixed with 7.2 kg of saturated water vapor at 200°C. What will be the quality of the resulting mixture? What will be its specific enthalpy?

4.17. A mass of 2.5 kg of superheated water vapor at 3000 kPa and 350°C is mixed with 0.8 kg of wet steam with a quality of 40% at 3000 kPa. Determine the resulting state and its specific internal energy.

4.18. Plot the saturation dome for Freon-12 on a *T-v* diagram and a *p-v* diagram using the property values in the appendix.

4.19. Determine the quality of ammonia at 10°C with a specific enthalpy of 1000 kJ/kg. What will be its specific volume?

4.20. Freon-12 exists in a saturated liquid state at 45°C. It is then expanded to a pressure of 308.6 kPa in such a process that its enthalpy remains constant. What is the quality of the Freon at the final pressure?

4.21. Determine the specific enthalpies of Freon-12 in each case.
    (a) −40°C and 300 kPa.
    (b) −40°C and 64.2 kPa in a liquid state.
    (c) −10°C and a quality of 50%.
    (d) 3344 kPa and saturated vapor.
    (e) 500 kPa and 70°C.

4.22. Water at 5000 kPa and 40°C is heated until it is completely vaporized in a constant-pressure heating process. If the mass of the water is 2.6 kg, determine the change in its internal energy and the amount of heat energy necessary to bring about the evaporation.

4.23. A cube of ice measuring 0.2 m on each side melts at 0°C while used to cool soft drinks at a picnic. What is the work done by the ice in the melting process? Is it useful work?

4.24. A large steam generator receives feed water in a saturated liquid state at 4.0 kPa. The pressure of the steam leaving the generator is 20 MPa and the quality is 0.98. If the rate of steam generation is 100,000 kg/h, determine the enthalpy gain of the water in the generator.

4.25. A superheater receives the steam from the generator in Problem 4.24 and raises its temperature to 460°C at a constant pressure of 20 MPa. How much enthalpy change is brought about in the superheater?

4.26. In a steam power plant, the steam leaving the steam turbine is condensed into liquid water in a condensor. The liquid water is then returned to the steam generator. In one such condenser the steam enters at 40°C with a quality of 89%. If it leaves the condenser as a saturated liquid at 40°C, how much energy is removed from the entering steam per kilogram of steam?

4.27. A container holds 14 kg of liquid water at 40°C. How much heat energy should be supplied to this body of water if the final condition is 10 kg of vapor in equilibrium with 4 kg of liquid at 170°C? Assume a constant pressure process.

4.28. A rigid container has a volume of 0.35 m³ and contains water vapor at 150°C and 100 kPa. The water vapor in the container is allowed to cool to 60°C. Determine the quantity of heat lost by the water and the final state of the water.

4.29. Saturated ice at −30°C sublimates into saturated water vapor at −30°C. If the mass of the ice is 6 kg, determine the amount of heat energy necessary to bring about the sublimation.

4.30. Plot the phase diagram for water showing the triple point and the critical point.

4.31. Plot a graph of volume against temperature for water vapor at 2500 kPa and determine the value of the coefficient of isobaric expansion for water vapor at (a) 250°C, (b) 500°C, and (c) 1200°C.

4.32. Determine the specific volume of water vapor at 4000 kPa and 500°C using (a) the steam table, (b) the ideal gas equation, and (c) the compressibility chart.

4.33. The pressure on 2 kg of a metal block is increased from 100 kPa to 100 MPa in a quasi-static isothermal process. The density and the isothermal compressibility are 5000 kg/m³ and 8.9×10^{10} Pa⁻¹, respectively. What is the work done in the process?

4.34. Hydrogen gas exists at 200°C and 800 kPa. Using the compressibility chart, determine if the gas can be treated as an ideal gas. If so, estimate the error in the specific volume calculated by using the ideal gas law.

4.35. Obtain expressions for the coefficients of isobaric expansion and isothermal compressibility for the van der Waals gas. (*Hint*: $(\partial v/\partial T)_p = -[(\partial p/\partial T)_v/(\partial p/\partial v)_T].$)

4.36. Do Problem 4.35 using the Redlich-Kwong equation of state.

4.37. Write a computer program to calculate a set of values of pressure, volume, and temperature from the Redlich-Kwong equation for water vapor. Compare the values with those from the superheated steam tables.

4.38. Methane gas at 1 MPa and 30°C is cooled at constant volume to saturated vapor and then condensed at constant pressure to saturated liquid state. The initial volume is 0.1 m³. Use the generalized compressibility chart to determine the final pressure, volume, and temperature.

4.39. Derive the following equation for a van der Waals gas:

$$c_p - c_v = \cfrac{R}{\left[1 - \cfrac{2a(v - b)^2}{RTv^3} \right]}$$

5

Open Systems
and the First Law

5.1 INTRODUCTION

The systems with which we have dealt with so far are closed systems. The mass contained in such systems remains fixed, and no mass crosses the system boundaries. In engineering applications, a majority of the problems are of such a nature that it is inconvenient to analyze them using a closed system approach. Air passing through a blower, water flowing in a pipeline or through a solar collector panel, steam expanding through a turbine, Freon vapor passing through the compressor of a refrigeration unit, and air expanding in the nozzle of a supersonic wind tunnel are some of the examples where it is cumbersome to use a closed-system approach. In these and many other situations we use an open-system approach. Intimately tied to this approach is the concept of control volume. In this chapter we shall present a discussion of control volume and the application of the principles of conservation of mass and energy to control volume.

We shall develop equations for steady-state and unsteady-state flow systems and then perform thermodynamic analyses of some typical pieces of apparatus, such as a compressor, a nozzle, and a heat exchanger. Finally, we shall examine some thermodynamic cycles that involve steady-flow processes.

5.2 CONTROL VOLUME

Recall from the definition of an open system (Section 1.6) that mass may enter or leave an open system and that the mass within it may or may not remain constant. In order to identify such a system, it is necessary to draw system boundaries. A system boundary may include some parts that are actual solid surfaces and other parts that are imaginary surfaces. The space enclosed by such a boundary is called a *control volume* (CV) and the bounding surface of a control volume is called a *control surface* (CS). A control volume may be fixed in space, moving through space, or changing its shape and size. We shall be concerned mostly with control volumes of fixed shape that are fixed in space.

Figure 5.1 shows some examples of control volume. In Figure 5.1(a) fluid passes through a right-angle bend, or an elbow, in a pipeline, and a dotted line identifies the boundary of the control volume to be used for a study of the fluid. The inside surface of the pipe at the elbow is part of the control surface. The circular areas at AB and CD are the imaginary, or fictitious, parts of the control surface. The boundary A–B–C–D–A encloses the control volume, and mass can cross the fictitious surfaces AB and CD. In Figure 5.1(b), a fluid turns around a corner and it then enters a nozzle. If we are interested in the flow in the nozzle, then we select a control volume, as shown in the figure. Here the fictitious surfaces are circular areas AB and CD. They are unequal in size.

A schematic of a steam generator is shown in Figure 5.1(c). Liquid water under pressure enters the steam generator at AB and steam leaves the generator at CD. In this example, the control volume A–B–B'–C–D–D'–A receives heat energy. Figure 5.1(d) shows a schematic of a compressor with low-pressure fluid entering the compressor at AB and high-pressure fluid leaving at CD. The control volume A–B–B'–C–D–D'–A is cut by the shaft of the compressor, and mechanical energy is transferred across the control volume at the aperture where the shaft penetrates the control volume. An ordinary balloon is depicted in Figure 5.1(e). If we select the boundaries of the balloon as a control surface, then an air mass crosses into or out of the balloon at its mouth. Thus we have an open system with mass either entering or leaving the control volume. Furthermore, the control volume itself does not remain fixed in shape and size.

These examples illustrate that mass may cross the boundaries of a control surface; this crossing occurs only at the fictitious boundary unless the solid boundary is porous. It is also possible for mass to accumulate in or be depleted from the control volume.

Heat energy, mechanical energy, or any other form of energy (such as solar or electrical), can cross the boundaries of a control volume. When mechanical energy crosses a control surface, it is usually via a shaft that cuts through the control surface. Kinetic and potential energies are carried into or out of a control volume, along with the mass that crosses the control surface. The kinetic and the potential energies of the fluid at the entrance to a control volume are not necessarily equal to those at the exit of the control volume. The potential energy of the fluid entering the vertical elbow at surface AB of the control volume in Figure 5.1(a) is less than the potential energy of the fluid leaving the control volume at CD. The same holds for the fluids entering and leaving the control volume in Figure 5.1(c). In Figure 5.1(b), the kinetic energies of the fluid entering and leaving the control volume (nozzle) are different.

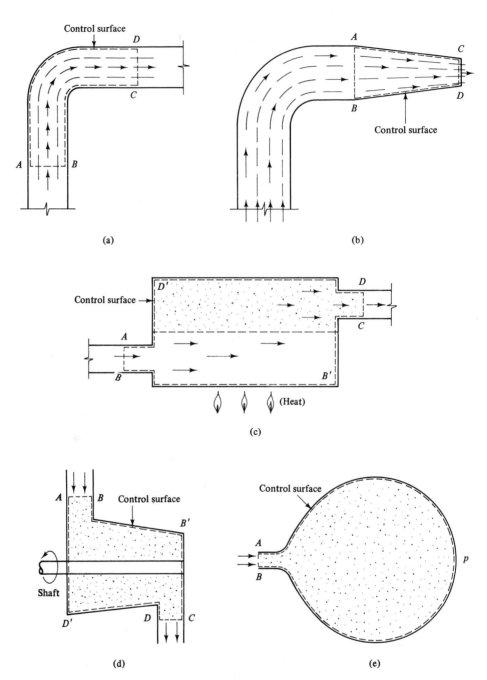

Figure 5.1 Control volume for flow. (a) Around a bend. (b) Through a converging section. (c) Through a steam generator. (d) Through a compressor. (e) Into a balloon.

5.2.1 Local Equilibrium in a Flow Process

Fluid properties at sections *AB* and *CD* in Figure 5.1 are, in general, nonuniform. For example, values of velocity, temperature, density, and pressure at point *A* (near the solid boundary) are different from those at a point halfway between *A* and *B* (midstream). Therefore thermodynamic equilibrium does not prevail, in the sense that it was defined in Chapter 1, at *AB* and *CD*. This is also true of the fluid in the control volume. It is still possible to analyze a general flow system using the control volume concept if we introduce the idea of local equilibrium.

The concept of equilibrium facilitates analysis of thermodynamic system, When we deal with a closed system containing a fixed mass, the phrase *a system in equilibrium* implies uniformity of properties throughout the system. When a system exchanges energy in the form of heat, work, or both with its environment, we assume for the purpose of analysis that the system remains essentially in a thermodynamic equilibrium throughout the energy transfer process. We refer to such a process as a *quasistatic process*. When we deal with a flow process in an open system, we find that—in general—the thermodynamic properties are not uniform throughout the flow field. In order to deal effectively with this situation, the concept of *local equilibrium* is used.

We consider a volume element of a fluid that is sufficiently small so that its properties, such as density and temperature, are uniform throughout this tiny volume. We also require that this volume element be large enough on a microscopic scale to contain a sufficiently large number of molecules. Under these conditions, a property such as density represents a statistical mean mass of the molecules in the elemental volume, and its value is not affected by the back-and-forth migration of molecules from the neighboring volume elements. Fortunately, these conditions are readily met in most engineering applications. A cube only 10^{-3} mm on a side contains as many as 10 billion molecules under ordinary conditions. We now consider two neighboring volume elements that also meet the above requirements on their sizes. If a fluid property measured in two such neighboring volume elements in a system shows change or variation that is very small compared with the average value of the property in the two elements, then local equilibrium exists.

Consider a flow of a viscous oil through a pipe. Due to the viscosity of the oil, the velocity of the oil at the pipe wall is zero, while it is maximum at the axis of the pipe. Although the velocity changes from zero to a maximum value, we do have local equilibrium since there are no abrupt changes in the velocity; the difference in the velocity of the oil in any two neighboring volume elements is extremely small in relation to the average velocity of the two fluid elements. Now consider air flowing through a pipe and then entering a tank. As the tank continues to receive air, we observe that the pressure in the tank builds up, but the pressure does not exceed the pressure of the air in the pipeline. If we now examine an area in the flow just upstream of the junction of the pipe and the tank, we find that a local equilibrium exists at this area. This does not preclude a sharp variation in the properties of the air at the junction of the pipe and the tank. In fact, by selecting a control volume large enough to include the junction and some flow upstream of it, we are able to ensure that we have a local equilibrium on the fictitious surface of the control volume. In practice, we find that the concept of local equilibrium can be employed effectively and advantageously.

5.3 CONSERVATION OF MASS AND CONTROL VOLUME

It is a common observation that mass can be neither destroyed nor created; that is, mass is conserved, if nuclear reactions are absent. For a closed system, the principle of conservation of mass can be expressed in equation form, namely,

$$dm_{\text{system}} = 0 \quad \text{or} \quad m_{\text{system}} = \text{constant}$$

The mathematical expression of mass conservation for an open system becomes somewhat more complicated because mass may cross the boundary of an open system at one or more points. With the use of vector algebra it is possible to write a single term that accounts for all of the mass crossing the control surface. We shall, however, use a physical approach to derive the equation for the principle of conservation of mass for an open system and then express it in a vector form.

We saw earlier that a control volume can be of any shape or size. A general representation of a control volume is shown in Figure 5.2. The section where the fluid enters the control volume is labeled section 1, and the section where the fluid leaves the control volume is labeled section 2. Typically, there may be both inflow into and outflow from a control volume. We shall assume that local equilibrium exists throughout the control volume.

We consider the fluid in the control volume and at the inlet and the outlet sections at two instants of time τ and $\tau + \Delta\tau$. On examining what happens in the short time interval $\Delta\tau$, we find the following:

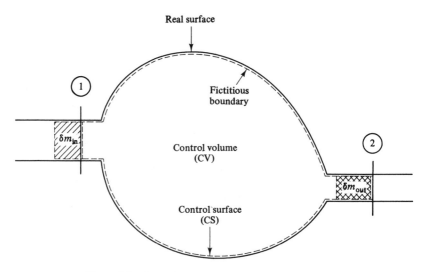

Figure 5.2   Control volume and conservation of mass.

1. Some fluid enters the control volume through the inlet pipe at section 1. We shall call this δm_{in}.
2. Some fluid leaves the control volume through the outlet pipe at section 2. We shall call this δm_{out}.
3. The mass contained in the control volume changes from $m_{CV,\tau}$ to $m_{CV,\tau+\Delta\tau}$.

Since mass is conserved, the net quantity of mass entering the control volume equals the change in the control-volume mass, or

$$\delta m_{in} - \delta m_{out} = m_{CV,\tau+\Delta\tau} - m_{CV,\tau}$$

Then division by $\Delta\tau$ gives

$$\frac{\delta m_{in}}{\Delta\tau} - \frac{\delta m_{out}}{\Delta\tau} = \frac{m_{CV,\tau+\Delta\tau} - m_{CV,\tau}}{\Delta\tau}$$

As we make the time step $\Delta\tau$ in the above equation very small, we find these facts.

1. $\delta m_{in}/\Delta\tau$ becomes the *mass flow rate*, $\dot{m}_{in}$, into the control volume*.
2. $\delta m_{out}/\Delta\tau$ becomes the *mass flow rate*, $\dot{m}_{out}$, out of the control volume.
3. The term on the right side of the equation becomes the *rate* of change of mass within the control volume, or or $(dm/d\tau)_{CV}$.

Thus the equation for conservation of mass for a control volume takes the following form (for the vector form of Equation 5.1, see *Notes* at the end of this chapter):

$$\dot{m}_{in} - \dot{m}_{out} = \left(\frac{dm}{d\tau}\right)_{CV} \tag{5.1}$$

If $\dot{m}_{in}$ is greater than $\dot{m}_{out}$, the mass within the control volume increases with time and $(dm/d\tau)_{CV}$ is positive. On the other hand, if $\dot{m}_{in}$ is equal to $\dot{m}_{out}$, then the mass within the control volume does not change and $(dm/d\tau)_{CV}$ is zero. Alternatively, if $\dot{m}_{in}$ is less than $\dot{m}_{out}$, then the mass within the control volume decreases with time and $(dm/d\tau)_{CV}$ is negative.

Let us now turn our attention to the method of determining the mass flow rate $\dot{m}$. We assume a one-dimensional flow; that is, fluid properties such as temperature, pressure, and velocity can vary only along the direction of the flow, but are uniform across any cross section. The direction of the flow does not have to be along a straight line. The flow can be along a smooth and continuous curve, such as the flow of water in a garden hose. Also, the cross-sectional area available for flow may or may not be constant along the flow direction. In the case of the garden hose it is constant; when the water in the garden hose passes through a nozzle, the cross-sectional area gradually decreases along the flow direction.

*Note that we do not write $dm/d\tau$ for the limiting value of $\delta m/\Delta\tau$. This is because the term represents the time rate at which the mass crosses the boundary of the control volume, denoted here by $\dot{m}$; it does not represent the time rate of change of mass at the inlet. In fact, there is no such entity as a time rate of change of mass at the inlet. These comments apply equally to the fluid leaving a control volume. The fluid within a control volume, however, can experience a change in its mass, and the use of the term $(dm/d\tau)_{CV}$ is proper.

Consider a *one-dimensional flow* of a fluid through a pipe. The uniform velocity of the fluid is in the direction normal to the cross-sectional area A of the pipe. If V is the uniform velocity of the flow normal to the area A and ρ is the density of the fluid, then the mass flow rate $\dot{m}$ for the fluid is given by

$$\dot{m} = \rho A V \qquad (5.2)$$

In reality, the velocity of a flowing fluid is not the same at every point in the flow cross section. It is usually zero at the wall of the duct carrying the fluid and reaches a maximum value at the center of the duct. In such cases we use the average, or the mean velocity, of the flow, V_{av}, over the cross section. It is defined by

$$V_{av} = \frac{1}{A} \iint_A V \, dx \, dy = \frac{1}{A} \int_A V \, dA$$

In the above equation, the local velocity V is a function of the xy-coordinates in the plane of the cross-sectional area A. In general, computation of the average velocity requires evaluation of a double integral and a knowledge of the functional relationship between the local velocity V and the coordinates x and y (or r and θ). The mass flow rate for an incompressible fluid (constant ρ) is then given by

$$\dot{m} = \rho A V_{av} \qquad (5.2a)$$

If the density of the fluid also varies, then the equation for the mass flow rate takes the following form:

$$\dot{m} = \iint_A \rho V \, dx \, dy = \int_A \rho V \, dA \qquad (5.2b)$$

where A is the surface area across which fluid moves into or out of a control volume. It is important to note that in the above equation, the outer normal vector to the elemental area dA and the velocity V must be in the same direction. In a one-dimensional flow, the cross-sectional area is usually normal to the flow direction.

The equation for conservation of mass in vector form is derived in the notes at the end of this chapter. The equation is

$$-\int_{cs} \rho \mathbf{V} \cdot d\mathbf{A} = \frac{d}{d\tau} \int_{cv} \rho \, dV \qquad (5.3)$$

where $\mathbf{V}$ is the velocity vector at a point on the control surface and $d\mathbf{A}$ is a vector representing an elemental area on the control surface. The symbol ρ in the left side of Equation 5.3 is the fluid density at $d\mathbf{A}$, whereas the symbol ρ in the right side is the fluid density in dV, an elemental volume within the control volume.

Example 5.1

Saturated steam at 150°C flows through a pipe of diameter 15 cm. The steam has an average velocity of 12 m/s. Determine the mass flow rate of the steam.

Solution:

Given: Saturated steam flows through a pipe with the following conditions:

$$t_1 = 150°C, \qquad d = 15 \text{ cm}, \qquad V = 12 \text{ m/s}$$

Objective: We are to determine the mass of the steam flowing through the pipe per unit time.

Assumptions: The density of the steam across a cross section of the pipe is uniform.

Relevant physics: Since the average velocity of steam is prescribed and we have assumed a uniform density, we can determine the mass flow rate using Equation 5.2(a).

$$\dot{m} = \rho A V_{av}$$

Analysis: We need to know the cross-sectional area of the tube, A, the density of steam, ρ, and the average velocity, V_{av}. We note that

$$A = \frac{\pi}{4}\left(\frac{15}{100}\right)^2 = 1.77 \times 10^{-2}\ m^2$$

and

$$V_{av} = 12\ m/s$$

From the steam tables (Appendix B2.1), we find that saturated steam at 150°C has a specific volume of 0.3928 m³/kg. Since the density is the reciprocal of specific volume, the mass flow rate can be expressed as

$$\dot{m} = \rho A V_{av} = \frac{A V_{av}}{v}$$

$$= \frac{1.77 \times 10^{-2} \times 12}{0.3928}$$

$$= 5.4\ kg/s,\ or\ 19,400\ kg/h$$

5.4 FLOW WORK AND TRANSFER OF ENERGY ACROSS CONTROL SURFACE

When a fluid flows across a control surface, there is a transfer of work energy across the surface. This transfer of work, known as *flow work*, is essential for maintaining a continuous movement of fluid across the control surface. To facilitate our understanding of flow work, we consider a fluid with uniform properties (p, T, and v) flowing through a pipe of uniform cross-sectional area A. We consider a fluid element A–B–C–D–A, shown in Figure 5.3, as our system. At time $\tau = 0$, the boundary AB

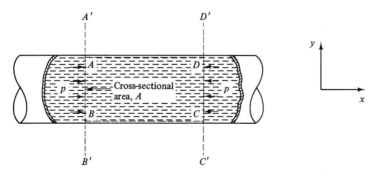

Figure 5.3   A fluid element in a flow through pipe.

of the element coincides with the line $A'B'$, which is fixed in space. We observe that the element experiences a pressure force pA along the x-axis. The pressure force acting on AB is in the positive x-direction, while that acting on face CD is in the negative x-direction. When the element travels a distance Δx (not shown in the figure) in time $\Delta \tau$ in the positive x-direction, the force pA on AB is displaced through a distance Δx. From the definition of work, we recognize that a quantity of work is done in this process. This is the work that the fluid to the left of the element $A–B–C–D–A$ does *on* the fluid within the element, since the directions of the force acting on AB and the displacement Δx of the element are both in the positive x-direction. As far as the fluid in the element $A–B–C–D–A$ is concerned, it receives this work energy, which is $-pA\Delta x$. The negative sign accounts for the system, the fluid in the element $A–B–C–D–A$, *receiving* work energy. The quantity $A\Delta x$ is the volume through which face AB has moved. It can be expressed as mv, where m is the mass of the fluid occupying the volume $A\Delta x$ and v is the specific volume of the fluid. Strictly speaking, m is the mass of the fluid on the left of $A'B'$ at time $\tau = 0$ that crosses over to the right of $A'B'$ in time $\Delta \tau$ as the fluid in the element $A–B–C–D–A$ moves to the right. The section $A'B'$ is fixed in space. It can be considered as a hypothetical gate through which fluid enters. Thus the expression for the work done by the fluid to the left of AB on the system, that is, on the fluid in the element $A–B–C–D–A$, is

$$W_{A'B'} = -pA\Delta x = -(pmv)_{A'B'}$$

On a unit-mass basis, the above becomes

$$w_{A'B'} = -(pv)_{A'B'}$$

Let us now consider the fluid on the right side of CD. We find that the fluid in the element $A–B–C–D–A$, as it moves through the distance Δx, pushes the fluid on the right side of CD. A quantity of work is done by the fluid in the element on the surroundings. Applying the same reasoning as that for $W_{A'B'}$, we find that this work is given by

$$W_{C'D'} = +pA\Delta x = (pmv)_{C'D'}$$

The positive sign tells us that the system—that is, the fluid in the element $A–B–C–D–A$—does work on the fluid to the right of $C'D'$. On a unit-mass basis, the above becomes

$$w_{C'D'} = (pv)_{C'D'}$$

The quantities of work, $W_{A'B'}$ and $W_{C'D'}$, are not available to an outsider for any use. They are necessary to maintain a flow. We can readily see that if the fluid is not moving, the work terms will be identically zero.

Let us now change our point of view and consider the control volume bounded by sections $A'B'$ and $C'D'$. As a fluid flows through this control volume, we observe that there is an inflow of work energy, $w_{A'B'}$ per unit mass, at $A'B'$ and an outflow of work energy, $w_{C'D'}$ per unit mass, at $C'D'$. This flow of work energy is called *flow work* and it is due to the fact that fluid is flowing and not stationary. Also, the flow work at the inlet and the outlet of a control volume is expressed by the term pv. To reiterate, the flow work is essential for maintaining a flow and is not available for outside use or application. Energy in the form of flow work crosses the boundaries that are fixed relative to a moving fluid, that is, the boundaries of a control volume. The quantity $-(pv)_{A'B'}$ can be interpreted as the amount of energy in the form of a flow work per unit mass that enters the control volume at section $A'B'$. Likewise, $(pv)_{C'D'}$ is the amount of energy in the form of a flow work per unit mass that leaves the control volume at section $C'D'$.

We now inquire into the different forms of energy that can be transported across the boundaries of a control volume. Some of the commonly encountered forms are flow work, internal energy, kinetic energy, potential energy, heat, and work. Of these, the first four are associated with fluid that crosses a control surface, and the last two are due to an interaction between control volume and its surroundings. In the preceding paragraphs we saw the role of flow work. We now consider the transfer of other forms of energy mentioned at the beginning of this paragraph. In considering internal energy we observe that with every unit mass of fluid is associated its specific internal energy u. Thus in Figure 5.3, for every unit mass of fluid crossing the boundary $A'B'$, there is a specific internal energy $u_{A'B'}$, that enters the control volume bounded by $A'B'$ and $C'D'$. Likewise, for every unit mass leaving the boundary $C'D'$, there is the specific internal energy, $u_{C'D'}$, leaving the control volume.

Associated with a moving mass is the kinetic energy. If $V_{A'B'}$, is the *average* velocity of the fluid entering the control volume in Figure 5.3, then the kinetic energy per unit mass, $V_{A'B'}^2/2$, enters the control volume along with each unit mass that crosses $A'B'$. Similarly, if $V_{C'D'}$, is the *average* velocity of the fluid leaving the control volume, then a specific kinetic energy, $V_{C'D'}^2/2$, leaves the control volume along with each unit mass that crosses $C'D'$.

Another form of energy that is transported by a mass across the boundaries of a control volume is the potential energy. It is given by the quantity gZ on a unit-mass basis. Thus the specific potential energy entering the control volume bounded by sections $A'B'$ and $C'D'$ is $(gZ)_{A'B'}$ and that leaving the control volume is $(gZ)_{C'D'}$.

To summarize, in most engineering applications there are four principal forms of energy that will be transported across (into or out of) the boundaries of a control volume by the fluid that flows across the control surface. They are the flow work pv, the specific internal energy u, the kinetic energy $V^2/2$, and the potential energy gZ. Any additional forms of energy transported along with the fluid crossing a control volume can be accounted for in the same manner as the internal energy. There are two other forms of energy that can enter a control volume or leave a control volume, independent of any mass crossing the boundaries of the control volume: heat and work. Transfer of heat requires a temperature difference between the control volume and its surroundings, while some mechanism is required to transfer work across the boundaries. The mechanism could be a mechanical shaft or electrical wires crossing the boundaries of the control volume. In the use of control volume analysis, it is important to identify the various forms of energy that are transported due to mass that crosses the boundaries. We are now ready to develop an energy equation for a control volume.

5.5 ENERGY EQUATION FOR A CONTROL VOLUME

The components of a steam power plant, such as a feedwater pump, a steam generator, a superheater, a turbine, or a condenser, involve energy transfer and flow processes. The power plant, as well as its components, can be analyzed by an appropriate choice of control volume. The same holds true for any device that processes energy in some manner. Some examples are a solar collector panel, an incinerator, a gasoline engine, a diesel engine, an electric battery, and a fuel cell. It is desirable to make an energy analysis of a control volume. We begin, therefore, with an arbitrary control volume such as the one shown in Figure 5.4.

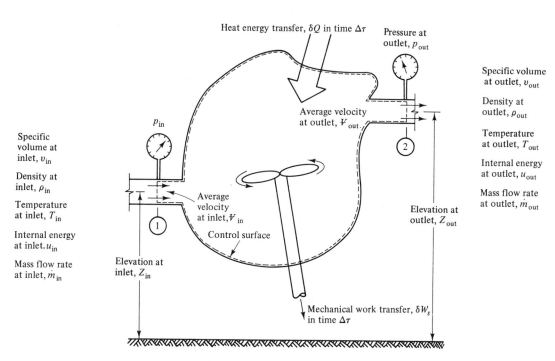

Figure 5.4 Control volume and the first law.

A fluid enters the control volume at section 1 with a mass flow rate $\dot{m}_{in}$ and leaves at section 2 with a mass flow rate $\dot{m}_{out}$. We designate fluid properties such as density, pressure, specific volume, temperature, velocity, and elevation at section 1 by the subscript *in* and those at section 2 by the subscript *out*. Also, heat energy δQ enters the control volume and mechanical work δW_s leaves the control volume. In order to emphasize that this useful work can cross the boundaries of a control volume only by suitable means, such as a shaft or electric wires, we use the subscript *s* in W_s.

If the status of the energy in the control volume and at sections 1 and 2 is considered at time τ and at time $\tau + \Delta\tau$, we find the following:

174

1. Some energy enters the control volume through the inlet pipe at section 1. We shall call this δE_{in}. The symbol E represents the sum total of energy in *all* forms.
2. Some energy leaves the control volume through the outlet pipe at section 2. We shall call this δE_{out}.
3. Some heat energy, δQ, enters the control volume.
4. Some mechanical work, δW_s, is transferred out of the control volume.
5. The energy contained *within* the control volume has changed. We shall denote this change of energy within the control volume by ΔE_{CV}. It equals the total energy within the control volume at time $\tau + \Delta\tau$ minus the total energy within the control volume at time τ. Or, $\Delta E_{CV} = E_{CV,\tau+\Delta\tau} - E_{CV,\tau}$

The first law requirement that energy be conserved can now be written as an equation.

(Energy entering the control volume in time $\Delta\tau$) $-$ (energy leaving the control volume in time $\Delta\tau$) $=$ (increase of energy within the control volume during time $\Delta\tau$)

Accounting for the various energy transfers into and out of the control volume, we can write

$$(\delta E_{in} + \delta Q) - (\delta E_{out} + \delta W_s) = \Delta E_{CV}$$

or

$$(\delta E_{in} - \delta E_{out}) + (\delta Q - \delta W_s) = \Delta E_{CV}$$

It is to be noted that in the above equations the E in E_{CV} and the E in E_{out} are not the same; E_{CV} represents the sum total of energy in all forms contained within the control volume, while E_{in} or E_{out} represent the sum total of energy in all forms associated with the fluid entering or leaving the control volume. We observe that

$$\delta E_{in} = e_{in}\delta m_{in} \quad \text{and} \quad \delta E_{out} = e_{out}\delta m_{out}$$

In the above equations, e represents the sum of all the energy forms associated with the mass crossing the boundary of the control volume on a unit mass basis. Typically, it is the sum of the integral energy u, the flow work pv, the kinetic energy $V^2/2$, and the potential energy gZ. Thus

$$e(\text{associated only with moving fluid}) = u + pv + \frac{V^2}{2} + gZ$$

Since the combination $u + pv$ is the enthalpy h, we can write

$$e(\text{associated only with moving fluid}) = h + \frac{V^2}{2} + gZ \tag{5.4}$$

The energy of the fluid in the control volume, E_{CV}, can be obtained by taking the product of a small mass within the control volume, $\rho\, dV$, and the specific energy, e_{CV}, associated with this mass and then integrating the product over the entire control volume. Typically, the specific energy of the fluid in the control volume is made of the internal energy u, the kinetic energy $V^2/2$, and the potential energy gZ. Observe that there is no flow-work term associated with the fluid contained in the control volume. Thus we can write

$$e(\text{associated with a control volume fluid}) = u + \frac{V^2}{2} + gZ$$

and

$$E_{\text{CV}} = \int_{\text{CV}} (e\rho)\, dV = \int_{\text{CV}} \left(u + \frac{V^2}{2} + gZ\right)\rho\, dV \qquad (5.4a)$$

Observe that E_{CV} involves u and not h.

In view of Equation 5.4, we can write

$$e_{\text{in}} = h_{\text{in}} + \frac{V_{\text{in}}^2}{2} + gZ_{\text{in}}$$

and

$$e_{\text{out}} = h_{\text{out}} + \frac{V_{\text{out}}^2}{2} + gZ_{\text{out}}$$

Also

$$\delta m_{\text{in}} = \dot{m}_{\text{in}}\, \Delta\tau \quad \text{and} \quad \delta m_{\text{out}} = \dot{m}_{\text{out}}\, \Delta\tau$$

Substitution of the various quantities in the energy equation and division by $\Delta\tau$ gives

$$\dot{m}_{\text{in}}\left(h_{\text{in}} + \frac{V_{\text{in}}^2}{2} + gZ_{\text{in}}\right) - \dot{m}_{\text{out}}\left(h_{\text{out}} + \frac{V_{\text{out}}^2}{2} + gZ_{\text{out}}\right) + \frac{\delta Q}{\Delta\tau} - \frac{\delta W_s}{\Delta\tau} = \frac{\Delta E_{\text{CV}}}{\Delta\tau} \qquad (5.5)$$

The terms that contain $\Delta\tau$ in their denominators must be carefully interpreted. We have seen that neither Q nor W_s are properties. They represent forms of energy in transit. Thus δQ and δW_s represent small quantities of energy transferred. When these are divided by $\Delta\tau$ and a limit is taken, they represent rates at which energy is transferred. This is somewhat analogous to the mass flow rate $\dot{m}$. Hence

$$\lim_{\Delta\tau \to 0} \frac{\delta Q}{\Delta\tau} = \dot{Q} \qquad (5.6)$$

and

$$\lim_{\Delta\tau \to 0} \frac{\delta W_s}{\Delta\tau} = \dot{W}_s \qquad (5.6a)$$

The quantity E, regardless of whether it represents the energy of the *fluid crossing a certain section* or that of the *fluid in a control volume*, is a property of a substance and is a function of its state. The limit of the ratio $\Delta E_{\text{CV}}/\Delta\tau$ is the time rate of change of total energy E function, or

$$\lim_{\Delta\tau \to 0} \frac{\Delta E_{\text{CV}}}{\Delta\tau} = \frac{d}{d\tau}(E_{\text{CV}}) = \left(\frac{dE}{d\tau}\right)_{\text{CV}} \qquad (5.6b)$$

Employing Equation 5.4(a), we have

$$\frac{d}{d\tau}(E_{\text{CV}}) = \frac{d}{d\tau}\int_{\text{CV}} \left(u + \frac{V^2}{2} + gZ\right)\rho\, dV \qquad (5.6c)$$

Substitution of Equations 5.6, 5.6(a), and 5.6(b) in Equation 5.5 gives

$$\dot{m}_{in}\left(h_{in} + \frac{V_{in}^2}{2} + gZ_{in}\right) - \dot{m}_{out}\left(h_{out} + \frac{V_{out}^2}{2} + gZ_{out}\right) + \dot{Q} - \dot{W}_s$$

$$= \frac{d}{d\tau}\int_{CV}\left(u + \frac{V^2}{2} + gZ\right)\rho\, dV \tag{5.7}$$

Equation 5.7 is the energy equation for unsteady flow through a control volume. The phrase *unsteady* implies that properties can change with time. If a steady flow exists, then all the derivatives with respect to time are identically zero. This makes the right side of Equation 5.7 zero for steady flow. Thus for a steady flow, the energy equation for a flow through a control volume becomes

$$\dot{Q} - \dot{W}_s = \dot{m}_{out}\left(h_{out} + \frac{V_{out}^2}{2} + gZ_{out}\right) - \dot{m}_{in}\left(h_{in} + \frac{V_{in}^2}{2} + gZ_{in}\right) \tag{5.7a}$$

Also, for steady flow, from Equation 5.1

$$\dot{m}_{in} = \dot{m}_{out} = \dot{m}$$

Equation 5.7(a) then becomes

$$\dot{Q} - \dot{W}_s = \dot{m}\left[(h_{out} - h_{in}) + \left(\frac{V_{out}^2}{2} - \frac{V_{in}^2}{2}\right) + (gZ_{out} - gZ_{in})\right] \tag{5.8}$$

Equation 5.8 is for a control volume that has only one entrance section and only one exit section. In practice, we often encounter multiple streams, both at entrance and exit. Equation 5.8 can be readily extended to such situations. It then takes the form

$$\dot{Q} - \dot{W}_s = \sum_{out}\left[\dot{m}\left(h + \frac{V^2}{2} + gZ\right)\right]_{out} - \sum_{in}\left[\dot{m}\left(h + \frac{V^2}{2} + gZ\right)\right]_{in} \tag{5.8a}$$

In the right side of the above equation, the first summation accounts for the energy leaving the control volume via *all* the outgoing streams of fluids and the second summation accounts for the energy entering the control volume via *all* the incoming streams of fluids.

If we divide Equation 5.8 by $\dot{m}$, we obtain

$$q - w_s = (h_{out} - h_{in}) + \left(\frac{V_{out}^2}{2} - \frac{V_{in}^2}{2}\right) + (gZ_{out} - gZ_{in}) \tag{5.8b}$$

where

$$q = \frac{\dot{Q}}{\dot{m}} \quad \text{and} \quad w_s = \frac{\dot{W}_s}{\dot{m}} \tag{5.8c}$$

The quantity q is the heat energy supplied to the control volume per unit mass of the fluid flowing through the control volume. Likewise, the quantity w_s is the mechanical work transmitted from the control volume along a shaft or electric wires per unit mass of the flowing fluid. You might be perplexed by the fact that Equation 5.8(b) applies to a flow process and yet does not contain any mass flow term. The explanation is that it is a normalized equation and represents an energy balance *on a unit mass* passing through a control volume under steady-state conditions.

5.6 APPLICATIONS OF THE STEADY-FLOW ENERGY EQUATION

The energy equation for a control volume, Equation 5.7, is highly useful in the analysis of a variety of devices. These devices include—

1. a feed pump that pressurizes liquid water and delivers it to a steam generator;
2. a compressor where a low-pressure fluid is compressed by utilizing mechanical work;
3. a turbine where a high-pressure, high-energy fluid (gas, vapor, or liquid) is expanded and mechanical work is produced;
4. a nozzle that converts some of the enthalpy of a fluid into kinetic energy;
5. a diffuser that converts part of the kinetic energy of a fluid into its enthalpy;
6. a heat exchanger that transfers energy from a high-energy fluid to a low-energy fluid (the steam condenser in a power plant is only a special type of heat exchanger);
7. a solar collector panel that receives heat energy from the sun and imparts it to the fluid flowing through the panel;
8. a combustion chamber into which fuel and air or oxygen are admitted, the fuel is burned, heat energy is released, and exhaust gases are discharged;
9. a steam generator that receives heat energy and converts liquid water into water vapor;
10. a superheater that raises the temperature of saturated vapor at constant pressure;
11. a garbage incinerator where garbage and fuel are mixed and combustion is brought about;
12. an internal combustion engine where an air-fuel mixture is ignited, hot, high-pressure gases are expanded, mechanical work is produced, and exhaust gases are rejected;
13. a fuel cell that converts chemical energy in fuel directly into electrical energy;
14. a valve that regulates the pressure of a fluid; and
15. a tank that is being filled with a fluid.

The above list, although long, is by no means complete. We can go on to include many other devices. The list is intended to suggest that the applications of the energy equation, Equation 5.7, are numerous.

When we wish to make an energy analysis of any device, we find the following procedure to be useful.

1. Draw a schematic diagram of the device, define a control volume, and draw the control surface.
2. Show clearly all the points where fluid enters and leaves the control surface. If there are no such points, then the system is closed.

3. If there is mechanical work input or output, show a shaft penetrating the control surface. Next to the shaft indicate the mechanical work, $\dot{W}_s$ or w_s, entering or leaving the system by an arrow pointing *away* from the control volume. The sign associated with the numerical value of $\dot{W}_s$ or w_s determines whether the mechanical work is leaving the control volume (positive sign) or entering it (negative sign).

4. Indicate the heat energy $\dot{Q}$ or q entering or leaving the control volume by an arrow pointing *into* the control volume. Once again, the sign associated with the numerical value of $\dot{Q}$ or q determines whether heat energy is entering the control volume (positive sign) or leaving the control volume (negative sign).

5. If there are forms of energy entering or leaving the control volume other than those carried by the mass of fluid crossing the control surface, identify them and show them on the control volume diagram with suitable arrows.

6. Make a list of the values of the properties of the fluid entering and leaving the control volume that are pertinent to the energy equation.

7. If any of the properties needed for the energy equation are missing, see if these can be determined by using (a) equation of state, (b) mass conservation, or (c) the equation for the process that the fluid experiences as it passes through the control volume.

8. If appropriate, make suitable assumptions consistent with the nature of the device so that some of the terms in the energy equation can be neglected. For example, the changes in the kinetic energy and the potential energy of a fluid as it passes through a feed pump, a steam generator, a combustion chamber, a superheater, or a heat exchanger can often be neglected.

9. Employ Equation 5.7 or Equation 5.8 to determine the unknown property or the unknown energy-transfer rate.

We shall now give several examples to demonstrate the above procedure for energy analyses of physical devices.

Example 5.2

Air at 100 kPa and 10°C enters a compressor and is compressed to 1000 kPa and 50°C. The specific heat of air is 1.01 kJ/kg K. If 15 kg of air are to be compressed every minute, determine the power requirement of the compressor. State your assumptions.

Solution:

Given: Air is compressed in a compressor.

$$\text{Inlet:} \quad p_{in} = 100 \text{ kPa}, \qquad t_{in} = 10°C$$
$$\text{Outlet:} \quad p_{out} = 1000 \text{ kPa}, \qquad t_{out} = 50°C$$

Also, $c_p = 1.01$ kJ/kg K and $\dot{m} = 15$ kg/min.

Objective: We are to determine the power requirement of the compressor, that is, the quantity $\dot{W}_s$ in Equation 5.8.

Diagram:

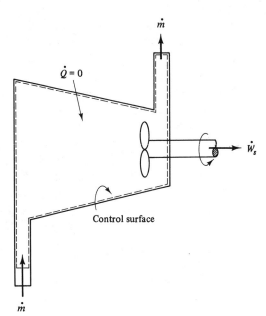

Figure 5.5 Schematic diagram of a compressor.

Assumptions: We assume the following about the air as it goes through the compressor.

1. Changes in the kinetic energy of the air are negligible.
2. Changes in the potential energy of the air are negligible.
3. No heat energy is gained or lost by the air in the compressor; that is, $\dot{Q} = 0$.
4. A steady flow prevails.

Relevant physics: We have an open system here with only one inlet and one outlet. With the assumptions made, equation 5.8 becomes

$$\overset{0}{\cancel{\dot{Q}}} - \dot{W}_s = \dot{m}\left[(h_{\text{out}} - h_{\text{in}}) + \left(\frac{V_{\text{out}}^2}{2} \overset{0}{\cancel{-\frac{V_{\text{in}}^2}{2}}}\right) + \overset{0}{\cancel{(gZ_{\text{out}} - gZ_{\text{in}})}}\right]$$

or

$$\dot{W}_s = \dot{m}(h_{\text{in}} - h_{\text{out}})$$

Also, for constant specific heat

$$h_{\text{in}} - h_{\text{out}} = c_p(t_{\text{in}} - t_{\text{out}})$$

so that

$$\dot{W}_s = \dot{m}c_p(t_{\text{in}} - t_{\text{out}})$$

Analysis: Since

$$\dot{m} = 15 \text{ kg/min} = 0.25 \text{ kg/s}$$

Substitution in the expression for $\dot{W}_s$ gives

$$\dot{W}_s = 0.25 \times 1.01(10 - 50) = -10.1 \text{ kJ/s}$$

The negative sign tells us that work is supplied to the control volume. Thus the power requirement of the compressor is 10.1 kW.

Let us next consider a steam turbine for the purpose of energy analysis. A steam turbine presents a slightly higher level of complexity compared to a gas turbine or a compressor because the quality of steam often enters into the picture. Tables of properties are also involved.

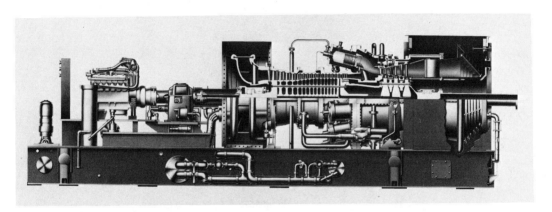

Plate 1   The cross section of a General Electric Company's MS6001 heavy duty gas turbine rated at 34 MW. (Courtesy of General Electric Company.)

Example 5.3

A steam turbine receives steam at 100 kPa and 350°C. The mass flow rate of the steam is 55 kg/s. The heat loss from the turbine is 50 kJ/kg of steam. It is known that the power output of the turbine is 40×10^3 kW. If the turbine exhausts at 10 kPa, determine the quality of the steam leaving the turbine.

Solution:

Given: Steam expands through a turbine and there is a mechanical work output.

$$\text{Inlet:} \quad p_{\text{in}} = 1000 \text{ kPa}, \quad t_{\text{in}} = 350°C$$

$$\text{Outlet:} \quad p_{\text{out}} = 10 \text{ kPa}$$

Also, $\dot{m} = 55$ kg/s, $\dot{W}_s = 40 \times 10^3$ kW, and $q = -50$ kJ/kg.

Objective: The quality of the turbine exhaust is to be determined.

Diagram:

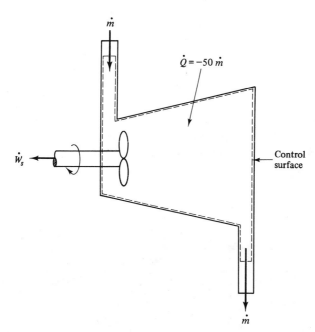

Figure 5.6 Schematic diagram of a turbine.

Assumptions: We make the following assumptions about the steam as it goes through the turbine.

1. Changes in its kinetic energy are negligible.
2. Changes in its potential energy are negligible.
3. A steady flow exists.

Relevant physics: With the assumptions made, the energy equation, Equation 5.8(a), for this open system becomes

$$\dot{Q} - \dot{W}_s = \dot{m}(h_{\text{out}} - h_{\text{in}})$$

with

$$\dot{Q} = \dot{m}q$$

Also, the enthalpy of the steam entering the turbine can be determined from the steam tables.

Analysis: The heat *loss* from the turbine is given to be 50 kJ/kg, or

$$\dot{Q} = -50\dot{m} = -50 \times 55 = -2750 \text{ kW}$$

and the rate of turbine work output is given to be

$$\dot{W}_s = 40,000 \text{ kW}$$

We find the saturation temperature for steam at 1000 kPa to be 179.91°C (Appendix B2.2), which is less than 350°C. Therefore the entering steam is superheated and its enthalpy (from Appendix B2.3) corresponding to 1000 kPa and 350°C is

$$h_{in} = 3157.7 \text{ kJ/kg}$$

Substituting for $\dot{Q}$, $\dot{W}_s$, $\dot{m}$, and h_{in} in the energy equation, we get

$$-2750 - 40{,}000 = 55(h_{out} - 3157.7)$$

so

$$h_{out} = 2380 \text{ kJ/kg}$$

On examining the saturated steam table (Appendix B2.2), we find that for $P_{out} = 10$ kPa,

$$h_{f,out} = 191.83 \text{ kJ/kg} \quad \text{and} \quad h_{g,out} = 2584.7 \text{ kJ/kg}$$

Since h_{out} has a value that is greater than $h_{f,out}$ but less than $h_{g,out}$ we conclude that the exhaust steam is wet. Its quality, x_{out}, is given by

$$
\begin{aligned}
x_{out} &= \frac{h_{out} - h_{f,out}}{h_{g,out} - h_{f,out}} \\
&= \frac{2380 - 191.83}{2584.7 - 191.83} \\
&= 0.91
\end{aligned}
$$

A compressor and a turbine have moving parts within them and mechanical work is involved in their operation. We now consider a nozzle in which there are no moving parts and neither mechanical work nor heat transfer is involved. Because of the contraction of the flow cross section occurring in a nozzle, there is a conversion of the enthalpy of the fluid into kinetic energy.

Example 5.4

Air with negligible velocity enters a nozzle. The inlet pressure and the temperature of the air are 400 kPa and 20°C, respectively. The pressure at the exit of the nozzle is maintained at 270 kPa. The cross-sectional area of the nozzle at the exit is 4000 mm². If the expansion of the air in the nozzle can be assumed to be quasistatic adiabatic, what is the mass flow rate of the air?

Solution:

Given: Air expands in a quasistatic adiabatic manner through a nozzle with the following conditions.

$$\text{Inlet:} \quad p_{in} = 400 \text{ kPa}, \quad t_{in} = 20°C, \quad V_{in} \simeq 0$$

$$\text{Outlet:} \quad p_{out} = 270 \text{ kPa}, \quad A_{out} = 4000 \text{ mm}^2$$

Objective: The mass flow rate of the air through the nozzle is to be determined.

Diagram:

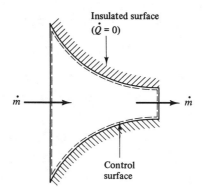

Figure 5.7 Control volume for a nozzle.

Assumptions:

1. The change in the potential energy of the air is negligible.
2. No mechanical work is done.
3. A steady flow exists.
4. Specific heat of the air is constant, and $c_p = 1.0035$ kJ/kg·K.
5. The air behaves like an ideal gas.

Relevant physics: In a nozzle, neither mechanical work nor a transfer of heat energy is involved.

With the use of our assumptions and the information given in the statement of the problem, the energy equation, Equation 5.8, becomes

$$\cancel{\dot{Q}}^{0} - \cancel{\dot{W}_s}^{0} = \dot{m}\left[(h_{out} - h_{in}) + \left(\frac{V_{out}^2}{2} - \cancel{\frac{V_{in}^2}{2}}^{0}\right) + (g\cancel{Z_{out}}^{0} - \cancel{g Z_{in}}^{0})\right]$$

or

$$(h_{out} - h_{in}) + \frac{V_{out}^2}{2} = 0$$

The above equation contains two unknowns, h_{out} and V_{out}, and therefore we need some additional information to solve for them. We were told that the air expands in a quasistatic adiabatic manner, so from the ideal gas law and Equation 3.14,

$$T_{in} p_{in}^{(1-\gamma)/\gamma} = T_{out} p_{out}^{(1-\gamma)/\gamma}$$

Rearrangement gives

$$T_{out} = T_{in}\left(\frac{p_{in}}{p_{out}}\right)^{(1-\gamma)/\gamma}$$

Once T_{out} is calculated, the change in the specific enthalpy can be calculated. Then V_{out} can be computed using the energy equation. The mass flow rate can be calculated from

$$\dot{m} = \rho_{out} A_{out} V_{out}$$

since ρ_{out} can be calculated by applying the ideal gas law at the outlet.

Analysis: For air

$$\gamma = 1.4$$

and we have

$$T_{out} = (273.15 + 20)\left(\frac{400}{270}\right)^{(1-1.4)/1.4}$$

$$= 262.01 \text{ K}$$

or

$$t_{out} = -11.14°C$$

We can now calculate the change in the enthalpy of the air as it expands through the nozzle.

$$(h_{out} - h_{in}) = c_p(t_{out} - t_{in}) = 1.0035(-11.14 - 20)$$

$$= -31.2 \text{ kJ/kg} = -31.2 \times 10^3 \text{ J/kg}$$

Substitution in the energy equation,

$$(h_{out} - h_{in}) + \frac{V_{out}^2}{2} = 0$$

gives

$$-31.2 \times 10^3 + \frac{V_{out}^2}{2} = 0$$

or

$$V_{out} = 250 \text{ m/s}$$

The mass flow rate of the air can now be calculated from

$$\dot{m} = \rho_{out} A_{out} V_{out}$$

$$= \left(\frac{p}{RT}\right)_{out} A_{out} V_{out}$$

$$= \frac{270 \times 10^3}{287 \times 262.01} \times 4000 \times 10^{-6} \times 250$$

$$= 3.6 \text{ kg/s}$$

We now consider a device in which more than one fluid is involved.

Example 5.5

A heat exchanger is designed to cool oil from 60°C to 30°C. The mass flow rate of the oil is 4500 kg/h. Water is used as a cooling agent and is available at 8°C. If the mass flow rate of the water is 6000 kg/h, determine the temperature of the water as it leaves the heat exchanger. The specific heat at constant pressure for water and oil are 4.18 kJ/ kg K and 3.8 kJ/kg K, respectively. Assume steady-flow operation.

Solution:

Given: Water is used in a heat exchanger to cool oil.

For oil: $t_{in} = 60°C$, $t_{out} = 30°C$, $\dot{m} = 4500 \text{ kg/h}$

For water: $t_{in} = 8°C$, $\dot{m} = 6000 \text{ kg/h}$, $c_p = 4.18 \text{ kJ/kg K}$

Objective: The temperature of the water leaving the heat exchanger is to be determined.

Diagram: We can draw a schematic diagram of the heat exchanger as a box with two tubes going in and two tubes coming out (Figure 5.8) of the box. A suitable control volume is also shown in the diagram.

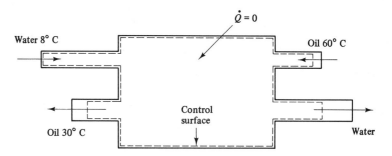

Figure 5.8 Control volume for a heat exchanger.

Assumptions:

1. Changes in the kinetic energy of oil, as well as of water, are negligible.
2. Changes in the potential energy of oil, as well as of water, are negligible.
3. No mechanical work is done.
4. Steady flow of oil and water prevail.
5. The specific heats of oil and water are constant.

Relevant physics: In this problem there is no transfer of heat or mechanical work across the control surface and there are four points where fluids cross the control surface. The transfer of heat from the hot oil to the cool water should not be confused with the term Q in Equation 5.8(c); the oil-to-water heat transfer is internal to the control volume. Consequently, Equation 5.8(a) takes the following form:

$$\Sigma \, (\dot{m}h)_{\text{out}} - \Sigma \, (\dot{m}h)_{\text{in}} = 0$$

or

$$[(\dot{m}h)_{\text{oil}} + (\dot{m}h)_{\text{water}}]_{\text{out}} - [(\dot{m}h)_{\text{oil}} + (\dot{m}h)_{\text{water}}]_{\text{in}} = 0$$

The above equation can be rewritten as

$$[\dot{m}(h_{\text{out}} - h_{\text{in}})]_{\text{oil}} + [\dot{m}(h_{\text{out}} - h_{\text{in}})]_{\text{water}} = 0$$

With our assumption of constant specific heats, $(h_{\text{out}} - h_{\text{in}}) = c_p(t_{\text{out}} - t_{\text{in}})$. The energy equation then becomes

$$[\dot{m}c_p(t_{\text{out}} - t_{\text{in}})]_{\text{oil}} + [\dot{m}c_p(t_{\text{out}} - t_{\text{in}})]_{\text{water}} = 0$$

Substituting the values of $\dot{m}$, c_p, and various temperatures in the energy equation, we have

$$\frac{4500}{60} \times 3.8(30 - 60) + \frac{6000}{60} \times 4.18(t_{\text{out}} - 8) = 0$$

Solving for t_{out}, we obtain

$$t_{\text{out}} = 28.45°C$$

Alternative procedure: This problem can also be handled by setting up two control volumes, one for the oil and one for the water. In such an approach, each control volume has an external heat transfer with $\dot{Q}_{\text{oil}} = -\dot{Q}_{\text{water}}$. It is generally easier to work with the approach requiring the use of a single control volume illustrated in this example when compared with the approach requiring two control volumes.

The next example involves an external heat transfer but no mechanical work.

Example 5.6

A solar collector panel (Figure 5.9) 20 m² in area receives solar energy at a rate of 750 W/m². It is estimated that 35% of the incident energy is lost to the surroundings. Water enters the panel at a steady flow rate of 0.05 kg/s and at 15°C. Calculate the temperature of the water leaving the solar collector panel. The specific heat of water may be taken to be 4.18 kJ/kg K.

Solution:

Given: The solar energy incident on a collector panel is 750 W/m². The area of the panel is 20 m². There is a loss of energy from the panel, which is 35% of the incident energy. For the water flowing through the panel, $\dot{m} = 0.05$ kg/s, $t_{in} = 14°C$, and $c_p = 4.18$ kJ/kg K.

Objective: We want to find the temperature of the water leaving the panel.

Diagram: We select a control volume to envelope the solar panel and to include a portion of the water pipes, as shown in Figure 5.9. We observe from the figure that there is a change in the elevation of the water as it enters and leaves the panel.

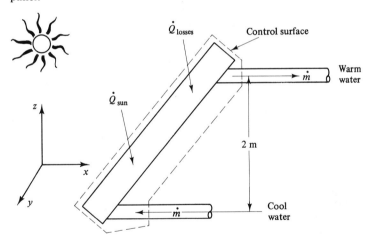

Figure 5.9   Schematic diagram of a solar collector.

Assumptions:

1. The change in the kinetic energy of the water is negligible.

2. No mechanical work is done.

3. The change in the density of water due to the temperature change of water can be ignored.

Relevant physics: A solar collector panel is a device that receives energy from the sun, absorbs a portion of it, and transfers the absorbed energy to a fluid flowing through it. The fluid is heated due to the transfer of energy taking place *within* the solar panel. The hot fluid then imparts its energy at the point of use and is then returned to the panel. In this example, water is the working fluid.

We note that there are two contributions to the term $\dot{Q}$ in the energy equation, namely, the incident solar energy, $\dot{Q}_{sun}$, and the thermal loss to the surroundings, $\dot{Q}_{losses}$. Equation 5.8 takes the following form

$$\dot{Q} - \cancel{\dot{W}_s}^{0} = \dot{m}\left[(h_{out} - h_{in}) + \left(\frac{\cancel{V^2_{out}}}{2} \Big/ \frac{\cancel{V^2_{in}}}{2}\right) + + (gZ_{out} - gZ_{in})\right]$$

or

$$\dot{Q}_{sun} + \dot{Q}_{losses} = \dot{m}[(h_{out} - h_{in}) + g(Z_{out} - Z_{in})]$$

Analysis: From the given information, we have

$$\dot{Q}_{sun} = 750 \times 20 = 15,000 \text{ W}$$

$$\dot{Q}_{losses} = -0.35\dot{Q}_{sun} = -5250 \text{ W}$$

$$\dot{m} = 0.05 \text{ kg/s}$$

$$h_{out} - h_{in} = c_p(t_{out} - t_{in}) = 4.18 \times 10^3(t_{out} - 14)$$

$$g(Z_{out} - Z_{in}) = 9.81(2) = 19.62 \text{ W}$$

We note that the gain in the potential energy is extremely small in comparison with the incident solar energy. Substitution gives

$$15,000 + (-5250) = 0.05[4.18 \times 10^3(t_{out} - 14) + 19.62]$$

Solving for t_{out}, we obtain

$$t_{out} = 60.6°C$$

Comments: The outlet temperature of 60.6°C is suitable for space heating but not for air conditioning.

5.6.1 Combustion Chamber

Although the process of combustion is complicated, it is possible to treat the energy-transfer process related to a combustion chamber using the developments in this chapter. In a combustion process, fuel is oxidized by the oxygen in the air and heat energy is released. Fuel and air may be preheated or they may enter a combustion chamber at room temperature. When a fuel burns, it releases a quantity of heat energy and produces the hot gases of combustion consisting mainly of carbon dioxide and water vapor. The temperature of the gases of combustion depends on the amount of heat energy released by the fuel and the ratio of the mass of the air used for every unit mass of the fuel. The temperature of these gases is usually much higher than the temperature of the incoming air. The hot gases themselves may do work, as in the internal combustion engine of an automobile, or they may transfer some of their energy to another fluid, as in the case of a steam generator or a home-heating furnace.

To facilitate analysis, we define the *heating value of a fuel* as the amount of heat energy released by the complete combustion of a unit mass of fuel. We also define the term *air-fuel ratio* as the actual mass of air used for combustion of a unit mass of fuel.

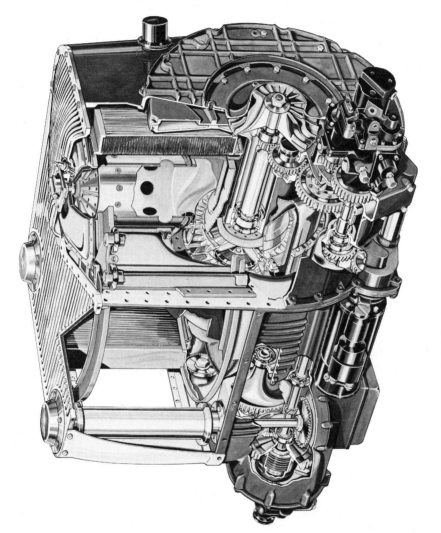

Plate 2 The cross-sectional drawing of the reverse-flow combustion system of a General Electric Company's MS6001 heavy-duty gas turbine. (Courtesy of General Electric Company.)

For an elementary analysis we can relate the heat released by a fuel in a combustion chamber to the quantity q in the energy equation, Equation 5.8(b). Also, the actual quantity of air used for combustion is often much more than the theoretically required quantity of air. For the purpose of energy analysis, the exhaust gases can be treated like the incoming air as a first approximation. Also, since the state for zero value of specific enthalpy of a substance is chosen somewhat arbitrarily (see Section 4.2.2), we may use 0°C as the temperature at which specific enthalpies of the gaseous fuel, combustion air, and products of combustion have a zero value.

Example 5.7

Natural gas with a heating value of 55,000 kJ/kg is to be burned in a home-heating furnace and the combustion gases used for heating air from -10°C to 40°C. The mass flow rate of the air to be heated is 40 kg/s. The temperature of the incoming air for combustion and of the natural gas is -20°C. The exhaust gases leave at 550°C. If the air-fuel ratio is 21, determine the rate at which the natural gas is consumed. The specific heat of the natural gas is 2.25 kJ/kg K and that of the air is 1.005 kJ/kg K.

Solution:

Given: The data on a home-heating furnace are as follows:

Fuel is natural gas:	$t_{in} = 20$°C,	$c_p = 2.25$ kJ/kg K
Heating value of the fuel:	55,000 kJ/kg	
Air to be heated:	$t_{in} = -10$°C,	$t_{out} = 40$°C
	$\dot{m} = 40$ kg/s,	$c_p = 1.005$ kJ/kg K
Combustion air:	$t_{in} = -20$°C	
Exhaust Gases:	$t_{out} = 550$°C	
Air-fuel ratio:	21	

Objective: The mass flow rate of the natural gas is to be determined.

Diagram: A schematic diagram of the combustion chamber and a suitable control volume are shown in Figure 5.10.

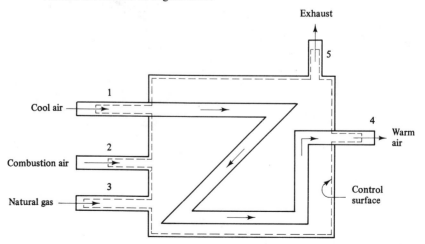

Figure 5.10 Control volume for a combustion chamber.

Assumptions: In order to solve the problem we make these assumptions.

1. No heat energy is lost from the walls of the combustion chamber.
2. Heat energy released by the combustion of the natural gas can be treated as a quantity of heat energy supplied from an external source.
3. No mechanical work is involved.
4. Changes in the kinetic energy and the potential energy are negligible.
5. Steady flow exists.
6. Specific heats are constant, and the specific heat of the products of combustion can be taken as equal to that for air.

Relevant physics: The energy equation, Equation 5.8(a), becomes

$$\dot{Q} - \overset{0}{\cancel{\dot{W}_s}} = [\Sigma\,(\dot{m}h)_{\text{out}} - \Sigma\,(\dot{m}h)_{\text{in}}] + \left[\Sigma\left(\dot{m}\frac{V^2}{2}\right)_{\text{oui}} - \overset{0}{\cancel{\Sigma\left(\dot{m}\frac{V^2}{2}\right)_{\text{in}}}}\right]$$

$$+ \overset{0}{[\cancel{\Sigma\,(\dot{m}gZ)_{\text{out}} - (\dot{m}gZ)_{\text{in}}}]}$$

or

$$\dot{Q} = \Sigma\,(\dot{m}h)_{\text{out}} - \Sigma\,(\dot{m}h)_{\text{in}}$$

where $\Sigma\,(\dot{m}h)_{\text{out}}$ is the sum of the $\dot{m}h$ products for all the outflow streams and $\Sigma\,(\dot{m}h)_{\text{in}}$ signifies a similar sum for all the inflow streams. We observe from Figure 5.10 that there are five locations where fluids cross the control surface. Air and fuel enter at points 1, 2, and 3, and hot air and exhaust gases leave at points 4 and 5. Thus

$$\dot{Q} = (\dot{m}_4 h_4 + \dot{m}_5 h_5) - (\dot{m}_1 h_1 + \dot{m}_2 h_2 + \dot{m}_3 h_3)$$

Analysis: From the statement of the problem

(Mass flow rate of air for combustion) $= 21 \times$ (mass flow rate of natural gas)

or
$$\dot{m}_2 = 21\dot{m}_3$$

Also, from Figure 5.10,
$$\dot{m}_1 = \dot{m}_4 = 40 \text{ kg/s}$$

For conservation of mass understeady-state condition

$$\underset{\text{in}}{\Sigma}\,\dot{m} = \underset{\text{out}}{\Sigma}\,\dot{m}$$

or
$$\dot{m}_1 + \dot{m}_2 + \dot{m}_3 = \dot{m}_4 + \dot{m}_5$$

Substituting for $\dot{m}_2$ and cancelling $\dot{m}_1$ against $\dot{m}_4$, gives

$$\dot{m}_5 = 22\dot{m}_3$$

We consider the enthalpy of the natural gas, air, and products of combustion to be zero at 0°C. We also assume that the specific heat of the products of combustion is the same as that of the air. The specific enthalpies of the various fluids crossing the boundaries of the control surface are then computed from

$$h = c_p t$$

They are

Incoming air: $\qquad\qquad\quad h_1 = 1.005 \times (-10) = -10.05 \text{ kJ/kg}$

Incoming air for

combustion: $\qquad\qquad\; h_2 = 1.005 \times (-20) = -20.1 \text{ kJ/kg}$

Incoming natural gas: $\quad h_3 = 2.25 \times (-20) = -45 \text{ kJ/kg}$

Outgoing air: $\qquad\qquad\;\; h_4 = 1.005 \times 40 = 40.2 \text{ kJ/kg}$

Outgoing exhaust: $\qquad\; h_5 = 1.005 \times 550 = 552.75 \text{ kJ/kg}$

Substitution in the reduced energy equation,

$$\dot{Q} = (\dot{m}_4 h_4 + \dot{m}_5 h_5) - (\dot{m}_1 h_1 + \dot{m}_2 h_2 + \dot{m}_3 h_3)$$

gives

$$55,000 \times \dot{m}_3 = [40 \times 40.2 + 22\dot{m}_3 \times 552.75]$$
$$- [40 \times (-10.05) + 21\dot{m}_3 \times (-20.1) + \dot{m}_3 \times (-45)]$$

or

$$(55,000 - 22 \times 552.75 - 21 \times 20.1 + 45)\dot{m}_3 = 40 \times 40.2 + 40 \times 10.05$$

Solving for $\dot{m}_3$, we obtain

$$\dot{m}_3 = 0.0473 \text{ kg/s, or } 170.4 \text{ kg/h}$$

The rate of natural gas consumption is 170.4 kg/h.

5.7 THROTTLE VALVE AND JOULE-THOMSON COEFFICIENT

Thus far we have considered a number of devices such as a compressor, a turbine, a heat exchanger, a solar collector panel, and a combustion chamber. These devices are encountered in a power plant or in a heating system. We employed the energy equation for a control volume in the analysis of these devices. We shall now discuss the throttle valve and make a first-law analysis of the flow through a throttle valve.

In many industrial situations, a fluid pressurized to a certain level at an upstream location is needed at a lower pressure level at a downstream location. This requires a reduction in the pressure of the fluid, and the reduction is usually accomplished by employing a throttle valve. Figure 5.11(a) shows the schematic of a throttle valve. A high-pressure fluid enters the left side of the valve at A. As the fluid passes through the restricted opening B in the valve, it experiences a decrease in the pressure and a flow at reduced pressure is obtained at C. By raising or lowering the valve stem D, the desired reduction in the fluid pressure can be obtained.

Although a throttle valve can reduce fluid pressure, it cannot control the fluctuations in the fluid pressure. In many situations the upstream pressure may fluctuate due to a variety of reasons, but the downstream pressure must remain constant to satisfy operational requirements. For example, the pressure of the steam produced in a steam geneartor may fluctuate due to variations in the demand for the steam; yet a steam turbine requires steam at a fairly constant value of pressure. As another example, in an experimental study of fluid flow it is important that the

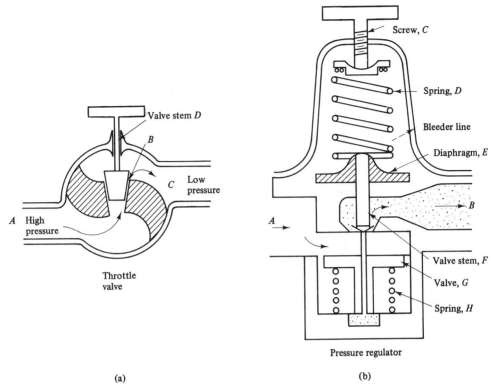

Figure 5.11
(a) Throttle valve.
(b) Pressure regulator.

pressure of the fluid be maintained constant in the test section, even though the supply pressure may fluctuate. Often the pressure for water in the supply line to a home is very high; under such conditions a faucet closed abruptly can cause water-hammer, or pounding sounds due to pressure waves and it is often desirable to reduce the pressure in the water line to a structurally safe level. In such and many other situations, a pressure regulator is used.

Figure 5.11(b) shows the schematic of a pressure regulator. High-pressure fluid enters the regulator at A and flows past the valve G. The fluid experiences a reduction in pressure and flows out through the exit port B. When the force in the spring D is greater than the force due to the downstream fluid pressure, the diaphragm E forces the valve stem F to open the main valve. If the upstream pressure increases, there is a momentary increase in the downstream pressure, which in turn relieves the force in the spring causing the stem and the valve G to rise. This then restricts the flow and causes the downstream pressure to return to its present level. If the upstream pressure continues to rise, the diaphragm will raise and the self-relieving disc seal will part from the valve stem F. This will then bleed the excess pressure from point B to atmosphere or to a return line.

The first-law analyses of a throttle valve and a pressure regulator are similar. The flow velocities in the immediate vicinity of the valve can be large due to restrictions in the flow path. If we select a control volume to include a large region on the upstream side, as well as on the downstream side, of the valve, then the flow velocities at the entrance and exit sections of the control surface are not influenced by the restriction in the flow. We note that under steady-state conditions, $\dot{m}_{\text{in}} = \dot{m}_{\text{out}}$ and that $\dot{m} = \rho A V$. If the fluid flowing through the valve can be regarded as incompressible and if the cross-sectional areas at the inlet and outlet sections are equal, then the flow velocities at the inlet and outlet section are equal; that is, $V_{\text{in}} = V_{\text{out}}$.

The process experienced by a fluid when it goes through a restriction, such as a throttle valve, a porous plug, or a plate with a very small hole in it, is called the *throttling process*. We can analyze the process by applying the energy equation for a steady-flow process, Equation 5.8. Consider a throttling process occurring across a porous plug, as shown in Figure 5.12(a). The cross-sectional areas upstream and downstream of the plug are equal, and the pipe with the porous plug is insulated. Under steady-state conditions

$$\dot{m}_{\text{in}} = \dot{m}_{\text{out}}$$

or

$$(\rho A V)_{\text{in}} = (\rho A V)_{\text{out}}$$

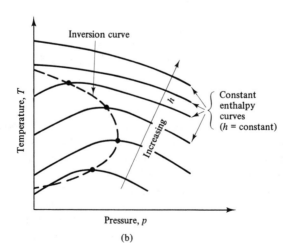

(a)

(b)

Figure 5.12
(a) Porous plug experiment.
(b) Inversion curve.

For a liquid, ρ_{in} and ρ_{out} have essentially the same values. Also, $A_{\text{in}} = A_{\text{out}}$. Thus for a liquid

$$V_{\text{in}} \simeq V_{\text{out}}$$

and the change in the kinetic energy is zero. Also

$$Z_{\text{in}} = Z_{\text{out}}$$

and the change in the potential energy is zero. With the absence of mechanical work and transfer of heat from the control volume, Equation 5.8 becomes

$$\cancel{\phi} - \cancel{W} = \dot{m}\left[(h_{\text{out}} - h_{\text{in}}) + \left(\frac{V_{\text{out}}^2}{2} - \frac{V_{\text{in}}^2}{2}\right) + (gZ_{\text{out}} \cancel{-} gZ_{\text{in}}) \right]^{0}$$

or

$$h_{\text{out}} - h_{\text{in}} = -\left(\frac{V_{\text{out}}^2}{2} - \frac{V_{\text{in}}^2}{2}\right)$$

As noted previously for a liquid flowing through the restriction,

$$V_{\text{in}} = V_{\text{out}}$$

and we obtain

$$h_{\text{out}} = h_{\text{in}}$$

If a gas were flowing through the porous plug, it would experience an increase in its surface volume as its pressure dropped. Since the flow rate is constant, V_{out} must increase. However, the resulting change in the kinetic energy of the gas is usually very small, so that we can write

$$h_{\text{out}} \simeq h_{\text{in}} \tag{5.9}$$

The throttling process is, therefore, sometimes known as an *isenthalpic process*. A *throttling process is nonquasistatic and involves nonequilibrium states*. It is, therefore, customary to represent a throttling process not by a solid line, but by a dashed or dotted line.

The change in pressure in a throttling process can be considered to be a measure of the extent of the throttling. For an ideal gas, $\Delta h = c_p \Delta T$, and in a throttling process an ideal gas does not experience any temperature change. In the case of a real gas, however, throttling can cause increase or decrease in the temperature of the gas. We define the change of temperature with respect to the change of pressure in a throttling process by

$$\mu_J = \left(\frac{\partial T}{\partial p}\right)_h \tag{5.9a}$$

The quantity μ_J is called the *Joule-Thomson coefficient*. The subscript h in Equation 5.9a is a reminder that the enthalpy remains constant in a throttling process. Figure 5.12(b) shows a number of constant-enthalpy curves on a T-p diagram for a typical real gas. Most of the curves have both positive and negative slopes, and they exhibit a point of zero slope. The locus of the points with zero slope—that is, $(\partial T/\partial p)_h = 0$—is called the *inversion curve*. A throttling process proceeds in the direction of decreasing pressure. Consequently, if a throttling operation is carried out in the domain to the right of the inversion curve, it results in an increase in the temperature of a gas. If it is carried out in the region to the left of the inversion curve, it causes a decrease in the temperature of a gas. Liquefaction of a gas is often accomplished by throttling the gas in the region that is to the left on the inversion curve.

5.8 FURTHER APPLICATIONS OF THE ENERGY EQUATION*

In the examples discussed so far in this chapter, we examined single components such as a heat exchanger and a compressor. We now briefly examine two systems, each of which employs a number of components: a gas turbine power plant and a heat pump. To facilitate the discussion of a turbine power plant we shall first describe the Brayton cycle.

5.8.1 Brayton Cycle

The Brayton cycle consists of two quasistatic adiabatic processes and two constant pressure heat transfer processes. Figure 5.13 shows the p-v diagram for such a cycle and the physical components that can be used to execute the processes in the cycle. The discussion here is intended to show how component flow processes can be combined together to produce a power cycle. A more detailed analysis of the Brayton cycle is presented later in the text.

The working medium in a Brayton cycle is usually a gas. The gas is compressed in a quasistatic adiabatic process from state 1 to state 2 in a compressor. The compressed gas is admitted into a heat exchanger, where it receives heat energy at constant pressure. The high-pressure, high-temperature gas (state 3) enters a turbine, where it expands in a quasistatic adiabatic manner to state 4. It is then admitted into a heat exchanger, where the gas rejects heat at a constant pressure and returns to its original state, namely, state 1.

It should be noted that we have four different pieces of hardware; each one, taken individually, operates as an open system. The hardware for the ideal Otto cycle (Chapter 3) operates as a closed system. We now obtain expressions for the thermal efficiency of the Brayton cycle.

We assume that the changes in the kinetic and potential energies are negligible as the gas goes through the compressor, the turbine, and the heat exchangers. For a unit mass flowing through the system under steady flow conditions, Equation 5.8(b) applies and we have

$$q - w_s = (h_{out} - h_{in}) + \left(\frac{V_{out}^2}{2} - \frac{V_{in}^2}{2}\right)^{\!\!0} + (gZ_{out} - gZ_{in})^{\!\!0}$$

For the compressor, $q_{1\text{-}2} = 0$, and

$$-w_{s,\,1\text{-}2} = h_2 - h_1$$

*Optional material for this chapter.

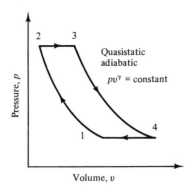

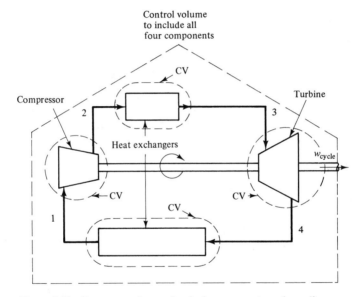

Figure 5.13 Brayton cycle: mechanical components and p-v diagram.

For the high-pressure heat exchanger, $w_{s,2\text{-}3} = 0$, and

$$q_{2\text{-}3} = h_3 - h_2$$

For the turbine, $q_{3\text{-}4} = 0$, so that

$$-w_{s,3\text{-}4} = h_4 - h_3$$

For the low-pressure heat exchanger, $w_{s,4\text{-}1} = 0$, and

$$q_{4\text{-}1} = h_1 - h_4$$

The net work output of the cycle is the *algebraic* sum of the work output of the turbine and the work output of the compressor.

$$W_{\text{cycle}} = W_{s,1-2} + W_{s,3-4}$$
$$= -(h_2 - h_1) - (h_4 - h_3)$$

or

$$W_{\text{cycle}} = (h_3 - h_4) - (h_2 - h_1)$$

Note that each of the parenthetical quantities is positive.

In the Brayton cycle, it is the high-pressure heat exchanger in which heat energy is supplied to the gas. Thus

$$q_{\text{in}} = q_{2\text{-}3} = h_3 - h_2$$

We recall from Chapter 3 that thermal efficiency is given by

$$\eta_t = \frac{\text{net work output}}{\text{heat energy supplied}} = \frac{W_{\text{cycle}}}{q_{\text{in}}} \tag{3.15a}$$

Substituting the expressions for W_{cycle} and q_{in}, we have

$$\eta_t = \frac{(h_3 - h_4) - (h_2 - h_1)}{(h_3 - h_2)}$$

This equation can be rearranged to read

$$\eta_t = \frac{(h_3 - h_2) - (h_4 - h_1)}{(h_3 - h_2)} = 1 - \frac{h_4 - h_1}{h_3 - h_2}$$

If we assume that the working medium behaves like an ideal gas and the specific heat at constant pressure, c_p, is constant, we can write

$$(h_4 - h_1) = c_p(T_4 - T_1) \quad \text{and} \quad (h_3 - h_2) = c_p(T_3 - T_2)$$

so that

$$\eta_t = 1 - \frac{T_4 - T_1}{T_3 - T_2}$$

The above equation can be further simplified by the introduction of the *pressure ratio* r_p. The pressure ratio is defined to be the ratio of the highest pressure to the lowest pressure experienced by the working medium in the ideal Brayton cycle.

$$r_p = \frac{p_2}{p_1} = \frac{p_3}{p_4} \tag{5.10}$$

Observing that processes 1–2 and 3–4 in Figure 5.13 are quasistatic adiabatic, we can write

$$\frac{T_2}{T_1} = \left(\frac{p_2}{p_1}\right)^{(\gamma-1)/\gamma} = r_p^{(\gamma-1)/\gamma}$$

and

$$\frac{T_3}{T_4} = \left(\frac{p_3}{p_4}\right)^{(\gamma-1)/\gamma} = r_p^{(\gamma-1)\gamma}$$

Hence

$$\frac{T_2}{T_1} = \frac{T_3}{T_4} \quad \text{or} \quad \frac{T_1}{T_4} = \frac{T_2}{T_3}$$

The equation for the efficiency can now be recast as

$$\eta_t = 1 - \frac{T_4(1 - T_1/T_4)}{T_3(1 - T_2/T_3)} = 1 - \frac{T_4}{T_3}$$

or

$$\eta_t = 1 - \frac{1}{r_p^{(\gamma-1)/\gamma}} \tag{5.11}$$

This last result tells us that the thermal efficiency of a Brayton cycle, using an ideal gas as the working medium, increases with an increase in the pressure ratio r_p.

Note that if we were to select a control volume large enough to envelope all four components (Figure 5.13), we would have a closed system. The expression for the thermal efficiency cannot be obtained by such an approach. You can verify that the only result that can be obtained by applying the equation

$$\oint \delta q = \oint \delta w$$

is

$$q_{2\text{-}3} + q_{4\text{-}1} - w_{\text{cycle}} = 0$$

5.8.2 A Gas Turbine Power Plant

A gas turbine power plant consists of a compressor, a combustion chamber, and a turbine (Figure 5.14). Atmospheric air is admitted into the compressor at point 1. The air is compressed adiabatically and then allowed to enter the combustion chamber. It is mixed with fuel and the mixture is combusted. The hot, high-pressure gases of combustion enter the turbine where they expand in an adiabatic process. The exhaust from the turbine consists of low-pressure, high-temperature gases that cannot be reused in the compressor. Attempts are often made to recover some of the energy from the hot waste gases by employing a heat exchanger (not shown in the figure).

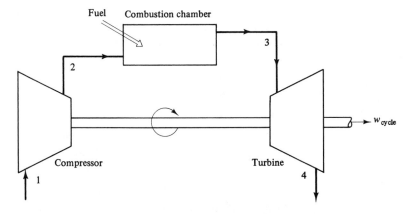

Figure 5.14   A schematic for gas turbine power plant.

If we compare the above system with the ideal Brayton cycle, we find that it does not contain a process that takes the working medium from state 4 back to state 1. However, this does not cause any difficulty when we want to analyze the actual operation. The equation for the thermal efficiency of the Brayton cycle can still be used, even though the actual operation is not a closed cycle because no work is involved and no heat is supplied to the system in the process 4–1. It should be noted that in practice the compressor and the turbine do not operate in a quasistatic adiabatic manner, and a constant pressure is not maintained between the compressor and the turbine. These considerations will be examined in detail in Chapter 7.

MS6001 COMBUSTOR

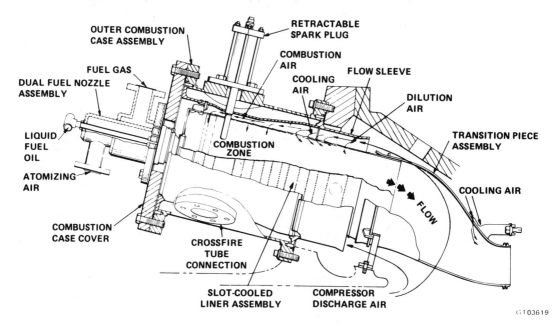

Plate 3 The experimental GT-309 gas turbine engine developed by the General Motors Research Laboratories. (Courtesy of General Motors Corporation.)

Example 5.8

The following data are available for a gas turbine power plant.

Compressor inlet:	air at 100 kPa and 11°C.
Compression:	quasistatic adiabatic.
Compressor outlet:	air at 1200 kPa.
Heat added in the combustion chamber:	850 kJ/kg of air.
Expansion in turbine:	from 1200 kPa to 100 kPa in a quasistatic adiabatic process.

Determine the net power plant work output and the thermal efficiency of the power plant if $c_p = 1.005$ kJ/kg K.

Solution:

Given: Using the nomenclature of Figure 5.14, we have for a gas turbine plant

$$\text{Air:} \qquad p_1 = 100 \text{ kPa}, \qquad t_1 = 11°C$$
$$p_2 = 1200 \text{ kPa}$$
$$q_{2\text{-}3} = 850 \text{ kJ/kg}$$
$$c_p = 1.005 \text{ kJ/kg K}$$

Compression and expansion processes: adiabatic

Objective: We need to determine the net work output from the turbine plant and the thermal efficiency.

Diagram:

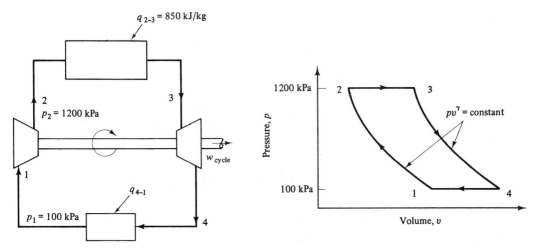

Figure 5.15

Assumptions:

1. Steady flow prevails.
2. The specific heat of air is constant.

Relevant physics: The material in Section 5.8.1 is directly applicable here.

Analysis: We have

$$p_1 = 100 \text{ kPa}, \ T_1 = 11 + 273.15 = 284.15 \text{ K}$$
$$p_2 = 1200 \text{ kPa}$$
$$\gamma \text{ for air} = 1.4$$

Since the compression is quasistatic adiabatic,

$$T_1 p_1^{(1-\gamma)/\gamma} = T_2 p_2^{(1-\gamma)/\gamma}$$

or

$$T_2 = T_1\left(\frac{p_1}{p_2}\right)^{(1-\gamma)/\gamma} = 284.15\left(\frac{100}{1200}\right)^{(1-1.4)/1.4}$$

$$= 578 \text{ K}$$

Following the development of the Brayton cycle in the preceding section, we can write for the compressor

$$-w_{s,1\text{-}2} = (h_2 - h_1) = c_p(T_2 - T_1)$$

$$= 1.005(578 - 284.15)$$

or

$$w_{s,1\text{-}2} = -295.3 \text{ kJ/kg}$$

For the combustion chamber

$$q_{2\text{-}3} = h_3 - h_2 = 850 \text{ kJ/kg}$$

Also,

$$c_p(T_3 - T_2) = h_3 - h_2 = 850 \text{ kJ/kg}$$

so that

$$T_3 = \left(\frac{850}{1.005}\right) + 578 = 1424 \text{ K}$$

For the turbine,

$$-w_{s,3\text{-}4} = h_4 - h_3 = c_p(T_4 - T_3)$$

The temperature of the exhaust from the turbine can be determined, as the process in the turbine is a quasistatic adiabatic one.

$$T_3 p_3^{(1-\gamma)/\gamma} = T_4 p_4^{(1-\gamma)/\gamma}$$

or

$$T_4 = 1424\left(\frac{1200}{100}\right)^{(1-1.4)/1.4} = 700 \text{ K}$$

and the turbine work output, $w_{s,3\text{-}4}$, is given by

$$-w_{s,3\text{-}4} = c_p(T_4 - T_3)$$

$$= 1.005(700 - 1424)$$

or

$$w_{s,3\text{-}4} = 727.6 \text{ kJ/kg}$$

The net work output of the power plant is

$$w_{\text{cycle}} = w_{s,1\text{-}2} + w_{s,3\text{-}4}$$

$$= -295.3 + 727.6$$

$$= 432.3 \text{ kJ/kg}$$

The thermal efficiency of the plant is then given by

$$\eta_t = \frac{w_{s,1\text{-}2} + w_{s,3\text{-}4}}{q_{2\text{-}3}}$$

$$= \frac{432.3}{850}$$

$$= 0.508, \text{ or } 50.8\%$$

5.8.3 Heat Pump

A heat pump transfers heat energy from a low-temperature source to a high-tempera-ture sink. In recent years, the heat pump has been increasingly used to provide winter heating in homes. The heat pump extracts energy from the atmosphere (or from a large body of water), which is at low temperature and supplies energy to a space to be maintained at a high temperature. It has the advantage that it can also be used as a cooling device for a summer air-conditioning system. Typically, a heat pump uses a vapor and not a gas as its working medium. The following discussion is intended only as an introduction to the subject. A more detailed description is presented in Chapter 8.

Figure 5.16(a) shows a schematic diagram of heat pump. A heat pump system consists of a compressor, a condenser, a throttle valve, and an evaporator. Also shown in the figure is a *p-h* diagram for the heat pump cycle. Experience has shown that it is advantageous to represent the thermodynamic processes in a heat pump system on a *p-h* diagram. The diagram shows the saturation dome for the vapor. It is desirable that point 1 of the heat pump cycle not be in the two-phase region, so that the com-pressor is required to handle only saturated or superheated vapor.

For convenience, we assume that saturated vapor at state 1 enters the compres-sor, where it is compressed to state 2 in a quasistatic adiabatic process. The high-pressure superheated vapor leaving the compressor (point 2) is then cooled in a condenser at a constant pressure until it is converted into a saturated liquid (point 3). The heat removed in the condensation is available for heating purposes in a heat pump application. This process appears as a horizontal line on a *p-h* diagram. The saturated liquid is then throttled, resulting in a low-temperature, low-pressure liquid-vapor mixture (point 4). This process appears as a vertical dotted line on a *p-h* diagram. The line is shown dotted because the throttling process is nonquasistatic and involves non-equilibrium states. The liquid-vapor mixture then passes through an evaporator where the mixture is transformed into a saturated or a superheated vapor. In an air-condi-tioning application, the cooling of water or air is accomplished in the evaporator part of the heat pump system.

In order to analyze the heat pump system, we draw control volumes around each piece of hardware. We also assume that the changes in kinetic and potential energies are negligible as the working fluid passes through the successive control volumes and that the thermodynamic state of the fluid does not change as it leaves one control volume and enters another. For steady flow, we can use Equation 5.8(b) for each one of the four control volumes. With the foregoing assumptions, the equation becomes

$$q - w_s = (h_{\text{out}} - h_{\text{in}}) + \underbrace{\left(\frac{V_{\text{out}}^2}{2} - \frac{V_{\text{in}}^2}{2}\right)}_{0} + \underbrace{(gZ_{\text{out}} - gZ_{\text{in}})}_{0}$$

Compressor. Although the reciprocating compressor shown in Figure 5.16 results in a pulsating flow at its exit, this unsteady flow effect becomes minimal if we

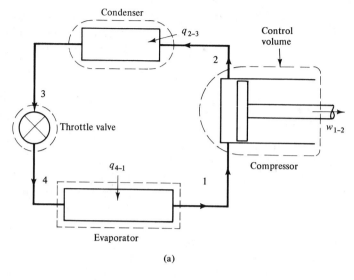

(a)

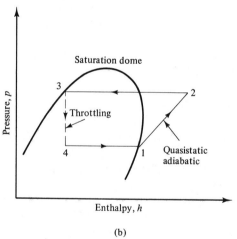

(b)

Figure 5.16 Vapor compression heat pump (refrigeration) cycle.
(a) Mechanical components.
(b) p-h diagram.

include large portions of the suction and delivery lines or damping chambers (not shown in the figure) near the suction and delivery ports in the control volume. Under ideal conditions, the compressor works in a quasistatic adiabatic manner, in which case $q_{1-2} = 0$ and we have

$$-w_{s,1-2} = h_2 - h_1$$

Condenser. A condenser is a heat exchanger in which sufficient energy is removed from a vapor to transform it completely into liquid. The energy removal is achieved by bringing a coolant into thermal contact with the vapor. The control volume shown around the condenser in Figure 5.16 includes only the pipeline that carries the working fluid of the process. With such a selection of the control volume, there is a transfer of heat to the control volume. No mechanical work is involved in a condenser, so that

$$q_{2\text{-}3} = h_3 - h_2$$

Note that for a condenser, $q_{2\text{-}3}$ has a negative numerical value; that is, heat is rejected by the working fluid.

Throttle valve. A throttle valve, as discussed earlier, is a simple way of reducing the pressure of a fluid resulting in an isenthalpic operation. Thus, from Equation 5.9

$$h_4 = h_3$$

You may ask why a device to produce mechanical work is not used instead of a throttle valve. The reason is that the quantity of mechanical work that can be obtained from such a device is very small since a liquid is involved and the operating pressure range is small.

Evaporator. An evaporator is a heat exchanger in which the working medium is vaporized. The source of heat—a relatively warm fluid—is not included in the control volume around the evaporator in Figure 5.16. No mechanical work is involved in an evaporator. Thus we have

$$q_{4\text{-}1} = h_1 - h_4$$

The performance of a heat pump is measured by the ratio of the amount of energy delivered by the working fluid in the condenser, $-q_{2\text{-}3}$, to the amount of mechanical work delivered to the the cycle. The only contribution to the cycle work is the compressor, $-w_{s,\,1\text{-}2}$. This ratio is called *coefficient of performance*, or COP. It is

$$(\text{COP})_{\text{HP}} = \frac{-q_{2\text{-}3}}{-w_{s,\,1\text{-}2}} = \frac{h_2 - h_3}{h_2 - h_1} \tag{5.12}$$

If a heat pump system is used to provide the cooling needed for an air-conditioning or a refrigerating system, then the desired benefit is the energy removed by the working medium in the evaporator from the fluid that is to be maintained at low temperature. This energy gain by the working medium is $q_{4\text{-}1}$. The coefficient of performance in such a case is given by

$$(\text{COP})_R = \frac{q_{4\text{-}1}}{-w_{s,\,1\text{-}2}} = \frac{h_1 - h_4}{h_2 - h_1} \tag{5.12a}$$

Example 5.9

A refrigeration system uses Freon-12 as the working medium. The evaporator and the condenser operate at $-10°C$ and $40°C$, respectively. The states of the Freon leaving the evaporator and the condenser are saturated vapor and saturated liquid, respectively.

If the enthalpy of the Freon leaving the compressor is 225 kJ/kg, determine the coefficient of performance of the refrigeration system.

Solution:

Given: A Freon-12 refrigeration system is prescribed. Using the nomenclature of Figure 5-17,

$$t_1 = t_4 = -10°C$$

$$t_3 = 40°C$$

State 1: saturated vapor

State 3: saturated liquid

$$h_2 = 225 \text{ kJ/kg}$$

Objective: The coefficient of performance of the system is to be calculated.

Diagram:

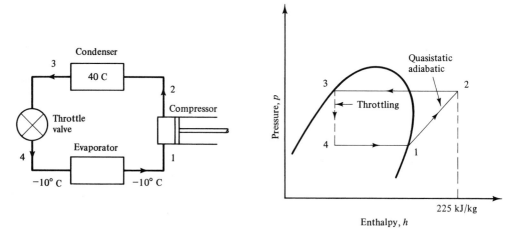

Figure 5.17

Assumptions:

1. Steady flow exists.

2. Enthalpy is a strong function of temperature but a weak function of pressure.

Relevant physics: The theory discussed in Section 5-8.3 is directly applicable here, except that we now have a refrigerator.

Analysis: Equation 5.12(a) tells us that the $(COP)_R$ is a function of the enthalpies at three state points. In order to determine an enthalpy value for a state, we must know the state uniquely. At point 1 in Figure 5.16, the temperature is $-10°C$ (evaporator temperature), and the fluid is in a saturated vapor state. Referring to Appendix C1 for saturated Freon-12, we read the value of h_g corresponding to $-10°C$:

$$h_1 = 183.058 \text{ kJ/kg}$$

The enthalpy value for state 2, that is, the exit point of the compressor, is given to be

$$h_2 = 225 \text{ kJ/kg}$$

The enthalpy value of the saturated liquid Freon leaving the condenser at point 3 can be determined by reading the value of h_f for 40°C and saturated Freon-12 from Appendix C1:

$$h_3 = 74.527 \text{ kJ/kg}$$

Since a throttling process occurs between points 3 and 4, we have

$$h_4 = h_3 = 74.527 \text{ kJ/kg}$$

Substituting in Equation 5.12(a), we get

$$(\text{COP})_R = \frac{h_1 - h_4}{h_2 - h_1} = \frac{183.058 - 74.527}{225 - 183.058}$$

$$= 2.6$$

Comments: This is a typical value of the COP for a refrigerator, especially when the evaporator temperature is of the order of -10°C. You can verify that if the system were used as a heat pump, its coefficient of performance would be 3.6.

5.9 ENERGY EQUATION FOR UNSTEADY FLOW PROCESSES

All the examples discussed up to this point in this chapter reflect steady-state operation; that is, properties such as pressure, density, temperature, velocity, and mass flow rate at any point in the control volume are constant. There are certain cases of open systems where some of the properties do not remain constant; rather they change with time. Under such conditions Equation 5.7, reproduce below, applies.

$$\dot{m}_{\text{in}}\left(h_{\text{in}} + \frac{V_{\text{in}}^2}{2} + gZ_{\text{in}}\right) - \dot{m}_{\text{out}}\left(h_{\text{out}} + \frac{V_{\text{out}}^2}{2} + gZ_{\text{out}}\right) + \dot{Q} - \dot{W}_s = \frac{d}{d\tau}(E_{\text{CV}})$$

where

$$E_{\text{CV}} = \int_{\text{CV}} \left(u + \frac{V^2}{2} + gZ\right)\rho \, dV$$

We now consider the application of the above equation to the process of filling a tank with a gas.

Consider a tank of volume V containing a gas at pressure p_i and temperature T_i. Let the tank (see Figure 5.18) be connected to a pipeline carrying gas of the same species as in the tank at pressure p_0 and temperature T_0. Let the valve be opened so that the high-pressure gas from the line enters the tank. We now wish to determine the final temperature of the gas in the tank.

We assume that the tank is insulated so that $Q = 0$ and that the gas in the tank is thoroughly mixed at all times so that the property values are uniform throughout the control volume at any given time. The final pressure of the gas in the tank equals the pressure p_0 in the pipeline. Let us draw a control volume around the tank, as shown in Figure 5.18. We observe that the gas enters the control volume at one point

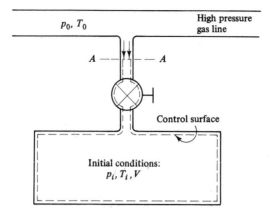

Figure 5.18 Filling of a tank.

and there is no point where it leaves the control volume, so that $\dot{m}_{out} = 0$. We shall assume that the kinetic energy and the potential energy are negligible when compared with the enthalpy of the gas entering the tank as well as that of the gas originally in the tank. Equation 5.7 then can be simplified to read

$$\dot{m}_{in}h_{in} = \frac{d}{d\tau}\left[\int_{cv} (u_{cv})\rho \, dV\right] = \frac{d}{d\tau}[mu]_{cv}$$

Multiplying both sides of this equation by $d\tau$ and integrating, we obtain

$$\int (\dot{m}_{in}h_{in}) \, d\tau = \int d\,(mu)_{cv}$$

The left side of the above equation represents the total enthalpy that enters the control volume through section A–A. It equals the product of the mass of the gas that enters the tank during the time it takes to fill the tank and the specific enthalpy of the gas at section A–A, which is the same as the enthalpy h_0 of the gas in the pipeline. Thus

$$\int (\dot{m}_{in}h_{in}) \, d\tau = (m_f - m_i)h_0$$

where m_f and m_i are the final mass and the initial mass of the gas in the tank, respectively.

The term $\int d\,(mu)_{cv}$ in the reduced energy equation represents the net change in the internal energy of the gas in the control volume. That is,

$$\int d\,[mu]_{cv} = [mu]_{cv,\,final} - [mu]_{cv,\,initial}$$

$$= m_f u_f - m_i u_i$$

If we assume the behavior of the gas to be ideal and c_v to be constant, $u = c_v T$ and if u is taken as zero at 0 K,

$$\int d\,[mu]_{cv} = c_v m_f T_f - c_v m_i T_i$$

208

where T_f and T_i are the final and initial temperatures, respectively. We can now write

$$\int (\dot{m}_{in}h_{in})\, d\tau = \int d\,[mu]_{CV}$$

or

$$(m_f - m_i)h_0 = c_v(m_f T_f - m_i T_i) \tag{5.13}$$

Since the initial conditions of the gas in the tank are known, we can determine m_i from the ideal gas equation. In the final state, the gas in the tank has pressure p_0 and volume V, so that

$$m_f T_f = \frac{p_0 V}{R} \tag{5.13a}$$

Equations 5.13 and 5.13a can be solved simultaneously for the two unknowns, T_f and m_f. If the tank is initially empty, $m_i = 0$, and Equation 5.13 becomes

$$m_f c_p T_0 = c_v m_f T_f$$

or

$$T_f = \gamma T_0 \tag{5.14}$$

The temperature of the gas in the tank will be somewhat greater than the temperature of the gas in the pipe since the value of γ is greater than one.

The problem of determining the final temperature of gas in a tank that is being emptied can be analyzed in a similar manner.

5.10 SUMMARY

In this chapter the first law of thermodynamics was applied to an open system. The concept of control volume was introduced to facilitate the analysis of open systems. Since thermal equilibrium as defined in Chapter 3 seldom exists in a flow system, we introduced the concept of local equilibrium to facilitate analysis of flow systems. A representation for conservation of mass was formulated for a flow system. We discovered that flow work is necessary to maintain a flow. The principle of conservation of energy applied to a control volume requires accounting of internal energy, flow work, kinetic energy, and potential energy for a fluid crossing a control surface; of internal energy, kinetic energy, and potential energy for fluid within the control volume; and of heat and mechanical work energies transferred across a control surface. The resulting equation appears more complex than the one for a closed system, but simplification is often possible because of the absence of certain energy terms. The energy equation was then applied to diverse situations of practical interest. The throttling process, in which enthalpy of fluid remains essentially unchanged, was described. The energy equation was then applied to a gas turbine plant and a heat pump system. Finally, the use of the unsteady-state equation was demonstrated via the problem of a tank being filled with a gas from a high-pressure line.

Notes

Equation for Conservation of Mass in Vector Form. Consider a control volume, across which flows a fluid with velocity $\mathbf{V}(x, y, z)$. If A represents a part of the control surface through which fluid leaves the control volume, then the outer normal vector to A, pointing away from the control volume, and the velocity vector $\mathbf{V}$ are in the *same* direction. This is considered to give a positive value for the integral $\int \rho V \, dA$ in Equation 5.2(b) and this integral then gives the rate of mass *outflow* from the control volume through area A. On the other hand, if A is a part of the control surface through which fluid enters the control volume, then the outer normal to A, again pointing away from the control volume and the velocity vector $\mathbf{V}$ are in the *opposite* direction. Consequently, the integral in the right side of Equation 5.2(b) is considered to be negative for the surface across which fluid enters the control volume, and the integral then gives the rate of mass *inflow* into the control volume through area A.

In two- and three-dimensional flows we write $\rho\mathbf{V} \cdot d\mathbf{A}$ for the integrand in Equation 5.2(b). The quantity $\mathbf{V} \cdot d\mathbf{A}$ is a dot or a scalar product between the local velocity vector $\mathbf{V}$ and the vector $d\mathbf{A}$, which represents an elemental area dA. The latter is always drawn as an *outward normal* to the area dA, that is, the outward normal points away from the control volume. The dot product represents the product of the magnitude of dA and the component of the vector $\mathbf{V}$ along the direction of $d\mathbf{A}$, that is, the normal to dA. The integral $\int_{A} \rho\mathbf{V} \cdot d\mathbf{A}$, gives the net outflow of a fluid from a control surface of area A because of the outward normal associated with $d\mathbf{A}$. Inasmuch as A now represents the entire control surface, we replace it by $\sum A$. The summation sign accounts for the various areas such as those through which fluid enters, those through which fluid leaves, and those that are solid, impervious surfaces. Thus Equation 5.2(b) becomes

$$\dot{m} = \int_{\Sigma A} \rho\mathbf{V} \cdot d\mathbf{A} \tag{5.14}$$

Thus we can visualize a control surface to consist of three parts: (1) one through which a fluid comes into the control volume; (2) a second through which the fluid leaves the control volume; and (3) the solid boundary across which there is usually no flow. The angle between the velocity vector and the area vector is between $\pi/2$ and π radians for (1), between 0 and $\pi/2$ radians for (2), and zero for (3). Since the dot product of two vectors is the product of the magnitudes of the two vectors and the cosine of the angle between the two vectors, the expression $\rho\mathbf{V} \cdot d\mathbf{A}$ will be negative for (1), positive for (2), and identically zero for (3). Thus we can write

$$\int_{CS} \rho\mathbf{V} \cdot d\mathbf{A} = \int_{(CS)_{in}} \rho\mathbf{V} \cdot d\mathbf{A} + \int_{(CS)_{out}} \rho\mathbf{V} \cdot d\mathbf{A} + \int_{\substack{solid \\ boundary}} \rho\mathbf{V} \cdot d\mathbf{A}$$

$$= -\dot{m}_{in} + \dot{m}_{out} + 0$$

$$= -(\dot{m}_{in} - \dot{m}_{out}) \tag{5.15}$$

The mass within a control volume at any time is $\int \rho \, dV$, and the rate of change of mass within the control volume can be written as

$$\frac{d}{d\tau}\left[\int_{CV} \rho \, dV\right]$$

The principle of conservation of mass for a control volume was expressed by Equation 5.1:

$$\dot{m}_{in} - \dot{m}_{out} = \left(\frac{dm}{d\tau}\right)_{CV} \tag{5.1}$$

Employing Equation 5.15, we obtain

$$-\int_{CS} \rho \mathbf{V} \cdot \mathbf{dA} = \frac{d}{d\tau}\int_{CV} \rho \, dV \tag{5.16}$$

Equation 5.16 expresses the principle of conservation of mass in a vector form.

PROBLEMS

Draw a control volume and show all relevant properties on it wherever applicable. If you neglect certain terms in the energy equation for a flow process, state your assumptions and reasons.

5.1. Air at 35°C and 1 atm flows through a horizontal pipe at a velocity of 30 m/s. Determine the diameter of the pipe and the mass flow rate of the air if the volumetric flow rate is 5 m³/s. Assume ideal gas behavior.

5.2. Do Problem 5.1 if the temperature of the air is −35°C.

5.3. Liquid water at 100°C flows through a 12-cm diameter pipe. If the flow velocity is 2 m/s, determine the mass flow rate of the water. What should be the velocity of saturated steam at 100°C through the same pipe for a flow rate that is 0.2 percent of the water flow rate?

5.4. Saturated Freon-12 vapor enters a compressor at 10°C with a mass flow rate of 0.1 kg/s. The velocity of the refrigerant must not exceed 12 m/s. Determine the smallest diameter of the tubing to carry the Freon vapor.

5.5. It was postulated by Joule that the temperature of water rises as the water comes down a waterfall. Consider the waterfall to be an open thermodynamic system and determine the height through which the water should fall for the temperature of the water to rise by 1°C. Assume that the c_p for water is 4.186 kJ/kg K. Write all the assumptions made in solving the problem.

5.6. Air flows through a control volume and it has the following properties at the inlet and the outlet.

	Inlet	Outlet
Pressure	15 Mpa	5 Mpa
Temperature	490°C	210°C
Velocity	350 m/s	50 m/s
Elevation	753 m	705 m
Mass Flow Rate	56 kg/s	56 kg/s

The rate at which heat energy is supplied to the control volume is -4 kJ/s and the rate at which mechanical work is delivered by the control volume is 6000 kJ/s. Determine the rate of accumulation of energy within the control volume.

5.7. A steady-flow system receives 50 kg/min of a gas at 200 kPa, 90°C, with negligible velocity, and discharges it at a point 25 m above the entrance section at a temperature of 300°C with a velocity of 2500 m/min. During this process 1.4 kW of heat is supplied from external sources and the increase in enthalpy is 8.4 kJ/kg. Determine the work done per kilogram of gas.

5.8. In a certain steady-flow system, 60 kJ of work is done by the system per kilogram of the fluid. The specific volume, pressure, and velocities at the inlet and exit are 0.45 m³/kg, 700 kPa, 15 m/s, and 0.067 m³/kg, 150 kPa, 250 m/s, respectively. The inlet is 50 m above the exit and the total heat loss is 10 kJ/kg of fluid. What is the change in the specific internal energy of the working fluid. ?

5.9. A pump located 5 m above the surface of well water supplies water to a tank that is 20 m above the pump. The pressure of the water in the suction pipe at the well-water level is 40 kPa, while that at the delivery to the tank is 100 kPa. The diameters of the suction and the delivery pipes are 20 cm and 15 cm, respectively. If the mass flow rate of the water is 3 kg/s determine the rate at which mechanical work should be supplied to the pump. The density of water is 1000 kg/m³ and is constant.

5.10. The inlet conditions of a water pump are 140 kPa, 15°C, and the exit pressure is 20 MPa. Assuming that the energy loss in the pump goes to heat the fluid, calculate the exit temperature. The rating of the pump is 50 kW and it pumps 10 m³ (at inlet conditions) of water per hour. Comment on the efficiency of the pump.

5.11. Water at 100 kPa and 40°C enters a pump at a rate of 60 kg/min. The power consumption of the pump is 60 kW, and the pump raises the pressure to 500 kPa. The water then passes through a boiler in which 2000 kJ/kg of heat is added. The diameter of the duct exiting the boiler is 20 cm. Neglecting the pressure drop in the boiler, determine the state at the exit of the boiler and the velocity at that point.

5.12. A venturimeter consists of converging-diverging pipes with a short constant-diameter section in between. The section with the smallest diameter is called the *throat* of the venturi. Water flows at a steady rate of 1500 kg/min through a horizontal venturimeter with inlet and throat diameters of 8.0 cm and 4.0 cm, respectively. If there is no transfer of heat or work and no change in internal energy and the density remains constant at 1000 kg/m³, what will be the pressure drop between the inlet and the throat in Pascals?

5.13. Air enters an air compressor and is pressurized to 1200 kPa and 80°C. The inlet conditions are 100 kPa and 10°C. The velocity at the outlet is eight times the velocity at the inlet, and the inlet diameter is 10 cm. If 0.16 kg of air are to be compressed every determine the diameter at the outlet and the rate at which work must be supplied to the compressor.

5.14. The compressor of a large gas turbine receives air from the surroundings at 95 kPa and 20°C. The air is compressed to 800 kPa, according to the relation $pv^{1.3}$ = constant. The inlet velocity is negligible and the outlet velocity is 100 m/s. The power input to the compressor is 2500 kW, 20% which is removed as heat from the compressor. What is the mass flow of the air?

5.15. A centrifugal air compressor receives a gas at a rate of 6 m³/min and compresses it from 90 kPa to 630 kPa. The initial and final specific volumes are 0.82 m³/kg and 0.21 m³/kg, respectively, and the diameters of the duct are 10 cm at the inlet and 5 cm at the exit. Determine (a) the flow work at the boundaries, (b) the mass flow rate, and (c) the change in velocity.

5.16. A ship propulsion system incorporates a compressor that receives steam at 240 kPa and a quality of 0.95. It delivers it dry and saturated at 600 kPa. It is known that the mass of steam handled is 5 kg/s, the diameters of the inlet and exit ducts are both 20 cm, the compression is adiabatic, and that the mechanical efficiency is 0.9. Calculate the necessary work input to the compressor.

5.17. A steam turbine receives saturated steam at 700 kPa. The exhaust from the turbine is 0.95 quality steam at 10 kPa and has a velocity of 75 m/s. Determine the steam flow rate necessary to produce 500 kW of power. Neglect the velocity of the steam at the entrance to the turbine.

5.18. A certain high-speed turbine operates on compressed air. Air at 600 kPa and 50°C enters the turbine, expands adiabatically, and leaves at 80 kPa. What should be the mass flow rate if the turbine output is to be 2.5 kW?

5.19. An inventor claims to have developed a turbine that will receive water from a reservoir located 500 m above the level of the turbine and produce 500 kW of power. The turbine would require a mass flow rate of 500 kg/s on a steady flow basis and a discharge at a velocity of 32 m/s. Verify the inventor's claim. The pressure of water at the intake from the reservoir and the outlet from the turbine are equal.

5.20. Steam with a specific enthalpy of 2.8 MJ/kg is supplied to a turbine. The specific enthalpy of the turbine exhaust is 2.2 MJ/kg. The entrance and exit velocities are 180 m/s and 300 m/s, respectively. If the heat loss is 20 kJ/kg of steam, what is the work done per kilogram of steam?

5.21. A steam turbine receives steam at a flow rate of 5000 kg/h and produces 500 kW. Neglecting the heat losses, find the change in the specific enthalpy of the steam flowing through the turbine in each case:
  (a) Entrance and exit velocities and the change in the entry and exit levels are negligible.
  (b) Entrance and exit velocities are 60 m/s and 250 m/s, respectively, and the inlet is 4 m above the exhaust.

5.22. The inlet conditions for the nozzle of a steam turbine are 5 MPa and 350°C. The exit conditions are 1 MPa and a quality of 90%. If the steam flow rate is 10,000 kg/h, determine the exit velocity and exit area.

5.23. Water at 500 kPa enters a horizontal nozzle at a velocity of 1.8 m/s. The diameter at the inlet of the nozzle is 9 cm and that at the outlet is 30 mm. Treating water as an incompressible liquid ($\rho = 1000$ kg/m³), determine the velocity of the water at the outlet. The temperature of the water may be assumed to be constant. Applying the energy equation for a flow process determine the pressure of the water at the outlet.

5.24. Air at a pressure of 300 kPa and a temperature of 15°C enters a nozzle with an inlet velocity of 20 m/s. It expands in a quasistatic adiabatic manner ($\gamma = 1.4$) to a pressure of 200 kPa as it reaches the exit of the nozzle. Calculate the temperature of the air at the exit of the nozzle. Then applying the energy equation, determine the exit velocity of the air. The air may be treated as an ideal gas.

5.25. The heating value of the fuel used in a gasoline engine is 1.03×10^7 J/kg. A stream of air and gasoline-vapor mixture in the ratio 15:1 by mass enters the engine at a temperature of

30°C and the exhaust gases leave at 170°C. The net heat transfer from the engine to the cooling water and surroundings is 2300 kJ/min. The shaft power delivererd is 26 kW. The mixture may be assumed to behave like an ideal gas with $c_p = 1.05$ kJ/kg K. Calculate the mass flow rate of the fuel and the increase in enthalpy.

5.26. Freon-12 is proposed as the working fluid for a solar heat pump. Calculate the surface area of the evaporator required to heat the fluid from 1 MPa, 30°C to 0.9 MPa, 100°C. The flow rate of the refrigerant is 10 kg/h, the solar heat flux 600 W/m², and the efficiency of the evaporator 60%.

5.27. Helium gas at -50°C and 100 kPa enters a diffuser at a mass flow rate of 0.25 kg/s. The inlet and the outlet areas of the diffuser are 36 cm² and 300 cm², respectively. The compression of the helium in the diffuser takes place in a quasistatic adiabatic process. The specific heats at constant pressure and constant volume for helium at 5.2 kJ/kg K and 3.123 kJ/kg K. Determine the velocity, pressure, and the temperature of the helium at the exit. (*Hint*: A trial-and-error solution is necessary.)

5.28. Air is heated in a tube of constant diameter of 6 cm. The air enters at 400 kPa and and 20°C with a velocity of 10 m/s. If the air exits at 350 kPa and 80°C, determine the velocity of the air at the exit and the quantity of heat supplied.

5.29. A constant area duct of diameter 12 cm carries a gas and is being heated over a 6-m length with a constant heat flux. The conditions at inlet and exit are:

Inlet: velocity, temperature and pressure uniform across the section, $T = 10$°C, $p = 100$ kPa.

Exit: $p = 100$ kPa, uniform, $V = 100(1 - r^2/R^2)$ m/s, $T = 300(1 + r^2/R^2)$°C.

The density and specific heat at constant pressure remain constant at 1 kg/m³ and 1.00 kJ/kg K, respectively. Determine the rate of heat supply in watts per square meter of the surface area.

5.30. At a particular time during the starting of a turbine, the inlet and exit conditions were noted as follows:

Inlet: $p_1 = 800$ kPa, $T_1 = 900$°C, $V_1 = 150$ m/s, $A_1 = 20$ cm²

Exit: $p_2 = 120$ kPa, $T_2 = 680$°C, $V_2 = 240$ m/s, $A_2 = 30$ cm²

At the instant considered, the output from the turbine was 50 kW. Assume the working fluid to be an ideal gas with $\gamma = 1.3$ and molar mass of 29.0 kg/kg-mol. Calculate the rate of change of the internal energy of the working fluid.

5.31. A certain gas turbine was found to have the following inlet and exit conditions averaged over a 1-min period of time:

	Inlet	Exit
Temperature (K)	1000	500
Pressure (MPa)	1.08	0.22
Velocity (m/s)	8.7	5.2

The diameters at the inlet and exit were observed to be 40 cm and 80 cm, respectively. The power output was 1 MW and the total heat loss in that minute was 8000 kJ. The working fluid was an ideal gas with a molar mass of 29 kg/kg-mol and $c_p = 1.0$ kJ/ kg K. Determine (a) the mass accumulated in the system, and (b) the increase in the internal energy of the system during the 1-min interval.

5.32. Sonic velocity of an ideal gas equals $\sqrt{\gamma RT}$, where R is the gas constant and T is the absolute temperature. An ideal gas is accelerated from rest and an initial temperature T_0. Determine the temperature of hydrogen under sonic conditions if T_0 is 300 K.

5.33. In a steady-one-dimensional flow through a nozzle, a gas enters the nozzle at negligible inlet velocity and emerges at a velocity V and a temperature T. At the exit the speed of sound is a where, for an ideal gas, $a^2 = \gamma RT$. If the Mach number M is defined as $M = V/a$, show that

$$\frac{T_0}{T} = 1 + \frac{\gamma - 1}{2} M^2$$

where T_0 is the temperature at the entrance.

5.34. Steam is used for heating buildings in winter. In one heating system, steam enters a pipe at ground level as saturated vapor at 150 kPa and leaves the top floor of the building, which is 300 m above the ground floor, at a pressure of 100 kPa. The heat transfer from the steam pipe is 125 kJ/kg. Determine the thermodynamic state of the steam as it leaves the top floor.

5.35. In a steam generator, saturated steam is produced at 8000 kPa at a rate of 8000 kg/h. The inlet conditions of water are 1200 kPa and 40°C. Determine the heat transfer rate to the water. If the efficiency of the generator is 81%, and if the heating value of the fuel is 2×10^6 J/kg, determine the fuel consumption.

5.36. Nitrogen at 400 kPa and 125°C enters a heat exchanger and leaves at 200 kPa and 50°C. The area at the entrance is 0.2 m² and that at the exit is 0.4 m². The inlet velocity of nitrogen is 200 m/s. The cooling agent is Freon-12, entering at 150 kPa and 20°C and leaving at 100 kPa and 90°C. Determine the mass flow rate of Freon-12 and the exit velocity of nitrogen. Assume nitrogen to behave as an ideal gas.

5.37. Saturated ammonia vapor at 1200 kPa is mixed with liquid ammonia at 1200 kPa and 10°C in a steady-flow process. The mass flow rate of liquid ammonia is twice the mass flow rate of the saturated vapor. It is found that after mixing, the pressure is 100 kPa and the quality is 75%. Determine the rate of heat transfer per kilogram of the resulting mixture in the mixing process.

5.38. The diameter of the incoming cold-water line and the outgoing hot-water line of a domestic hot water heater is 12 mm. The water is known to flow at a uniform velocity of 3 m/s. The temperature of the water at the inlet is 4°C. The gas used for the water heater produces 40,000 kJ/kg of gas. It is known that 20% of the heat released by the combustion of the gas is lost through the stack and another 4% is lost from the body of the heater to the room in which the heater is located. If the temperature of the water leaving the heater is 65°C, determine the rate of gas consumption in kilogram per hour.

5.39. In a gas-cooled breeder reactor, helium gas enters the steam generator at 700°C with a flow rate of 300 kg/s and leaves at 300°C. Water enters the steam generator at 30°C and 20,000 kPa, and the steam leaves at 500°C. The water inlet pipe is 12 cm in diameter, and the steam exit pipe is 20 cm and 5 m above the inlet pipe. Determine the inlet and exit velocities of water and the mass flow rate of steam.

5.40. Determine the gain in the energy of the water flowing through the solar panel of Example 5.6. If the losses to the surroundings are reduced to 10% of the incident energy and if the heating requirement for a home is 15,000 kJ/h, determine the size of the storage tank where the hot water returning from the solar panel can be stored. Assume that the tank is insulated, and the energy stored in the tank water should be sufficient to heat the home for three days in the absence of any solar radiation.

5.41. Freon-12 is throttled from a saturated liquid condition at 50°C to a pressure of 100 kPa. The inlet and exit velocities can be considered to be equal. Determine the exit temperature and quality.

5.42. Steam at 800 kPa is throttled to 500 kPa. It is observed that the temperature of the steam after throttling is 160°C. Determine the quality of steam before throttling.

5.43. A simple steam power plant consists of a feed pump, a steam generator, a turbine and a condenser. Water enters the pump at 14 kPa and 45°C. The pump work is 5 kJ/kg. Steam leaves the steam generator at 2000 kPa and 300°C and enters the turbine at 1900 kPa and 290°C. Steam with a quality of 90% at a pressure of 15 kPa leaves the turbine and enters the condensor. Determine each of the following.
  (a) Heat transfer in the pipeline between the steam generator and the turbine.
  (b) Turbine work output.
  (c) Heat transfer to the condenser.
  (d) Heat transfer to the steam generator.
  (e) Thermal efficiency of the power plant.

5.44. In a Freon-12 refrigeration system, saturated Freon vapor at 150 kPa enters the compresser and leaves at 1000 kPa and 90°C. It then enters an air-cooled condenser and leaves as saturated liquid at 0.95 MPa. The mass flow rate of Freon is 0.02 kg/s and the power input to the compressor is 1.4 kW. Determine each of the following.
  (a) Heat transfer rates from the compressor and the condenser.
  (b) Coefficient of performance for the refrigeration system.

5.45. A steam turbine receives steam at 1000 kPa and 400°C. The turbine exhausts into a large tank having a volume of 100 m³. The tank is initially evacuated. The turbine can operate until the pressure in the tank reaches 100 kPa. If the steam temperature at this point is 250°C, determine the work done by the turbine during this process. Assume the entire process to be adiabatic.

5.46. Write a computer program to solve Example 5.8. Then, holding the temperature at the exit from the combustion chamber constant, vary the compressor outlet pressure from 200 kPa to 2000 kPa, with increments of 200 kPa. For each value of the compressor outlet pressure, determine the net work output from the system and the thermal efficiency of the plant. Plot the network output as a function of the pressure ratio and hence determine the optimum pressure ratio for maximum work output from the system.

5.47. Show that the pressure ratio yielding the most work per unit mass for the idealized Brayton cycle of Figure 5.13 is

$$r_{p,\text{optimum}} = \left(\frac{T_3}{T_1}\right)^{\gamma/2(\gamma-1)}$$

5.48. Air enters the compressor of a Brayton cycle at 100 kPa and 15°C and leaves at 500 kPa. If the maximum temperature in the cycle is 1000°C, determine the following.
  (a) Pressure and temperature at each point in the cycle.
  (b) Net work and efficiency of the cycle.
  (c) Mass flow rate necessary to produce 10 kW of power.

5.49. A vapor compression refrigeration system uses ammonia as the refrigerant. The condenser outlet conditions are saturated liquid at 50°C and the evaporator outlet conditions are saturated vapor at −10°C. Determine the COP of the system and the cooling effect that can be produced with a mass flow rate of 8 kg/s.

5.50. If the system in Problem 5.49 were to be used as a heat pump, determine its COP and the heating effect.

5.51. An evacuated tank is connected to a pipe line carrying air at a pressure of 1000 kPa and a temperature of 25°C. A valve regulates the flow of air from the line to the tank. Determine the final temperature of the air in the tank as its pressure reaches 1000 kPa.

5.52. If the air in the tank in Problem 5.51 is allowed to exchange heat with the surroundings after it is pressurized and sealed, the tank eventually returns to the surroundings' temperature of 20°C. What is the final pressure inside the tank?

5.53. A tank of volume 50 m³, containing air at 600 kPa and 25°C, is rapidly bled until its pressure drops to 580 kPa. Determine the final temperature of the air in the tank.

6 —————————————

The Second Law
of Thermodynamics
and Entropy

6.1 INTRODUCTION

The first law of thermodynamics is the principle of conservation of energy. It forbids creation or destruction of energy, but permits transformation of one form of energy into another. It does not put any restriction on the direction in which the transformation of energy may occur. Let us amplify this point by means of a few examples. Imagine a liquid in an insulated container with a paddle wheel immersed in the liquid. If we rotate the paddle wheel and thereby do some work on the liquid, we find that the temperature of the liquid increases. Thus a transfer of work energy to the liquid system has increased its internal energy. This transformation of energy is in accordance with the first law, and we can indeed verify such a transformation in a laboratory. Now suppose that we want a decrease in the internal energy of the liquid and a corresponding output of work energy. The first law does not prohibit such transformation. In practice, we find that we can cause a decrease in the internal energy of the liquid by permitting some heat transfer, but we are not able to produce a corresponding work output.

Next, consider a high-temperature body in thermal contact with a low-temperature body. It is a common experience that there is a transfer of heat energy from the high-temperature body to the low-temperature body until both the bodies have the same temperature. If we ignore the amount of work associated with the changes in the volumes of the two bodies, we find that the increase in the internal energy of the cold body equals the decrease in the internal energy of the hot body. This transformation of energy is in accordance with the first law.

218

Now, suppose we wish to restore the two bodies to their original temperatures. The first law does not prohibit the restoration in any way. However, we find that such a restoration does not occur naturally. The restoration can be brought about only with the aid of an external device, such as a heat pump. Note that such an external device was not necessary to cause the heat transfer from the hot body to the cold body. In a similar fashion, we find that electrical work supplied to a heating coil in an electric range produces a heating effect and raises the internal energy of the coil, but the converse—that is, a drop in the internal energy of the coil, producing electric current and doing corresponding electrical work—is not observed in practice. A river flows from a mountain to the sea, bringing about a conversion of the potential energy of water into kinetic energy and finally dissipating this energy as heat, everything occuring in accordance with the first law of thermodynamics. However, we never see a river originating from the sea and going up to a mountaintop. This would require a reduction in the internal energy of water and a corresponding increase in its potential energy, all of which would not violate the first law. Yet it does not happen. Such experiences and observations force one to think that there must be a hierarchy associated with energy, and that natural processes degenerate energy from higher to lower grades. These characteristics of energy are not addressed by the first law.

A *high-grade energy* can be converted to virtually any form of energy. Work energy is a high-grade energy, since it can be entirely and naturally converted into thermal energy, kinetic energy, potential energy and many other forms of energy. On the other hand, the internal energy of a body at atmospheric temperature is a low-grade energy because none of it can be readily converted into work. This conversion would be possible only if a sink of a temperature less than the atmospheric temperature were available, but such a sink is not generally available. A high-grade energy such as the internal energy of gases in a combustion chamber has a potential to do work. When this high-grade energy becomes low-grade energy due to a heat transfer process, it loses its potential. It is similar to the water in an elevated lake flowing down to sea level and losing its potential to produce work, for example, by driving a hydraulic turbine. Such degradation of energy is the subject matter of the second law of thermodynamics. In order to study the second law, it is necessary to understand the concepts of a reversible and an irreversible process. We shall now discuss these concepts.

6.2 REVERSIBLE AND IRREVERSIBLE PROCESSES

When a thermodynamic process can be reversed so that at the end of the reversal the thermodynamic states of the system and its surroundings are exactly the same as they were at the start of the original process, then the process is said to be a *reversible process*. If during the reversal, we have to use energy from the surroundings that was not originally delivered by the system to the surroundings, the process is *irreversible*.

Let us examine the heat transfer process between two bodies at different temperatures in the light of the foregoing definitions. Let body A at temperature T_{high} be brought in thermal contact with body B at temperature T_{low}. We soon find that the two bodies attain an equilibrium temperature T_{eq} (see Figure 6.1(a)). The process involved is irreversible. In order to restore the two bodies to their original temperatures, we would need to supply heat energy to body A and to remove energy from body B. We can use a heat pump (See Section 5.8.3) to bring this about, as shown in Figure 6.1(b). We can operate the heat pump until body A attains its original temperature, T_{high}. In so doing we would supply exactly the same quantity of heat energy to body A as it originally (Figure 6.1(a)) supplied to body B. However, the energy delivered by the heat pump must equal the sum total of the amount of work delivered to the heat pump from the surroundings and the amount of heat transferred from body B to the heat pump. Consequently, body B will not be restored to T_{low} in this operation. We could continue to operate the heat pump for an additional period of time until body B reaches a temperature of T_{low}. We would then find that body A is at a temperature that is higher than T_{high}.

We may argue that body A can then be restored to T_{high} by allowing a transfer of heat energy between body A and its surroundings. The magnitude of this heat transfer is exactly equal to the work energy supplied to the heat pump by the surround-

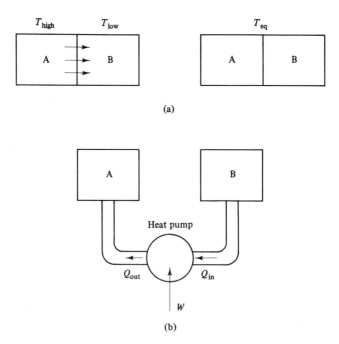

(a)

(b)

Figure 6.1
(a) Heat flows from high temperature to low temperature.
(b) Work input from surroundings is necessary to restore bodies A and B to their initial states.

ings. In order to restore the surroundings to its original status, all that we have to do is to convert this heat energy into an equal amount of work. However, this cannot be done. Man has not been able to convert heat into work with 100 percent efficiency using a cyclic device. Thus no matter how hard we try, there is no way to restore the two bodies in Figure 6.1 to their original states without affecting the surroundings. Therefore we conclude that *heat transfer across a finite temperature difference is an irreversible process.*

Consider a frictionless piston-cylinder system (see Figure 6.2) enclosing a gas and receiving heat energy. For convenience, let us assume that the gas undergoes a constant pressure expansion. As heat is supplied to the gas, it does some work. This work results in an increase in the potential energy of the discs that are raised by the expanding gas. In a real piston-cylinder system, friction is present at the area of contact between the piston and the cylinder. Part of the work available from the expanding gas is spent in overcoming the frictional force. Consequently, the actual increase in the potential energy of the discs in a real situation is less than that in an ideal frictionless case. This is shown in Figure 6.2 by a greater expansion of the gas for the frictionless case than for the case with friction. If we try to restore the gas to its original state, we find that we must use work from the surroundings that was not delivered by the gaseous system in the first instance. Thus the expansion of a gas in a real piston-cylinder system is irreversible, since friction is present.

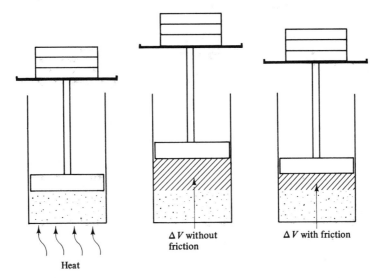

Figure 6.2   Work output from a system with friction is less than the work output from a frictionless system.

If a system does not pass through a series of equilibrium states—that is, if the process is not quasistatic—then the process is irreversible. Consider a gas held under pressure in a piston-cylinder system with suitable restraints. If the restraints on the piston are removed abruptly, the gas undergoes a rapid expansion. The gas is not in equilibrium at any point during the ensuing expansion process, and such a process is irreversible. The same thing holds true when a gas expands into an initially evacuated chamber.

Let us consider a few other processes. Imagine an insulated, partitioned box, with one side evacuated and the other containing a gas under pressure. If the partition is ruptured, the entire box will be filled with the gas and the pressure of the gas will decrease. Now suppose that we are interested in a reversal of this process, that is, restoration of all the gas to one side of the partition to its original pressure. The only way the reversal of the process can be accomplished is by bringing an external agency —a pump—to evacuate one side and pressurize the other. Such an external agency would require mechanical work. Thus the original process is irreversible. A brine solution can be formed with little expenditure of energy by mixing salt and water. The constituents of the brine solution, namely, salt and water, do not separate spontaneously; the separation can be effected only by an expenditure of a significant quantity of heat energy to vaporize all the water. This raises the temperature and internal energy of the salt and the water. The salt and the water can be restored to their initial states by cooling them. However, the effect of this process on the surroundings will be a transfer of energy from a high-temperature source, used for heating the brine solution, to a low-temperature sink, used for cooling the separated salt and water vapor. Thus the mixing of salt and water is an irreversible process.

A heat transfer process requires a temperature difference. A reversible process precludes a temperature difference. Does this mean that there can be no heat transfer in a reversible process? The answer is that the transfer of heat in a reversible process requires an infinitesimal temperature difference. Under such conditions the heat transfer process between an energy reservoir and a system is essentially an isothermal process. Furthermore, the rate of heat transfer in the presence of an infinitesimal temperature difference is also infinitesimal, and such a slow rate of heat transfer can be rarely be achieved in practice.

There are other phenomena where the processes involved are irreversible. The expansion or compression of a fluid occurring at a finite rate is irreversible. The flow of electric current through a resistor is irreversible. All finite-rate chemical reactions are irreversible, as are all mixing processes. The phenomenon of diffusion is migration of a solute from areas of high concentration to areas of low concentration, and it is analogus to conduction of heat. If the difference in concentrations is infinitesimal, the ensuing diffusion is reversible. If it is finite, the diffusion is irreversible. Magnetic hysteresis is an irreversible phenomenon. The free fall of a body in a gravitational field is irreversible. A pressurized gas filling an evacuated chamber is an irreversible process. To restore these systems to their original states, it is necessary to use mechanical work from the surroundings. A pendulum swinging in air soon comes to rest. This is also an irreversible process; the energy of the pendulum is dissipated due to the friction of the air molecules.

We can make the following statements with regard to irreversible processes.

1. All natural processes are irreversible.

2. Whenever a process takes a system through nonequilibrium states, the process is irreversible. A direct consequence of this is that any process proceeding at a finite rate is irreversible.

3. Whenever there is heat transfer due to a finite temperature difference in a given process, the process is irreversible.

4. Whenever there is friction in a given process, the process is irreversible.

We might ask at this point whether there is any reason to consider the reversible process if there is no way to achieve it in practice. The answer is that the concept of a reversible process is as important in thermodynamics as is the concept of a frictionless device in mechanics. All real devices have friction. Yet a first analysis of these devices assumes no friction. In a similar fashion the reversible process provides a first approximation for evaluation of actual processes. Also, it is a useful concept in dealing with the ideal power and refrigeration cycles in thermodynamics. While considering applications of thermodynamics, we do account for the fact that real processes are not reversible. This is analogous to accounting for frictional forces in the applications of mechanics.

Often irreversibility is classified as *internal irreversibility* and *external irreversibility*. When a system undergoes an irreversible process with finite temperature or pressure gradients just outside the system boundary, and yet the system is *essentially* at thermal equilibrium, the process is said to be externally irreversible but internally reversible. Consider saturated water at 100°C receiving heat energy from combustion gases at 600°C. As long as a complete change of state does not occur, the water is essentially at thermal equilibrium. On the other hand, at the boundary of the system (water), there is a finite temperature difference. Hence the process is internally reversible, but with external irreversibility. If a system is not in thermal equilibrium during a process, then the process is internally irreversible. Rapid expansion of a gas falls in this category.

From this discussion it should be evident by now that the quasistatic processes discussed in Chapter 1 are reversible, and that the words *quasistatic* and *reversible* can be used interchangeably.

6.3 THE SECOND LAW OF THERMODYNAMICS

In the preceding sections we saw that natural processes occur in a certain preferred direction and that they are irreversible. The second law of thermodynamics specifies in what direction a process may proceed. There are two classical statements, both of which are called the second law of thermodynamics. One is due to Clausius and the other is due to Kelvin. As we shall show later, these statements lead to a mathematical expression of the second law and to a new property called *entropy*.

The **Clausius statement**:   It is impossible to construct a device that executes a thermodynamic cycle so that the sole effect is to produce a transfer of heat energy from a body at a low temperature to a body at a high temperature.

The **Kelvin statement**, also known as the **Kelvin-Planck statement**: It is impossible to construct a device that executes a thermodynamic cycle, exchanges heat energy with a *single* reservoir, and produces an equivalent amount of work.

It is important to note that *both statements apply to cyclic processes only*. The Clausius statement is a formal form of the observation that in nature heat transfer does not occur from low temperature to high temperature. The Kelvin-Planck statement means that thermal energy cannot be converted into work by a cyclic process with 100 percent efficiency; that is, only a part of the thermal energy received by a cyclic device can be converted into mechanical work. It is not possible to prove these statements. However, it can be shown that these two statements are equivalent. The proof consists of showing that a violation of one statement leads to a violation of the other.

Let us assume that the Clausius statement can be violated. We begin with a Carnot heat engine, shown schematically in Figure 6.3. The heat engine operates between two reservoirs of energy, one at T_{high} and the other at T_{low}. The engine delivers a quantity of work W and rejects a quantity of heat Q_{low} (considered positive here) to the reservoir at T_{low}. Let there be a device that operates between the two reservoirs and violates the Clausius statement. Let this device draw a quantity of heat, Q_{low}, from the reservoir at T_{low} and supply it to the reservoir at T_{high}. Next, consider a system consisting of the heat engine and the device. Examining Figure 6.3 we find that the net effect is a transfer of a quantity of heat energy equal to $Q - Q_{low}$ from a high temperature reservoir to the system and a conversion of all of this heat energy into work

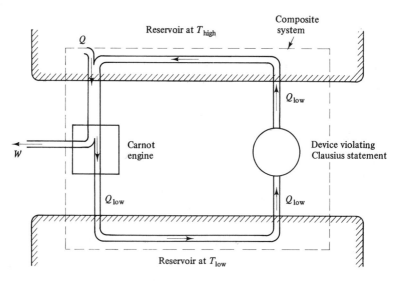

Figure 6-3   A composite system to demonstrate that the Clausius statement cannot be violated.

W. This is a violation of the Kelvin-Planck statement, since we have just created a cyclic device with 100 percent thermal efficiency. Thus a violation of the Clausius statement leads to a violation of the Kelvin-Planck statement. The converse can also be readily proved.

A *perpetual motion machine of the first kind* is a machine whose energy output is greater than its energy input. Such a machine violates the first law. A *perpetual motion machine of the second kind* is a machine that receives thermal energy from a single source and converts all of it into mechanical work. Such a machine does not create energy and therefore does not violate the first law. However, it does violate the second law since it is in direct conflict with the Kelvin-Planck statement. The second law is sometimes expressed by the statement that a *perpetual motion machine of the second kind is impossible.* If such a machine were possible, one could produce work by extracting energy from the atmosphere, a river, or a sea.

There are two important corollaries of the second law.

1. Any heat engine operating between two constant temperature reservoirs cannot have an efficiency that is greater than the efficiency of a reversible engine operating between the same two reservoirs.
2. All reversible heat engines operating between two constant temperature reservoirs have the same efficiency regardless of the operating media used by these engines. The efficiency is solely a function of the temperatures of the reservoirs.

There are a number of reversible engines that may be considered in examining the above corollaries. Of these, the Carnot engine was historically the first engine to be used for thermodynamic analysis. Also, it is conceptually elegant and simple.

Let us recapitulate what we know about the Carnot cycle. It is a cycle that exchanges heat energy with only two reservoirs. It receives heat energy from a high-temperature reservoir and rejects heat energy to a low-temperature reservoir in a reversible (quasistatic) manner. The remaining two processes in the cycle are reversible (quasistatic) adiabatic. The Carnot cycle discussed in Section 3.6.1 operated as a heat engine, had a net work output, and was traced clockwise on a *p-V* diagram. All processes in a Carnot cycle are reversible, so a Carnot cycle can be reversed. It is then traced counterclockwise on a *p-V* diagram and is said to function as a refrigerator or as a heat pump. A Carnot refrigerator receives heat energy from a low-temperature reservoir and supplies heat energy to a high-temperature reservoir with the aid of external mechanical work.

The two corollaries stated earlier mean that it is impossible to have an engine operating between two energy reservoirs whose thermal efficiency is greater than that of a Carnot engine operating between the same two reservoirs. We begin our demonstration of the validity of the two corollaries by first assuming that the corollaries can be violated.

Figure 6.4(a) shows two reservoirs with a Carnot engine operating between them, as well as another device with thermal efficiency η_a greater than that of the Carnot engine. We shall show that such a device violates the Clausius statement.

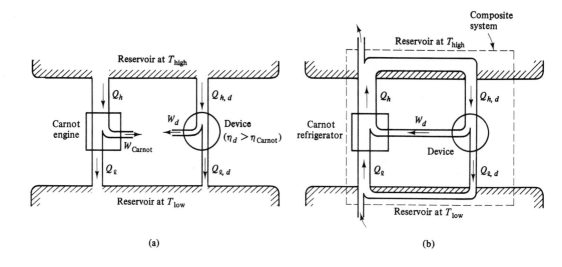

(a)

(b)

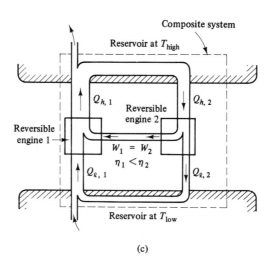

(c)

Figure 6.4
(a) A device with thermal efficiency greater than the Carnot efficiency.
(b) A composite system to demonstrate that the device in (a) violates Clausius statement.
(c) A composite system to demonstrate that the thermal efficiencies of reversible engines operating between same temperatures must be equal.

Our assumption is

$$\eta_d > \eta_{\text{Carnot}}$$

or

$$\frac{W_d}{Q_{h,d}} > \frac{W_{\text{Carnot}}}{Q_h}$$

where the subscript h on Q represents the heat energy extracted from the high-temperature reservoir and the subscript d refers to the device. Let the Carnot engine and the device have equal outputs of work. Then

$$W_d = W_{Carnot}$$

so that

$$Q_{h,d} < Q_h$$

Also, the various Q's and W's are related by the first law,

$$Q_{h,d} - Q_{\ell,d} = W_{device} = W_{Carnot} = Q_h - Q_\ell$$

where Q_ℓ represents the heat energy delivered to the low-temperature reservoir. Note that the quantities $Q_{h,d}$, $Q_{\ell,d}$, Q_h, and Q_ℓ are treated as positive in the foregoing equations.

A rearrangement gives

$$Q_{h,d} + Q_\ell = Q_{\ell,d} + Q_h$$

or

$$Q_{\ell,d} < Q_\ell$$

The above inequalities mean that the heat energy $Q_{h,d}$ drawn by the device from the high-temperature reservoir is less than the quantity of heat energy Q_d drawn from the same reservoir by the Carnot engine. Likewise, the quantity of heat energy $Q_{\ell,d}$ rejected by the device is less than the quantity of heat energy Q_ℓ rejected by the Carnot engine. Since the Carnot engine is reversible, we can operate it as a refrigerator and couple it to our device, as shown in Figure 6.4(b). Since the work output of the device equals the work input for the Carnot refrigerator, the overall effect of the coupled system is a net transfer of heat energy, $Q_\ell - Q_{\ell,d}$, from a low-temperature reservoir to a high-temperature reservoir, which is a violation of the Clausius statement. Thus our assumption that there can be a device whose efficiency is greater than that of a Carnot engine is invalid.

The truth of the second corollary can be proved by a similar reasoning. Consider two reversible engines, both operating between two reservoirs with temperatures T_h and T_ℓ and having unequal efficiencies. Also, let the work output of both the engines be equal. Since both the engines are reversible, we can reverse the one with lower efficiency, engine 1, and operate it as a refrigerator by supplying the work output from the more efficient engine, engine 2, as shown in Figure 6.4(c). The combined system results in a transfer of heat energy from a low temperature reservoir to a high temperature reservoir, without using work from the surroundings. Thus the Clausius statement is violated. Consequently, neither engine can have an efficiency that is greater than the efficiency of the other. Thus we have shown that the efficiencies of reversible engines operating between two reservoirs are equal.

6.3.1 The Kelvin Temperature Scale

If we have two energy reservoirs at temperatures T_h and T_ℓ, respectively, then of all the engines operating between these two reservoirs, the Carnot engine has the highest thermal efficiency. This statement, which follows from the earlier discussion, can be used to establish a temperature scale.

We recall that the thermal efficiency of a cycle is defined by

$$\eta_t = \frac{\text{work done}\,(W_{\text{cycle}})}{\text{heat supplied}\,(Q_{\text{in}})}$$

A Carnot cycle receives heat energy, Q_{in}, from a reservoir at T_h and rejects heat energy, Q_{out}, to a reservoir at T_ℓ. The work output W_{cycle} equals the quantity $Q_{\text{in}} - Q_{\text{out}}$. We therefore write

$$\eta_t = \frac{Q_{\text{in}} - Q_{\text{out}}}{Q_{\text{in}}} = 1 - \frac{Q_{\text{out}}}{Q_{\text{in}}}$$

or

$$\frac{Q_{\text{out}}}{Q_{\text{in}}} = 1 - \eta_t$$

In view of the corollaries discussed earlier in this section, the efficiency of a Carnot cycle is strictly a function of the temperatures of the reservoirs. After taking the reciprocal of both sides of the last equation, we can write

$$\frac{Q_{\text{in}}}{Q_{\text{out}}} = \frac{1}{1 - \eta_t} = g(T_h, T_\ell) \qquad (6.1)$$

where g is an unknown function.

Consider three energy reservoirs at temperatures T_h, T_i, and T_ℓ. Consider three Carnot engines, the first operating between T_h and T_ℓ (see Figure 6.5), the second operating between T_h and T_i, and the third operating between T_i and T_ℓ. The first two engines receive the same quantity of heat energy, Q_{in}, from the reservoir at T_h, and the first and the third engines reject the same quantity of heat energy, Q_{out}, to the reservoir at T_ℓ. We can write the following functional relations for each of the three engines.

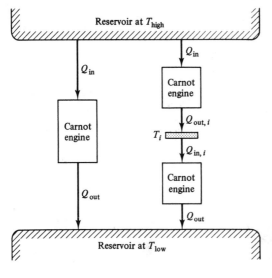

Figure 6.5 A scheme of Carnot engine operations to establish the Kelvin temperature scale.

$$\frac{Q_{in}}{Q_{out}} = g(T_h, T_\ell)$$

$$\frac{Q_{in}}{Q_{out,i}} = g(T_h, T_i)$$

$$\frac{Q_{in,i}}{Q_{out}} = g(T_i, T_\ell)$$

Since

$$Q_{out,i} = Q_{in,i}$$

we can write

$$\frac{Q_{in}}{Q_{out}} = \frac{Q_{in}}{Q_{out,i}} \times \frac{Q_{in,i}}{Q_{out}}$$

or

$$g(T_h, T_\ell) = g(T_h, T_i) \times g(T_i, T_\ell)$$

In the above equation the left side is a function of T_h and T_ℓ. The right side has all three temperatures, T_h, T_i, and T_ℓ. Therefore T_i must somehow cancel out of the right-hand side. The only way in which this can be accomplished is for the function g to be a ratio of two functions, both these functions having the same functional form. That is,

$$g(T_h, T_i) = \frac{f(T_h)}{f(T_i)} \quad \text{and} \quad g(T_i, T_\ell) = \frac{f(T_i)}{f(T_\ell)}$$

Thus we have

$$g(T_h, T_\ell) = \frac{f(T_h)}{f(T_i)} \times \frac{f(T_i)}{f(T_\ell)}.$$

or

$$g(T_h, T_\ell) = \frac{f(T_h)}{f(T_\ell)} \tag{6.2}$$

There are many functional forms that will satisfy Equation 6.2. One rather simple form, also suggested by Lord Kelvin, is

$$g(T_h, T_\ell) = \frac{T_h}{T_\ell}$$

Employing this last form for the function g in Equation 6.1, we obtain

$$\frac{Q_{in}}{Q_{out}} = \frac{T_h}{T_\ell} \tag{6.3}$$

where Q_{in} and Q_{out} are the *magnitudes* of the heat energy involved. From this and with

$$\eta_t = 1 - \frac{Q_{out}}{Q_{in}}$$

for a Carnot engine, it follows that

$$\eta_{Carnot} = 1 - \frac{T_\ell}{T_h} \tag{6.4}$$

Equation 6.3 is now a *definition* of the ratio of two temperatures. This ratio equals the ratio of the quantities of heat energy exchanged by a Carnot engine with two reservoirs at the temperatures T_h and T_ℓ. The temperature scale based on Equation 6.3 is known as the *Kelvin scale*. In determining an unknown temperature, it requires measurement of a quantity of heat rather than of such factors as pressure or volume. A quantity of heat may be measured by measuring input of electrical energy to a reservoir. The Kelvin scale can be completely defined by assigning an arbitrary value to the difference between any two temperatures. For instance, we may select two reservoirs, one at ice point and the other at steam point, operate a Carnot engine between these two reservoirs, and measure the quantities of heat exchanged. We may further arbitrarily assign a numerical value of 100 to the difference between the steam point temperature and the ice point temperature. Equation 6.3 will then enable us to determine the temperatures at the ice point and the steam point on the Kelvin scale.

In practice it would be very difficult to devise a reversible Carnot engine operating between two reference temperatures such as the ice point and the steam point. Instead the ideal gas thermometer, discussed in Chapter 1, is used, and the triple point of water is assigned a value of 273.16 K. On this scale the ice point and the steam point are 0°C and 100°C, or 273.15 K and 373.15 K, respectively.

An interesting conclusion can be drawn from the second law and the expression for the efficiency of the Carnot cycle. According to the second law there cannot be a device which converts thermal energy into work with 100 percent efficiency. For a Carnot cycle, this means

$$\eta_{Carnot} = 1 - \frac{T_{low}}{T_{high}} < 1$$

or

$$T_{low} > 0$$

In other words it is impossible to have an energy reservoir with its temperature equal to absolute zero. The unattainability of absolute zero temperature is sometimes called the *third law of thermodynamics*.

You may wonder at this point if there is a mathematical form of the second law. Such a form does exist, and the property of entropy (S) is involved in that form. The entropy of a substance is a property of that substance and its value is determined by the state of the substance. The Clausius inequality is the basis for establishing entropy as a property.

6.4 CLAUSIUS INEQUALITY

The Clausius inequality is concerned with the quantities of heat energy (Q) that a system undergoing a cyclic process exchanges with reservoirs and with the temperatures (T) of these reservoirs. It states that the *algebraic* sum of the ratios of heat energy extracted to temperature (that is, Q/T) for each of the various reservoirs exchanging heat energy with a system undergoing a cyclic process is less than or equal to zero.

The basis for proving the Clausius inequality is the observation that heat transfer always takes place from high temperatures to low temperatures (the Clausius statement).

Figure 6.6 shows three reservoirs of energy at temperatures T_h, T_i, and T_ℓ. Also shown in the figure is a system S that undergoes a cyclic process, exchanges heat energy with the three reservoirs, and has a specified work output W. We assume that the system receives heat energy from the two reservoirs at T_h and T_i and rejects heat energy to the third reservoir at T_ℓ. We shall prove the Clausius inequality for this system and then extend the analysis to a system exchanging heat energy with a number of reservoirs.

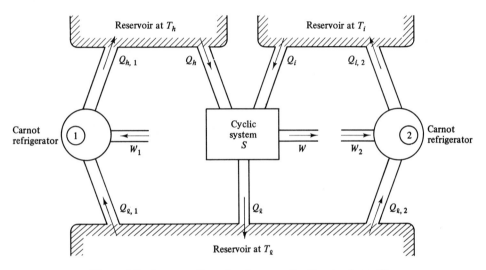

Figure 6.6   A composite system to prove the Clausius inequality.

We construct two Carnot refrigerators, one to operate between T_ℓ and T_h and the other to operate between T_ℓ and T_i. Refrigerator 1 operates in such a way that the amount of heat rejected by refrigerator 1 to the reservoir at temperature T_h numerically equals the amount of heat energy drawn by the system S from the same reservoir. That is,

$$Q_h + Q_{h,1} = 0 \qquad (6.5)$$

and there is no change in the energy level of the reservoir at T_h. Recall that Q is considered positive if heat energy enters a system. With this sign convention, a quantity of heat given up by a reservoir is positive. Thus if numerical values for Q_h and $Q_{h,1}$ were to be supplied, the former will be positive while the latter will be negative. In this paragraph, as well as in the following material, the sign convention for various Q's is determined from the point of view of the system.

In a like manner we have refrigerator 2 working in such a way that

$$Q_i + Q_{i,2} = 0 \qquad (6.5a)$$

Equation 6.5(a) means that the reservoir at T_i gives up as much energy to system S as it gains from refrigerator 2.

Let us examine the effects of one complete cycle executed by the system S and the two refrigerators on the reservoirs and the rest of the surroundings. As far as the two reservoirs at T_h and T_i, respectively, are concerned, there is no change in their thermodynamic states by virtue of Equations 6.5 and 6.5(a). On the other hand, the reservoir at T_ℓ exchanges a net quantity of heat energy, $Q_{\ell,\mathrm{net}}$, with system S and the two refrigerators.

$$Q_{\ell,\mathrm{net}} = Q_\ell + Q_{\ell,1} + Q_{\ell,2}$$

If $Q_{\ell,\mathrm{net}}$ is positive, the reservoir at T_ℓ has lost heat energy on balance, and the combination of system S and the two refrigerators has received heat energy on balance. There is net work, W_{net}, from this combination.

$$W_{\mathrm{net}} = W + W_1 + W_2$$

We now have a situation where there is, in effect, heat exchange with only one reservoir in a cyclic process and there is a net quantity of work. According to the first law

$$Q_{\ell,\mathrm{net}} = W_{\mathrm{net}}$$

As far as the first law is concerned, $Q_{\ell,\mathrm{net}}$ and W_{net} can be positive, negative, or zero. However, the second law will not tolerate all three possibilities. A positive sign for $Q_{\ell,\mathrm{net}}$ and W_{net} means heat energy taken from the reservoir at T_ℓ is completely converted into mechanical work, which is a violation of the Kelvin-Planck statement. Therefore the only acceptable situations are those where the reservoir at T_ℓ neither gains nor loses heat—that is,

$$Q_{\ell,\mathrm{net}} = 0$$

and where the reservoir at T_ℓ receives heat on balance—that is,

$$Q_{\ell,\mathrm{net}} < 0$$

Thus the second law dictates that

$$Q_{\ell,\mathrm{net}} = Q_\ell + Q_{\ell,1} + Q_{\ell,2} \leq 0 \tag{6.6}$$

We now proceed to eliminate $Q_{\ell,1}$ and $Q_{\ell,2}$ from Equation 6.6.

Rewriting Equation 6.3 for the two Carnot refrigerators and noting that the quantities $Q_{h,1}$ and $Q_{i,2}$ are negative, we have

$$\frac{-Q_{h,1}}{Q_{\ell,1}} = \frac{T_h}{T_\ell} \quad \text{and} \quad \frac{-Q_{i,2}}{Q_{\ell,2}} = \frac{T_i}{T_\ell}$$

Solving for $Q_{\ell,1}$ and $Q_{\ell,2}$, we obtain

$$Q_{\ell,1} = -\frac{T_\ell}{T_h} Q_{h,1} \quad \text{and} \quad Q_{\ell,2} = -\frac{T_\ell}{T_i} Q_{i,2}$$

Employing Equations 6.4 and 6.4(a), we have

$$Q_{\ell,1} = +\frac{T_\ell}{T_h} Q_h \quad \text{and} \quad Q_{\ell,2} = +\frac{T_\ell}{T_i} Q_i$$

Inserting these expressions for $Q_{\ell,1}$ and $Q_{\ell,2}$ in Equation 6.6, we get

$$Q_\ell + \frac{T_\ell}{T_h}Q_h + \frac{T_\ell}{T_i}Q_i \leq 0$$

Finally, division by T_ℓ gives

$$\frac{Q_\ell}{T_\ell} + \frac{Q_h}{T_h} + \frac{Q_i}{T_i} \leq 0$$

The above inequality has been established for a system exchanging heat energy with three reservoirs. This was done by introducing two Carnot refrigerators in such a manner that the new system, consisting of the Carnot refrigerators and the original system, was effectively exchanging heat energy with a single reservoir—namely, the one at the lowest temperature—and by invoking the second law.

If we have a system executing a cyclic process and thermally interacting with N reservoirs, we can devise $N - 1$ Carnot refrigerators, apply the same procedure, and show that

$$\sum_{i=1}^{N} \frac{Q_i}{T_{i,\text{reservoir}}} \leq 0 \tag{6.7}$$

When the number of reservoirs becomes infinitely large and a system exchanges a very small quantity of heat energy with each one of them, Equation 6.7 takes the form

$$\oint \frac{\delta Q}{T_\text{reservoir}} \leq 0 \tag{6.8}$$

Equations 6.7 and 6.8 represent the Clausius inequality. The inequality relates to the quantities of heat energy received by a system from energy reservoirs and *their* temperatures when a system thermally interacts with them. The system temperatures do not enter the inequality. Also, reversibility or irreversibility of the interactions is of no consequence to Equations 6.7 or 6.8.

6.5 ENTROPY

The Clausius inequality now can be used to lead to a new property, entropy. Consider a system undergoing a reversible cyclic process, path 1–A–2–B–1 (Figure 6.7) and interacting with a number of energy reservoirs. Inasmuch as the cyclic process is reversible, the system and the reservoirs have essentially identical temperatures at every point in the cycle. In such a case, $T_\text{reservoir}$ in Equation 6.8 can be replaced by the system temperature T. We can then write the Clausius inequality for the system as

$$\oint_{1-A-2-B-1} \frac{\delta Q}{T} \leq 0$$

Now let the system reverse the cycle, that is, follow the path 1–B–2–A–1. For this path we can write

$$\oint_{1-B-2-A-1} \frac{\delta Q}{T} \leq 0$$

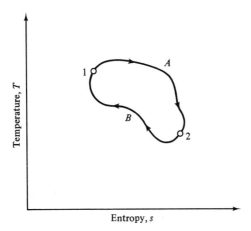

Entropy, s

Figure 6.7 A reversible cyclic process.

The only way to satisfy both these equations is to have

$$\oint_{\substack{\text{reversible} \\ \text{cycle}}} \frac{\delta Q}{T} = 0 \tag{6.9}$$

Equation 6.9 holds only for a system undergoing a reversible cyclic process and has been obtained from the Clausius inequality, which applies to a system interacting with a number of reservoirs.

Next, we note that the cyclic path 1–A–2–B–1 is made of two segments, 1–A–2 and 2–B–1. Equation 6.9 can then be written as

$$\oint_{1-A-2-B-1} \frac{\delta Q}{T} = \int_{1-A-2} \frac{\delta Q}{T} + \int_{2-B-1} \frac{\delta Q}{T} = 0$$

Reversing path 2–B–1 to 1–B–2 and changing the sign of the integrand accordingly gives

$$\int_{1-A-2} \frac{\delta Q}{T} - \int_{1-B-2} \frac{\delta Q}{T} = 0$$

or

$$\int_{1-A-2} \frac{\delta Q}{T} = \int_{1-B-2} \frac{\delta Q}{T}$$

This result means that if we have a *reversible process* taking a system from point 1 to point 2, then the integral of the quantity, $\delta Q/T$, is the same regardless of the path chosen to go from point 1 to point 2, and the value of the integral is dependent solely on states 1 and 2. We saw in Chapter 3 that if a function, when integrated, has the same value regardless of the path of integration so long as the two end points are the same, we are dealing with a point function; the integrand in such case is an exact differential. We represent the quantity $\delta Q/T$ by dS, an exact differential. It is the differential of a new thermodynamic property, called *entropy* (S), of a system.

$$dS = \left(\frac{\delta Q}{T}\right)_{\text{reversible}} \tag{6.10}$$

and

$$S_2 - S_1 = \int_{1 \atop \text{reversible}}^{2} \frac{\delta Q}{T} \tag{6.10a}$$

Entropy, S, is an extensive property and has units of J/K. We use the symbol s to denote specific entropy (J/kg K). Entropy is a function of the thermodynamic state of a system and it is defined only for equilibrium states. Only changes in entropy can be calculated at this point. Fortunately, in most applications, we are concerned only with changes in entropy. Equation 6.10(a) allows us to calculate changes in entropy, provided the process is reversible. We can imagine any number of reversible paths connecting two given state points for a system and the change in entropy would be the same for all those paths. When a system undergoes an irreversible process taking it from state 1 to state 2, we determine the change in the entropy by devising a reversible process, taking the system from state 1 to state 2, and computing ΔS for this reversible process.

6.5.1 T-s *Diagram*

We can rewrite Equation 6.10 on a unit-mass basis to read

$$\delta q = T \, ds \tag{6.10b}$$

This is comparable with the equation for the reversible work done by a gaseous system.

$$\delta W = p \, dv$$

We notice that we can plot a graph of a reversible process on a T-s diagram, with absolute temperature T and specific entropy s as the ordinate and the abscissa, respectively. Such a graph for an arbitrary reversible process is shown in Figure 6.8(a). If we take a narrow vertical strip between the curve and the abscissa, we find that its area is $T \, ds$. By virtue of Equation 6.10(b), the area also represents a small quantity of heat δq supplied to the system. The area under the entire curve between points 1 and 2 is $\int_1^2 T \, ds$, or the total quantity of heat energy supplied to the system per unit mass during the reversible process. This interpretation for the area under a curve on a T-s diagram is useful in applications. Note that for an area on a T-s diagram to represent the amount of heat supplied, the process must be reversible.

In Figure 6.8(b) there is a vertical line connecting points 1 and 2. Since the entropy does not change in process 1–2—that is, $\Delta s = 0$—the area under the line is zero. Therefore, from Equation 6.10(b), we conclude that $q = 0$, or no heat transfer takes place during the process. We know that such a process is called an *adiabatic process*. But an adiabatic process may or may not be reversible, while the process 1–2 in Figure 6.8(b) is reversible. Therefore we call a reversible adiabatic process a *constant entropy process*, or *an isentropic process*.

All of the processes that we represented earlier on p-v and T-v diagrams can be represented on a T-s diagram. Curve 3–4 in Figure 6.8(b), for example, represents a constant volume process. Recall that it appears as a vertical straight line on a p-v

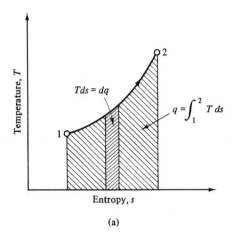

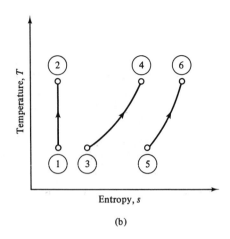

Figure 6.8
(a) Interpretation of the area under a curve on a *T-s* diagram.
(b) Reversible adiabatic, constant volume, and constant pressure processes on a *T-s* diagram.

diagram (see Figure 1.14(a)). Likewise, curve 5–6 represents a constant pressure expansion, which appears as a horizontal line on a *p-v* diagram (see Figure 1.14(a)). A constant pressure line on a *T-s* diagram has a steeper slope than a constant volume line on the same diagram.

We shall next reconstruct the Carnot cycle on a *T-s* diagram.

6.5.2 The Carnot Cycle

The Carnot cycle consists of two reversible constant temperature processes and two reversible adiabatic, or isentropic, processes. The *T-s* diagrams for the Carnot cycle are shown in Figure 6.9.

Consider a piston-cylinder system containing a gas for the execution of the Carnot heat engine [Figure 6.9(a)]. The thermodynamic state of this gas is represented by point 1 on the *T-s* diagram. Let an energy reservoir at temperature T_h supply heat to the gas reversibly so that the gas now has a new state, state 2. Then allow the gas to undergo an isentropic expansion ($Q = 0$) to state 3. Now let the gas reject heat to another reservoir at temperature T_ℓ in a reversible constant temperature process until state 4 is reached. Finally, we return the gas to its initial state 1 by subjecting the gas to isentropic compression ($Q = 0$).

The thermal efficiency of a power cycle is the ratio of the net work of the cycle to the heat energy supplied to the cycle. In view of the first law, the net work output of a cycle equals the net heat input to the cycle. Or

$$W_{\text{cycle}} = Q_{\text{in}} - Q_{\text{out}}$$

where Q_{in} and Q_{out} are the magnitudes of the heat energy supplied to and rejected by the cycle, respectively.

236

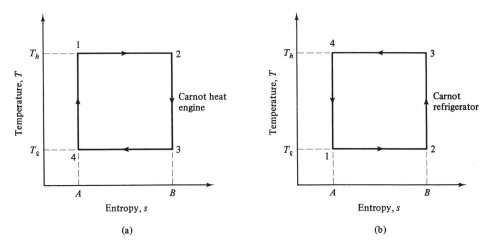

Figure 6.9 *T-s* diagram for (a) Carnot heat engine, and (b) Carnot refrigerator.

In view of Equation 6.10(b)

$$Q_{in} = \text{area } 1\text{–}2\text{–}B\text{–}A\text{–}1 \text{ in Figure 6.9(a)}$$
$$= T_h(S_2 - S_1)$$

and

$$Q_{out} = \text{area } 3\text{–}B\text{–}A\text{–}4\text{–}3 \text{ in Figure 6.9(a)}$$
$$= T_\ell(S_3 - S_4)$$

Since processes 2–3 and 4–1 are isentropic, we can write

$$S_3 = S_2 \quad \text{and} \quad S_4 = S_1$$

so that

$$Q_{out} = T_\ell(S_2 - S_1)$$

The net work of the cycle is then given by

$$W_{cycle} = Q_{in} - Q_{out}$$
$$= T_h(S_2 - S_1) - T_\ell(S_2 - S_1)$$
$$= (T_h - T_\ell)(S_2 - S_1)$$

The thermal efficiency of the cycle, η_t, is

$$\eta_t = \frac{\text{net work output}}{\text{gross heat supply}} = \frac{W_{cycle}}{Q_{in}}$$
$$= \frac{(T_h - T_\ell) \times (S_2 - S_1)}{T_h(S_2 - S_1)}$$

or

$$\eta_{Carnot} = \frac{T_h - T_\ell}{T_h} = 1 - \frac{T_\ell}{T_h}$$

which is exactly the same expression as the one derived earlier in this chapter (Equation 6.4) and the one obtained in Chapter 3 (Equation 3.17(a)).

Earlier in this chapter we also used the Carnot cycle as a refrigerator. It is of interest to derive an expression for the coefficient of performance (COP) of a Carnot refrigerator, extracting heat from a reservoir at T_ℓ and rejecting heat to a reservoir at T_h. We have

$$(\text{COP})_R = \frac{\text{heat extracted from the low–temperature reservoir}}{\text{work supplied to the refrigerator}} \qquad (6.11)$$

For a Carnot refrigerator, the cycle (see Figure 6.9(b)) is traced counterclockwise. Considering only the magnitudes of quantities of heat involved, we have

$$\text{Heat extracted} = T_\ell(S_2 - S_1)$$

$$\text{Heat rejected} = T_h(S_3 - S_4) = T_h(S_2 - S_1)$$

$$\text{Work supplied} = \text{heat rejected} - \text{heat extracted}$$

$$= T_h(S_2 - S_1) - T_\ell(S_2 - S_1)$$

$$= (T_h - T_\ell)(S_2 - S_1)$$

The COP is then given by

$$(\text{COP})_R = \frac{T_\ell(S_2 - S_1)}{(T_h - T_\ell) \times (S_2 - S_1)}$$

or

$$(\text{COP})_{R,\text{Carnot}} = \frac{T_\ell}{T_h - T_\ell} \qquad (6.11a)$$

The thermodynamic principle of a heat pump is the same as that of a refrigerator. The practical objective of a heat pump, however, is to supply heat to a high-temperature reservoir, in contrast to that of a refrigerator, which is to extract heat from a low-temperature reservoir. This is accomplished by using mechanical work to extract heat from a low temperature reservoir. The coefficient of performance of a heat pump is given by

$$(\text{COP})_{\text{HP}} = \frac{\text{heat supplied to the high-temperature reservoir}}{\text{work supplied to the heat pump}} \qquad (6.12)$$

Thus with reference to Figure 6.9

$$(\text{COP})_{\text{HP}} = \frac{T_h(S_2 - S_1)}{(T_h - T_\ell) \times (S_2 - S_1)}$$

or

$$(\text{COP})_{\text{HP,Carnot}} = \frac{T_h}{T_h - T_\ell} \qquad (6.12a)$$

Example 6.1

Calculate the change in entropy of a gas of mass 8 kg when 100 kJ of heat energy is supplied to the gas. The temperature of the gas remains constant at 100°C during the heat transfer process.

Solution:

Given: A gaseous system receives heat energy in a constant temperature process; $m = 8$ kg, $Q_{1-2} = 100$ kJ, $t_1 = 100°C$, $t_2 = 100°C$.

Objective: The change in the entropy of the system is to be determined.
Diagram:

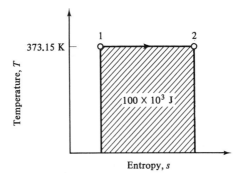

Figure 6.10   *T-s* diagram.

Assumptions: The heat transfer process is reversible.

Relevant physics: The solution to this problem involves a direct application of Equation 6.10(a):

$$S_2 - S_1 = \int_1^2 \frac{\delta Q}{T}$$ (6.10a)

For the given process (Figure 6.10), T is constant and

$$S_2 - S_1 = \frac{1}{T} \int_1^2 \delta Q = \frac{Q_{1-2}}{T}$$

Analysis: We are given that $t_1 = t_2 = t = $ constant $= 100°C$, or $T = 373.15$ K. Substituting the values of Q_{1-2} and T, we obtain

$$S_2 - S_1 = \frac{100 \times 10^3}{373.15}$$

$$= 268 \text{ J/K}$$

The change in the specific entropy is given by

$$\Delta s = \frac{S_2 - S_1}{m}$$

$$= \frac{268}{8}$$

$$= 33.5 \text{ J/kg K}$$

Comments: This example illustrates that a supply of heat to a system is accompanied by an increase of its entropy.

Example 6.2

A mass of 4 kg of liquid water is cooled from $90°C$ to $10°C$ under a constant pressure. Determine the change in its entropy, assuming the process to be reversible.

Solution:

Given: A liquid-water system is cooled in a constant pressure process; $m = 4$ kg, $t_1 = 90°C$, $t_2 = 10°C$.

Objective: The change of entropy of the water is to be determined.

Diagram:

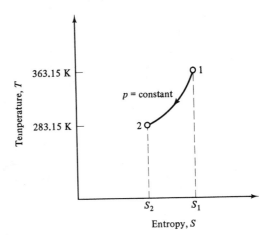

Figure 6.11

Assumptions:

1. The process is reversible.
2. The specific heat of liquid water is constant at 4.18 kJ/kg K.

Relevant physics: For a constant pressure process, from Equation 3.9(a),

$$\delta Q_p = mc_p \, dT$$

Hence, for the entropy change, we can write

$$S_2 - S_1 = \int_{T_1}^{T_2} \frac{\delta Q}{T} = \int_{T_1}^{T_2} \frac{mc_p dT}{T}$$

or

$$S_2 - S_1 = mc_p \int_{T_1}^{T_2} \frac{dT}{T} = mc_p \ln \frac{T_2}{T_1}$$

Analysis: With

$$m = 4 \text{ kg}, \qquad T_2 = 273.15 + 10 = 283.15 \text{ K},$$
$$\text{and} \quad T_1 = 273.15 + 90 = 363.15 \text{ K}$$

we have

$$S_2 - S_1 = (4)(4.18 \times 10^3) \ln \left(\frac{283.15}{363.15}\right) = -4160.6 \text{ J/K}$$

Comments: This example demonstrates that the cooling of a system causes a decrease in its entropy. This is true for reversible, as well as irreversible, cooling processes.

Example 6.3

A mass of 10 kg of liquid water at 90°C is mixed with 20 kg of liquid water at 10°C. Determine the change in the entropy of water.

Solution:

Given: Our system consists of 10 kg of water initially at 90°C and 20 kg initially at 10°C.

$$m_1 = 10 \text{ kg}, \qquad T_1 = 273.15 + 90 = 363.15 \text{ K}$$
$$m_2 = 20 \text{ kg}, \qquad T_2 = 273.15 + 10 = 283.15 \text{ K}$$

Objective: The entropy change due to mixing is to be calculated.

Diagram:

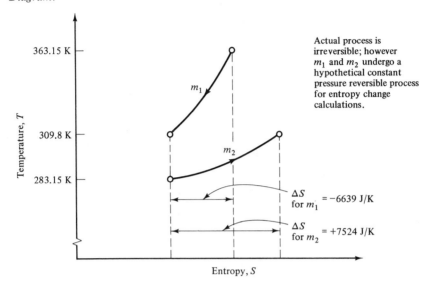

Figure 6.12

Assumptions:

1. The work done in the mixing process is negligible.
2. There is no heat transfer from the system consisting of m_1 and m_2 to the surroundings.
3. The specific heat of liquid water is constant at 4.18 kJ/kg K.
4. The value of the internal energy of water at 0 K is zero.

Relevant physics: We have a system undergoing an irreversible mixing process with known initial state. The final state can be determined by applying the first law. In order to calculate the entropy change, we must devise a reversible process taking the system from the same initial state to the same final state. The change in entropy for the reversible process equals the change in entropy for the actual process. The T-S diagram shown in Figure 6.12 depicts the history for the reversible process.

Analysis: In view of the assumptions made,

$$\cancel{Q}^{0} - \cancel{W}^{0} = \Delta U = U_f - U_i$$

where U_f and U_i represent the final and the initial internal energy of our system. Thus

$$U_f = U_i$$

If T_f is the final temperature of the mixture, we can write

$$\int_0^{T_f} m_1 c_v \, dT + \int_0^{T_f} m_2 c_v \, dT = \int_0^{T_1} m_1 c_v \, dT + \int_0^{T_2} m_2 c_v \, dT$$

or

$$m_1 c_v T_f + m_2 c_v T_f = m_1 c_v T_1 + m_2 c_v T_2$$

Solving for the final temperature T_f, we obtain

$$T_f = \frac{m_1 T_1 + m_2 T_2}{m_1 + m_2} \tag{6.13}$$

Substitution of the known values gives

$$T_f = \frac{10 \times 363.15 + 20 \times 283.15}{10 + 20} = 309.8 \text{ K}$$

Next, we assume that the masses of water, m_1 and m_2, are brought to their final temperature of T_f by reversible constant pressure processes with the aid of suitable energy reservoirs. We can then write (see Example 6.2)

$$\text{For } m_1: \quad (\Delta S) = m_1 c_p \ln \frac{T_f}{T_1}$$

$$= 10 \times 4.18 \times 10^3 \ln \left(\frac{309.8}{363.15} \right)$$

$$= -6639 \text{ J/K}$$

$$\text{For } m_2: \quad (\Delta S) = m_2 c_p \ln \frac{T_f}{T_2}$$

$$= 20 \times 4.18 \times 10^3 \ln \left(\frac{309.8}{283.15} \right)$$

$$= +7524 \text{ J/K}$$

If we were to compute the change in the entropy of the reservoirs that helped us bring about a reversible change in the temperature of m_1, we would find that the change is $+6639$ J/K. Likewise, the change in the entropy of the reservoirs that helped us bring about a reversible change in the temperature of m_2 would be -7524 J/K. The total change in the entropy of the universe would be

$$(\Delta S)_{\text{universe. rev.}} = (-6639 + 7524) + (6639 - 7524)$$

$$= 0$$

In the actual problem on hand, the change in temperature is brought about in an irreversible manner, and there are no reservoirs. The change in the entropy of the universe is then

$$(\Delta S)_{\text{universe, irrev.}} = (-6639) + (7524)$$

$$= 885 \text{ J/K}$$

Comments: In this problem both mass m_1 and mass m_2 are part of the system. There is transfer of heat from m_1 to m_2, but it is internal to the system. Since there is no heat transfer across the boundary of the system, the process is adiabatic. This example illustrates that an irreversible process results in an increase in the entropy of the universe or in a production of entropy. For the first time you have come across a property that is not conserved, but is created in a process.

6.5.3 Entropy and an Irreversible Process

We remarked previously that in order to calculate the change in the entropy of a system undergoing an irreversible process, we devise a reversible process connecting the same end states and calculate the change in entropy for the devised reversible process. Whenever there is a reversible heat transfer, and it is present in all reversible processes except the adiabatic one, there is at least one energy reservoir involved. A reversible heat transfer causes changes in entropy of both the system and the reservoir. These changes are equal in magnitude but opposite in sign. In an irreversible heat transfer process, there is a finite source of energy instead of an energy reservoir, and the temperature difference during the heat transfer process is finite. The finite temperature difference results in a greater numerical entropy change for the body that receives heat energy than that for the body imparting its thermal energy. This is due to the fact that temperature appears in the denominator of the integrand, $\delta Q/T$ in Equation 6.10(a). Since the body receiving heat has temperature that is less than the temperature of the body imparting heat, the entropy change for the body receiving heat is positive, and the entropy change for the body imparting heat is negative. Also, the former is numerically greater than the latter. Consequently, the change in the entropy of the universe, which is the algebraic sum of these two changes, is positive in an irreversible heat transfer process. This is proved as follows.

Consider a reversible cycle (see Figure 6.13) taking a system from state 1 to state 2 along path 1–A–2 and then bringing it back to state 1 along path 2–B–1. The Clausius inequality for this reversible cycle becomes

$$\oint_{1-A-2-B-1} \frac{\delta Q}{T} = 0$$

or

$$\int_{1-A-2} \frac{\delta Q}{T} + \int_{2-B-1} \frac{\delta Q}{T} = 0$$

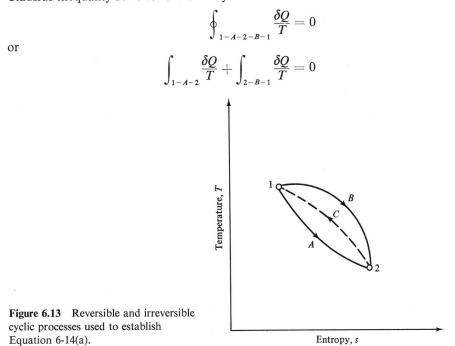

Figure 6.13 Reversible and irreversible cyclic processes used to establish Equation 6-14(a).

Now suppose that the system is brought back from state 2 to state 1 by some irreversible process. This is represented by the dotted line 2–C–1 in Figure 6.13. The Clausius inequality for the irreversible cycle 1–A–2–C–1 becomes

$$\oint_{1-A-2-C-1} \frac{\delta Q}{T_{\text{reservoir}}} < 0$$

or

$$\int_{1-A-2} \frac{\delta Q}{T} + \int_{2-C-1} \frac{\delta Q}{T_{\text{reservoir}}} < 0$$

If we subtract the last equation on page 243 from the above inequality, we obtain

$$\int_{2-C-1} \frac{\delta Q}{T_{\text{reservoir}}} - \int_{2-B-1} \frac{\delta Q}{T} < 0$$

The second integral in the above inequality is for the reversible process 2–B–1 and therefore can be replaced by ΔS or $S_1 - S_2$. Note that the starting point for this path is 2 and the terminal point is 1. Then the foregoing inequality becomes

$$\int_{2-C-1} \frac{\delta Q}{T_{\text{reservoir}}} - (S_1 - S_2) < 0$$

or

$$(S_1 - S_2) > \int_{2-C-1} \frac{\delta Q}{T_{\text{reservoir}}}$$

The above inequality means that if we have two processes taking a system from one point to another, one process reversible and the other irreversible, then the change of entropy in the ensuing change of state is greater than the integral of the quantity $(\delta Q/T)$ for the irreversible process. It should be noted that the T in the integrand is the temperature of the external heat source (or sink). In fact, since the entropy change is a function of end states only, this result means that in any irreversible process, the change in entropy is always greater than the integral of the quantity $\delta Q/T$. Thus for an irreversible process we write

$$dS > \frac{\delta Q}{T} \tag{6.14}$$

For a system undergoing an irreversible process from state 1 to state 2,

$$S_2 - S_1 > \int_1^2 \frac{\delta Q}{T} \qquad \text{irreversible} \tag{6.14a}$$

For a reversible process we know that

$$dS = \frac{\delta Q}{T} \tag{6.10}$$

and that

$$S_2 - S_1 = \int_1^2 \frac{\delta Q}{T} \qquad \text{reversible} \tag{6.10a}$$

Next, consider an isolated system. An isolated system is one that does not interact with its surrounding in any manner; that is, $\delta Q = 0$ for an isolated system. Combining these results we can write, for an isolated system,

$$(S_2 - S_1) \geq 0$$

or

$$(\Delta S)_{\substack{\text{isolated} \\ \text{system}}} \geq 0 \qquad\qquad (6.14b)$$

This equation is considered yet another expression of the second law of thermodynamics. If the entropy of an isolated system does not change, then its state does not change, and it undergoes a reversible process. If the isolated system undergoes an irreversible process, then its entropy increases.

6.5.4 Lost Work in an Irreversible Process

When heat energy is transferred with a finite temperature difference, an irreversible process results. When a gas expands in an unrestricted fashion, an irreversible process also results. Irreversible processes of the first type were discussed in the earlier sections. Now we shall examine the effect of the latter type of irreversibility. In our discussions of work in Chapter 2, we always required a restraining force on a spring, a gas, or a piston equal in magnitude to the internal force. Motion was permitted due to an infinitesimal imbalance between the internal force and the external force, and we had a quasistatic process producing work. The system was essentially in thermodynamic equilibrium during the work interactions. Now we ask this question: How is the work interaction affected if the system is not in equilibrium during a process? This question was answered in Section 2.4 by introducing lost work. There it was stated that when a system undergoes a nonquasistatic process, a potential to do work is lost forever. We are now better equipped to discuss lost work in view of the second law and entropy.

Consider a gas confined in the left part of a closed chamber and separated from the right part by a partition such as a membrane (see Figure 6.14(a)). We shall treat the entire chamber as our system. Let the right part of the chamber be evacuated. Then let the partition be removed. After a while the entire system is filled with the gas. We observe that the system boundary does not move, and there is no work done. An application of the first law to this irreversible process gives

$$\delta Q - \cancel{\delta W}^{\;0} = dU$$

We assume that sufficient time elapses so that the temperature of the gas returns to the original value due to heat transfer between the system and its surroundings.

Now suppose that the same gas is held in place by a frictionless piston (see Figure 6.14(b)) with vacuum on the right side of the piston. Let there be a reservoir at $T + dT$ supplying heat to the gas and causing it to expand isothermally. As the piston moves out very slowly, with the restraining force on the piston appropriately decreasing to accomodate the reduced gas pressure, we have a reversible isothermal

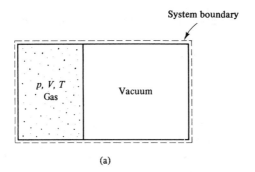

System boundary

(a)

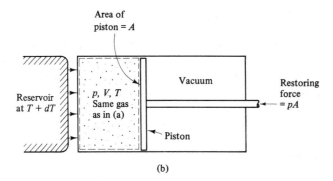

Area of piston = A

(b)

Figure 6.14
(a) Free expansion of a gas.
(b) Controlled expansion of a gas.

process with some work output. For this process we have

$$\delta Q - \delta W = dU$$

Since this is a reversible process,

$$(\delta Q)_{rev} = T\,dS \quad \text{and} \quad (\delta W)_{rev} = p\,dV$$

In both of these processes, the initial and the final temperatures are the same and the initial and the final volumes are the same, so the pressures are the same. With identical initial and final states, the irreversible process produces no work, and the reversible process produces maximum possible work. The amount of work that is irretrievably lost is called *lost work* (W_{lost}). The relationship between the reversible work, the actual work in an irreversible process, and the lost work is

$$(\delta W)_{rev} = (\delta W)_{irrev} + \delta(W_{lost}) \tag{6.15}$$

In Figure 6.10(b) if the piston were not frictionless, the actual work output would be less than $(\delta W)_{rev}$, the difference being the lost work. The lost work is dissipated as heat energy and transferred to the surroundings.

From the first law, we have

$$\delta Q - \delta W = dU$$

From the second law for a reversible process we have

$$T\,dS = (\delta Q)_{rev}$$

Combining the two equations, we get

$$T\,dS = (\delta W)_{\text{rev}} + dU \tag{6.16}$$

The above equation is known as the *combined first and second law* for a reversible process. Substituting Equation 6.15 in Equation 6.16, we obtain

$$T\,dS = (\delta W)_{\text{irrev}} + \delta(W_{\text{lost}}) + dU$$

In an actual irreversible process between the same end states as those for a reversible process, the energy balance is again governed by the first law and we have

$$(\delta Q)_{\text{irrev}} - (\delta W)_{\text{irrev}} = dU$$

Eliminating dU from the two preceding equations, we obtain

$$T\,dS = (\delta Q)_{\text{irrev}} + \delta(W_{\text{lost}})$$

and

$$dS = \frac{(\delta Q)_{\text{irrev}}}{T} + \frac{\delta(W_{\text{lost}})}{T} \tag{6.17}$$

The above result gives us an equality for an irreversible process. For a reversible process, the lost work is zero and the subscript *irrev* on δQ is replaced by the subscript *rev*, and we have

$$dS = \frac{(\delta Q)_{\text{rev}}}{T}$$

In our future use of Equation 6.17, we shall drop the subscript on δQ, since the term $\delta(W_{\text{lost}})$ will serve to remind us that we are dealing with an irreversible process.

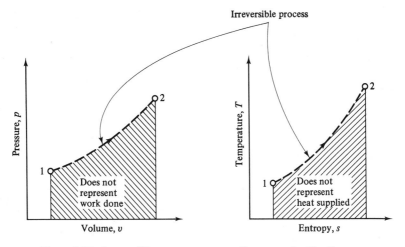

Figure 6.15   Irreversible process on a *p-v* diagram and a *T-s* diagram.

6.5.5 Representation of an Irreversible Process

When a system undergoes a reversible process, the system is in thermal equilibrium at every state of the process. A solid line is usually drawn on a p-v or a T-s diagram to represent such a process, with every point on the line representing a thermodynamic state through which the system actually passes. In an irreversible process, on the other hand, we are sure only of the initial and the final thermodynamic states. At no point is the system in a thermodynamic equilibrium during an irreversible process. A solid line on a p-v or a T-s diagram cannot represent the actual state of affairs during an irreversible process. Therefore it is customary to show an irreversible process by a dotted or dashed line connecting the two end states (see Figure 6.15). The area under such a dotted curve has no physical meaning. It represents neither work on a p-v diagram nor heat energy on a T-s diagram. It is only for a reversible process that the areas under p-v or T-s diagrams represent work or heat energy.

Example 6.4

A mass of 2.5 kg of saturated liquid water at atmospheric pressure and 100°C is completely evaporated into saturated vapor. The heat energy is supplied to the water by a source of heat that is at 1000°C. Determine the change in the entropy of the water.

Solution:

Given: A mass of 2.5 kg of saturated liquid water at 100°C is vaporized using a high temperature heat source at 1000°C.

Objective: The entropy change of water is to be determined.

Diagram:

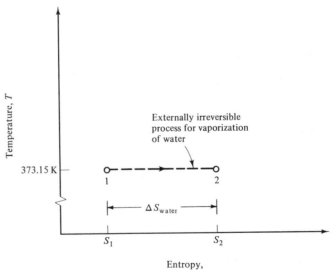

Figure 6.16

Assumptions: None.

Relevant physics: The process is externally irreversible, since there is a temperature difference of 1000 − 100 or 900°C between the water and the source of heat.

248

In order to calculate the change of entropy of water, we hypothesize a reversible process where an energy reservoir with a temperature infinitesimally higher than 100°C supplies heat to the water during the process of evaporation. This is a constant pressure process and

$$Q = \Delta H$$

Then we have

$$(\Delta S)_{water} = \int \frac{\delta Q}{T} = \frac{Q}{T}$$

Analysis: From Appendix B2.1, we find the latent heat of vaporization for water at atmospheric pressure and 100°C is

$$h_{fg} = 2257 \times 10^3 \text{ J/kg}$$

Also

$$\Delta H = mh_{fg} = 2.5 \times 2257 \times 10^3 \text{ J}$$

and

$$Q = \Delta H = 5642.5 \times 10^3 \text{ J}$$

For our *hypothesized* reversible constant pressure, constant temperature process

$$(\Delta S)_{water} = \frac{Q}{T} = \frac{5642.5 \times 10^3}{273.15 + 100} = 15,121 \text{ J/K}$$

This is also the actual change of entropy of the water since the *end states* for the *hypothesized* and actual processes are the same.

Comments: In this assumed reversible process, the change in the entropy of the heat source, which is essentially at 100°C, is

$$(\Delta S)_{\substack{source \\ ideal}} = \int \frac{\delta Q}{T} = \frac{-5642.5 \times 10^3}{273.15 + 100}$$

$$= -15,121 \text{ J/K}$$

In the *actual* irreversible process the temperature of the source is 1000 C

and

$$(\Delta S)_{\substack{source \\ actual}} = \int \frac{\delta Q}{T} = \frac{-5642.5 \times 10^3}{273.15 + 1000} = -4432 \text{ J/K}$$

which is less than the ΔS for the water. Note that δQ in the above integral is the quantity of heat received by the reservoir which is a negative quantity and T is the temperature of the heat source.

6.6 ENTROPY CHANGE FOR IDEAL GAS

For a gaseous system undergoing a reversible process

$$(\delta W)_{rev} = p \, dV$$

and Equation 6.16 becomes

$$T \, dS - p \, dV = dU$$

For an ideal gas

$$dU = mc_v \, dT$$

and the $T\,dS$ equation becomes

$$T\,dS - p\,dV = mc_v\,dT$$

A rearrangement gives

$$dS = mc_v\frac{dT}{T} + \frac{p}{T}\,dV$$

Noting that $p/T = mR/V$ for an ideal gas, we obtain

$$dS = mc_v\frac{dT}{T} + mR\frac{dV}{V}$$

and

$$S_2 - S_1 = \Delta S = m\int_1^2 c_v\frac{dT}{T} + mR\int_1^2\frac{dV}{V} \qquad (6.18)$$

For constant properties, Equation 6.18 can be readily integrated to yield

$$S_2 - S_1 = \Delta S = mc_v\ln\left(\frac{T_2}{T_1}\right) + mR\ln\left(\frac{V_2}{V_1}\right) \qquad (6.18a)$$

This equation expresses the change in entropy of an ideal gas with constant properties in terms of the initial and final values of volumes and temperatures. Since pressure, volume, and temperature are related by the equation of state, we can express the right side of Equation 6.18 in terms of temperature and pressure or volume and pressure. For an ideal gas,

$$pV = mRT$$

Taking the natural logarithm of both sides of this equation, we get

$$\ln p + \ln V = \ln(mR) + \ln T$$

For differential changes in p, V, and T the above becomes

$$\frac{dp}{p} + \frac{dV}{V} = \frac{dT}{T} \qquad (6.19)$$

Substituting for dT/T from Equation 6.19 in Equation 6.18 gives

$$\Delta S = mR\int_1^2\frac{dV}{V} + m\int_1^2 c_v\left(\frac{dp}{p} + \frac{dV}{V}\right)$$

or

$$\Delta S = m\int_1^2 c_p\frac{dV}{V} + m\int_1^2 c_v\frac{dp}{p} \qquad (6.20)$$

where $R + c_v$ has been replaced by c_p. For constant properties, Equation 6.20 becomes

$$S_2 - S_1 = \Delta S = mc_p\ln\left(\frac{V_2}{V_1}\right) + mc_v\ln\left(\frac{p_2}{p_1}\right) \qquad (6.20a)$$

If we eliminate dV/V from Equations 6.18 and 6.20, we can readily obtain the following result

$$\Delta S = m\int_1^2 c_p\frac{dT}{T} - mR\int_1^2\frac{dp}{p} \qquad (6.21)$$

Once again for constant properties, Equation 6.21 takes the form

$$S_2 - S_1 = \Delta S = mc_p \ln \left(\frac{T_2}{T_1}\right) - mR \ln \left(\frac{p_2}{p_1}\right) \tag{6.21a}$$

We note that in Equations 6.18, 6.20, and 6.21, the specific heat appears inside the integral signs. Functional relationships between specific heat at constant pressure at zero pressure and absolute temperature are given in Appendix H for a number of ideal gases. These can be used while evaluating the integrals in the above equations. Evaluating these integrals in this manner, however, is a tedious task. To circumvent this difficulty, the following procedure is used.

Entropy at a reference temperature T_0 is taken as zero. Pressure is assumed to be constant at 100 kPa. Then the second integral in Equation 6.21 vanishes, and the equation for a unit mass reduces to

$$\Delta s = \int_{T_0}^{T} c_p \frac{dT}{T}$$

Since the pressure is very low (100 kPa), we can replace c_p by c_{p0}, the zero-pressure, constant pressure specific heat in the above equation. Also, Δs becomes $s - 0$, or s. Since we are using c_{p0}, we use a superscript 0 on s.

The equation then becomes

$$s^0(T) = \int_{T_0}^{T} c_{p0} \frac{dT}{T} \tag{6.22}$$

The function $s^0(T)$, which is a function of temperature alone, is known as *standard-state entropy*; it is tabulated in Appendix I1 for a number of gases, including air. The values listed are on a molal basis. When we want to determine the entropy change between any two state points involving moderately low pressures, we have (from Equation 6.21) for a unit mass

$$\Delta s = \int_{T_1}^{T_2} c_p \frac{dT}{T} - R \int_{p_1}^{p_2} \frac{dp}{p}$$

$$= \int_{T_0}^{T_2} c_{p0} \frac{dT}{T} - \int_{T_0}^{T_1} c_{p0} \frac{dT}{T} - R \int_{p_1}^{p_2} \frac{dp}{p}$$

or

$$s_2 - s_1 = s^0(T_2) - s^0(T_1) - R \ln \left(\frac{p_2}{p_1}\right) \tag{6.23}$$

Equation 6.23, together with the values of $s^0(T)$ listed in Appendix I1 for different gases, enables us to calculate the change in entropy for an ideal gas.

Example 6.5

Air is compressed from 100 kPa to 500 kPa in a reversible isothermal process. Determine the change in the specific entropy of the air.

Solution:

Given: Air undergoes a reversible isothermal process.

$$\text{Initial state: } p_1 = 100 \text{ kPa}$$

$$\text{Final state: } p_2 = 500 \text{ kPa}$$

Objective: The change in the specific entropy of the air is to be determined.
Diagram:

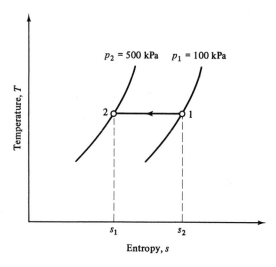

$p_2 = 500$ kPa $p_1 = 100$ kPa

2   1

s_1 s_2

Entropy, s **Figure 6.17**

Temperature, T

Assumptions: The air can be treated as an ideal gas.

Relevant physics: To calculate the change in entropy, we could start with the equation

$$ds = \delta q/T$$

We could calculate the amount of heat supplied by noting that $\Delta u = 0$ for an ideal gas undergoing an isothermal process, by calculating the work done, and by applying the first law. However, since the temperature of the isothermal process is not specified, this procedure cannot be used. For this problem, we can advantageously use Equation 6.21, which for our problem becomes

$$\Delta s = \int_1^2 c_p \frac{\overset{0}{dT}}{T} - R \ln \frac{p_2}{p_1}$$

Analysis: Noting that for air $R = 0.287$ kJ/kg K, we have

$$\Delta s = -0.287 \times 10^3 \ln \left(\frac{500 \times 10^3}{100 \times 10^3}\right)$$

$$= -461.9 \text{ J/kg K}$$

Comments: Isothermal compression involves removal of heat energy and, therefore, a decrease in the entropy of the gas being compressed. The reservoir which receives this heat energy under reversible isothermal conditions undergoes a change in entropy that is positive and equal in magnitude to the change experienced by the gas. The quantity of heat removed will be a function of the temperature of the reservoir or the temperature at which compression is carried out.

Example 6.6

Carbon dioxide gas expands from an initial state at 750 kPa and 30°C in a polytropic process, $pV^{1.3} = $ constant. The final pressure is 120 kPa. Determine the change in the specific entropy of the gas.

Solution:

Given: Carbon dioxide gas expands polytropically.

Initial state: $p_1 = 750 \text{ kPa}$, $t_1 = 30°C$

Final state: $p_2 = 120 \text{ KPa}$

Process equation: $pV^{1.3} = \text{constant}$

Objective: The change in the specific entropy of the carbon dioxide gas is to be evaluated.

Diagram:

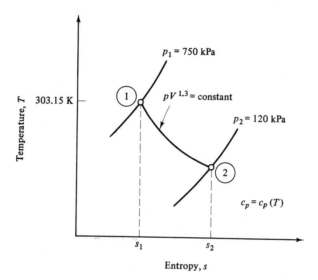

Figure 6.18

Assumptions:

1. The polytropic process is reversible.

2. The gas behaves ideally.

Relevant physics: The equation for the polytropic process and the given properties of the gas can be used to determine the final temperature. With the pressures and temperatures at the end states known, we can use Equation 6.21 to determine Δs by assuming a constant c_p-value. Alternately, Equation 6.23, which permits a variation in c_p, may be used to calculate Δs.

Analysis: The ideal gas and polytropic process relationships for the end states can be written as

$$\frac{p_1 V_1}{p_2 V_2} = \frac{T_1}{T_2} \quad \text{and} \quad p_1 V_1^{1.3} = p_2 V_2^{1.3}$$

or

$$\frac{V_1}{V_2} = \frac{T_1}{T_2} \cdot \frac{p_2}{p_1} \quad \text{and} \quad \frac{V_1}{V_2} = \left(\frac{p_2}{p_1}\right)^{1/1.3}$$

Combining the two equations, we obtain

$$\left(\frac{p_2}{p_1}\right)^{1/1.3-1} = \frac{T_1}{T_2} \quad \text{or} \quad \frac{T_2}{T_1} = \left(\frac{p_2}{p_1}\right)^{1-1/1.3}$$

We are given that

$$p_1 = 750 \text{ kPa} \qquad T_1 = 273.15 + 30 = 303.15 \text{ K}$$
$$p_2 = 120 \text{ kPa}$$

so that

$$T_2 = 303.15\left(\frac{120}{750}\right)^{0.3/1.3} = 198.6 \text{ K}$$

From Equation 6.21,

$$\Delta s = \int_{303.15}^{198.6} c_p \frac{dT}{T} - R \int_{750}^{120} \frac{dp}{p}$$

For carbon dioxide, CO_2, assuming c_p to be constant,

$$c_p = 841.8 \text{ J/kg K} \quad \text{and} \quad R = 188.92 \text{ J/kg K}$$

and

$$\Delta s = 841.8 \ln\left(\frac{198.6}{303.15}\right) - 188.92 \ln\left(\frac{120}{750}\right)$$
$$= -9.82 \text{ J/kg K}$$

Alternately, if we wish to account for the variation in specific heat with temperature, we can proceed with Equation 6.23:

$$s_2 - s_1 = s^0(T_2) - s^0(T_1) - R \ln\left(\frac{p_2}{p_1}\right) \tag{6.23}$$

$$= \frac{s^0(198.6) - s^0(303.15)}{44.01} - 188.92 \ln\left(\frac{120}{750}\right)$$

$$= \frac{(199.68 - 214.38)10^3}{44.01} - 188.92 \ln(0.16)$$

$$= +12.23 \text{ J/kg K}$$

Comments: It is interesting to observe that, in this example, the assumption of constant specific heat gives a decrease in entropy, or a heat removal, while accounting for the variation in specific heat gives an increase of entropy, or a supply of heat to the system.

6.6.1 Entropy of a Pure Substance

The property tables for various substances such as water, Freon, and ammonia list the entropy values for these substances. For convenience, the entropy of water at 0.01°C is considered zero. For most refrigerants the entropy at −40°C is considered zero. The entropy of a liquid-vapor mixture can be calculated in the same way that the enthalpy or the specific volume of a mixture is calculated (See Section 4.2.2). The entropy of a liquid-vapor mixture is given by

$$s = xs_g + (1-x)s_f = s_f + xs_{fg} \tag{6.24}$$

where x is the quality of the mixture. The subscripts f, g, and fg in Equation 6.24 represent saturated liquid state, saturated vapor state, and the change from the saturated liquid state to the saturated vapor state. Values of entropies for a number of substances such as water, Freon, and ammonia are listed in the appendix.

Entropies of liquids and vapors will be extensively used in the chapter on steam power plants and refrigeration. Also, we shall find temperature-entropy diagrams and enthalpy-entropy diagrams (Mollier chart) of saturated liquids and vapors useful.

6.6.2 Gas Tables

In the design and analyses of thermal systems, calculations often have to be repeated a number of times by changing the state points. The gas tables simplify some of these calculations. The gas tables by Keenan and Kaye contain data on properties of air, carbon dioxide, and several other gases of practical importance. These tables assume that the ideal gas law, $pv = RT$, can be used for the pressure ranges commonly encountered in practice. These tables are particularly useful when isentropic processes are involved.

The key property in using the gas tables is the temperature T in degrees Kelvin. Also, relative pressure p_r and relative volume v_r are two additional properties providing a link between the two end states of a process. The relative pressure is the ratio of the pressure at a given state to the pressure at a reference state. The relative volume is similarly defined. Denoting the reference state by subscript 0, we have

$$p_r = \frac{p}{p_0} \quad \text{and} \quad v_r = \frac{v}{v_0}$$

For an isentropic process, Equation 6.21 becomes

$$0 = \int c_p \frac{dT}{T} - R \int \frac{dp}{p}$$

Integrating between the reference state and a given state and rearranging, we obtain

$$\ln \frac{p}{p_0} = \frac{1}{R} \int_{T_0}^{T} \frac{c_p}{T} dT$$

or

$$p_r = \exp\left[\frac{1}{R} \int_{T_0}^{T} \frac{c_p}{T} dT \right]$$

Similarly, starting with Equation 6.18, we can obtain the following expression for v_r:

$$v_r = \exp\left[-\frac{1}{R} \int_{T_0}^{T} \frac{c_v}{T} dT \right]$$

Since the relationships between c_p and T and between c_p and c_v for an ideal gas are known, we can evaluate the right-hand sides of the above equations to determine the values of p_r and v_r corresponding to a given T. The values of h in the gas tables are calculated from the specific heat data, while those of u are derived from the relation $u = h - RT$. The values of the standard-state entropy s^0 are calculated by the method given in Section 6.6. For convenience, the values of h, u, and s^0 are taken as zero at absolute zero temperature. The values of p_r, v_r, h, u, and s^0 are listed in the gas tables by Keenan and Kaye, and those for air are given in Appendix I2.

The following example illustrates the use of the air tables.

(a) Air expands isentropically.

$$\text{Initial state:} \quad T_1 = 400 \text{ K}, \quad p_1 = 1000 \text{ kPa}$$

$$\text{Final state:} \quad T_2 = 250 \text{ K}$$

We wish to determine the pressure after the expansion, p_2.

Looking in Appendix I2 at 400 K, we find

$$p_{r1} = 3.806$$

Likewise, for 250 K,

$$p_{r2} = 0.7329$$

Then

$$p_2 = p_{r2} \times p_0 = p_{r2} \times \left(\frac{p_1}{p_{r1}}\right)$$

$$= 0.7329 \times \frac{1000}{3.806}$$

$$= 192.6 \text{ kPa}$$

If v_1 is prescribed instead of p_1, we can determine v_2 from

$$v_2 = v_{r2} \times v_0 = v_{r2} \times \left(\frac{v_1}{v_{r1}}\right)$$

since the values of v_{r1} and v_{r2} can be read from the table.

If we use Appendix I2 to find the values of h_1 and h_2 corresponding to 400 K and 250 K, respectively, we find that the enthalpy change in this illustration is

$$\Delta h = h_2 - h_1$$

$$= 400.98 - 250.05$$

$$= 150.93 \text{ kJ/kg}$$

(b) Air undergoes a process with the end states given below.

$$\text{Initial state:} \quad p_1 = 150 \text{ kPa}, \quad T_1 = 300 \text{ K}$$

$$\text{Final state:} \quad p_2 = 750 \text{ kPa}, \quad T_2 = 400 \text{ K}$$

We wish to know the change of entropy per unit mass.

From Equation 6.21

$$\Delta s = \int_1^2 c_p \frac{dT}{T} - R \int_1^2 \frac{dp}{p}$$

$$= (s_2^0 - s_1^0) - R \ln\left(\frac{p_2}{p_1}\right)$$

$$= (2.8052 - 2.5153) - 0.287 \ln\left(\frac{750}{150}\right)$$

$$= -0.1720 \text{ kJ/kg K}$$

6.7 ENTROPY CHANGE FOR A CONTROL VOLUME

Thus far we have considered a fixed mass or a closed system while discussing the second law of thermodynamics. The law, of course, is universally valid and applies to open systems as well. We shall find it convenient to make use of Equation 6.17, developed for a fixed mass system undergoing an irreversible process, to obtain expressions for the entropy change in an open system undergoing either a reversible or an irreversible process. Beginning with Equation 6.17

$$dS = \frac{\delta Q}{T} + \frac{\delta(W_{\text{lost}})}{T} \tag{6.17}$$

and integrating for a finite change, we obtain

$$\Delta S = \int \frac{\delta Q}{T} + \int \frac{\delta(W_{\text{lost}})}{T} \tag{6.25}$$

Consider a control volume such as the one shown in Figure 6.19, with a fluid entering and leaving the control volume. The properties at the inlet and the outlet are designated by the subscripts *in* and *out*, respectively. Let us examine a fixed mass of the fluid at time τ. This mass occupies all of the control volume and a small portion of the inlet pipe. The masses in these two portions are designated by $m_{\text{CV},\tau}$ and δm_{in}, respectively. Since the fluid is moving continuously, the same mass of the fluid occupies a different space at time $\tau + \Delta\tau$. The fluid is now occupying the control volume and a small portion of the outlet pipe. The masses in these two portions are designated by $m_{\text{CV},(\tau+\Delta\tau)}$ and δm_{out}, respectively. We are interested in analyzing the change of entropy of this mass of fluid during a period of time $\Delta\tau$ and in extending the analysis to a control volume. For the left side of Equation 6.25, we can write

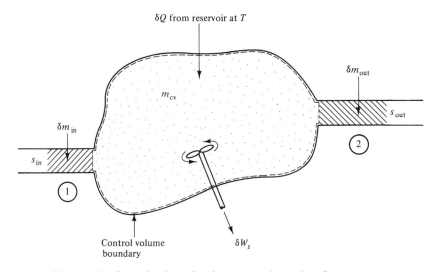

Figure 6.19 Control volume for the entropy change in a flow process.

$$\Delta S = (\text{entropy of the mass at time } \tau + \Delta \tau)$$
$$- (\text{entropy of the mass at time } \tau)$$

so that

$$\Delta S = (S_{\text{CV}, (\tau + \Delta \tau)} + s_{\text{out}} \delta m_{\text{out}}) - (S_{\text{CV}, \tau} + s_{\text{in}} \delta m_{\text{in}})$$

or

$$\Delta S = (S_{\text{CV}, (\tau + \Delta \tau)} - S_{\text{CV}, \tau}) + (s \, \delta m)_{\text{out}} - (s \, \delta m)_{\text{in}}$$

where S_{CV} is the entropy of the entire mass in the control volume.

Next, we have to interpret the right side of Equation 6.25 in a suitable manner. For simplicity, we assume that there is only one source supplying heat to our mass. In the quantity $\int (\delta Q/T)$, the integrand represents the ratio of a small amount of heat supplied to the mass under consideration to the temperature of the system receiving the heat. The integral of $\delta Q/T$ represents the cumulative value of the ratio $\delta Q/T$ over the time period $\Delta \tau$. In a similar fashion the quantity $\int (\delta [W_{\text{lost}}]/T)$ represents the cumulative ratio of the lost work to the temperature of the system receiving heat over the time period, $\Delta \tau$. Substituting for ΔS and employing the foregoing interpretation for the right-hand members of Equation 6.25, we can rewrite the equation to read

$$S_{\text{CV}, (\tau + \Delta \tau)} - S_{\text{CV}, \tau} + (s \, \delta m)_{\text{out}} - (s \, \delta m)_{\text{in}} = \int_{\tau}^{\tau + \Delta \tau} \frac{\delta Q}{T} + \int_{\tau}^{\tau + \Delta \tau} \frac{\delta (W_{\text{lost}})}{T}$$

We can divide this equation by $\Delta \tau$ to yield

$$\frac{S_{\text{CV}, (\tau + \Delta \tau)} - S_{\text{CV}, \tau}}{\Delta \tau} + \frac{(s \, \delta m)_{\text{out}}}{\Delta \tau} - \frac{(s \, \delta m)_{\text{in}}}{\Delta \tau} = \frac{1}{\Delta \tau} \int_{\tau}^{\tau + \Delta \tau} \frac{\delta Q}{T} + \frac{1}{\Delta \tau} \int_{\tau}^{\tau + \Delta \tau} \frac{\delta (W_{\text{lost}})}{T}$$

If we now take the limit as $\Delta \tau$ tends to zero of each of the terms in the above equation, we find

$$\lim_{\Delta \tau \to 0} \frac{S_{\text{CV}, (\tau + \Delta \tau)} - S_{\text{CV}, \tau}}{\Delta \tau} = \frac{dS_{\text{CV}}}{d\tau}$$

$$\lim_{\Delta \tau \to 0} \frac{(s \, \delta m)_{\text{out}}}{\Delta \tau} = (\dot{m}s)_{\text{out}}$$

$$\lim_{\Delta \tau \to 0} \frac{(s \, \delta m)_{\text{in}}}{\Delta \tau} = (\dot{m}s)_{\text{in}}$$

$$\frac{1}{\Delta \tau} \int_{\tau}^{\tau + \Delta \tau} \frac{\delta Q}{T} = \frac{\dot{Q}}{T}$$

and

$$\frac{1}{\Delta \tau} \int_{\tau}^{\tau + \Delta \tau} \frac{\delta (W_{\text{lost}})}{T} = \frac{(\dot{W}_{\text{lost}})}{T}$$

Thus the equation for change of entropy for an open system exchanging heat with a single reservoir at temperature T becomes

$$\frac{dS_{\text{CV}}}{d\tau} + (\dot{m}s)_{\text{out}} - (\dot{m}s)_{\text{in}} = \frac{\dot{Q}}{T} + \frac{(\dot{W}_{\text{lost}})}{T} \tag{6.26}$$

In the left side of this equation, the first term is the rate of change of entropy within the control volume. If there are local variations of the properties within a control volume, this term is expressed as

$$\frac{dS_{CV}}{d\tau} = \frac{d}{d\tau} \int_{CV} s\rho \, dV$$

The term $(\dot{m}s)_{out}$ represents the rate at which entropy leaves the control volume due to an outflow of mass. Likewise, $(\dot{m}s)_{in}$ represents the rate at which entropy enters the control volume due to an inflow. There are situations in practice where more than one point of inflow and more than one point of outflow exist. In such cases, $(\dot{m}s)_{out}$ and $(\dot{m}s)_{in}$ are replaced by $\sum\limits_{out} \dot{m}s$ and $\sum\limits_{in} \dot{m}s$, respectively.

In the right side of Equation 6.26, the term $\dot{Q}/T$ is the ratio of the rate of heat supply on the control volume to the temperature at which the heat supply occurs. There can be situations where the rate of heat supply and the temperature at which heat is supplied are nonuniform over the entire control surface. If $\mathcal{Q}$ denotes the rate of heat supply per unit surface area of the control surface CS, $\mathcal{Q}$ being a function of the surface coordinates, we can write

$$\frac{\dot{Q}}{T} = \int_{CS} \frac{\mathcal{Q}}{T} \, dA$$

The lost-work term in Equation 6.26 is difficult to evaluate. Therefore Equation 6.26 is often rewritten as an inequality by dropping the lost-work term, which is always positive. Thus we can write

$$\frac{dS_{CV}}{d\tau} + (\dot{m}s)_{out} - (\dot{m}s)_{in} \geq \frac{\dot{Q}}{T} \tag{6.26a}$$

or

$$\frac{dS_{CV}}{d\tau} + \sum_{out} \dot{m}s - \sum_{in} \dot{m}s \geq \frac{\dot{Q}}{T} \tag{6.26b}$$

or

$$\frac{d}{d\tau} \int_{CV} s\rho \, dV + \sum_{out} \dot{m}s - \sum_{in} \dot{m}s \geq \int_{CS} \frac{\mathcal{Q}}{T} \, dA \tag{6.26c}$$

The equality sign in Equations 6.26(a), 6.26(b), and 6.26(c) applies only to a reversible process, since the lost-work term is identically zero for a reversible process.

The principle of the increase of entropy was expressed mathematically when we dealt with a closed system. It is possible to obtain a similar expression for the principle in relation to an open system.

Consider a control volume with fluid entering and leaving it. Let the control volume receive heat energy at a rate $\dot{Q}$ from the surroundings at temperature T_0. Then the rate of change of entropy of the control volume is given by Equation 6.26(a). It is

$$\frac{dS_{CV}}{d\tau} \geq (\dot{m}s)_{in} - (\dot{m}s)_{out} + \frac{\dot{Q}}{T}$$

with the greater than sign applying for an irreversible heat transfer and the equals sign for a reversible heat transfer.

We observe that the fluid entering the control volume depletes the surroundings entropy at a rate of $(\dot{m}s)_{in}$ and the fluid exiting the control volume brings in entropy at a rate of $(\dot{m}s)_{out}$ into the surroundings. Also, as the heat energy $\dot{Q}$ leaves the surroundings—which are at T_0—the surroundings experience a decrease in entropy equal to $\dot{Q}/T_0$. Thus the change of entropy of the surroundings is given by

$$\frac{dS_{surr}}{d\tau} = (\dot{m}s)_{out} - (\dot{m}s)_{in} - \frac{\dot{Q}}{T_0} \tag{6.27}$$

When we add these expressions for the time rate of change of entropy of the control volume and of the surroundings, we have

$$\frac{dS_{CV}}{d\tau} + \frac{dS_{surr}}{d\tau} \geq \frac{\dot{Q}}{T} - \frac{\dot{Q}}{T_0}$$

In the foregoing equation, when $T_0 > T$, $\dot{Q}$ is positive; and when $T_0 < T$, $\dot{Q}$ is negative. Consequently, the right side of the above equation is always positive. The equation then takes the form

$$\frac{dS_{CV}}{d\tau} + \frac{dS_{surr}}{d\tau} \geq 0 \tag{6.28}$$

Equation 6.28 is the general principle of the increase of entropy.

Example 6.7

Figure 6.20 shows an insulated variable-area pipe. It is known that steam flows through this pipe and has the following properties: $p_1 = 40$ kPa, $t_1 = 75.87°C$, $p_2 = 75$ kPa, and $t_2 = 91.78°C$. It is also known that the steam at section 1 has a quality of 0.98 and the steam at section 2 is saturated vapor. Determine the direction of the steam flow.

Solution:

Given: Steam flows through an insulated pipe of variable cross-sectional area.

$$\text{Section 1:} \quad p_1 = 40 \text{ kPa}, \quad t_1 = 75.87°C, \quad x_1 = 0.98$$

$$\text{Section 2:} \quad p_2 = 75 \text{ kPa}, \quad t_2 = 91.78°C, \quad x_2 = 1.00$$

Objective: The direction of the steam flow is to be ascertained.
Diagram:

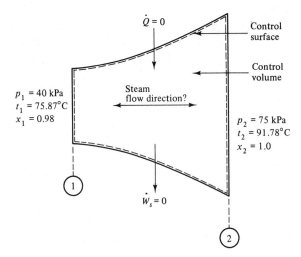

Figure 6.20   Control volume.

Assumptions: None.

Relevant physics: There is a temptation to say that the steam flows from high pressure to low pressure, but such temptation is to be resisted since the variable-area pipe can function as a nozzle (drop in pressure) or a diffuser (rise in pressure). Instead, we have to be guided by the second law or the principle of the increase of entropy, Equation 6.26(a). We note that for the control volume in Figure 6.20, $\dot{Q} = 0$ and $W_{shaft} = 0$. Assuming a steady state, we have $\dot{m}_{in} = \dot{m}_{out} = \dot{m}$, and $(dS/d\tau)_{CV} = 0$. Equation 6.26(a) then reduces to

$$\cancelto{0}{\frac{dS_{cv}}{d\tau}} + \dot{m}(s_{out} - s_{in}) \geq \cancelto{0}{\frac{\dot{Q}}{T}}$$

or

$$(s_{out} - s_{in}) \geq 0$$

Analysis: We first determine the entropies at sections 1 and 2.

Section 1: Steam here has a quality of 98%. From Equation 6.24

$$s = s_f + x s_{fg}$$

From Appendix B2.1, for $p_1 = 40$ kPa,

$$s_f = 1.0259 \text{ kJ/kg K} \quad \text{and} \quad s_{fg} = 6.6441 \text{ kJ/kg K}$$

so that

$$s_1 = 1.0259 + 0.98 \times 6.6441 = 7.537 \text{ kJ/kg K}$$

Section 2: The steam here is known to be saturated vapor. Hence

$$s_2 = s_g \text{ at } 75 \text{ kPa} = 7.4564 \text{ kJ/kg K}$$

Since the entropy at section 1 is greater than the entropy at section 2, we conclude that section 1 is the outlet and section 2 is the inlet. In other words, the variable-area section is functioning as a nozzle.

6.8 AVAILABLE WORK

The first law of thermodynamics tells us that energy may be transformed from one form into another, and that energy is always conserved. The second law tells us that heat energy cannot be converted into mechanical energy with 100 percent efficiency when a cyclic device is used or the entropy of an isolated system either remains constant or increases. An increase of entropy of the universe is synonymous with a degradation of energy, which implies lowering of the grade of energy. We also noted previously that mechanical energy is the highest grade energy, and that heat energy at a high temperature is superior to heat energy at a low temperature.

When a reversible cycle operates between two temperatures and produces mechanical work, in accordance with the second law some heat energy is always rejected to the reservoir at the lower temperature. Although there is a transfer of heat energy to the low-temperature reservoir, there is no degradation of energy, since only reversible processes are involved and there is no change in the entropy of the universe.

In an irreversible process, there is always some degradation of energy. In an irreversible heating process that produces no work, there is a maximum degradation of energy. Now consider a mass undergoing an irreversible process that produces some work and changes the state of the mass. There would be a degradation of energy in such a process, since the actual work obtained is less than the reversible work that can be obtained with the same change of state of the mass.

The degradation of energy occurring in a given process can be determined if the actual work obtained can be compared with the work available from a reversible process. The available work of a mass is that portion of the thermal energy available from the mass for conversion into work. In such a conversion, heat energy is rejected to a sink using a cyclic process. As the temperature T_0 of the sink is lowered, the portion of the thermal energy converted into work increases.

We define the *available work*, AW, of a system as the reversible work that can be obtained from a system when it undergoes a given change of state. The available work is variously called the *reversible work*, the *available energy*, and the *available portion of heat*. The term *available work* is preferred since the word *reversible* has been liberally used throughout this chapter, while the word *available* can be used only when we are dealing with the reversible work that can be extracted from a given system. Also, the word *work* in *available work* signifies the specific form of energy that is available from a system. We shall now determine the available work from different systems.

6.8.1 An Energy Reservoir at T_h

Consider a reservoir at T_h, from which a quantity of heat Q can be removed to produce work. We are interested in determining the available work in this process. We assume that a reservoir at T_ℓ is available, so that a reversible Carnot cycle can be operated between the two reservoirs. This cycle will receive heat Q from the reservoir at T_h and produce work.

The heat energy Q received by the cycle from the reservoir at T_h is represented by the area 1–2–B–A–1 in Figure 6.21(a). All this heat energy is not available for conversion into work. The amount of heat energy rejected by the Carnot cycle is $T_\ell(S_4 - S_3)$ and is represented by area 3–4–A–B–3 in Figure 6.21(a). The energy available for conversion into work is $Q + T_\ell(S_4 - S_3)$. The quantity $S_4 - S_3$ represents the change in the entropy of the working medium of the cycle during the isothermal heat transfer process 3–4. Since the quantity $(S_4 - S_3)$ is equal and opposite to the increase in the entropy ΔS of the reservoir at T_ℓ, we can write the following expression for AW.

$$AW = Q - T_\ell(\Delta S)_{\text{reservoir}} \tag{6.29}$$

The available work can be increased by selecting a reservoir with very low temperature. From a practical viewpoint it would be difficult to find a reservoir whose temperature is less than that of the local atmosphere. If T_0 is the temperature

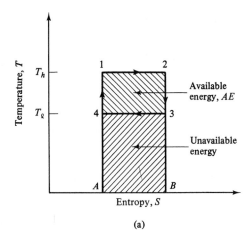

(a)

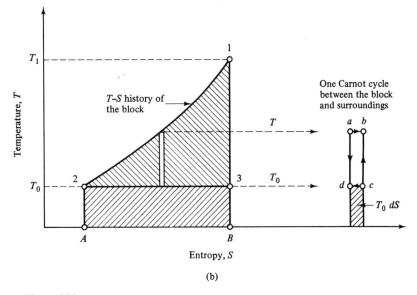

(b)

Figure 6.21
(a) Available and unavailable energies from a reservoir at T_h and a sink at T_ℓ.
(b) Available work from a finite source—reversible heat transfer between a block and the surroundings.

of the local atmosphere acting as the low temperature reservoir, the available work is

$$AW = Q - T_0 \, \Delta S$$

where ΔS is the increase in the entropy of the local atmosphere. The quantity $T_0 \, \Delta S$ is the amount of energy that now resides with the atmosphere and it represents the *unavailable energy*.

263

6.8.2 A Finite Source of Energy

Now consider the available work from a *finite* source of energy, such as a hot metal block. As long as the temperature of this block is higher than the temperature T_0 of the available sink, it would be possible to extract work by operating a reversible Carnot engine between the sink and the block.

With the execution of each cycle of the reversible engine, a small quantity of heat is extracted from the block and some work is done. Also, the temperature and the entropy of the block decrease while the entropy of the sink increases. The typical cycle, a–b–c–d–a, is shown in the right side of Figure 6.21(b). During the heat supply process (a–b), the entropy of the working medium increases, and the entropy of the source (block) decreases, precisely by the same magnitude. Also, during the heat rejection process, the entropy decrease of the working medium is equal in magnitude to the entropy increase of the sink. Since the processes b–c and d–a are isentropic, we conclude that the entropy changes $S_b - S_a$ and $S_d - S_c$ for the reversible cycle a–b–c–d–a, are equal and opposite. Consequently, the change in the entropy of the sink is equal and opposite to the change in the entropy of the block.

After a sufficiently large number of cycles are executed, the temperature of the block is the same as the temperature T_0 of the sink. In Figure 6.21(b), curve 1–2 shows the T-S history of the block, while line 2–3 represents the T-S history of the sink. It is to be noted that the initial states of the block and the sink are represented by points 1 and 2, respectively, and the corresponding final states are represented by points 2 and 3, respectively.

In Figure 6.21(b), area 1–2–A–B–3–1 represents the heat removed from the block, area 1–2–3–1 represents the available work, and area 2–3–B–A–2 represents the unavailable energy. For a single cycle of the Carnot engine, (a–b–c–d–a in Figure 6.21(b)) operating between the source at temperature T and the sink at temperature T_0, the available work is given by

$$\delta \text{AW} = \delta Q - T_0 \, dS$$

When the block changes state from 1 to 2, the state of the surroundings changes from 2 to 3 and the available work is given by

$$\text{AW} = \int \delta \text{AW} = \int_1^2 \delta Q - \int_2^3 T_0 \, dS$$

$$\text{AW} = Q_{1-2} - T_0(S_3 - S_2)_{\text{surr}} \tag{6.30}$$

Note that the quantity Q_{1-2} is the total heat energy supplied to the numerous Carnot cycles that operate between the finite source of energy and the surroundings and is represented by area 1–2–A–B–1 in Figure 6.21(b). We also note that the quantity $(S_3 - S_2)_{\text{surr}}$ represents the increase in the entropy of the surroundings. Thus both the quantities in the right side of Equation 6.30 are positive. We may rewrite Equation 6.30 to read

$$\text{AW} = Q - T_0(\Delta S)_{\text{surr}} \tag{6.30a}$$

Equation 6.30(a) can also be used in situations where the temperature of the finite source is not reduced to T_0.

Example 6.8

A block of copper has a mass of 9 kg and is at temperature 500 K. The c_p for copper is 0.383 kJ/kg K. If the surroundings are at 27°C, determine the available work of the block.

Solution:

Given: A block of copper has the following properties.

$$m = 9 \text{ kg}, \qquad T_1 = 500 \text{ K}, \qquad c_p = 0.383 \text{ kJ/kg K}$$

Also, $t_0 = 27°C$.

Objective: The available work of the block is to be determined.
Diagram:

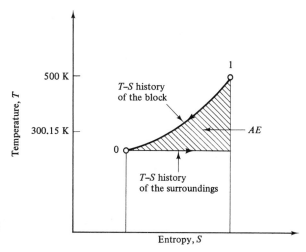

Figure 6.22 Available energy in a reversible heat transfer between a block and the surroundings.

Assumptions: The specific heat of copper is constant.

Relevant physics: In order to apply Equation 6.30, we need to calculate the amount of energy, Q, that can be extracted from the block in the form of heat. As energy is extracted from the block its temperature will fall; the limit for this falling temperature is the surroundings' temperature T_0. Thus if T_1 is the initial temperature of the block, the quantity Q_{1-2} in Equation 6.30 becomes Q_{1-0} and is given by

$$Q_{1-0} = mc_p(T_1 - T_0)$$

Next, as noted earlier, when a reversible engine operates between the block and the surroundings, the change in the entropy of the surroundings is equal and opposite to the change in the entropy of the block. Thus

$$(\Delta S)_{\text{surr}} = -(\Delta S)_{\text{block}} = -\int_1^0 \frac{\delta Q}{T}$$

For a constant pressure heat transfer from the block, $\delta Q = mc_p\, dT$. Integration gives

$$(\Delta S)_{\text{surr}} = -mc_p \ln\left(\frac{T_0}{T_1}\right)$$

The available work is given by

$$AW = Q_{1-0} - T_0(\Delta S)_{surr}$$

Analysis:

$$Q_{1-0} = 9 \times 0.383 \times 10^3[500 - (273.15 + 27)]$$

$$= 688,880 \text{ J, or } 689 \text{ kJ}$$

$$(\Delta S)_{surr} = -9 \times 0.383 \times 10^3 \times \ln(300.15/500)$$

$$= 1759.1, \text{ or } 1.76 \text{ kJ/k}$$

$$AW = 688,880 - (273.15 + 27)1759.1 \text{ J}$$

$$= 161 \text{ kJ}$$

Example 6.9

The copper block of Example 6.8 is brought in thermal contact with 5 kg of water at 27°C. Determine the loss of available work and the increase in the entropy of the universe, where c_p for water = 4.18 kJ/kg K.

Solution:

Given: The copper block of Example 6.8 is in thermal contact with water.

For water: $c_{p,w} = 4.18$ kJ/kg K, $T_{1,w} = 300.15$ K, $m_w = 5$ kg

For copper block: $c_{p,b} = 0.383$ kJ/kg K, $T_{1,b} = 500$ K, $m_w = 9$ kg

Objective: The loss of available work due to the irreversible heat transfer from the block to the water and the entropy change of the universe are to be determined.

Diagram:

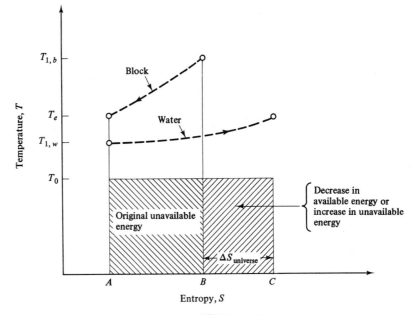

Figure 6.23   *T-s* diagram for block and water.

Assumptions: The specific heats of the block and the water are constant.

Relevant physics: Now we have an irreversible energy exchange between the block and the water. Assuming constant properties, the equilibrium temperature T_e is given by the application of the first law.

$$m_b c_{p,b}(T_e - T_{1,b}) + m_w c_{p,w}(T_e - T_{1,w}) = 0$$

where the subscripts b and w refer to the block and the water, respectively. The quantity of energy that can be removed from the composite system of the block and the water by operating a series of Carnot cycles is

$$Q_{e-0} = m_b c_{p,b}(T_e - T_{\text{surr}}) + m_w c_{p,w}(T_e - T_{\text{surr}})$$

If a series of Carnot cycles are operated between the composite system at T_e and the surroundings at T_0, then

$$(\Delta S)_{\text{surr}} = -(\Delta S) \text{ for the composite system}$$

$$= -(\Delta S_{\text{block}} + \Delta S_{\text{water}})$$

$$= -\left(\int_e^0 m_b c_{p,b} \frac{dT}{T} + \int_e^0 m_w c_{p,w} \frac{dT}{T} \right)$$

$$= -\left[m_b c_{p,b} \ln \frac{T_0}{T_e} + m_w c_{p,w} \ln \frac{T_0}{T_e} \right]$$

The available work for the composite system is

$$\text{AW} = Q_{e-0} - T_0 (\Delta S)_{\text{surr}}$$

The decrease in the available work due to the irreversible heat transfer between water and block is

$$(\text{AW})_{\text{block}} - (\text{AW})_{\text{composite}}$$

The increase in the entropy of the universe in the irreversible process is

$$(\Delta S)_{\text{uni}} = m_b c_{p,b} \ln \frac{T_e}{T_{1,b}} + m_w c_{p,w} \ln \frac{T_e}{T_{1,w}}$$

Analysis: Inserting the appropriate numerical values, we obtain the following equation for T_e.

$$9 \times 0.383 \times 10^3 (T_e - 500) + 5 \times 4.18 \times 10^3 (T_e - 300.15) = 0$$

or

$$T_e = 328.4 \text{ K}$$

Next,

$$Q_{e-0} = 9 \times 0.383 \times 10^3 (328.4 - 300.15) + 5 \times 4.18 \times 10^3 (328.4 - 300.15)$$

$$= 688,880 \text{ J, or } 689 \text{ kJ}$$

This quantity is the same as the Q_{1-0} of the preceding example, since energy is conserved. Also, the entropy change for the surroundings is given by

$$(\Delta S)_{\text{surr}} = -[9 \times 0.383 \times 10^3 + 5 \times 4.18 \times 10^3] \ln \left(\frac{300.15}{328.4} \right)$$

$$= 2190 \text{ J/K}$$

$$\text{AW} = 688,880 - 300 \times 2190$$

$$= 31,880 \text{ J, or } 32 \text{ kJ}$$

Decrease in AW is

$$161 - 32 = 129 \text{ kJ}$$

$$(\Delta S)_{uni} = 9 \times 0.383 \times 10^3 \ln \frac{328.4}{500} + 5 \times 4.18 \times 10^3 \ln \frac{328.4}{300}$$

$$= 421.3 \text{ J/K}$$

This is shown in Figure 6.23 as distance BC. Also shown in the figure are T-S histories of the block and the water.

Comments: If we multiply $(\Delta S)_{uni}$ by T_0, we find that

$$T_0(\Delta S)_{uni} = 431.3 \times 300 \text{ J}$$

$$= 129 \text{ kJ}$$

which is exactly equal to the decrease in the available work. This quantity is referred to as an *increase in the unavailable energy*, or degradation of energy due to the irreversible process.

It is not a coincidence that the quantity $T_0(\Delta S)_{uni}$ also gives the decrease in the available work. It was observed earlier in the chapter that the opportunity to extract mechanical work is lost forever when a process is irreversible. A measure of this loss is the increase in the entropy of the universe. The quantity $T_0(\Delta S)_{uni}$ can be regarded as the amount of energy gained by the surroundings at temperature T_0. This quantity represents energy that was originally of a high grade, but which has now become low-grade energy due to the irreversible process. This low-grade energy can essentially be written off as far as the possibility of converting it to mechanical work is concerned. The only way in which some of this energy can be retrieved as work is to have a sink with temperature less than T_0.

6.8.3 A Closed System

To derive a general expression for the available work for a closed system, consider a closed system such as a gas enclosed in a piston-cylinder system. Figure 6.24 shows the system, a reservoir at T_0, and the boundary of the surroundings. We are interested in determining the maximum available work from the system. Some of this work may be expended in pushing the atmosphere, and to that extent may not be useful work; however it is still a part of the available work. If the system receives a small quantity of heat, δQ, from the reservoir, the system would in general experience a small change in its internal energy and would deliver some work, δW. In accordance with the first law

$$\delta Q - \delta W = dU$$

or

$$\delta W = -dU + \delta Q \qquad (6.31)$$

The right side of this equation represents the work obtainable when a system undergoes a change of state. The inexact differential δQ can be eliminated by an appropriate consideration of the second law. When a system receives heat energy, it experiences a change in entropy, dS. If the heat transfer process is reversible, the reservoir supply-

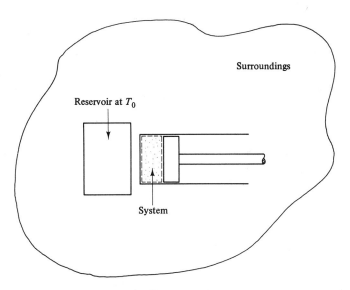

Figure 6.24 Available work in a closed system.

ing the heat energy δQ undergoes a change of entropy dS_0, which is negative and numerically equal to dS. If the heat transfer process is irreversible, dS_0 is numerically less than dS (see Equation 6.14a). Thus we can write

$$dS + dS_0 \geq 0$$

in accordance with Equation 6.14(b). The entropy change for the reservoir is $-\delta Q/T_0$ and we have

$$dS - \frac{\delta Q}{T_0} \geq 0$$

or

$$\delta Q \leq T_0\, dS$$

Substituting the foregoing upper limit of δQ in Equation 6.31, we obtain

$$\delta W \leq T_0\, dS - dU$$

Since we are interested in the available work, we drop the inequality sign and replace δW by $d(\text{AW})$.

$$d(\text{AW}) = T_0\, dS - dU \tag{6.32}$$

In Equation 6.32, we have used $d(\text{AW})$ and not $\delta(\text{AW})$, since the right side is an exact differential.

If the system were to undergo a finite change in its state, an integration of Equation 6.32 would give the maximum avialable work due to such a change.

$$\text{AW}_{1-2} = \int_1^2 T_0\, dS - \int_1^2 dU$$
$$= T_0(S_2 - S_1) - (U_2 - U_1)$$

269

A rearrangement gives

$$AW_{1-2} = (U_1 - U_2) - T_0(S_1 - S_2) \tag{6.32a}$$

The quantity on the right side of Equation 6.32 is sometimes called *reversible work*. A comparison of Equation 6.32(a) with the equation for AW developed earlier in this section shows that the expressions for AW are essentially the same.

6.8.4 An Open System

In practice we more often than not encounter open systems interacting with surroundings. For determining the available work in an open system, we consider a control volume such as the one shown in Figure 6.25. The control volume receives

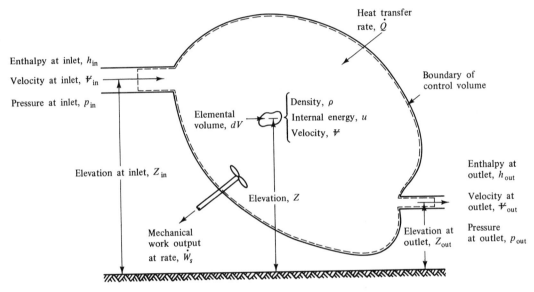

Figure 6.25 Available work in an open system.

heat energy at a rate $\dot{Q}$ from the surroundings at T_0 and the control volume delivers work at a rate $\dot{W}_s$. The properties of fluids entering and leaving the control volume are identified by the subscripts *in* and *out*, respectively. Application of the first law to this control volume (Equation 5.7) gives

$$\dot{m}_{in}\left(h_{in} + \frac{V_{in}^2}{2} + gZ_{in}\right) - \dot{m}_{out}\left(h_{out} + \frac{V_{out}^2}{2} + gZ_{out}\right) + \dot{Q} - \dot{W}_s$$

$$= \frac{d}{d\tau} \int_{CV} \left(u + \frac{V^2}{2} + gZ\right) \rho \, dV \tag{6.33}$$

From Equation 6.26(a)

$$\frac{\dot{Q}}{T_0} \leq \frac{dS_{CV}}{d\tau} + (\dot{m}s)_{out} - (\dot{m}s)_{in}$$

Since T_0 is constant, we can write

$$\dot{Q} \le \frac{d(T_0 S_{CV})}{d\tau} + \dot{m}_{out} T_0 s_{out} - \dot{m}_{in} T_0 s_{in}$$

Substituting the above expression for the upper limit of $\dot{Q}$ in Equation 6.33 and rearranging gives

$$\dot{W}_s \le \dot{m}_{in}\left(h_{in} - T_0 s_{in} + \frac{V_{in}^2}{2} + gZ_{in}\right) - \dot{m}_{out}\left(h_{out} - T_0 s_{out} + \frac{V_{out}^2}{2} + gZ_{out}\right)$$
$$- \frac{d}{d\tau}\int_{CV}\left(u - T_0 s + \frac{V^2}{2} + gZ\right)\rho\, dV$$

Since we are interested in the time rate of available work that would be obtainable only in a reversible process, we drop the inequality sign, replace $\dot{W}_s$ by $A\dot{W}_s$, and write

$$A\dot{W}_s = \left[\dot{m}\left(h - T_0 s + \frac{V^2}{2} + gZ\right)\right]_{in} - \left[\dot{m}\left(h - T_0 s + \frac{V^2}{2} + gZ\right)\right]_{out}$$
$$- \frac{d}{d\tau}\int_{CV}\left(u - T_0 s + \frac{V^2}{2} + gZ\right)\rho\, dV \qquad (6.34)$$

If there is more than one point for both entry and exit, Equation 6.34 takes the form

$$A\dot{W}_s = \sum_{in}\dot{m}\left(h - T_0 s + \frac{V^2}{2} + gZ\right) - \sum_{out}\dot{m}\left(h - T_0 s + \frac{V^2}{2} + gZ\right)$$
$$- \frac{d}{d\tau}\int_{CV}\left(u - T_0 s + \frac{V^2}{2} + gZ\right)\rho\, dV \qquad (6.34a)$$

Under steady-flow conditions and with only one inlet and one outlet, the equation reduces to the following form:

$$A\dot{W}_s = \dot{m}\left[(h_{in} - h_{out}) - T_0(s_{in} - s_{out}) + \frac{V_{in}^2 - V_{out}^2}{2} + g(Z_{in} - Z_{out})\right] \qquad (6.35)$$

6.8.5 Irreversibility

The actual work output from a system—closed or open—is always less than the available work as given by Equations 6.32 or 6.34. This difference is called *irreversibility*. Irreversibility, I, for a process is measured by the excess of the available work, AW, over the actual work output, W.

$$I = AW - W \qquad (6.36)$$

For a closed system, AW is given by Equation 6.32(a) and W is given by an integration of Equation 6.31. Substitution for AW and W in Equation 6.36 then gives

$$I_{1-2} = [(U_1 - U_2) - T_0(S_1 - S_2)] - [Q - (U_2 - U_1)]$$

or

$$I_{1-2} = T_0(S_2 - S_1) - Q \qquad (6.37)$$

For an open system we consider the time rate of change of irreversibility. This is given by

$$\dot{I} = A\dot{W}_s - \dot{W}_s \tag{6.38}$$

The rate of available work, $A\dot{W}_s$ is given by Equation 6.34 while the rate of actual work output, $\dot{W}_s$, is given by Equation 5.7. Substitution of these equations in Equation 6.38 gives

$$\dot{I}_{1-2} = \left\{ \left[\dot{m}\left(h - T_0 s + \frac{V^2}{2} + gZ\right) \right]_{\text{in}} - \left[\dot{m}\left(h - T_0 s + \frac{V^2}{2} + gZ\right) \right]_{\text{out}} \right.$$
$$- \frac{d}{d\tau}\int_{\text{CV}} \left(u - T_0 s + \frac{V^2}{2} + gZ\right) \rho \, dV \Big\}$$
$$- \left\{ \dot{Q} + \left[\dot{m}\left(h + \frac{V^2}{2} + gZ\right) \right]_{\text{in}} - \left[\dot{m}\left(h + \frac{V^2}{2} + gZ\right) \right]_{\text{out}} \right.$$
$$- \frac{d}{d\tau}\int_{\text{CV}} \left(u + \frac{V^2}{2} + gZ\right) \rho \, dV \Big\}$$

or

$$\dot{I}_{1-2} = \left[(\dot{m}T_0 s)_{\text{out}} - (\dot{m}T_0 s)_{\text{in}} \right] + \left[\frac{d}{d\tau}\int_{\text{CV}} (T_0 s) \rho \, dV \right] - \dot{Q} \tag{6.39}$$

Since T_0 is constant, the above equation can be rewritten as

$$\dot{I}_{1-2} = T_0 \left[(\dot{m}s)_{\text{out}} - (\dot{m}s)_{\text{in}} - \frac{\dot{Q}}{T_0} + \frac{dS_{\text{CV}}}{d\tau} \right] \tag{6.39a}$$

The first three quantities in the brackets on the right side, taken together, represent $(dS_{\text{surr}}/d\tau)$ (see Equation 6.27).
Thus

$$\dot{I}_{1-2} = T_0 \left[\frac{dS_{\text{surr}}}{d\tau} + \frac{dS_{\text{CV}}}{d\tau} \right] \tag{6.40}$$

For a real process, the quantity on the right side of Equation 6.40 is always positive and is the product of the surroundings' temperature and the rate of change of entropy of the universe. Thus the rate of change of irreversibility is directly related to the second law of thermodynamics.

It can be readily seen that under steady-flow conditions, Equation 6.39(a) reduces to

$$\dot{I}_{1-2} = \dot{m}T_0 s_{\text{out}} - \dot{m}T_0 s_{\text{in}} - \dot{Q} \tag{6.41}$$

Example 6.10

A rigid tank with a volume of 1.8 m³ contains air at 1000 kPa and 150°C. The air is cooled to 20°C by transferring heat to the surrounding air which is also at 20°C. Calculate the available work and the irreversibility of the process.

Solution:

Given: Air in a tank loses heat energy to the surrounding air.

Initial state:	$V_1 = 1.8$ m³,	$p_1 = 1000$ kPa,	$t_1 = 150°C$
Final state:	$V_2 = 1.8$ m³,	$t_2 = 20°C$,	
Surroundings:		$t_0 = 20°C$	

Objective: The available work and the irreversibility of the process are to be calculated.
Diagram:

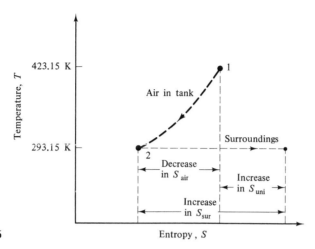

Figure 6.26

Assumptions: Specific heat of air at constant volume is constant; $c_v = 716.5$ J/kg K
Relevant physics: In relation to the ground state (20°C), the air in the tank at state 1
has a certain potential to deliver work. Likewise, there is a certain potential for
work at state 2. The difference between these two potentials of work is the
available work for the process, and this can be calculated from Equation 6.32(a).
The irreversibility of the process is the difference between the available work,
AW_{1-2} and the actual work W. The latter is zero since the volume of the air in
the tank remains constant.
Analysis: Use of Equation 6.32(a) to calculate available work requires computation of
the change of internal energy and the change of entropy for the air in the tank.
To this end, we first calculate the mass of the air in the tank.

$$m = \frac{p_1 V_1}{RT_1} = \frac{1000 \times 10^3 \times 1.8}{287(273.15 + 150)} = 14.82 \text{ kg}$$

Next

$$U_2 - U_1 = mc_v(T_2 - T_1) = 14.82 \times 716.5(20 - 150)$$

or

$$U_2 - U_1 = -1{,}380{,}400 \text{ J, or } -1380.4 \text{ kJ}$$

The change in entropy of the air is given by Equation 6.18(a). It is

$$S_2 - S_1 = mc_v \ln\left(\frac{T_2}{T_1}\right) + mR \ln\left(\frac{V_2}{V_1}\right)$$

$$= 14.82 \times 716.5 \ln\left(\frac{273.15 + 20}{273.15 + 150}\right) + 0$$

$$= -3897 \text{ J/K}$$

$$= -3.897 \text{ kJ/K}$$

From Equation 6.32(a), the available work is

$$AW_{1-2} = (U_1 - U_2) - T_0(S_1 - S_2)$$
$$= (1380.4) - (273.15 + 20) \times (3.897)$$
$$= 238 \text{ kJ}$$

In this constant volume process, there is no work done and the irreversibility is

$$I = AW_{1-2} - W$$
$$= 238 - 0 = 238 \text{ kJ}$$

Example 6.11

In order to regulate the power output from a certain steam turbine, the supply steam—at 1000 kPa and 400°C—is throttled to 800 kPa. Determine the irreversibility per unit mass of the throttling process if the surroundings are at 25°C.

Solution:

Given: Steam is throttled with the following conditions in a flow process.

Inlet:	$p_{in} = 1000$ kPa,	$t_{in} = 400$°C
Outlet:	$p_{out} = 800$ kPa	
Surroundings:	$t_0 = 25$°C	

Objective: The irreversibility per unit mass in the throttling process is to be determined.

Diagram:

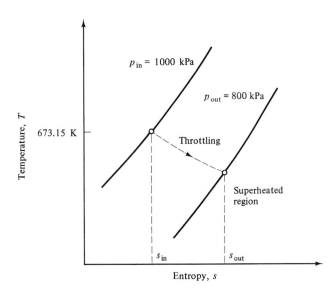

Figure 6.27

Assumptions: Steady flow exists.

Relevant physics: We have an open system here with no work output and no heat transfer. Equation 6.41 then takes the following form:

$$\dot{I}_{1-2} = \dot{m}T_0(s_{out} - s_{in})$$

Analysis: From Appendix B3, we find that the supply steam is superheated and has

$$s_{in} = 7.4651 \text{ kJ/kg K}$$
$$h_{in} = 3263.9 \text{ kJ/kg K}$$

Also, in a throttling process the enthalpy remains constant so that

$$h_{out} = 3263.9 \text{ kJ/kg K}$$

For $p_{out} = 800$ kPa and with the above value of h_{out}, we find (from Appendix B3)

$$s_{out} = 7.5667 \text{ kJ/kg K}$$

so that

$$\frac{\dot{I}_{1-2}}{\dot{m}} = (273.15 + 25) \times (7.5667 - 7.4651)$$

$$= 30.29 \text{ kJ/kg}$$

Example 6.12

A steam turbine expands steam from 4000 kPa and 500°C to 100 kPa in an adiabatic manner. The actual work output is found to be 800 kJ/kg. Determine the rate of available work and the rate of irreversibility of the expansion process. The surroundings are at 25°C.

Solution:

Given: Steam expands adiabatically in a turbine.

$$\text{Inlet:} \quad p_1 = 4000 \text{ kPa}, \qquad t_{in} = 500°C$$
$$\text{Outlet:} \quad p_2 = 100 \text{ kPa}, \qquad w_s = 800 \text{ kJ/kg},$$
$$\text{Surrounding:} \qquad\qquad t_0 = 25°C$$

Objective: The rates of available work and irreversibility are to be determined.

Diagram:

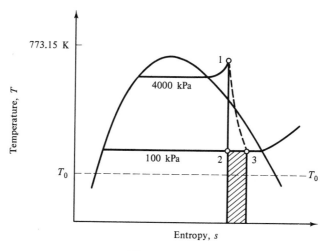

Figure 6.28

Assumptions:

1. Steady flow exists.
2. The changes in the kinetic and potential energies are negligible.

Relevant physics: This example is a straightforward application of Equation 6.35. With the assumptions made, the equation reduces to

$$A\dot{W}_s = \dot{m}[(h_{in} - h_{out}) - T_0(s_{in} - s_{out})]$$

Analysis: For the given inlet conditions, we have (from Appendix B3)

$$h_{in} = 3445.3 \text{ kJ/kg} \quad \text{and} \quad s_{in} = 7.0901 \text{ kJ/kg K}$$

We can now determine the actual state of the exiting steam from Equation 5.7, which for our problem takes the form

$$\frac{-\dot{W}_s}{\dot{m}} = h_{out} - h_{in}$$

$$-800 = h_{out} - 3445.3$$

$$h_{out} = 2645.3 \text{ kJ/kg}$$

Corresponding to the exit pressure of 100 kPa

$$h_{g,out} = 2675.5. \text{ kJ/kg} > h_{out}$$

Therefore the steam at exit is wet. Its quality is given by

$$x_{out} = \frac{h_{out} - h_{f,out}}{h_{fg}}$$

$$= \frac{2645.3 - 417.46}{2258}$$

$$= 0.9867$$

and

$$s_{out} = s_{f,out} + x_{out}s_{fg,out}$$

$$= 1.3026 + 0.9867 \times 6.0568$$

$$= 7.2788 \text{ kJ/kg K}$$

Then

$$\frac{A\dot{W}_s}{\dot{m}} = (h_{in} - h_{out}) - T_0(s_{in} - s_{out})$$

$$= (3445.3 - 2645.3) - (273.15 + 25) \times (7.0901 - 7.2788)$$

$$= 856.26 \text{ kJ/kg}$$

From Equation 6.41,

$$\frac{\dot{I}_{1-2}}{\dot{m}} = T_0(s_{out} - s_{in})$$

$$= (273.15 + 25) \times (7.2788 - 7.0901)$$

$$= 56.26 \text{ kJ/kg}$$

Alternatively,

$$\frac{\dot{I}}{\dot{m}} = \frac{A\dot{W}_s - \dot{W}_s}{\dot{m}}$$

$$= 856.26 - 800$$

$$= 56.26 \text{ kJ/kg}$$

which is the same result as before.

Comments: If the expansion in the turbine were to be isentropic, the exit condition would be at point 2, as shown on the *T-s* diagram in Figure 6.28. It can be readily shown that the work output for a reversible expansion from state 1 to state 2 would be 870.3 kJ/kg. This operation involves no change of entropy. The maximum reversible work of 856.26 kJ/kg corresponds to a reversible process between the same initial state, state 1, and a different terminal state, state 3.

We can devise a two-step reversible process between states 1 and 3; an isentropic process, 1–2, and an isothermal process, 2–3. The latter process would require a heat transfer to the steam, which is represented by the shaded area in Figure 6.28, after the steam has expanded isentropically to point 2. To bring about this transfer of heat, we would have to supply work to a heat pump, the magnitude of this work being equal to $(T_2 - T_0)(s_0 - s_2)$. For our problem, this work, w_{pump}, is

$$w_{\text{pump}} = (99.63 - 25)(7.2788 - 7.0901)$$

$$= 14.04 \text{ kJ/kg}$$

This quantity of work is also given by

$$w_{\text{pump}} = \frac{A\dot{W}_{s,1-2}}{\dot{m}} - \frac{A\dot{W}_{s,1-3}}{\dot{m}}$$

$$= 870.3 - 856.26$$

$$= 14.04 \text{ kJ/kg}$$

which is exactly the same as before.

6.9 AVAILABILITY

In the preceding sections we considered available work for a given change in the state of a closed system and the rate of available work for a control volume with fluid entering and leaving it. The determination of available work required a knowledge of the initial and the final states of the inlet and outlet conditions. We now consider availability, which is the maximum *useful* work available from a system of fixed mass due to its given initial state. In the European literature, availability is called *exergy* and this word is gaining popularity in this country also. Clearly, exergy would be a function of the end state of the system. An examination of Equation 6.32(a) shows that if the internal energy of a system at the end state is minimum and if the entropy at the end state is maximum, we shall have maximum available work from the system. When a system is in equilibrium with its surroundings, there is no possibility of extracting any work from the system. This equilibrium implies equilibrium of all types—that is, thermal, chemical, electrical, magnetic, mechanical, and so on. A state in which a system is in equilibrium with the surroundings is designated by subscript 0. Maximum work from a system in a given state will then be obtained when the system attains equilibrium with its surroundings.

For a closed system, the maximum useful available work may be calculated by using Equation 6.32(a) after allowing for the fact that a certain quantity of work will be expended in pushing the atmosphere. When a given system expands from its initial volume V to its final volume V_0, this work of expansion is given by $p_0(V_0 - V)$ and is not available, so it must be subtracted from the right side of Equation 6.32(a). When this is done we obtain an expression for the availability of a system, Φ, or the maximum useful available work. If contributions of kinetic and potential energies are ignored, the availability of a system is given by

$$\Phi = \text{(available work)} - \text{(work done on surroundings)}$$

or

$$\Phi = (U - U_0) - T_0(S - S_0) + p_0(V - V_0) \tag{6.42}$$

The above expression can be readily modified if the system has certain amount of initial kinetic energy and potential energy.

If a closed system does not come to equilibrium with its surroundings but instead goes from state 1 to state 2, the useful available work, AW_{1-2}, can be expressed as

$$AW_{1-2} = \Phi_1 - \Phi_2 - p_0(V_1 - V_2) \tag{6.43}$$

In case of an open system, we deal with a control volume that is fixed. The rate of available work from an open system, given by Equation 6.35, will be a maximum if the outlet field is in equilibrium with its surroundings and has zero potential and kinetic energies. We of course recognize that for a system to be open, there must be a flow and, therefore, the velocity is nonzero; however, we can visualize a vanishingly small flow velocity at outlet and hence a kinetic energy that approaches zero. We note that there is no work done in pushing the atmosphere. The *time rate of availability*, $\dot{\Psi}$, or the rate of maximum available work for a steady flow open system, is given by Equation 6.35 after replacing the properties designated by the subscript *out* by the properties that correspond to atmospheric conditions.

$$\dot{\Psi} = \dot{m}\left[(h - h_0) - T_0(s - s_0) + \frac{V^2}{2} + gZ\right] \tag{6.44}$$

We can define availability ψ for a unit mass in an open system as

$$\psi = (h - h_0) - T_0(s - s_0) + \frac{V^2}{2} + gZ \tag{6.44a}$$

If the outlet state is not in equilibrium with the surroundings, we can write the rate of available work from a flow system with inlet conditions 1 and outlet conditions 2, $A\dot{W}_{s,1-2}$:

$$A\dot{W}_{s,1-2} = \dot{m}\psi_1 - \dot{m}\psi_2 \tag{6.45}$$

If there are multiple inlet points and exit points,

$$\text{A}\dot{\text{W}}_{s,\,\text{in-out}} = \sum_{\text{in}} \dot{m}\psi - \sum_{\text{out}} \dot{m}\psi \qquad (6.46)$$

Example 6.13

Freon-12 at 20°C and 100 kPa is compressed to 100°C and 800 kPa. What is the availability per unit mass at the outlet in a steady flow process? The ambient temperature is 20°C.

Solution:

Given: Freon-12 is compressed in a steady flow process.

$$\text{Inlet:} \quad p_1 = 100 \text{ kPa}, \quad t_1 = 20°\text{C}$$

$$\text{Outlet:} \quad p_2 = 800 \text{ kPa}, \quad t_2 = 100°\text{C}$$

Objective: The availability per unit mass at the outlet is to be determined.

Diagram:

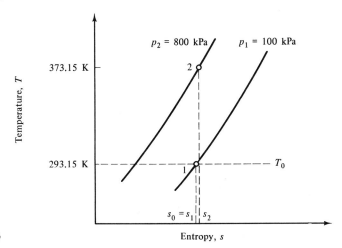

Figure 6.29

Assumptions: The potential and kinetic energies at the outlet are negligible.

Relevant physics: This example involves an application of Equation 6.44(a). With the assumptions made, the equation becomes

$$\psi_2 = (h_2 - h_0) - T_0(s_2 - s_0)$$

Analysis: From Appendix C we find that for 800 kPa and 100°C,

$$h_3 = 249.26 \text{ kJ/kg K}, \qquad s_2 = 0.8283 \text{ kJ/kg K}$$

To determine h_0 and s_0 we look up these properties for Freon-12 corresponding to atmospheric pressure (about 100 kPa) and 20°C. From Appendix C, we find

$$h_0 = 203.707 \text{ kJ/kg K}, \qquad s_0 = 0.8275 \text{ kJ/kg K}$$

The availability per unit mass is given by

$$\psi_2 = (h_2 - h_0) - T_0(s_2 - s_0)$$

$$= (249.26 - 203.707) - (273.15 + 20) \times (0.8283 - 0.8275)$$

$$= 45.32 \text{ kJ/kg}$$

6.10 SUMMARY

We can make the following statements to summarize the essence of the topics studied in this chapter.

1. A reversible process is an idealization that cannot be attained in practice. However, the concept provides an excellent tool for the analysis of a thermodynamic system. All natural processes are irreversible.

2. The Clausius statement of the second law is a formal expression of the everyday observation that heat transfer always occurs from a high-temperature body to a low-temperature body. The Kelvin-Planck statement negates a cyclic device that can convert thermal energy into mechanical work with 100 percent efficiency.

3. The Clausius inequality relates to a device employing a cyclic process and exchanging heat energy with a number of reservoirs at different temperatures. It is

$$\oint \frac{\delta Q}{T_{reservoir}} \leq 0$$

It is based on the Clausius statement. The less than sign applies to an irreversible process, while the equals sign applies to a reversible process.

4. The application of the Clausius inequality to a reversible cycle results in the important relation

$$dS = \left(\frac{\delta Q}{T}\right)_{reversible}$$

and the definition of entropy. Entropy is a property of a substance and is a function of the state of the substance. Only changes in entropy can be calculated. Since entropy is a function of state, the changes in entropy for an irreversible process are calculated by replacing the irreversible process in question by a reversible process joining the same end states.

5. When a system undergoes an irreversible process, maximum possible work is not obtained, some work is irretrievably lost. The change in the entropy of the universe is a measure of such lost work.

6. When the second law is applied to a control volume, we find for a steady state

$$\dot{m}(S_{out} - S_{in}) \geq \frac{\dot{Q}}{T}$$

7. The second law leads to the concepts of available work and availability. These involve reversible work and entropy. Available work, AW, is the reversible work that can be obtained when a system undergoes a given change of state. Irreversibility is a measure of the amount by which actual work falls short of the reversible work. Availability is the maximum useful work that can be obtained by operating a reversible cyclic device between a given body and atmosphere, which is the lowest-temperature sink available from a practical point of view.

PROBLEMS

6.1. State if the following processes are reversible, only internally reversible, or irreversible, and give your reasons.
  (a) Going up and down in a roller-coaster.
  (b) Grinding a tool bit.
  (c) Transferring heat from a baseboard heater to room air.
  (d) Filling an air balloon.
  (e) Boiling water in a teakettle.
  (f) Transferring heat from a source of energy at 100.01°C to a gas at 100°C and 50 kPa.
  (g) Expanding of gas in (f).
  (h) Condensing saturated Freon vapor at 5°C due to heat transfer to ice at 0°C.

6.2. Show that a violation of the Kelvin-Planck statement leads to a violation of the Clausius statement.

6.3. It is proposed to produce mechanical work by taking advantage of the thermal gradients that are present in an ocean. A source temperature of 25°C and a sink temperature of 5°C are available. What is the highest possible thermal efficiency of a heat engine operating between these two temperatures?

6.4. A person claims to have designed a heat engine that will operate between 30°C and 5°C, receive 500 kJ of heat energy, and produce 55 kJ of work. Verify the claim.

6.5. A certain heat engine is designed to undergo a cyclic process and exchange heat energy with four different reservoirs. The quantities of the heat energy to be exchanged and the temperatures of the reservoirs with which the exchange is to occur are (a) +9000 J at 1200°C, (b) +1800 J at 900°C (c) −3000 J at 100°C, and (d) −2500 J at 50°C. The engine is claimed to have a work output of 5300 J. Comment on the feasibility of the engine.

6.6. A modification is proposed to the engine in Problem 6.5 to increase the work output. It consists of eliminating the exchange of heat energy with the reservoir at 100°C. Comment on the feasibility of the modification.

6.7. A Carnot engine absorbs 4000 J of heat from a reservoir at the saturation temperature of water at 100 kPa and rejects it to a reservoir at the temperature of the triple point of water. Determine the heat rejected, the work done, and the efficiency of the engine.

6.8. A Carnot engine operating between 400°C and 0°C produces 16 kJ of work. Determine each of the following.
  (a) Heat supplied.
  (b) Change of entropy during heat rejection.
  (c) Thermal efficiency of the engine.

6.9. A Carnot power cycle operates on 1.5 kg of air between the limits of 21°C and 260°C, completing one cycle each second. The pressures at the beginning and end of isothermal expansion are 30 kPa and 15 kPa, respectively. Determine each of the following.
  (a) Volume at the end of the isothermal compression.
  (b) Changes in entropy during the isothermal processes.
  (c) Heat absorbed and heat rejected.
  (d) Power developed.
  (e) Efficiency.

6.10. Which is the more effective way to increase the efficiency of a Carnot engine: to increase the temperature of the hot reservoir, keeping that of the cold reservoir constant, or to decrease the temperature of the cold reservoir, keeping that of the hot reservoir constant. Justify your answer.

6.11. A body at 200°C undergoes an isothermal reversible process. The heat energy removed in the process is 7875 J. Determine the change in the entropy of the body. What should be the temperature of the reservoir receiving the heat energy from the body?

6.12. A mass of 7 kg of liquid water at 0°C is heated till it reaches a temperature of 100°C. Assuming the process to be reversible, determine the change in the entropy of the water. If the water is now completely vaporized, what would be the change of entropy? (*Note:* $h_{fg} = 2257.0\,\text{kJ/kg.}$)

6.13. A mass of 0.5 kg ice at $-10°C$ is mixed with 18 kg of water at 20°C. Determine the change in the entropy of the universe.

6.14. If the water in Problem 6.12 was converted into vapor by a reservoir at 400°C, determine the change in the entropy of the water, of the reservoir, and of the universe.

6.15. Air at 600 kPa and 40°C expands adiabatically to 100 kPa. The mass of the air is 2 kg. The work of expansion was measured to be 167 kJ. Determine if the adiabatic expansion was reversible or not; if not, determine the lost work and give your reasons for the possible causes of irreversibility. Represent the process on p-V and T-s diagrams.

6.16. Air expands from 650 kPa and 85°C to 50 kPa and 40°C. If the mass of the air is 0.8 kg, determine the change in the entropy of the air. Assume a constant specific heat.

6.17. Work Problem 6.16 using the air tables in appendix.

6.18. Nitrogen is compressed from a volume of 0.3 m³ to a volume of 0.05 m³. The pressure changes from 1 atm to 7 atm. What is the entropy change of the gas? Initial temperature of the air is 20°C.

6.19. A chamber is partitioned into two equal compartments. On one side there is oxygen at 850 kPa and 14°C. On the other side there is also oxygen, but at 100 kPa and 14°C. The chamber is insulated and has a volume of 7500 cc. The partition is abruptly removed. Determine the change in the entropy of the universe.

6.20. A perfect conductor connects two reservoirs at temperatures of 1000 K and 400 K. The steady heat flow rate from the hot to the cold reservoir is 100 W. What will be the rate of entropy production in the conductor?

6.21. An electric current of 10 A is maintained for 1 s in a resistor of 25 Ω, while the temperature of the resistor is kept constant at 310 K by keeping it in contact with a reservoir at 300 K. What is the entropy change of (a) the resistor, (b) the reservoir, and (c) the universe?

6.22. In the resistor of Problem 6.21, the same current is maintained for the same length of time. But the resistor is now thermally insulated and has an initial temperature of 310 K. If the resistor has a mass of 20 g and a specific heat of 0.74 kJ/kg K, what is the entropy change of (a) the resistor, and (b) the universe?

6.23. The temperature of 1 kg of water is raised from 300 K to 400 K by bringing it in contact with a heat reservoir at 400 K. Determine the entropy change of (a) the water, (b) the heat reservoir, and (c) the universe.

6.24. If the water in Problem 6.23 is heated from 300 K to 400 K by first keeping it in contact with a reservoir at 350 K and then with a reservoir at 400 K, what will be the change in entropy of the universe?

6.25. Air is contained in a piston-cylinder arrangement. Initially it has a volume of 0.015 m^3, a pressure of 100 atm, and a temperature of 15°C. The gas expands in a constant pressure process to a volume of 0.09 m^3. Calculate the change in entropy for the end state and for five intermediate points using Equation 6.23. Plot the points on a T-s diagram and draw a suitable curve.

6.26. Calculate the change in entropy for the air in Problem 6.25 using Equation 6.23, but now assume that the air undergoes a reversible constant volume heating to a temperature of 200°C. Plot a graph on a T-s diagram by calculating at least six data points. If the air were heated by a reservoir at 201°C, what would be the change in entropy of the air? Show the process on the same T-s plot.

6.27. A piston-cylinder arrangement containing 1 kg of saturated liquid water is initially at 500 kPa. Energy is slowly added to it in the form of heat until the water is completely evaporated and the piston moves in such a way that the pressure remains constant. Make a second-law analysis of the system.

6.28. A thermally insulated cylinder closed at both ends is fitted with a frictionless heat-conducting piston, which divides the cylinder into two parts. Initially, the piston is is clamped in the center with 4 liters (L) of air at 500 K and 2 atm on one side, and 4 L of air at 500 K and 1 atm on the other side. The piston is released and reaches equilibrium in pressure and temperature at a new position. Compute the final pressure, temperature, and the total increase of entropy.

6.29. A rigid, insulated container holds 5 kg of an ideal gas. The gas is stirred so that its state changes from 5 kPa and 300 K to 15 kPa. Assuming $c_p = 1.0 \text{ kJ/kg K}$ and $\gamma = 1.4$, determine the change in the entropy of the system.

6.30. An insulated chamber of volume $2V$ is divided by a rigid partition into two parts of volume V each. One chamber contains an ideal gas at a pressure p and temperature T. The other is evacuated. The partition is suddenly removed. Show that the temperature of the gas is unchanged. What is the change in entropy?

6.31. A piston-cylinder arrangement containing 4 kg of dry, saturated steam at 4 MPa is brought into thermal contact with a heat sink at 200°C. The steam rejects 1700 kJ of heat during a constant pressure process. Determine each of the following.
(a) Final state of steam.
(b) Change in entropy of steam.
(c) Entropy produced during the process.

6.32. A mass of 4 kg of saturated liquid water at 100 kPa is mixed in an insulated vessel with 8 kg of steam at 100 kPa and 250°C in such a way that the pressure remains constant. Determine the final state and the change in entropy.

6.33. A mass of 1 kg of an ideal gas at 21 kPa and 550 K is mixed with 1 kg of the same gas at 7 kPa and 320 K. The mixing takes place at constant volume, and during the process the system rejects 100 kJ of heat to the environment, which is at 300 K. It is known that the molar mass of the gas is 32 kg/kg-mol and $\gamma = 1.33$. Determine each of the following.
(a) Final volume, temperature, and pressure.
(b) Change in the entropy of the universe.

6.34. A mass of 1 kg of a liquid at 330 K is mixed with 1 kg of the same liquid at 450 K in an adiabatic calorimeter with rigid walls. Assume that there is no phase change and that the average specific heat of the liquid is constant at 6 kJ/kg K. Calculate the change in entropy due to the mixing process.

6.35. Water at 1 atm and 15°C is converted into wet steam ($x = 0.7$) at 250°C. Determine the change in entropy of the water.

6.36. Plot a T-s diagram for saturated liquid water and saturated water vapor using the values of s_f and s_g in the steam tables in the appendix. The temperature values should range from 50°C to the critical temperature for water. Use a minimum of 12 data points.

6.37. An operator reported that the steam in a turbine expanded adiabatically from 600 kPa and 400°C to 50 kPa. The final quality was found to be 0.98. What is the change in the entropy of the steam? Is the process possible?

6.38. Saturated vapor of Freon-12 at 10°C is compressed to 50°C and 1000 kPa. Determine the change in entropy. If the process is to be represented by the equation $pv^n =$ constant, determine the value of the exponent n.

6.39. A person claims to have developed a steam turbine that delivers 2950 kW. The steam enters the turbine at 600 kPa, 250°C and leaves the turbine at 15 kPa with a mass flow rate of 4.3 kg/s. Make a second-law analysis of this claim.

6.40. Determine the change in entropy of the gas in Example 2.5 of Chapter 2 where the gas undergoes a controlled expansion.

6.41. One kilogram of Freon-12 at 80°C and 2.3046 MPa is condensed at constant pressure to saturated liquid at 80°C by the transfer of heat to the surrounding air at 26°C. Determine the change of entropy of the universe and the highest air temperature consistent with the second law.

6.42. The conditions at two sections in a pipe carrying air are

$$t_1 = 19°C \qquad t_2 = 20°C$$
$$p_1 = 190 \text{ kPa} \qquad p_2 = 180 \text{ kPa}$$
$$V_1 = 285 \text{ m/s} \qquad V_2 = 290 \text{ m/s}$$

Determine the direction of flow assuming that the pipe is insulated.

6.43. The inlet conditions for a steam turbine are 400°C and 100 kPa, and the steam exhausts at 10 kPa. What is the maximum work output possible? State your assumptions.

6.44. The turbine exhaust of Problem 6.43 has a quality of 0.92 and is condensed to a saturated liquid in a condensor. The lowest available temperature is 17°C. What fraction of the energy rejected to the condensor is available energy?

6.45. Determine the available work and unavailable energy of a large body of water of mass 1,000,000 kg at a temperature of 45°C. The surroundings temperature is 10°C.

6.46. A metal block of mass 8 kg and specific heat of 0.4 kJ/kg K experiences a drop in temperature from 250°C to 80°C by a reversible process. Determine the reduction in the available work if the atmospheric temperature is 25°C.

6.47. Consider an irreversible process taking a system from state 1 to state 2, involving a supply of heat to the system and resulting in T_2 greater than T_1. How much of this heat can be converted into work if the surroundings are at T_0, a temperature less than T_1 and T_2?

6.48. A mass of 7 kg of water is contained in an insulated tank. Work of 12 kJ is performed on the water by a paddle wheel. If the initial temperature of the water is 26°C and the temperature of the surroundings is 20°C, determine the entropy change of the universe.

6.49. A rigid tank contains saturated water vapor at 100 kPa. Heat energy is added to the tank until the pressure reaches 200 kPa. If the environmental temperature is 20°C, determine the available portion of the heat added.

6.50. A pressure vessel with a volume of 0.8 m³ contains air at 200 kPa and 150°C. Due to heat transfer to the surrounding air, which is at 25°C, the temperature of the air in the tank drops to 25°C. Determine the change in the availability of the air and the irreversibility of the process.

6.51. Express the irreversibility of an ideal gas expanding through a turbine in terms of c_p, T_0, η_{turb}, and pressure ratio across the turbine.

6.52. The heat source for a power cycle ususally consists of a mixture of fuel and air, at atmospheric pressure and temperature, combusting irreversibly to produce high temperature gases. As heat is transferred out, the temperature of the hot gases drops to that of the atmosphere. State and discuss the maximum work available.

6.53. Combustion gases leave a stack at 280°C and 1 atm pressure. If the temperature of the surroundings is 22°C, determine the maximum quantity of work that can be obtained. Assume that the gases have a constant pressure specific heat of 1.05 kJ/kg K.

6.54. A processing plant requires a supply of hot water at 5 kg/s at 92°C. This supply is effected by mixing a supply of water at 11°C with steam from a boiler as wet vapor of 90% quality at 100 kPa. What is the rate of irreversibility of the process?

6.55. An economizer is a heat exchanger that utilizes the energy in the exhaust gases from a boiler to preheat the feedwater. For a certain economizer, the following data are available.

	Gas	Water
Mass flow rate, kg/s	190.7	124.7
c_p, kJ/kg K	1.0043	4.187
Inlet temperature, K	701	395
Outlet temperature, K	469	457

The ambient conditions are 101 kPa and 290 K. Determine the change of the availability of the water and of the gas and the rate of entropy production.

6.56. A closed feedwater heater consists of a heat exchanger, where the feedwater flows through tubes and steam extracted from the turbine condenses on the outside surface of the tubes. The condensate is pumped and mixed with the feedwater leaving the heater. In one such feedwater heater, the feedwater enters at 2000 kPa and 35°C and leaves at 1900 kPa and 95°C. The steam entering the heater is saturated at 150 kPa. Determine the irreversibility per kilogram of feedwater leaving the heater.

7 ~·—·—·—·—·—·—·—·—·

Energy Conversion—
Gas Cycles

7.1 INTRODUCTION

In this chapter, we shall consider some thermodynamic cycles for energy conversion. For the most part, we shall be concerned with conversion of thermal energy into mechanical energy, physical devices used for the conversion, and the thermal efficiency of such conversion. The cycles that we shall analyze in this chapter are called *air-standard cycles*. In an air-standard cycle, regardless of the mechanical hardware used for implementing the cycle, the cycle is treated as a closed cycle. The working medium of an air-standard cycle is air with constant properties. Since the working medium for the cycles to be discussed will be a gas, we shall trace the cycles in that region of the p-V and T-s diagrams, which is well above the critical point.

The physical device that converts thermal energy into mechanical energy is generally called a *prime mover*, or an engine. The prime mover can be of a reciprocating type or a rotary type. In the reciprocating type, a piston moves back and forth inside a cylinder, which is equipped with intake and exhaust valves. This piston is connected to the crankshaft via a connecting rod and a crank. The Otto cycle, discussed in Section 3.6.2, is generally executed in a reciprocating device. In a rotary device, the fluid may flow radially or axially past blades attached to a shaft. The gas turbine in a Brayton cycle, discussed in Chapter 5, is a rotary device for power production.

We shall first introduce the relevant terminology for receprocating machines. Then we shall present a discussion of the reciprocating compressor and of various power producing cycles.

7.2 NOMENCLATURE FOR A RECIPROCATING DEVICE

Figure 7.1 shows a schematic of a reciprocating device, which may be a compressor or an engine. Its principal parts are a crankshaft, a crank, a crank pin, a connecting rod, a crankcase, a piston, a wrist pin, a cylinder, and inlet (suction) and exhaust (delivery) ports with a suitable valve mechanism. The crank and the connecting rod convert rotary motion of the crankshaft into reciprocating motion of the piston, or vice versa. When the piston is at its extreme position, so that the volume enclosed between the cylinder walls and the piston top is a minimum, the piston is said to be at its *top dead center* (TDC), or *inner dead center* (IDC). The volume enclosed by the piston top and the cylinder walls when the piston is at TDC is called the *clearance volume*. When the piston is at its other extreme position, the volume enclosed is a maximum. This position of the piston is called *bottom dead center* (BDC), or *outer dead center* (ODC).

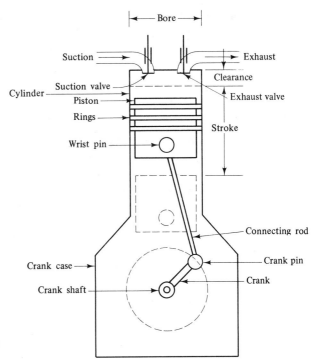

Figure 7.1 Nomenclature for a reciprocating device.

The volume displaced by the piston as it moves from TDC to BDC is called the *piston displacement* (PD). The maximum distance traveled by the piston in one direction is called the *stroke* (L) of the piston. The inner diameter of the cylinder is called the *bore* (D) of the cylinder. The piston displacement is then given by

$$PD = \left(\frac{\pi}{4}\right) D^2 L \tag{7.1}$$

The compression ratio r of a reciprocating device is based on the ratio of the volumes enclosed by the piston and the cylinder walls when the piston is at its two extreme positions.

$$r = \frac{\text{volume at the bottom dead center}}{\text{volume at the top dead center}} \qquad (7.2)$$

or

$$r = \frac{\text{clearance volume} + \text{piston displacement}}{\text{clearance volume}} \qquad (7.2a)$$

The clearance volume is sometimes expressed as a fraction of the piston displacement. It is then called *clearance factor*, CF. Thus,

$$\text{CF} = \frac{\text{clearance volume}}{\text{piston displacement}} \qquad (7.2b)$$

A reciprocating device can be single-acting or double-acting. In a single-acting device, the working fluid is only on the top side of the piston and the device has one pair of valves. In a double-acting device, the working fluid is on both the top and the bottom of the piston, and the bottom end of the cylinder is sealed from the crank case. Also, there are two pairs of inlet and exhaust valves. Double-acting devices are used in steam engines, air compressors, and feed pumps for water. Single-acting devices are used in compressors and internal conbustion engines.

7.3 RECIPROCATING COMPRESSOR

The air-standard reciprocating compressor is a two-stroke device; that is, a complete thermodynamic cycle is executed in two strokes of the piston, the suction stroke and the compression stroke. Consider the reversible cycles sketched in Figure 7.2. Figure 7.2(a) shows the $p\text{-}V$ diagram for a reciprocating compressor with no clearance volume, and Figure 7.2(b) shows the $p\text{-}V$ diagram for a compressor with clearance volume. Referring to Figure 7.2(a), the suction stroke begins at 1 and the low-pressure gas enters the cylinder. When the piston reaches the BDC, the compression stroke begins. During part of the compression stroke, the inlet and exhaust ports (not shown in the figure) are closed, and the gas is compressed in process 2–3. This process is usually polytropic. The exhaust port opens at point 3, and during the 3–4 part of the compression stroke, the delivery of the compressed gas takes place. At point 4, *all* the gas is removed from the cylinder and, as the piston begins its suction stroke, the thermodynamic state changes instantly from 4 to 1, state 1 being the thermodynamic state of the gas in the suction line. This cycle, although desirable, is not practical since a reciprocating device has to have some clearance volume.

Figure 7.2(b) shows the $p\text{-}V$ diagram for a compressor with clearance volume. Suction occurs during the constant pressure process 1–2 while process 2–3 is the same as in Figure 7.2(a). At the end of the delivery stroke (point 4) there is a residual high-pressure gas in the cylinder. This gas begins to expand as soon as the suction stroke begins. The expansion of this gas appears as a polytropic curve (4–1) in the

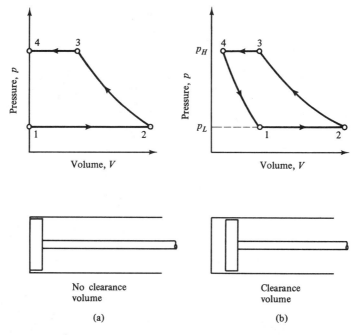

Figure 7.2 p-V diagram for a reciprocating compressor (a) with no clearance volume, and (b) with clearance volume.

figure. In practical situations, the exponents in the polytropic expansion and compression processes are not the same. It is only when the pressure within the cylinder equals the suction pressure that a fresh charge of gas can enter the cylinder. Thus with the presence of clearance volume, induction of the gas takes place only during a part of the suction stroke. If there is a pressure drop at the inlet and exhaust ports, the actual pressure p_d in the delivery line will be less than p_3 or p_4 and the actual pressure p_s in the suction line will be higher than p_1 or p_2.

7.3.1 Volumetric Efficiency

Consider a reciprocating compressor, with the piston at its TDC. In this position of the piston the valves are closed and the compressed fluid is at the highest pressure in the cycle. The volume of this residual fluid at this point equals the clearance volume. As the piston begins to move downward, we would like a fresh charge to enter the cylinder. However, this cannot happen immediately since the pressure within the cylinder is greater than that in the suction line. The residual fluid must expand during the downward stroke; only when its pressure falls below that in the suction line can a fresh charge enter the cylinder. Consequently, the actual mass of the fluid processed per cycle by the device is less than the theoretical mass of the fluid that can be processed per cycle. The ratio of the two masses is called *volumetric efficiency*. It is

$$\eta_v = \frac{\text{actual mass of fluid compressed per cycle}}{\text{theoretical mass of fluid compressed per cycle}} \qquad (7.3)$$

$$= \frac{(V_2 - V_1)/v_2}{(V_2 - V_4)/v_s} \qquad (7.4)$$

where v_2 is the specific volume of the gas in the cylinder during the theoretical suction stroke, v_s is the specific volume of the gas in the suction line, and V_1, V_2, V_3, and V_4 are defined in Figure 7.2(b).

We now proceed to express η_v in terms of the clearance factor CF, polytropic exponent n, and the pressure ratio.

We can write

$$\frac{V_2 - V_1}{V_2 - V_4} = \frac{V_2 - V_4 + V_4 - V_1}{V_2 - V_4}$$

$$= 1 + \frac{V_4}{V_2 - V_4} - \frac{V_4}{V_2 - V_4} \cdot \frac{V_1}{V_4}$$

$$= 1 + CF - CF\left(\frac{p_4}{p_1}\right)^{1/n}$$

In the last step, we have used the relation

$$p_4 V_4^n = p_1 V_1^n$$

Since $p_4 = p_H$ and $p_1 = p_L$, Equation 7.4 can then be rewritten as

$$\eta_v = \left[1 + CF - CF\left(\frac{p_H}{p_L}\right)^{1/n}\right]\frac{v_s}{v_2} \qquad (7.5)$$

The volumetric efficiency is less than unity because of a number of factors, such as pressure drop at the valves, heat transfer between the surroundings and the device, and the presence of residual fluid in the cylinder at the end of the delivery stroke. As the pressure ratio increases, the quantity in the square bracket becomes small. For a sufficiently large pressure ratio, the volumetric efficiency for a given clearance factor would be zero.

7.3.2 Compressor Work

For a reversible cycle such as the one shown in Figure 7.2(b), the area 1–2–3–4–1 represents the work input for one cycle. An expression for this work can be obtained by considering a closed system where a mass m undergoes a sequence of processes.

$$W_{\text{cycle}} = W_{1\text{-}2} + W_{2\text{-}3} + W_{3\text{-}4} + W_{4\text{-}1}$$

$$= \int_1^2 p \, dV + \int_2^3 p \, dV + \int_3^4 p \, dV + \int_4^1 p \, dV$$

For simplicity in the analysis, we assume that the exponent n in the polytropic processes 2–3 and 4–1 is the same. Noting that processes 1–2 and 3–4 are isobaric, we then obtain

$$W_{\text{cycle}} = p_2(V_2 - V_1) + \frac{p_3 V_3 - p_2 V_2}{1 - n} + p_3(V_4 - V_3) + \frac{p_1 V_1 - p_4 V_4}{1 - n}$$

If m is the mass of the gas entering the cylinder per cycle, and if v_L and v_H are the specific volumes at states 2 and 3, we can write

$$(V_2 - V_1) = mv_L \quad \text{and} \quad (V_3 - V_4) = mv_H$$

Also

$$p_L = p_1 = p_2 \quad \text{and} \quad p_H = p_3 = p_4$$

so that

$$W_{\text{cycle}} = p_L mv_L - p_H mv_H + \frac{(p_H V_3 - p_L V_2) + (p_L V_1 - p_H V_4)}{1 - n}$$

$$= m\left(p_L v_L - p_H v_H + \frac{-p_L v_L + p_H v_H}{1 - n}\right)$$

$$= \frac{-mn}{n-1}(p_H v_H - p_L v_L) \tag{7.6}$$

The work per unit mass of the gas is

$$w_{\text{cycle}} = \frac{-n}{n-1}(p_H v_H - p_L v_L) \tag{7.6a}$$

$$= \frac{-np_L v_L}{n-1}\left[\left(\frac{p_H}{p_L}\right)^{(n-1)/n} - 1\right] \tag{7.6b}$$

This result can also be obtained by considering a steady-flow process in which a fluid at p_2, v_2 enters the compressor and leaves at p_3, v_3. If changes in the kinetic and potential energies are neglected, for steady flow the conservation of energy requires

$$-\delta w_s = dh - \delta q$$

where the subscript s in w_s denotes shaft work. For a reversible process, $\delta q = T\,ds$, so that*

$$-\delta w_s = (du + p\,dv + v\,dp) - T\,ds$$

Since

$$T\,ds = du + p\,dv$$

we can write

$$-\delta w_s = v\,dp$$

and

$$W_{\text{cycle}} = -\int_2^3 v\,dp = -\int_L^H v\,dp$$

For the reciprocating compressor

$$pv^n = C \quad \text{(constant)}$$

so that

$$w_{\text{cycle}} = -\int_L^H C^{1/n}\,p^{-1/n}\,dp$$

$$= -C^{1/n}(p_H^{1-1/n} - p_L^{1-1/n})\frac{n}{n-1}$$

$$= \frac{-n}{n-1}(p_H v_H - p_L v_L) \tag{7.6a}$$

which is the same as obtained earlier. The work input per unit mass for a substance continuously undergoing a change of state along a specified process has a unique

*See Section 6-5.

value. It is irrevelent of the model used to calculate it, whether an open system or a closed system.

Inasmuch as the expression for work per unit mass (Equation 7.6(b)) does not contain the clearance factor, we conclude that the clearance volume has no effect on the work per unit mass. It does, however, have an effect on the work per cycle. If the clearance volume is increased, the volumetric efficiency, as well as the work per cycle, decreases, and it takes a greater number of cycles to compress a given mass of gas than before.

7.3.3 Isothermal Compression Efficiency

If the compression process in a compressor were reversible and isothermal ($pv = C$), then

$$w_{s,2-3} = -\int_L^H Cp^{-1}\, dp$$

$$= -p_L v_L \ln\left(\frac{p_H}{p_L}\right) \tag{7.7}$$

It can be shown that, for the same pressure ratio, the work of compression for a polytropic process is greater than the work for an isothermal process. Figure 7.3(a) and (b) shows p-V and T-s diagrams for three processes of compression, isentropic, polytropic, and isothermal. The three different areas enclosed on the p-V diagram, 1–2–3–4, 1–2–3A–4, and 1–2–3B–4, respectively, represent the work inputs for the isentropic, the polytropic, and the isothermal compression. It is clear from the relative sizes of these areas that the work input will be the least for isothermal compression. Therefore, if we wish to compress a gas with minimum work input, the compression should be isothermal. We define the *isothermal compression efficiency* as

$$\eta_{C,t} = \frac{\text{the isothermal work of compression from } p_L \text{ to } p_H}{\text{the actual work of compression from } p_L \text{ to } p_H} \tag{7.8}$$

Example 7.1

A single-acting, reciprocating air compressor takes in air at 100 kPa and delivers at 700 kPa. The compression is polytronic with $n = 1.28$. The temperature of the air at the inlet is 20°C. There is a pressure drop of 5 kPa at the suction port and of 10 kPa at the discharge port. If the clearance factor is 8%, determine the volumetric efficiency. If the piston displacement is 500 cm³ and if the compressor operates at 900 rpm, determine the power required to drive the compressor and the mass of air compressed per hour.

Solution:

Given: A single-acting reciprocating compressor has the following characteristics.

Intake: $p_1 = 100 \text{ kPa} = p_2$, $t_1 = 20°C$

Delivery: $p_3 = 700 \text{ kPa} = p_4$

Pressure drop at suction: 5 kPa

Pressure drop at discharge: 10 kPa

Compression: $pV^n = \text{constant}$, $n = 1.28$

PD $= 500 \text{ cm}^3$,   CF $= 0.08$, $N = 900 \text{ rpm}$

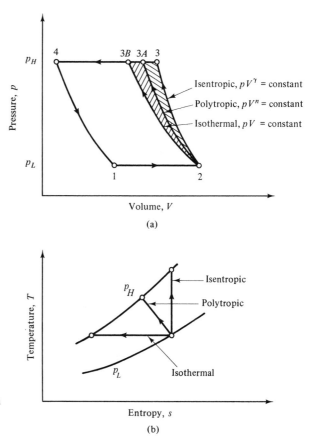

Figure 7.3 Three possible compression processes (a) on a *p-V* diagram, and (b) on a *T-s* diagram.

Objective: We are to determine the volumetric efficiency, the power input to the compressor, and the mass flow rate through the compressor.

Diagram:

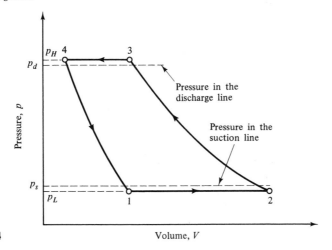

Figure 7.4

Assumptions:

1. The compression and expansion processes are both polytropic with $n = 1.28$.
2. The suction and discharge occur at their respective constant pressures.
3. The air behaves like an ideal gas.
4. The processes are reversible.

Relevant physics: This problem involves a straightforward application of Equation 7.5 for volumetric efficiency, η_v, and Equation 7.6(b) for specific work. The specific volume at suction can be determined from the ideal gas equation.

Analysis: Referring to Figure 7.4, we have

p_H = pressure in the discharge line + pressure drop at the discharge port

$\quad = 700 + 10 = 710 \text{ kPa}$

p_L = pressure in the suction line $-$ pressure drop at the intake port

$\quad = 100 - 5 = 95 \text{ kPa}.$

The specific volume (v_s) of the air in the suction line is given by

$$v_s = \frac{RT_s}{p_s}$$

$$= \frac{287 \times (20 + 273.15)}{100 \times 10^3}$$

$$= 0.841 \text{ kg/m}^3$$

Also

$$v_L = \frac{287(273.15 + 20)}{95 \times 10^3}$$

$$= 0.885 \text{ kg/m}^3$$

From Equation 7.5, we obtain

$$\eta_v = \left[1 + CF - CF\left(\frac{p_H}{p_L}\right)^{1/n}\right]\frac{v_s}{v_L}$$

$$= \left[1 + 0.08 - 0.08\left(\frac{710}{95}\right)^{1/1.28}\right]\left(\frac{0.841}{0.885}\right)$$

$$= 0.66, \text{ or } 66\%$$

The quantity of mass handled per unit time is

$$\dot{m} = (PD)\,(\text{number of cycles per unit time}) \times \frac{\eta_v}{v_s}$$

Since the compressor is single-acting, the number of cycles equals the number of revolutions and

$$\dot{m} = \frac{(500 \times 10^{-6})(900) \times 0.66}{0.841}\text{ kg/min}$$

$$= 0.3532 \text{ kg/min}$$

The work per unit mass is given by Equation 7.6(b) and is

$$w = \frac{-n\,p_L v_L}{n-1}\left[\left(\frac{p_H}{p_L}\right)^{(n-1)/n} - 1\right]$$

$$= -\frac{1.28 \times 95 \times 10^3 \times 0.885}{1.28 - 1}\left[\left(\frac{710}{95}\right)^{(1.28-1)/1.28} - 1\right]$$

$$= -212.42 \text{ kJ/kg}$$

The required power input is

$$P = \dot{m}w_s$$

$$= \left(\frac{0.3532}{60}\right)(212.42)$$

$$= 1.25 \text{ kW}$$

The temperature of the discharge air is

$$T_H = T_L\left(\frac{p_H}{p_L}\right)^{(n-1)/n}$$

$$= (20 + 273.15)\left(\frac{710}{95}\right)^{(1.28-1)/1.28}$$

$$= 455.17 \text{ K, or } 182°\text{C}$$

7.3.4 Intercooling

For a given intake pressure, as the discharge pressure is raised the temperature of the discharge also rises. In the preceding example, the discharge temperature was found to be 182°C for a pressure ratio of 7.5. Excessively high temperatures are detrimental to the seals, valves, and the lubricating oil and should, therefore, be avoided. Also, with large pressure ratios the volumetric efficiency decreases, as does the isothermal compression efficiency. When a gas is to be compressed through a large pressure ratio, the compression is carried out in more than one stage. Staging offers an opportunity to cool the discharge from the low-pressure stage before the gas enters the high-pressure stage. This is called *intercooling*, and it results in an improved isothermal compression efficiency. This is shown in Figure 7.5. The polytropic process 1–a–2 involves no intercooling, while process 1–b–3 is isothermal. Process 1–a is the polytropic compression in the low pressure stage, followed by the intercooling process a–b and the polytropic compression b–c in the high-pressure stage.

Figure 7.5 shows that there is a saving in work input to the compressor as a result of intercooling at the intermediate pressure, p_I. It can be seen from Figure 7.5 that if p_I is too close to either p_H or p_L, the saving in work input due to intercooling will be very small. There exists an optimum value of p_I for which the savings in the work of compression is maximum. Noting that $p_L v_1 = p_I v_b$, the total work of compression is given by

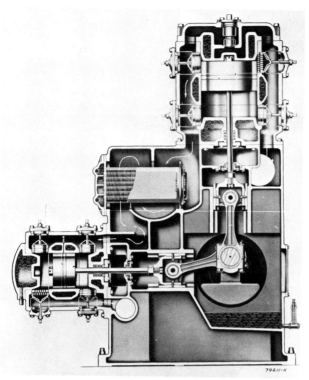

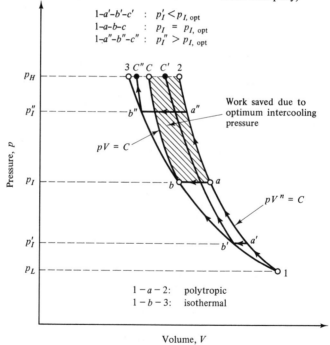

$1-a'-b'-c'$: $p_I' < p_{I,\,opt}$
$1-a-b-c$: $p_I = p_{I,\,opt}$
$1-a''-b''-c''$: $p_I'' > p_{I,\,opt}$

Work saved due to optimum intercooling pressure

$pV = C$

$pV^n = C$

$1-a-2$: polytropic
$1-b-3$: isothermal

Figure 7.5 Effect of the intermediate pressure on the work input for a two-stage compression with intercooling.

$$w = \frac{-n}{n-1}\left\{p_L v_1\left[\left(\frac{p_I}{p_L}\right)^{(n-1)/n} - 1\right] + p_I v_b\left[\left(\frac{p_H}{p_I}\right)^{(n-1)/n} - 1\right]\right\}$$

$$= \frac{-n p_L v_1}{n-1}\left[\left(\frac{p_I}{p_L}\right)^{(n-1)/n} + \left(\frac{p_H}{p_I}\right)^{(n-1)/n} - 2\right]$$

Since p_H, p_L, v_1, and n are fixed, the expression in the square bracket can be differentiated with respect to p_I and the result equated to zero. When this is done, we obtain the optimum value of the intermediate pressure

$$p_{I,\text{opt}} = \sqrt{p_H p_L} \tag{7.9}$$

It can be verified that the above value of $p_{I,\text{opt}}$ when substituted in the equation for work yields a minimum value of w. If there are three stages, then the two optimum intermediate pressures are found to be

$$p_{I1,\text{opt}} = \sqrt[3]{p_H^2 p_L} \quad \text{and} \quad p_{I2,\text{opt}} = \sqrt[3]{p_H p_L^2} \tag{7.9a}$$

7.4 GAS POWER CYCLES—OTTO CYCLE

The objective of a gas power cycle is to convert thermal energy into mechanical energy. In a gas power cycle, the working medium remains in a gaseous state throughout the cyclic process. The medium receives heat energy from a source, delivers work, and rejects heat energy to a sink. The supply of heat energy during the cycle may be external to the work output device as in Brayton cycle, or the heat supply may be due to release of heat energy from the combustion of the fuel within the work output device, as in the Otto cycle. The engines for gas power cycles are sometimes classified as *external combustion engines* and *internal combustion engines*. The term *engine* generally implies a reciprocating engine, although it may also include a rotary engine such as the Wankel engine. We note that the gas turbine, a rotary device, accomplishes only one segment of the Brayton cycle. Strictly speaking, a complete thermodynamic cycle is not executed within an internal combustion engine since the working medium undergoes a chemical change. However, one can analyze the thermodynamics of an internal cumbustion engine by considering an air-standard cycle, discussed in Section 7.1, taking place within a reciprocating device.

The Otto cycle discussed in Section 3.6.2 was an air-standard cycle. We briefly recapitulate its features here. It consists of two isentropic processes and two constant volume processes (Figure 7.6). Heat energy is supplied in the constant volume process 2–3 and heat energy is extracted in the constant volume process 4–1. The expression for the thermal efficiency of the Otto cycle was obtained in Chapter 2 by considering the work done and the heat energy supplied in each segment of the Otto cycle. We can obtain the same result by considering the heat supplied to and rejected by a system of mass m undergoing the Otto cycle.

Assuming an ideal gas behavior and constant specific heats, the heat supplied is given by

$$Q_{\text{in}} = \int_2^3 mc_v\, dT = mc_v(T_3 - T_2)$$

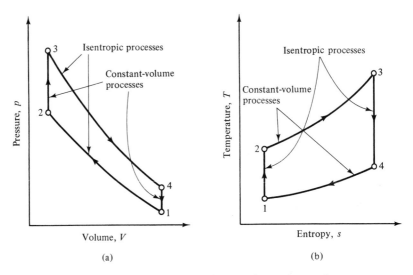

Figure 7.6 *p-V* and *T-s* diagrams for an Otto cycle.

The heat rejection by the system is also at constant volume, and we can write

$$Q_{out} = -Q_{4-1} = -\int_4^1 mc_v \, dT = mc_v(T_4 - T_1)$$

The thermal efficiency is then given by

$$\eta_t = \frac{Q_{in} - Q_{out}}{Q_{in}}$$

$$= \frac{mc_v(T_3 - T_2) - mc_v(T_4 - T_1)}{mc_v(T_3 - T_2)}$$

$$= 1 - \frac{T_4 - T_1}{T_3 - T_2}$$

Introducing the compression ratio

$$r = \frac{V_1}{V_2} = \frac{V_4}{V_3}$$

and noting that for the isentropic processes 1–2 and 3–4

$$T_1 V_1^{\gamma-1} = T_2 V_2^{\gamma-1} \quad \text{and} \quad T_3 V_3^{\gamma-1} = T_4 V_4^{\gamma-1}$$

we obtain

$$\frac{T_2}{T_1} = \frac{T_3}{T_4} = r^{\gamma-1} \quad \text{or} \quad \frac{T_2}{T_3} = \frac{T_1}{T_4}$$

The expression for the thermal efficiency then becomes

$$\eta_t = 1 - \frac{T_4[1 - (T_1/T_4)]}{T_3[1 - (T_2/T_3)]}$$

$$= 1 - \frac{T_4}{T_3}$$

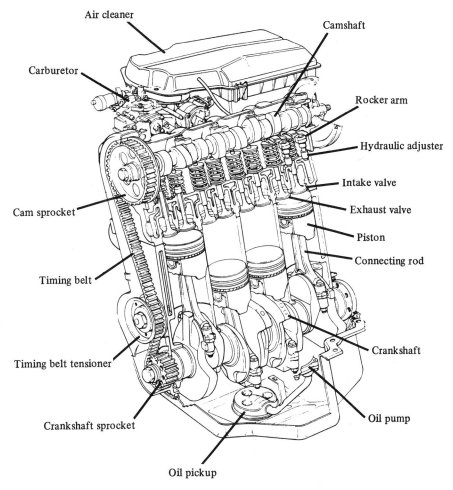

Plate 5 Cross-section of a Chrysler Corporation 2.2 litre, four cylinder Otto cycle engine. (Courtesy of Chrysler Corporation.)

or

$$\eta_{\text{Otto}} = 1 - \frac{1}{r^{\gamma-1}}$$ (7.10)

Equation 7.10 shows that for a given gas as the working medium, the thermal efficiency of an Otto cycle increases with the compression ratio. Figure 7.7 shows the effect of an increase in the compression ratio r on the efficiency for three different values of γ. For $r > 10$, the increase in η with an increase in r is less pronounced, and the curves tend to flatten out. For a given compression ratio, a monatomic gas ($\gamma = 1.67$) as the working fluid results in the highest thermal efficiency. The middle curve represents the theoretical limit for the internal combustion engines. The actual efficiency curve of an internal combustion engine operating on the Otto cycle is much below the middle curve in Figure 7.7. The curve for $\gamma = 1.2$ in the figure would be for a working

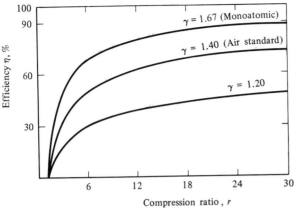

Figure 7.7 Effect of changing the compression ratio on the Otto cycle efficiency.

medium such as ethane. It should be pointed out that as the pressure ratio is increased to improve the efficiency, the pressure at the end of the compression stroke increases dramatically, thus putting a limitation on the design of the engine.

7.4.1 Actual Otto Cycle

In the analysis of the air-standard Otto cycle it was assumed that the working medium behaves as an ideal gas, that specific heats are constant, and that all the processes comprising the cycle are quasistatic or reversible. In practice, none of these assumptions are strictly fulfilled. Due to the deviations from these assumptions, the p-V diagram of an actual cycle looks much different from that in Figure 7.6.

Typically, the cycle is executed in a piston-cylinder arrangement, equipped

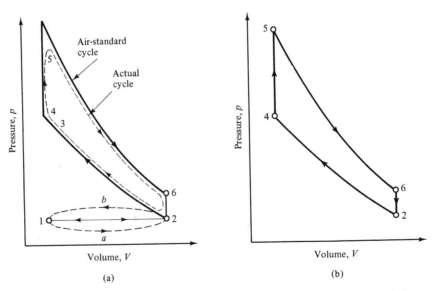

Figure 7.8 Indicator diagram of (a) an actual Otto cycle, and (b) a theoretical Otto cycle.

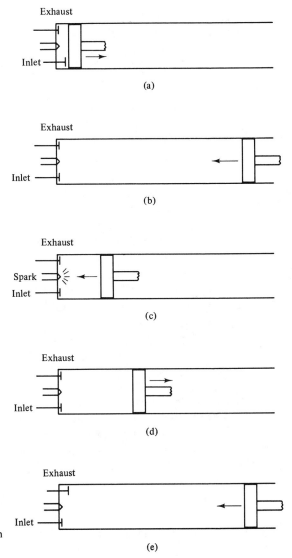

Figure 7.9 Sequence of events in an actual Otto cycle.

with inlet and exhaust valves and a spark plug. Referring to Figures 7.8(a) and 7.9(a), the suction stroke begins at point 1 as the piston moves away from the cylinder head. At this point the inlet valve is open and the air-fuel (gasoline vapor) mixture is taken into the cylinder. Under ideal conditions, this process would be reversible and appear as a constant pressure, solid, horizontal line, line 1–2 in Figure 7.8(a). In practice, due to the pressure drop necessary for the air-fuel mixture to flow into the cylinder and the irreversibilities in the process, the actual process would appear as a dotted line, line 1–a–2. At point 2, the piston reaches the BDC and the suction stroke is completed.

The compression stroke begins with the inward motion of the piston, that is, toward the cylinder head [Figure 7.9(b)]. At the start of the compression stroke, both the valves are closed and as the compression stroke proceeds, the air-fuel charge is compressed with a corresponding rise in the gas temperature. Shortly before the TDC is reached, a spark from the spark plug Figure 7.9(c) ignites the compressed charge, point 3 in Figure 7.8(a), and in a very short time the pressure within the cylinder rises to point 5. This entire process appears as curve 2–3–4–5, shown in Figure 7.8(a). In the air-standard Otto cycle, there is no spark ignition; instead, the ideal gas in the cylinder is compressed from state 2 to state 4 [Figure 7.8(b)] and then heat energy is supplied to the gas in a reversible constant volume process, raising its pressure until state 5 is reached.

The expansion, or power, stroke begins at point 5 with the outward movement of the piston, and the products from the combustion of the fuel expand [Figure 7.9(d)]. In practice, this appears as the dotted curve 5–6, shown in Figure 7.8(a). The curve deviates somewhat from the insentropic expansion curve 5–6 in Figure 7.8(b) for the air-standard Otto cycle.

The exhaust valve opens before point 6, and the exhaust stroke begins (Figure 7.9(e)). This serves to remove the combustion gases from the cylinder. Since a pressure difference is necessary for the flow of the exhaust gases, the exhaust stroke appears as curve 6–2–b–1, as in Figure 7.8(a). The pressure in the cylinder during the exhaust stroke is higher than the pressure at the end of the stroke. In the air-standard Otto cycle, heat energy is reversibly removed from the ideal gas at constant volume. This appears as process 6–2, shown in Figure 7.8(b). In practice, the inlet valve opens before the TDC and the exhaust valve opens before the BDC. Also, since we used four stroke to execute the actual Otto cycle, it is known as a *four-stroke cycle*.

It is possible to execute all the operations of a four-stroke engine in two strokes, in which case the engine is known as a *two-stroke engine*. In such an engine, the crankcase is sealed and the outward movement of the piston is used to compress the charge in the crankcase. In the latter half of the power stroke, the inlet and the exhaust ports —which are on the side of the cylinder—are opened so that the incoming charge drives the exhaust gases out of the cylinder.

There are a number of additional deviations from the air-standard in the actual Otto cycle. The working medium in an actual cycle is not standard air. For part of the cycle, it is an air-fuel mixture, and for the rest of the cycle, it is the products of combustion, namely, carbon dioxide, carbon monoxide, oxides of nitrogen, unburnt fuel. water vapor, oxygen, and nitrogen. In the air-standard cycle we assume the specific heats to be constant. Even for an ideal gas, the specific heats change with temperature, although the difference between c_p and c_v remains constant. With such behavior the value of γ is less at high temperatures than it is at low temperatures. Consequently, to reach the same peak temperature, more heat energy is required than for the case where the specific heats remain constant. Another effect is that, for a given compression ratio, the expansion in the power stroke results in a higher exhaust temperature and a greater heat rejection. Furthermore, the high temperatures reached inside the cylinder in the combustion process, enhance molecular dissociation. Molecular dissociation is endothermic, that is, it requres thermal energy. All these factors contribute to a greater than the ideal heat input.

The processes in an actual engine are not reversible. Also, they are neither adiabatic nor constant volume. Heat energy is constantly removed from the engine by the coolant. Finally there are pumping losses in an actual cycle that are nonexistent in an air-standard cycle. During the suction stroke, there must be a negative gauge pressure in the cylinder, and during the exhaust stroke, there must be positive gauge pressure. All such deviations from the air-standard cycle result in a significantly less than ideal performance of the engine.

The equation for the thermal efficiency of the air-standard Otto cycle, Equation 7.10, was obtained by the application of the first law *only*, and reversible processes were used throughout the cycle.

In the air-standard Otto cycle, there is no degradation of energy, and the work output of the cycle can always be used to restore the heat source and the sink to their original conditions. This is not so with the actual Otto cycle. The combustion of the fuel and the accompanying release of heat energy is an irreversible process, resulting in a large increase in the entropy of the universe. The degradation of the energy in the process is significant. We can calculate the decrease in the availability of the system in all the processes of actual Otto cycle using Equation 6.42. Such a calculation would show that in the actual cycle, high-grade energy is converted into low-grade energy. In most cases the work output (high-grade energy) of the actual cycle is *ultimately* dissipated. It is this feature of the actual cycles that contributes to the "energy crisis." The degeneration of energy is not evident from the first-law analysis of a cycle.

7.4.2 Performance of An Actual Engine

Referring to the *p-V* diagrams for the air-standard and actual Otto cycle, Figure 7.8(b) and (a), respectively, we notice that the air-standard Otto cycle has only one loop. This loop is traced clockwise and its area represents the net work output of the cycle. The actual Otto cycle has two loops, one traced clockwise (loop 2–3–4–5–6–2) and the other (loop 2–*b*–1–*a*–2) counter clockwise; the former represents work output and the latter represents work input. The net work output of the cycle is the algebraic sum of the areas of the two loops. For a given pressure ratio and peak temperature, the net work output of an actual cycle is less than that of an air-standard cycle.

When a thermodynamic cycle is traced on a *p-V* diagram, the resulting loop is called *indicator diagram*. When steam engines were in use in the late nineteenth century, the indicator diagram was the principal means of guaranteeing the performance of a steam engine. The diagram is obtained by a device called an *indicator*, which senses the pressure inside the cylinder and the movement of the piston. Since the area of an indicator diagram for a cycle represents the net work output of the cycle, the mean height of the diagram—that is, the area of the diagram when multiplied by the appropriate scale factor and divided by the piston displacement—gives the average pressure within the cylinder during the cycle. This average pressure is called *mean effective pressure* (mep). If an engine operates at n thermodynamic cycles per minute, then the indicated work per minute is

$$\text{Indicated work} = (n)(\text{mep})(PD)$$

For a four-stroke engine, because there is one thermodynamic cycle requiring four strokes which are executed in two revolutions,

$$n = \frac{\text{number of revolutions per minute}}{2} = \frac{N}{2}$$

where N is the engine rpm; and for a two-stroke engine,

$$n = \text{number of revolutions per minute} = N$$

Also,

$$PD = (\text{number of cylinders}) \times (\text{area of the piston face}) \times (\text{stroke})$$

$$= N_c \frac{\pi}{4} D^2 L$$

where N_c is the number of cylinders of the engine.

In Figure 7.8(b) the area of the p-V diagram of the air-standard Otto cycle gives the *indicated theoretical work ouput* of the cycle. In Figure 7.8(a), the net area of the p-V diagram—that is, the algebraic sum of the areas of the two loops—gives the *actual indicated work output*. This actual work output is the net work delivered to the piston face. This output is greater than the work output from the engine shaft, which is often called the *brake work output*. The term *brake* is used because in general a band brake is applied to the flywheel on the shaft, and the output torque T of the engine is determined. The product of the torque and the angular speed of the shaft gives the brake work output per unit time. The difference between the actual indicated work output and the brake work output is expended in overcoming the friction within the engine and the friction between the moving parts of the engine and the surrounding air.

We can now define the follwing terms for an engine.

$\dot{W}_b$: Brake work output per unit time, which equals $2\pi NT$, where N is the engine rpm and T is the torque measured at the shaft.

$\dot{W}_i$: Actual indicated work output per unit time, which equals $n(\text{mep})(PD)$.

$\dot{Q}$: Energy supplied per unit time to produce $\dot{W}_i$ or $\dot{W}_b$. It is the same as the rate of heat supply, or $\dot{Q}_{in}$, which equals $\dot{m}_f(\text{HV})$, where $\dot{m}_f$ is the mass flow rate of the fuel (per minute) and HV is the heating value of the fuel.

Figure 7.10 shows the relationship among $\dot{Q}$, $\dot{W}$, $\dot{W}_i$, and $\dot{W}_b$. We define the indicated thermal efficiency η_i by

$$\eta_i = \frac{\text{actual indicated work output per unit time}}{\text{energy supplied per unit time}}$$

$$= \frac{\dot{W}_i}{\dot{Q}} \tag{7.11}$$

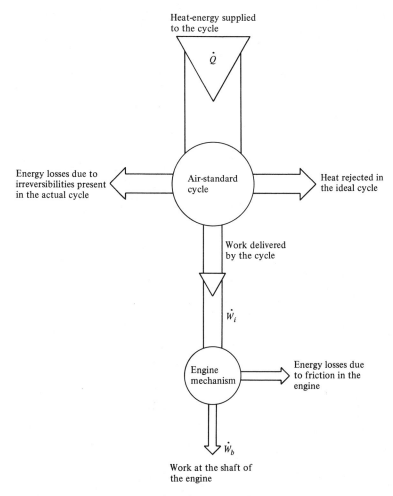

Heat-energy supplied
to the cycle

$\dot{Q}$

Energy losses due to
irreversibilities present
in the actual cycle

Air-standard
cycle

Heat rejected in
the ideal cycle

Work delivered
by the cycle

$\dot{W}_i$

Engine
mechanism

Energy losses due
to friction in the
engine

$\dot{W}_b$

Work at the shaft of
the engine

Figure 7.10 Distribution of the energy input to a power cycle.

Similarly, the brake thermal efficiency, η_b, is defined as

$$\eta_b = \frac{\text{brake work output per unit time}}{\text{energy supplied per unit time}}$$

$$= \frac{\dot{W}_b}{\dot{Q}} \tag{7.11a}$$

Example 7.2

An air-standard, four-stroke, four-cylinder Otto engine has a compression ratio of 8.6 and a piston displacement of 1000 cc.. The conditions at the beginning of the compression stroke are 100 kPa and 18°C. The amount of heat supplied per cylinder per cycle is 135 J. Determine the thermal efficiency and the pressure and temperature at the end of the heat supply process.

305

Solution:

Given: An air-standard Otto cycle operates with the following conditions: $r = 8.6$, $PD = 1000$ cm³, $N_c = 4$, $p_1 = 100$ kPa, $t_1 = 18°C$ or $T_1 = 291.15$ K, $Q_{2-3} = 135$ J/cycle.

Objective: We are to calculate η_{Otto}, p_3, and t_3.

Diagram:

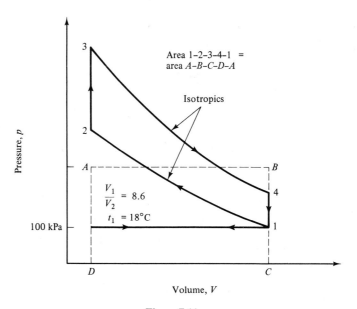

Figure 7.11

Assumptions: The assumptions are those for an air-standard cycle.

1. Air behaves like an ideal gas.
2. Specific heats are constant.
3. The cyclic process is reversible.
4. There is no dissociation at the end of the heating process.

Relevant physics: The thermal efficiency of the cycle can be determined from Equation 7.10:

$$\eta_{Otto} = 1 - \frac{1}{r^{\gamma-1}} \tag{7.10}$$

Since r and T_1 are given, the equation for the isentropic compression can be employed to evaluate T_2. Once this is done, T_3 can be determined from

$$Q_{2-3} = mc_v(T_3 - T_2)$$

since Q_{2-3} is prescribed.

To calculate p_3, we can use the equations for the isentropic process between states 1 and 2 and the constant volume process between states 2 and 3.

Analysis: Noting that $\gamma = 1.4$ for air,

$$\eta_{\text{Otto}} = 1 - \frac{1}{r^{\gamma-1}} = 1 - \frac{1}{8.6^{1.4-1}}$$

$$= 57.7\%$$

For the isentropic process 1–2, we have

$$T_1 V_1^{\gamma-1} = T_2 V_2^{\gamma-1}$$

so that the temperature at the end of the isentropic compression is

$$T_2 = T_1(r)^{\gamma-1}$$

$$= 291.15(8.6)^{1.4-1}$$

$$= 688.5 \text{ K}$$

Also

$$V_1 = \text{clearance volume} + \text{piston displacement per cylinder}$$

$$= \left[\left(\frac{1}{r}\right) + 1\right]\left(\frac{PD}{N_c}\right)$$

$$= \left(\frac{1}{8.6} + 1\right)\left(\frac{10^{-3}}{4}\right)$$

$$= 0.28 \times 10 \text{ m}^3$$

The mass of the air in one cylinder may be calculated from

$$m = \frac{p_1 V_1}{RT_1}$$

$$= \frac{100 \times 10^3 \times 0.28 \times 10^{-3}}{287 \times 291.15}$$

$$= 0.334 \times 10^{-3} \text{ kg}$$

We are given that the heat supplied is 135 J per cylinder, per cycle, so that for a constant specific heat

$$Q_{2\text{-}3} = mc_v(T_3 - T_2)$$

or

$$135 = 0.334 \times 10^{-3} \times 0.7165 \times 10^3(T_3 - 688.5)$$

which yields

$$T_3 = 1252.5 \text{ K, or } 979.4°C$$

Since 2–3 is a constant volume process, pressure p_3 is found from

$$p_3 = p_2\left(\frac{T_3}{T_2}\right)$$

Thus

$$p_3 = p_1(r)^{\gamma}\left(\frac{T_3}{T_2}\right)$$

$$= 100(8.6)^{1.4}\left(\frac{1252.5}{688.5}\right)$$

$$= 3700 \text{ kPa}$$

Example 7.3

Determine the theoretical mean effective pressure for the engine of Example 7.2. If the engine operates at 3000 rpm, determine the theoretical rate of work output.

Solution:

Given: The air-standard engine of Example 7.2 is given.

$$Q_{\text{in}} = 135 \text{ J/cycle/cylinder}, \qquad N = 3000 \text{ rpm}, \qquad N_c = 4$$

Objective: The theoretical mep and the rate of work output of the engine are to be determined.

Diagram: See Figure 7.11.

Assumptions: Same as those for Example 7.2.

Relevant physics: We need to know the work per cycle to calculate the mean effective pressure, as well as to find the theoretical power output of the engine. The work output per cycle per cylinder can be calculated from

$$W_{\text{cycle}} \text{ per cylinder} = \eta_t Q_{\text{in}} \text{ per cylinder}$$

Then the mep can be determined from

$$\text{mep} = \frac{W_{\text{cycle}} \text{ per cylinder}}{PD \text{ of one cylinder}}$$

Analysis:

$$W_{\text{cycle}} = 0.577 \times 135 = 77.9 \text{ J/cylinder}$$

The mean effective pressure is given by

$$\text{mep} = \frac{77.9}{10^{-3}/4} = 311.6 \text{ kPa}$$

The theoretical rate of work output is

W_{cycle} per cylinder × number of cylinders × number of cycles per second

$$= 77.9 \times 4 \times \left(1/2 \times 300 \times \frac{1}{60} \right)$$

$$= 7788 \text{ W, or } 7.8 \text{ kW}$$

Example 7.4

The area of an indicator diagram taken for an engine measures 6.5 cm². The scale for the volume axis is 1 cm to 250 cm³. The pressure in the cylinder is sensed by springs, which move the indicator pen by 1 cm for every 50 kPa change in pressure. If the engine operates at 1200 cycles per minute, determine the rate of work output.

Solution:

Given: The data on the area of an indicator diagram for an engine is as follows.

Area of the diagram: 6.5 cm².

Volume-axis scale: 1 cm to 250 cm³.

Pressure-axis scale: 1 cm to 50 kPa.

$$n = 1200 \text{ cycles per minute}$$

Objective: The power output of the engine is to be determined.

Diagram:

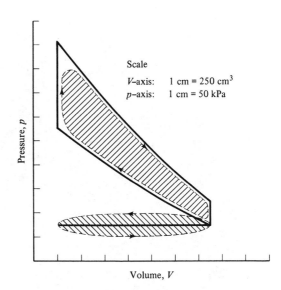

Scale

V-axis: 1 cm = 250 cm^3
p-axis: 1 cm = 50 kPa

Pressure, p

Volume, V

Figure 7-12

Assumptions: None

Relevant physics: The given area of the indicator diagram is the net area; that is, it equals the area of the loop traced clockwise on the diagram minus the area traced counterclockwise on the p-V diagram. This net area is proportional to the work output per cycle. The proportionality factor is reflected by the scale factors for the pressure and the volume areas.

Analysis: The work output per cycle is represented by the 6.5-cm² area of the indicator diagram. This area must be multiplied by the scale factors. Then

$$W_{\text{cycle}} = 6.5 \text{ cm}^2 \times \frac{250 \text{ cm}^3}{\text{cm}} \times \frac{10^{-6} \text{ m}^3}{\text{cm}^3} \times \frac{50 \times 10^3 \text{ Pa}}{\text{cm}}$$

$$= 81.25 \text{ J}$$

$$\text{Power} = 81.25 \times \left(\frac{1200}{60}\right)$$

$$= 1625 \text{ W, or } 1.6 \text{ kW}$$

7.5 THE DIESEL CYCLE

The air-standard Diesel cycle consists of two isentropics, one constant pressure process and one constant volume process. The ideal cycle is shown on p-V and T-s diagrams in Figure 7.13. At the end of the isentropic compression process 1–2, heat energy is supplied to the working medium at constant pressure, process 2–3, until the volume increases to V_3. The supply of heat energy is terminated and the volume V_3 at this point is called *cutoff volume*. The volume ratio (V_3/V_2) is called the *cutoff*

Plate 6 Cutaway of a John Deere diesel engine. (Courtesy of the John Deere Product Engineering Center.)

ratio, r_c. The working medium then isentropically expands from state 3 to state 4. Finally, heat is rejected by the working medium in the constant volume process, process 4.1.

In practice, the working medium does not behave as an ideal gas, the specific heats change with temperature, and irreversibilities are present. An actual Diesel cycle requires suction and exhaust strokes, which are absent in an air-standard Diesel cycle. The p-V diagram for an actual Diesel cycle is shown in Figure 7.14(a) as the dotted loop 1–a–2–3–4–5–6–2–b–1. The solid lines in the figure represent idealized processes.

The cycle is generally executed in a piston-cylinder system equipped with inlet and exhaust valves and a fuel injection system. Various positions of the piston as the cycle is executed are shown in Figure 7.14. The suction stroke begins at point 1, Figure 7.14(a). At this point, the inlet valve is open. As the piston moves out, Figure 7.14(b), the air is drawn into the cylinder. Ideally, this process would be reversible and would appear as the constant pressure solid horizontal line 1–2 in Figure 7.14(a). In practice, due to the pressure drop necessary for the air to flow into the cylinder

Plate 7 Cutaway of a 1980 GM Oldsmobile diesel engine for passenger cars. (Courtesy of General Motors Corporation.)

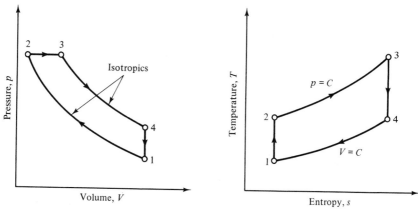

Figure 7.13 p-V and T-s diagrams for the Diesel cycle.

and the irreversibilities in the process, the actual process appears as a dotted line 1-a-2. At point 2, the piston reaches the BDC and the suction stroke is completed.

At the start of the compression stroke, Figure 7.14(c), both the valves are closed. As the compression proceeds along the dotted curve 2–3–4 in Figure 7.14(a), the temperature and the pressure of the air in the cylinder increase. At point 4, a position before the TDC, the temperature of the air is sufficiently high that atomized fuel injected into this air ignites without a spark from a spark plug, Figure 7.14(d). The fuel supply is continued as the piston reaches TDC and begins to move outward until the contents of the cylinder expand to the cutoff volume, V_5, Figure 7.14(e). The piston continues its outward travel until BDC is reached and the combustion gases expand adiabatically along the dotted curve 5–6 in Figure 7.14(a). This stroke is the power stroke of the cycle. Shortly before reaching the BDC, the exhaust valve is opened.

Finally, in its fourth stroke the piston begins to move inward as shown in Figure 7.14(f) and the combustion gases are removed from the cylinder. Since a pressure difference is necessary for the flow of the exhaust gases, the exhaust stroke appears as curve 6–b–1. Ideally, the exhaust stroke would appear as a horizontal straight line. Thus we see that the actual Diesel cycle is executed in four strokes. The indicator diagram of an actual Diesel cycle, like the one for an actual Otto cycle, has two loops —one positive and one negative. The net work output of the cycle is then given by the algebraic sum of the areas enclosed by the loops 2–3–4–5–6–2 and 1–a–2–b–1.

The thermal efficiency of the air-standard Diesel cycle can be determined by considering the heat supplied and the heat rejected. Referring to Figure 7.13, we have,

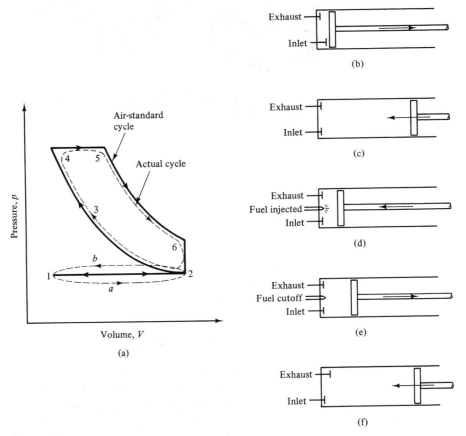

Figure 7.14 Actual Diesel cycle.
(a) Indicator diagram.
(b)–(f) Sequence of events.

for the constant pressure process 2–3,

$$Q_{in} = mc_p (T_3 - T_2)$$

For the constant volume process 4–1

$$Q_{out} = mc_v(T_4 - T_1)$$

The thermal efficiency is then given by

$$\eta_t = \frac{Q_{in} - Q_{out}}{Q_{in}}$$

$$= \frac{mc_p(T_3 - T_2) - mc_v(T_4 - T_1)}{mc_p(T_3 - T_2)}$$

$$= 1 - \frac{1}{\gamma} \frac{T_4 - T_1}{T_3 - T_2}$$

$$= 1 - \frac{1}{\gamma} \frac{T_1[(T_4/T_1) - 1]}{T_2[(T_3/T_2) - 1]}$$

Noting that

$$T_1 V_1^{\gamma-1} = T_2 V_2^{\gamma-1} \quad \text{and} \quad T_3 V_3^{\gamma-1} = T_4 V_4^{-\gamma 1}$$

and

$$r_c = \frac{V_3}{V_2} \quad \text{and} \quad r = \frac{V_1}{V_2} = \frac{V_4}{V_2}$$

we obtain

$$\frac{T_2}{T_1} = r^{\gamma-1} \quad \text{and} \quad \frac{T_4}{T_3} = \left(\frac{V_3}{V_2} \times \frac{V_2}{V_4}\right)^{\gamma-1} = \left(\frac{r_c}{r}\right)^{\gamma-1}$$

Also, since 2–3 is a constant pressure process,

$$\frac{T_3}{T_2} = \frac{V_3}{V_2} = r_c$$

Hence

$$\frac{T_4}{T_1} = \frac{T_4}{T_3} \cdot \frac{T_3}{T_2} \cdot \frac{T_2}{T_1} = \left(\frac{r_c}{r}\right)^{\gamma-1} \cdot r_c \cdot r^{\gamma-1} = r_c^\gamma$$

The expression for the thermal efficiency,

$$\eta_t = 1 - \frac{1}{\gamma} \frac{T_1}{T_2} \frac{[T_4/T_1 - 1]}{[T_3/T_2 - 1]}$$

then becomes

$$\eta_{\text{Diesel}} = 1 - \frac{1}{\gamma} \left(\frac{1}{r}\right)^{\gamma-1} \left[\frac{r_c^\gamma - 1}{r_c - 1}\right] \qquad (7.12)$$

For the same compression ratio, the Otto cycle has a greater work output than the Diesel cycle, and the two cycles reject the same quantity of heat energy. This can be seen by comparing the Otto cycle, 1–2–3*A*–4–1, and the Diesel cycle, 1–2–3*D*–4–1, on the *T-s* diagrams in Figure 7.15(a). Consequently, for the same *r*-value, the Otto cycle is more efficient than the Diesel cycle. If the maximum temperature is the same for both cycles, then the Diesel cycle has greater work output than the Otto cycle, they both reject the same quantity of heat energy, and the Diesel cycle is more efficient than the Otto cycle. The T-s diagrams in Figure 7.15(b) for the Otto cycle, 1–2A–3–4–1, and the Diesel cycle, 1–2D–3–4–1, illustrate this fact. Also, the thermal efficiency of the Diesel cycle decreases with an increase in the cutoff ratio.

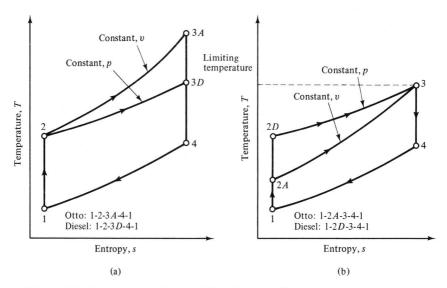

Figure 7.15 Comparison of Otto and Diesel cycles with reference to T-s diagrams.

Example 7.5

An air-standard Diesel cycle has a compression ratio of 20 and heat is transferred to the working medium at a constant pressure for 4% of the power stroke. The pressure and the temperature at the beginning of the compression stroke are 100 kPa and 20°C, respectively. Calculate the temperature and the pressure at the beginning of each process in the cycle. Determine the thermal efficiency and the mean effective pressure; $c_p = 1.0031$ kJ/kg K.

Solution:

Given: An air-standard Diesel cycle has the following characteristics.

$$p_1 = 100 \text{ kPa} \qquad T_1 = 20 + 273.15 = 293.15 \text{ K}$$

$$r = 20 \qquad \qquad \frac{v_3 - v_2}{v_1 - v_2} = 0.04$$

$$c_p = 1.0031 \text{ kJ/kg} \cdot \text{K}$$

Objective: We have to calculate T_2, p_2; T_3, p_3; T_4, p_4; η_t; and mep.

Diagram: The *p-V* and *T-s* diagrams for the cycle are shown in Figure 7.16.

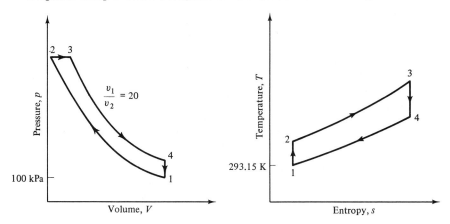

Figure 7.16

Assumptions:

1. Air behaves like an ideal gas.
2. Specific heats are constant.
3. The cyclic process is reversible.
4. There is no dissociation at the end of the heating process.
5. The mass of the air handled per cycle is 1 kg.

Relevant physics: The temperature and pressure at point 1 in the cycle are known. From the ideal gas law, v_1 can be calculated. The isentropic process equation and the prescribed compression ratio permit computation of properties at point 2. In a similar manner, the properties at points 3 and 4 can be calculated.

The thermal efficiency can be calculated from Equation 7.12 and mep equals $W_{\text{cycle}}/\text{PD}$.

Analysis:

$$p_1 = 100 \text{ kPa}, \qquad T_1 = 20 + 273.15 = 293.15 \text{ K}$$

$$v_1 = \frac{RT_1}{p_1} = \frac{287 \times 293.15}{100 \times 10^3} = 0.841 \text{ m}^3/\text{kg}$$

$$r = \frac{v_1}{v_2} = 20 \qquad v_2 = \frac{0.841}{20} = 0.042 \text{ m}^3/\text{kg}$$

For the isentropic compression process 1–2,

$$T_2 = T_1 \left(\frac{v_1}{v_2}\right)^{\gamma - 1} \qquad \text{and} \quad p_2 = p_1 \left(\frac{v_1}{v_2}\right)^{\gamma}$$

$$= 293.15(20)^{1.4-1} \qquad\qquad = 100(20)^{1.4}$$

$$= 971.63 \text{ K} \qquad\qquad\qquad = 6628.9 \text{ kPa}$$

From the data supplied,

$$\frac{v_3 - v_2}{v_1 - v_2} = 0.04$$

or

$$v_3 - v_2 = 0.04 v_2 \left(\frac{v_1}{v_2} - 1\right)$$

$$= 0.04(20 - 1) v_2$$

$$= 0.76 v_2$$

and

$$v_3 = 1.76 v_2 \quad \text{or} \quad r_c = 1.76$$

For the constant pressure process 2–3,

$$\frac{T_3}{T_2} = \frac{v_3}{v_2}$$

so that

$$T_3 = 971.63 \times 1.76$$

$$= 1710 \text{ K}$$

$$p_3 = p_2 = 6628.9 \text{ kPa}$$

For the isentropic expansion process 3–4,

$$T_4 = T_3 \left(\frac{v_3}{v_4}\right)^{\gamma - 1} \qquad \text{and} \quad p_4 = p_3 \left(\frac{v_3}{v_4}\right)$$

$$= 1710 \left(\frac{1.76}{20}\right)^{1.4 - 1} \qquad\qquad = 6628.9 \left(\frac{1.76}{20}\right)^{1.4}$$

$$= 646.8 \text{ K} \qquad\qquad\qquad\quad = 220.6 \text{ kPa}$$

where we have used the fact that $v_4 = v_1$.

The thermal efficiency can now be calculated from Equation 7.12.

$$\eta_{\text{Diesel}} = 1 - \frac{1}{\gamma r^{\gamma - 1}} \frac{r_c^\gamma - 1}{r_c - 1}$$

$$= 1 - \frac{1}{1.4 \times 20^{1.4 - 1}} \cdot \frac{1.76^{1.4} - 1}{1.76 - 1}$$

$$= 0.658, \text{ or } 65.8\%$$

It is instructive to obtain the efficiency from the considerations of heat supplied and removed. We then have

$$\eta_t = 1 - \frac{Q_{\text{out}}}{Q_{\text{in}}}$$

$$= 1 - \frac{c_v (T_4 - T_1)}{c_p (T_3 - T_2)}$$

$$= 1 - \frac{1}{1.4} \cdot \frac{646.8 - 293.15}{1710 - 971.63}$$

$$= 0.658, \text{ or } 65.8\%$$

which is the same result as before.

The mean effective pressure is given by

$$\text{mep} = \frac{W_{\text{cycle}}}{PD}$$

$$= \frac{\eta_t c_p (T_3 - T_2)}{v_1 - v_2}$$

$$= \frac{0.658 \times 1.0031 \times (1710 - 971.63)}{0.841 - 0.042}$$

$$= 610 \text{ kPa}$$

7.6 THE DUAL CYCLE

In the Otto cycle, the working medium receives heat energy at constant volume; in the Diesel cycle, the heat supply is at constant pressure. In reality, the heat supply occurs partly at constant volume and partly at constant pressure. To represent this behavior, the *dual*, or *mixed*, *cycle* has been proposed. The cycle is shown in Figure 7.17.

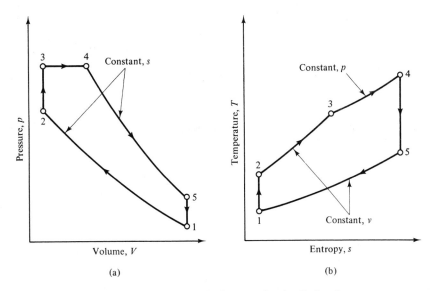

Figure 7.17   *p-V* and *T-s* diagrams for the dual cycle.

The thermal efficiency of the air-standard dual cycle is given by

$$\eta_{\text{dual}} = 1 - \frac{1}{r^{\gamma-1}} \left\{ \frac{\left(\frac{p_3}{p_2}\right)(r_c^{\gamma}) - 1}{\left(\frac{p_3}{p_2} - 1\right) + \gamma \frac{p_3}{p_2}(r_c - 1)} \right\} \qquad (7.13)$$

where r is the compression ratio V_1/V_2, and r_c is the cutoff ratio V_4/V_3.

In practice, the Otto engine uses gasoline as a fuel. However, alcohol or an alcohol-gasoline mixture can also be used as a fuel for the Otto engine. For a given power output, the Otto engine is lighter than the Diesel engine. Also, at low speeds, the former provides a higher torque than the latter. The Otto engine has widespread applications, for example, in automobiles, propeller-driven aircraft, motorcycles, lawnmowers, and small farm implements.

The Diesel engine uses diesel fuel, which is virtually the same as number 2 home-heating oil. Because of the high compression ratio necessary for the auto-ignition in an actual Diesel cycle, the Diesel engine is more efficient than the Otto engine. With the need for fuel economy, some automobile manufacturers are intro-ducing Diesel engines in some of their products. The high compression ratio of the Diesel means high pressure, which in turn means a heavy engine. The Diesel engine has applications in heavy earth-moving machinery, trucks, and marine engines.

7.7 THE STIRLING CYCLE

The air-standard Stirling cycle consists of two isothermal processes and two constant volume processes. The p-V and T-s diagrams of the cycle are shown in Figure 7.18. The heat supply occurs during the isothermal process at T_H and the heat is rejected during the isothermal process at T_L. There is also heat transfer in the constant volume processes 2–3 and 4–1. Heat is transferred from the working medium to a regenerator during process 4–1 and the same regenerator transfers heat to the working medium during the constant volume process 2–3.

A regenerator—a device which alternately stores and releases heat energy—is an essential component of a Stirling-cycle engine. The thermal mass (product of

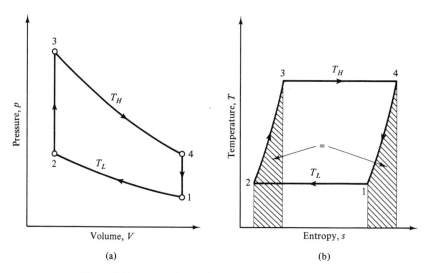

Figure 7.18   p-V and T-s diagrams for the Stirling cycle.

Fuel Nozzle

Cooled Exhaust
Outlet

Preheater Spiral
Passages

Preheater
Assembly

Hot Exhaust

Hot Space

Regenerator

Cylinder

Cooler Tubes

Cold Space

Power Piston

Rhombic Drive

Power Piston
Connecting Rod

Timing Gears

Combustion Chamber

Heater Tubes

Hot Combustion Air

Displacer Piston

Combustion Air
Inlet

Cooling Water
Connections

Seal

Buffer Space

Seal Assembly

Displacer Piston Rod

Power Piston Rod

Power Piston Yoke

Power Piston Yoke
Pin

Displacer Piston
Connecting Rod

Displacer Piston
Yoke

Plate 8 (a) The Stirling engine developed by General Motors Corporation for the Stir-Lec hybrid vehicle. The hybrid arrangement allows the use of a 6 kW engine to provide electric power for moderate constant speed driving, with reserve power stored in a bank of batteries. Courtesy of the General Motors Research Laboratories.

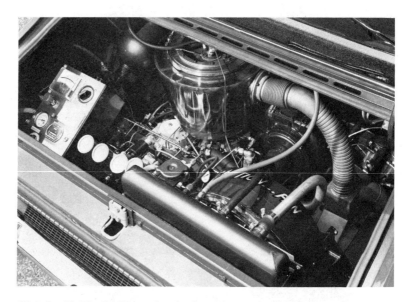

mass and specific heat) of a regenerator is usually high, thus enabling it to store and release large amounts of thermal energy with only a modest change in its temperature. First, a hot fluid is passed through the regenerator so that there is a heat transfer from the hot fluid to the regenerator. This operation stores energy in the regenerator. When the energy stored in the regenerator is to be utilized, a cold fluid is passed through it. This brings about a heat transfer from the regenerator to the cold fluid.

In the Stirling cycle shown in Figure 7.18, the heat transfer in processes 4–1 and 2–3 is internal. Heat transfer to the system from an external source occurs only during process 3–4, and heat rejection to an external sink occurs only during process 1–2. During these processes the temperatures are held constant at T_H and T_L, respectively. Consequently, the thermal efficiency of the Stirling cycle is the same as that of the Carnot cycle working within identical temperature limits. That is,

$$\eta_{\text{Stirling}} = 1 - \frac{T_L}{T_H} \qquad (7.14)$$

Figure 7.19 shows a hypothetical system that can execute the Stirling cycle consisting of a long cylinder equipped with two pistons facing each other. At the center of the cylinder is an ideal regenerator with negligible volume, but with large heat capacity. The left side of the cylinder is at T_H, while the right side is at T_L. At point 1 in Figure 7.19, the right and the left pistons are at their extreme right positions. In process 1–2, the right piston moves to the left, while the left piston is stationary, thus compressing the working medium. During this process heat is removed from the right cylinder, while maintaining the contents at T_L. In the constant volume process

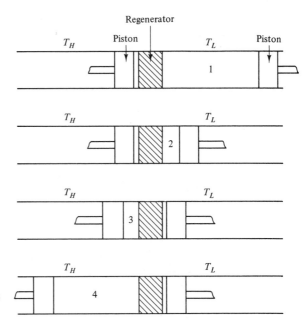

Figure 7.19 A hypothetical system to execute the Stirling cycle.

2–3, both the pistons move to the left, the working medium passes through the regenerator, and it receives heat from the regenerator raising the pressure to 3. Then, in the isothermal process 3–4, the right piston is stationary and the working medium receives heat energy. The medium expands pushing the left piston to the far left position, state 4 in Figure 7.19. In the final segment of the cycle, both the pistons move to the right, while maintaining the volume constant. In this process the working medium once again passes through the regenerator, storing heat in it.

It should be noted that it would be difficult to build a device of the type described in the preceding paragraph to execute the Stirling cycle. This is due to the discontinuous motion of the pistons in the device.

Like the actual Otto and Diesel cycles, in a practical Stirling cycle there would be irreversibilities and accompanying losses in work, resulting in an efficiency that is less than the air-standard efficiency. Furthermore, the regenerator in the cycle would not operate at 100 percent efficiency in practice: that is, all the energy given up by the working medium in process 4–1 (Figure 7.18) to the regenerator would not be available in process 2–3.

In an engine operating on Stirling cycle, it is possible to burn the fuel externally and supply heat at constant temperature to the engine. This is not the case in the Otto and Diesel engines, where fuel is burned inside the cylinder, requiring special types of fuels. With the external combustion for a Stirling engine, it is possible to use a variety of fuels and to insure relatively complete combustion. This makes it relatively easy to control the emissions. The "hot-air," external-combustion Stirling engine was patented in 1916. It competed successfully with the steam engine in certain applications but soon lost ground to the internal-combustion engine. With the world-wide concern for energy and pollution, the interest in the Stirling engine has revived in the last two decades. The Philips Research Laboratories of the Netherlands have successfully developed a Stirling engine suitable for trucks and buses. It has an efficiency of over 50 percent.

7.8 THE ERICSSON CYCLE

The air-standard Ericsson cycle has two isotherms, during which heat is supplied and removed. Instead of the two constant volume processes, as in the Stirling cycle, the Ericsson cycle uses two constant pressure processes, 1–2 and 3–4, during which internal heat transfer occurs. The internal heat transfer involves regeneration similar to the Stirling cycle. The cycle is shown in Figure 7.20. Since the external heat exchange between the surroundings and the system undergoing the Ericsson cycle occurs only at two temperatures, T_H and T_L, the thermal efficiency of the Ericsson cycle is the same as that of a Carnot cycle operating between T_H and T_L (Equation 7.12). The Ericsson cycle was not a commercially attractive cycle until gas turbine power plants came into increasing usage. A gas turbine plant with reheat, intercooling, and regeneration executes a cycle that is close to the Ericsson cycle.

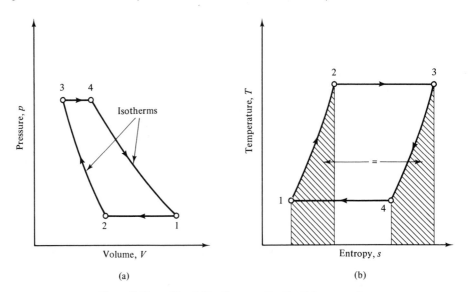

Figure 7.20 p-V and T-s diagrams for the Ericsson cycle.

7.9 THE ATKINSON CYCLE

The Atkinson cycle is similar to the Otto cycle, except that the heat rejection occurs at constant pressure instead of at constant volume. This means that the isentropic expansion continues until the pressure at the end of the expansion, p_4, equals the pressure at the beginning of the compression process, p_1 (see Figure 7.21). The increase in the work output per cycle relative to the output of the corresponding Otto cycle is shown by the shaded area. It can be shown that the thermal efficiency of the air-standard Atkinson cycle is given by

$$\eta_{\text{Atkinson}} = 1 - \frac{\gamma}{(V_1/V_2)^{\gamma-1}}\left[\frac{(p_3/p_2)^{1/\gamma}-1}{(p_3/p_2)-1}\right] \tag{7.15}$$

In Equation 7.15, p_3/p_2 is the constant volume pressure ratio and V_1/V_2 is the isentropic compression ratio. The cycle executed by the free-piston engine–gas turbine system is close to the Atkinson cycle. A free-piston engine consists of two opposed pairs of piston-cylinder systems, each system consisting of a large piston and a small piston. All the pistons move on the same axis and no rotary motion is involved.

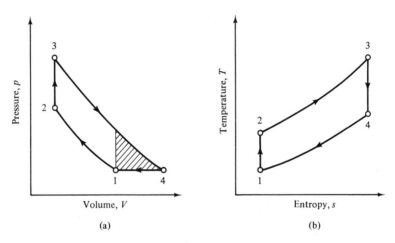

Figure 7.21 p-V and T-s diagrams for the Atkinson cycle.

7.10 THE BRAYTON CYCLE

The air-standard Brayton cycle and its thermal efficiency were discussed in Section 5.8.1. To recapitulate, this cycle consists of two isentropic and two constant pressure processes with heat transfer occuring in the constant pressure processes. The p-v and the T-s diagrams for the cycle are shown in Figure 7.22. There are four components for the execution of the air-standard Brayton cycle—namely, a compressor, two heat exchangers, and a turbine—and together they constitute an external combustion system. Unlike the Otto and the Diesel cycles, the Brayton cycle involves flow processes. The thermal efficiency of the Brayton cycle is given by (see Section 5.8.1)

$$\eta_t = \frac{W_{\text{cycle}}}{q_{\text{in}}}$$

$$= \frac{(h_3 - h_4) - (h_2 - h_1)}{h_3 - h_2}$$

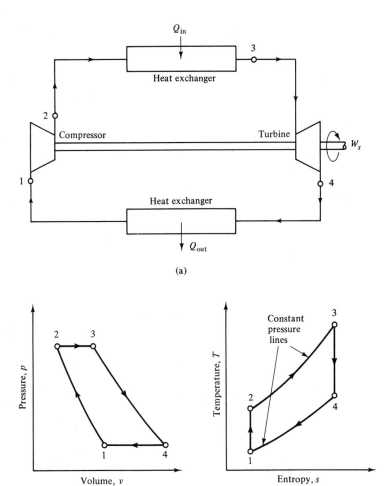

Figure 7.22 A closed Brayton cycle.
(a) Mechanical components.
(b) *p-V* diagram.
(c) *T-s* diagram.

or

$$\eta_{\text{Brayton}} = 1 - \frac{1}{r_p^{(\gamma-1)/\gamma}} \tag{7.16}$$

where r_p is the pressure ratio and is given by

$$r_p = \frac{p_2}{p_1} = \frac{p_3}{p_4} \tag{7.16a}$$

The cycle, as depicted in Figure 7.22(a), is called a *closed cycle*.

When the Brayton cycle is put into practice, the heat supply is provided by burning a fuel in a combustor which receives compressed air from a compressor. The hot, high-pressure gases of combustion are then expanded in a turbine. Consequently, the exhaust from the turbine cannot be reused and must necessarily be discarded, thus obviating the second heat exchanger in Figure 7.22(a). A Brayton cycle operating in this manner is called an *open cycle*. If the exhaust is hot enough, then some of the thermal energy of the exhaust can be recovered and used to heat the compressed air before the air is admitted into the combustor.

Plate 9 The compressor and turbine rotor of General Electric Company's MS6001 heavy duty gas turbine rated at 34 MW suspended above the bottom-half casing. (Courtesy of General Electric Company.)

If a graph of efficiency versus pressure ratio (see Figure 7.23) is plotted for the Brayton cycle, it shows increasing efficiency with increasing pressure ratio. An examination of Equation 7.16(a) also reveals the same trend. The increase in the efficiency is quite rapid for $r_p < 10$. For larger values of r_p, the slope of the efficiency curve decreases. In a practical situation, the pressure ratio cannot be increased indefinitely. For a given heat supply, an increase in r_p also increases the peak temperature in the cycle. There are structural and metallurgical limitations on the peak temperature, which is not usually allowed to exceed 1200°C.

It is interesting to study the effect of change of pressure ratio on the net work output of the cycle when the maximum and minimum temperatures are fixed. Three Brayton cycles, A, B, and C, are shown in Figure 7.24. In cycle A, the pressure ratio

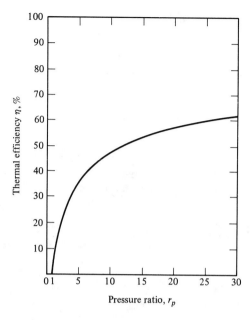

Figure 7.23 Effect of pressure ratio on the Brayton cycle efficiency.

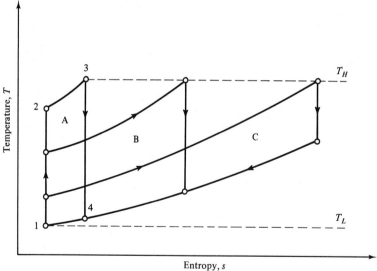

Figure 7.24 Effect of pressure ratio on the work output of a Brayton cycle with fixed T_H and T_L.

is very high, resulting in a significant temperature increase in the compression process and requiring relatively little heat supply to reach the fixed peak temperature, T_H. In cycle B, the pressure ratio is moderate and the heat supply is greater than that for cycle A. Cycle B produces a greater work output, as measured by the area enclosed on the T-s diagram, than that for cycle A. In cycle C, the pressure ratio is very small, requiring a large supply of heat to reach T_H. The net work of this cycle is less than that of cycle B. A comparison of the three cycles leads us to conclude that there must be an optimum pressure ratio for which the net-work output of the Brayton cycle is a maximum. For a unit mass of the working medium this optimum pressure ratio, $r_{p,opt}$ can be determined as follows.

$$w_{cycle} = (h_3 - h_4) - (h_2 - h_1)$$

$$= c_p T_1 \left(\frac{T_3}{T_1} - \frac{T_4}{T_1} - \frac{T_2}{T_1} + 1 \right)$$

Since 1–2 and 3–4 are isentropic processes,

$$\frac{T_4}{T_3} = r_p^{(1-\gamma)/\gamma} \quad \text{and} \quad \frac{T_2}{T_1} = r_p^{(\gamma-1)/\gamma}$$

Also, $T_3 = T_H$ and $T_1 = T_L$ so that

$$\frac{T_4}{T_1} = \frac{T_4}{T_3} \cdot \frac{T_3}{T_1} = r_p^{(1-\gamma)/\gamma} \cdot \frac{T_H}{T_L}$$

and

$$w_{cycle} = c_p T_L \left(\frac{T_H}{T_L} - \frac{T_H}{T_L} \cdot r_p^{(1-\gamma)/\gamma} - r_p^{(\gamma-1)/\gamma} + 1 \right)$$

Setting the derivative of w_{cycle} with respect to r_p equal to zero and solving the resulting equation, we obtain

$$r_{p,opt} = \left(\frac{T_H}{T_L} \right)^{\gamma/(2\gamma-2)} \tag{7.17}$$

and

$$w_{cycle,max} = c_p T_L \left(\sqrt{\frac{T_H}{T_L}} - 1 \right)^2 \tag{7.17a}$$

For an air-standard Brayton cycle ($\gamma = 1.4$) operating between 1200 K and 300 K, the optimum pressure ratio for maximum work works out to be 11.3.

7.10.1 Regenerative Brayton Cycle

It was noted earlier that if the exhaust from the turbine is hot enough, it is possible to recover some of the thermal energy of the turbine exhaust. This possibility is depicted in Figure 7.25, which shows that the temperature of the turbine exhaust (state 4) is higher than the temperature of the air leaving the compressor (state 2). Therefore, in principle, it is possible to preheat the compressed air to a temperature T_{2R}, equal to T_4, by employing a counter-flow heat exchanger. In a counter-flow heat exchanger, the hot and cold fluids move in opposite directions, making it possible to bring up the cold fluid exit temperature close to the hot fluid entrance temperature. Such a heat exchanger in a Brayton cycle would function as a regenerator. The regeneration does not alter the area enclosed by the cycle on the T-s diagram and the net work output of the cycle remains unchanged. However, regeneration

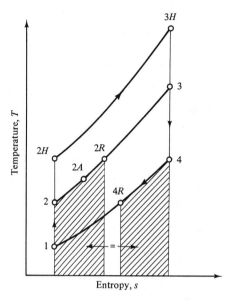

Figure 7.25 Regenerative Brayton cycle.

does reduce the external supply of heat. Thus a regenerator can increase the thermal efficiency of a Brayton cycle.

A careful examination of Figure 7.25 reveals that if the pressure ratio is increased, point 2 will move straight up. Since the turbine exhaust must be at pressure $p_4 = p_1$, point 4 would remain as it is. A sufficient rise in the pressure ratio can move the state of the compressor discharge up to $2H$, so that the temperatures at $2H$ and 4 are identical. For pressure ratios greater than p_H/p_1, regeneration would not be possible.

The regenerator discussed in this section is an ideal one. In practice, a finite temperature difference across the walls of the heat exchanger (regenerator) is necessary for rapid heat transfer. Alternately, a large surface area of the heat exchanger can also be used for rapid heat transfer. However, this would cause large pressure drops along the flow path of the fluid. Thus in a practical regenerator, it would not be possible to preheat the compressed air to state $2R$ (Figure 7.25), but only to $2A$. We then define the efficiency of a regenerator by

$$\eta_R = \frac{h_{2A} - h_2}{h_{2R} - h_2} \tag{7.18}$$

If the specific heat is assumed to be constant, the expression for η_R becomes

$$\eta_R = \frac{T_{2A} - T_2}{T_{2R} - T_2} = \frac{T_{2A} - T_2}{T_4 - T_2} \tag{7.18a}$$

7.10.2 Deviations from the Air-Standard Brayton Cycle

Due to fluid friction and the finite temperature difference between the compressor and the surroundings, the compression process is irreversible. The same is true of the expansion process in the turbine. Consequently, the actual work of compression

$(h_2 - h_1)$ is greater than the isentropic work of compression $(h_{2s} - h_1)$. We define the isentropic compressor efficiency (Figure 7.26(a)) by

$$\eta_{i,C} = \frac{h_{2s} - h_1}{h_2 - h_1} \tag{7.19}$$

In case of the turbine, the actual work output $(h_3 - h_4)$ is less than the isentropic work output $(h_3 - h_{4s})$. We define the isentropic turbine efficiency by

$$\eta_{i,T} = \frac{h_3 - h_4}{h_3 - h_{4s}} \tag{7.19a}$$

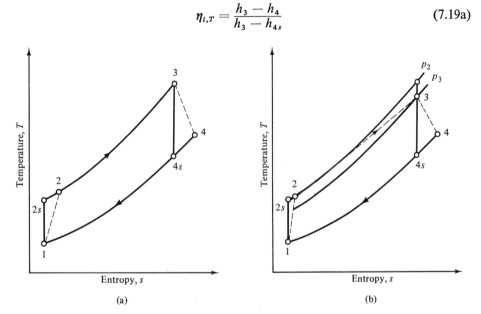

Figure 7.26   Deviations from the ideal Brayton cycle.

Typically the isentropic compressor and turbine efficiencies are of the order of 85 percent. As the compressor and the turbine performances deviate from the ideal, progressively larger work input to the compressor is required, the turbine output diminishes, and the net work output of the cycle rapidly decreases. If the isentropic efficiencies were to drop below 60 percent, hardly any net work output would be available from the cycle.

In actual practice, there is a pressure drop in the combustion chamber due to fluid friction. This means that pressure p_3 at the turbine inlet is less than the pressure p_2 at the compressor discharge. This is shown in Figure 7.26(b). In this figure, path 1–2–3–4 represents the actual open cycle with irreversibilities. Note that the area enclosed by the dotted path does not represent net work output of the cycle, since the processes are not reversible.

Example 7.6

In a gas turbine cycle, the compressor compresses air from 100 kPa and 22°C to 600 kPa. The turbine inlet temperature is 800°C. It is known that a regenerator with

80% efficiency is available. Determine the improvement in the efficiency resulting from the installation of the regenerator. Assume $\gamma = 1.4$ and $c_p = 1.03$ kJ/kg K.

Solution:

Given: A regenerative gas turbine plant has the following characteristics.

$$p_1 = 100 \text{ kPa} \qquad T_1 = 22 + 273.15 = 295.15 \text{ K}$$

$$p_2 = 600 \text{ kPa} \qquad T_3 = 800 + 273.15 = 1073.15 \text{ K}$$

$$r_p = \frac{p_2}{p_1} = 6 \qquad c_p = 1.03 \text{ kJ/kg K}$$

$$\eta_R = 0.80 \qquad \gamma = 1.4$$

Objective: We have to determine the efficiencies with and without the regenerator and the percent improvement in the efficiency due to the regenerator.

Diagram:

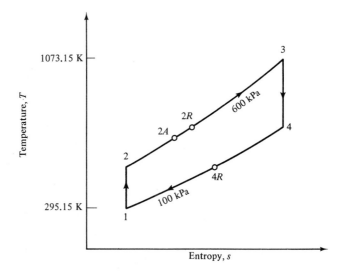

Figure 7.27

Assumptions:

1. Ideal gas behavior can be assumed for the air and the combustion gases.
2. Properties of the combustion gases are the same as those of air.
3. Specific heats are constant.
4. The processes are reversible.

Relevant physics: In order to determine the efficiencies before and after the installation of the regenerator, we need to determine T_2, T_{2R}, T_{2A}, T_4, and T_{4R}. These temperatures can be determined from the equations for the isentropic processes 1–2 and 3–4.

Analysis: For the isentropic processes 1–2 and 3–4,

$$T_2 = T_1 r_p^{(\gamma-1)/\gamma}$$
$$= 295.15(6)^{(1.4-1)/1.4}$$
$$= 492.46 \text{ K}$$

Likewise

$$T_4 = T_3 r_p^{-(\gamma-1)/\gamma}$$
$$= 1073.15(6)^{-(1.4-1)/1.4}$$
$$= 643.18 \text{ K}$$

For a constant specific heat,

$$\eta_R = \frac{T_{2A} - T_2}{T_4 - T_2} \tag{7.16a}$$

or

$$T_{2A} = \eta_R(T_4 - T_2) + T_2$$
$$= 0.8(643.18 - 492.46) + 492.46$$
$$= 613.04 \text{ K}$$

Thermal efficiency without regeneration:

$$\eta_t = \frac{c_p(T_3 - T_4) - c_p(T_2 - T_1)}{c_p(T_3 - T_2)}$$
$$= 1 - \frac{T_4 - T_1}{T_3 - T_2}$$
$$= 1 - \frac{643.18 - 295.15}{1073.15 - 492.46}$$
$$= 0.40, \text{ or } 40\%$$

Thermal efficiency with regeneration:

$$\eta_t = \frac{c_p(T_3 - T_4) - c_p(T_2 - T_1)}{c_p(T_3 - T_{2A})}$$
$$= \frac{(1073.15 - 643.18) - (492.46 - 295.15)}{1073.15 - 613.04}$$
$$= 0.506, \text{ or } 50.6\%$$

The improvement in the efficiency due to the installation of the regenerator is

$$\frac{0.506 - 0.40}{0.40}, \text{ or } 26\%$$

Example 7.7

Rework Example 7.6 if the isentropic efficiencies of the compressor and the turbine are 0.90 and 0.85, respectively.

Solution:

Given: A regenerator gas turbine plant has the following data.

$$p_1 = 100 \text{ kPa} \qquad T_1 = 295.15 \text{ K}$$
$$p_2 = 600 \text{ kPa} \qquad T_3 = 1073.15 \text{ K}$$
$$r_p = 6 \qquad c_p = 1.03 \text{ kJ/kg K}$$
$$\eta_R = 0.8 \qquad \gamma = 1.4$$
$$\eta_{i,c} = 0.90 \qquad \eta_{i,T} = 0.85$$

Objective: The thermal efficiencies with and without the regenerator are to be determined, and then the improvement in the efficiency due to the regenerator is to be calculated.

Diagram:

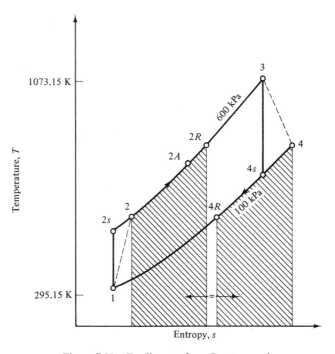

Figure 7.28   *T-s* diagram for a Brayton cycle.

Assumptions:

1. Ideal gas behavior can be assumed for the air and the combustion gases.
2. Properties of the combustion gases are the same as those of air.
3. Specific heats are constant.

Relevant physics: The actual compression takes path 1–2 in Figure 7.28, and the actual expansion occurs as process 3–4. Exhaust at point 4 is available for regeneration. Since the regenerator is only 80% efficient, the compressed air would leave the regenerator at state 2A. The various temperatures can be calculated from the process equations and Equations 7.19 and 7.19(a).

Analysis: From the data and the solution of Example 7.6, we have

$$r_p = 6$$

$$T_1 = 295.15 \text{ K} \qquad T_3 = 1073.15 \text{ K}$$

$$T_{2s} = 492.46 \text{ K} \qquad T_{4s} = 643.18 \text{ K}$$

For a constant c_p, Equation 7.19 for the isentropic compressor efficiency becomes

$$\eta_{i,c} = \frac{T_{2s} - T_1}{T_2 - T_1}$$

Upon rearrangement, we obtain

$$T_2 = T_1 + \frac{T_{2s} - T_1}{\eta_{i,c}}$$

$$= 295.15 + \frac{(492.46 - 295.15)}{0.9}$$

or

$$T_2 = 514.38 \text{ K}$$

For a constant c_p, Equation 7.19(a) for the isentropic turbine efficiency becomes

$$\eta_{i,T} = \frac{T_3 - T_4}{T_3 - T_{4s}}$$

A rearrangement gives

$$T_4 = T_3 - \eta_{i,T}(T_3 - T_{4s})$$

$$= 1073.15 - 0.85(1073.15 - 643.18)$$

or

$$T_4 = 707.68 \text{ K}$$

Also

$$T_{2A} = \eta_R(T_4 - T_2) + T_2$$

$$= 0.8(707.68 - 514.38) + 514.38$$

$$= 669.02 \text{ K}$$

For thermal efficiency without regeneration,

$$\eta_t = \frac{c_p(T_3 - T_4) - c_p(T_2 - T_1)}{c_p(T_3 - T_2)}$$

$$= \frac{(1073.15 - 707.68) - (514.38 - 295.15)}{1073.15 - 514.38}$$

$$= 0.262, \text{ or } 26.2\%$$

For thermal efficiency with regeneration,

$$\eta_t = \frac{c_p(T_3 - T_4) - c_p(T_2 - T_1)}{c_p(T_3 - T_{2A})}$$

$$= \frac{(1073.15 - 707.68) - (514.38 - 295.15)}{1073.15 - 669.02}$$

$$= 0.362, \text{ or } 36.2\%$$

The improvement in the thermal efficiency due to the installation of the regenerator is

$$\frac{0.362 - 0.262,}{0.262}, \text{ or } 38\%$$

Comments: Although the thermal efficiency with regeneration for this problem is comparable with the thermal efficiency without regeneration for Example 7.6, the net work outputs of the cycles are far apart, namely, 151 kJ/kg and 293 kJ/kg, respectively.

7.10.3 Intercooling and Reheat

In order to increase the thermal efficiency of a Brayton cycle, we can increase the pressure ratio, r_p. However, as the pressure ratio is increased, the temperature at the outlet of the compressor increases, causing problems with seals and metal fatigue. Also, the physical size of a compressor increases with the increase in the pressure ratio. To minimize such problems, compression is accomplished in two or more stages, as shown in Figure 7.29(a) with intercooling between stages. Ideally, in intercooling the compressed air at the exit of one stage is cooled to the inlet temperature of that stage and then compressed in the next stge. The total work of compression in the cycle is the sum of the work inputs for each compression stage. For minimum work input, the pressure at the intermediate stage in a two-stage operation is the square root of the product of the high pressure and the low pressure (see Section 7.3.2).

When a large pressure ratio is involved, the expansion process occurs in more than one stage. This permits reheating of the exhaust from one stage before that exhaust is admitted to the next stage. This is shown in Figure 7.29(a). Ideally, at the end of the reheating, the temperature is the same as that at the inlet to the previous stage. The *T-s* diagram for the Brayton cycle with intercooling and reheat is shown in Figure 7.29(b). As can be seen from the *T-s* diagram in the figure, both intercooling and reheat increase the net work available from the cycle. For the cycle shown in Figure 7.29(b),

$$w_{in} = (h_4 - h_3) + (h_2 - h_1)$$
$$w_{out} = (h_5 - h_6) + (h_7 - h_8)$$
$$q_{in} = (h_5 - h_4) + (h_7 - h_6)$$

The thermal efficiency of the cycle with intercooling and reheat can then be calculated in the usual manner. If the compressor and the turbines have isentropic efficiencies of less than 100 percent and if there is also regeneration, the analysis becomes a bit more complex but presents no extraordinary difficulty.

We may ask this question: What effect would there be if there were numerous stages of compression and expansion with intercooling and reheat? The *T-s* diagram for such a cycle is shown in Figure 7.30. If in the limit the pressure change across each stage is infinitesimal, we can replace the two sawtooth curves in the figure by constant temperature processes 2–3 and 4–1. With 1–2 and 3–4 as constant pressure processes, cycle 1–2–3–4 would represent the Ericsson cycle, provided there is ideal regeneration between 1–2 and 3–4. We know that the thermal efficiency of the Ericsson

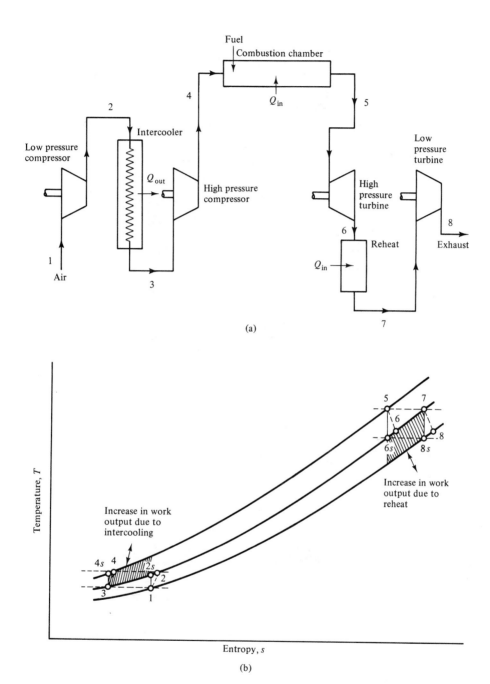

(a)

(b)

Figure 7.29 Brayton cycle with intercooling and reheat.
(a) Mechanical components.
(b) *T-s* diagram.

Plate 10 Experimental "AGT-5" gas turbine passenger car engine developed by General Motors engineering staff. (Courtesy of General Motors Corporation.)

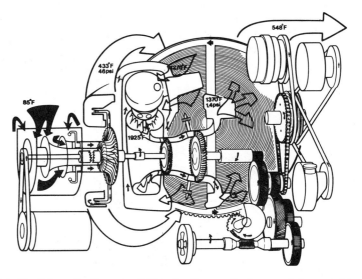

Plate 11 Schematic diagram of Chrysler Corporation's upgraded gas turbine engine. (Courtesy of the Chrysler Corporation.)

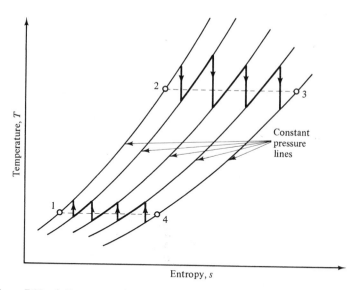

Figure 7.30 A Brayton cycle with a large number of intercooling and reheat stages approaches the Ericsson cycle.

cycle equals that of a Carnot cycle operating between the same temperature limits. Thus a multistage gas turbine cycle with intercooling, reheat, and regeneration approaches the Ericsson cycle as the number of stages is increased. In practice, because of the cost of the devices for intercooling and reheat, the number of stages are limited, and a trade-off between an improvement in operating efficiency and increase in initial cost is achieved.

Example 7.8

In a two-stage gas turbine cycle with ideal intercooling and reheat, the pressure ratio in each stage is 3.5. The inlet conditions are 300 K and 100 kPa and the temperature at the inlet to the turbines is 1300 K. A regenerator with an efficiency of 70% is used to improve the efficiency. Determine the compressor work, the turbine work, and the thermal efficiency of the cycle; $\gamma = 1.4$ and $c_p = 1.03$ kJ/kg K.

Solution:

Given: A two-stage regenerative gas turbine cycle with ideal intercooling and reheat is prescribed.

$$T_1 = T_3 = 300 \text{ K} \qquad r_{p1} = r_{p2} = r_p = 3.5$$
$$T_5 = T_7 = 1300 \text{ K}$$
$$\eta_R = 0.70, \quad \gamma = 1.4 \qquad c_p = 1.03 \text{ kJ/kg K}$$

Objective: We are to determine the compressor work, the turbine work, and the thermal efficiency.

Diagram:

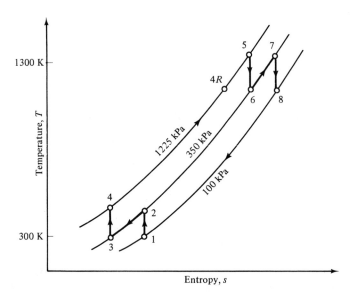

Figure 7.31   *T-s* diagram for a Brayton cycle.

Assumptions:

1. Ideal gas behavior can be assumed for the air and the combustion gases.
2. Properties of the combustion gases are the same as those of air.
3. Specific heats are constant.
4. With the exception of the regeneration process, all other processes in the cycle are reversible.

Relevant physics: This example involves a direct application of Section 7.7.3.
Analysis:

$$T_4 = T_2 = T_1 r_p^{(\gamma-1)/\gamma}$$
$$= 300 \times (3.5)^{(1.4-1)/1.4}$$
$$= 429.11 \text{ K}$$

Likewise,

$$T_6 = T_8 = T_5 r_p^{-(\gamma-1)/\gamma}$$
$$= 1300(3.5)^{-(1.4-1)/1.4}$$
$$= 908.85 \text{ K}$$

The compressor input is

$$w_{in} = c_p[(T_4 - T_3) + (T_2 - T_1)] = 2c_p(T_2 - T_1)$$
$$= 2 \times 1.003(429.11 - 300)$$
$$= 259.0 \text{ kJ/kg}$$

The turbine output is

$$w_{\text{out}} = c_p[(T_5 - T_6) + (T_7 - T_8)] = 2c_p(T_5 - T_6)$$
$$= 2 \times 1.003(1300 - 908.85)$$
$$= 784.64 \text{ kJ/kg}$$

For the regenerator,

$$0.7 = \frac{T_{4A} - T_4}{T_{4R} - T_4}$$

or

$$T_{4A} = 0.7(908.85 - 429.11) + 429.11$$
$$= 764.93 \text{ K}$$

The heat supplied is given by

$$q_{\text{in}} = (h_5 - h_{4A}) + (h_7 - h_6)$$
$$= 1.003(1300 - 764.93 + 1300 - 908.85)$$
$$= 929 \text{ kJ/kg}$$

We can now calculate the thermal efficiency.

$$\eta_t = \frac{w_{\text{out}} - w_{\text{in}}}{q_{\text{in}}}$$
$$= \frac{784.64 - 259}{929}$$
$$= 0.566, \text{ or } 56.6\%$$

Comments: If the entire compression were to take place in a single stage and if there were no regeneration, the efficiency would be 51.1% (Equation 7.14).

Figure 7.22—used for the study of the gas turbine cycle— shows a single shaft connecting the compressor and the turbine. In practice, there may be two turbines and one compressor, with one turbine coupled to the compressor and the other turbine connected to the load. This arrangement is shown in Figure 7.32(a). In such an arrangement, the load fluctuation does not affect the compressor performance. In a two-stage compression operation, a single turbine may provide the work input for both the compressors, the second turbine being free to respond to the load. Alternatively there may be two separate turbine-compressor sets, each with its own shaft. These arrangements are shown in Figure 7.32(b) and (c).

7.10.4 Jet Propulsion

Jet propulsion is based on creating an unbalanced force by exhausting gases at high velocity from a nozzle. This unbalanced force is called *thrust* and it is used for jet propulsion which is based on an open Brayton cycle. An open Brayton cycle consists of a compressor, a combustion chamber, and a turbine. In an open Brayton cycle for a stationary power plant, the turbine produces power in excess of the compressor requirements, the excess being available for driving the load. In a jet propulsion system, the turbine produces just enough power to meet the needs of the compressor.

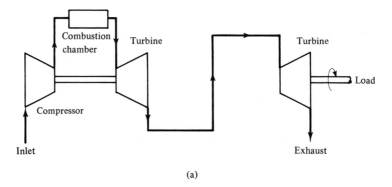

(a)

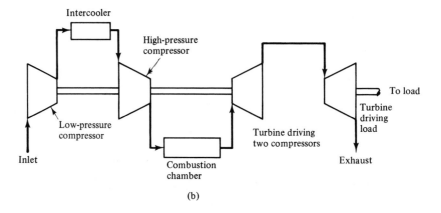

(b)

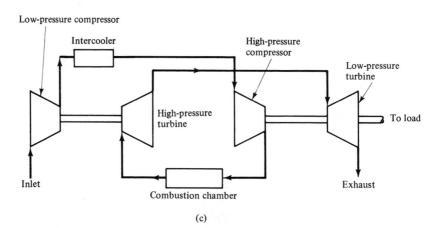

(c)

Figure 7.32 Some possible arrangements of the mechanical components in single-stage and double-stage Brayton cycles.

The thrust is obtained by partially expanding the high-temperature, high-pressure combustion gases in a turbine that is used to drive the compressor, and then expanding the exhaust from the turbine in a nozzle to produce high-velocity gases. The velocity of the gases exiting from the nozzle can be determined if the nozzle inlet and outlet pressures and the inlet temperature are known. It is usually assumed that the expansion in the nozzle is isentropic. Then the thrust can be calculated by an application of the momentum equation.

7.11 SUMMARY

Reciprocating and rotary devices are used for implementing thermodynamic cycles. The terminology for a reciprocating device was first introduced in this chapter. Then various energy conversion systems to compress gases and to produce power were discussed. In the discussion of reciprocating compressors, volumetric efficiency and isothermal compression efficiency were introduced. The air-standard Otto cycle, the Otto cycle in practice, and the deviations from the ideal cycle were discussed. Also discussed were the Diesel cycle, the dual cycle, the Stirling cycle, the Ericsson cycle, the Atkinson cycle, and the Brayton cycle. The Brayton cycle forms the basis for today's gas turbine power plants. For the Brayton cycle, expressions for the thermal efficiency and optimum pressure ratio for maximum net work output were obtained. Regeneration for improving the Brayton cycle efficiency was discussed and regeneration efficiency was defined. In studying the deviations from the ideal Brayton cycle, isentropic efficiencies for the compressor and the turbine were discussed. Modifications to the simple Brayton cycle such as regeneration, intercooling, and reheat, were presented. Also, the basis for jet propulsion was discussed.

PROBLEMS

(Draw p-V and T-s diagrams wherever applicable.)

7.1. A reciprocating single-acting compressor has a piston displacement of 200 cc and a clearance factor of 5%. The compressor is used to compress air from 1 atm to 8 atm. The exponent for compression is 1.34.

(a) Determine the volumetric efficiency of the compressor.

(b) If there is a drop of 7 kPa at the inlet port and 9 kPa at the discharge port, what would be the volumetric efficiency?

7.2. Derive an expression for the pressure ratio of a compressor for a zero volumetric efficiency. For the compressor in Problem 7.1(b), determine the pressure ratio that would result in a zero volumetric efficiency.

7.3. A single-acting reciprocating compressor is needed to compress 50 kg of air per hour from 100 kPa and 25°C to 900 kPa. The compression is expected to follow $pv^{1.25} =$ constant.

(a) Determine the power required to drive the compressor under ideal conditions.

(b) If the compressor has a clearance factor of 4% and if it is expected to operate at 1200 rpm, what is the requisite piston displacement?

(c) Determine the isothermal efficiency of the compressor.

7.4. Carbon dioxide gas is compressed in a reciprocating compressor with a clearance factor of 6%. The inlet conditions are 100 kPa and 270 K, and the delivery pressure is 500 kPa. Calculate each of the following. Assume isentropic compression.

(a) The volumetric efficiency.

(b) The peak temperature.

(c) The rate of work input for compressing 10 kg of the gas per minute.

7.5. Rework Problem 7.4 if the compression follows the relation $pv^{1.1} = $ constant.

7.6. Rework Problem 7.4 if there is a pressure drop of 5 kPa and 10 kPa at the suction and discharge valves, respectively.

7.7. A single-cylinder, double-acting, reciprocating compressor is designed to handle 3000 cm³ of air at 99 kPa per revolution. The air is to be compressed to 594 kPa according to the relation $pv^{1.32} = $ constant. The clearance factor is to be 6% and the stroke to bore ratio is to be 1.1. Calculate the cylinder dimensions.

7.8. Air is to be compressed at a rate of 0.24 m³/sec from 101 kPa and 289 K to 850 kPa. Calculate the piston displacement and the power requirement for a single stage isentropic compression. Assume a clearance factor of 7.8% and neglect the pressure drop at the valves.

7.9. Rework Problem 7.8 assuming the compressor to be a two-stage unit with equal pressure ratio in the two stages. Further, assume perfect intercooling.

7.10. Derive an expression for the work input for a two-stage compressor, with intercooling where the intermediate pressure is given by Equation 7.9. What will be the saving in power if the compression in Problem 7.3 was carried out in two stages with an optimum intermediate pressure?

7.11. Air at 100 kPa and 20°C enters a two stage compressor. The pressure ratio across each stage is same. Air leaves the compressor at 1000 kPa. With ideal intercooling and isentropic compression in each stage, what is the work required per kilogram of air? Compare the work with that in a single-stage compressor.

7.12. The volumetric efficiency η_v of a reciprocating compressor is a function of the pressure ratio, the polytropic exponent n, and the clearance factor CF. Write a computer program to calculate η_v by using increments of 0.05 for n from 1.05 to 1.45 and by using increments of CF from 1% to 10% for a pressure ratio of 6. Then plot curves of η_v against CF with n as a parameter.

7.13. An air-standard cycle consisting of isothermal compression, constant pressure heat supply, isentropic expansion, and constant pressure heat removal is proposed. The conditions at the beginning of the isothermal compression are 98 kPa and 20°C. The volumes at the beginning and the end of the heating process are 0.04 m³/kg and 0.20 m³/kg, respectively. Draw p-v and T-s diagrams of the cycle and calculate the state points at the end of each process. Also, calculate each of the following.

(a) Amount of heat added.

(b) Net work output per kilogram of the working medium.

(c) Thermal efficiency of the cycle.

7.14. Calculate the mean effective pressure for the cycle in Problem 7.13.

7.15. In an air-standard Otto cycle with a compression ratio of 8.5, the conditions at the beginning of the compression stroke are 100 kPa and 27°C. If 200 kJ/kg of heat are supplied, determine the following, assuming a piston displacement of 43,000 cc.
(a) Temperature and pressure at the end of each process.
(b) Thermal efficiency.
(c) Mean effective pressure.

7.16. An Otto cycle with a compression ratio of 7.5 has suction conditions of 98 kPa and 285 K. Find the pressure and temperature at the end of the compression stroke and the air-standard efficiency for a working substance whose γ-value is (a) 1.4, and (b) 1.3.

7.17. The air-fuel ratio of an ideal Otto engine with a clearance factor of 8.7% is 16 and the fuel has a heating value of 4.5 MJ/kg. The engine operates on 0.20 kg/s of air and the suction conditions are 100 kPa and 20°C. Determine each of the following, if $c_v = 0.71$ kJ/kg K.
(a) Pressure and temperature at each point of the cycle.
(b) Power developed.
(c) Efficiency.
(d) Mean effective pressure.

7.18. The compression ratio of an Otto cycle is 6 and the suction temperature and pressure are 300 K and 100 kPa, respectively. Heat supplied in the constant volume process is 540 kJ/kg. The air-flow rate is 100 kg/h. Assume $\gamma = 1.4$ and $c_v = 0.71$ kJ/kg K, and determine each of the following.
(a) Power output.
(b) Mean effective pressure.
(c) Efficiency.

7.19. In an air-standard Otto cycle, the conditions at the beginning of the compression stroke are 100 kPa and 15°C. The pressure at the end of the compression stroke is 800 kPa. Calculate the temperatures at the end of the compression and expansion strokes if the peak temperature in the cycle is 1400°C. Assume $\gamma = 1.4$ and constant specific heats. Also, determine the thermal efficiency and the mean effective pressure of the cycle.

7.20. A four-cylinder four-stroke Otto engine with a bore of 8 cm and a stroke of 10 cm operates at 3000 rpm and delivers 25 kW of power at the shaft. The brake thermal efficiency of the engine is measured to be 28%. Calculate the heat energy supplied per cycle. An indicator diagram obtained from one cylinder of the engine has an area of 28 cm². The scale factor for the pressure is 1 cm to 50 kPa and that for the volume is 1 cm to 240 cm³. Determine the indicated thermal efficiency of the engine. What is the quantity of heat supplied to one cylinder per cycle?

7.21. The pressure at the beginning of the compression stroke in an air standard Diesel cycle is 90 kPa. The corresponding temperature is 10°C. The compression ratio is 18 and the maximum temperature in the cycle is 2100°C. Determine the thermal efficiency and the thermodynamic states at the beginning and at the end of the adiabatic expansion stroke.

7.22. Prove that for the same compression ratio the thermal efficiency of a Diesel cycle is always less than that of an Otto cycle.

7.23. A diesel cycle has a compression ratio of 16. The temperature at the beginning of the compression stroke is 288 K and the temperature at the end of expansion is 940 K. Assume $\gamma = 1.4$. Calculate the efficiency of the cycle.

7.24. The energy released in a six-cylinder, four-stroke, 45-cm bore, 65-cm stroke Diesel engine is 0.28 MJ/kg of the charge in the cylinder. The other data are as follows.

Inlet conditions: 300 K and 100 kPa

Heat released by the combustion of the fuel: 4.5 MJ/kg of fuel

Compression ratio: 17

Volumetric efficiency: 80%

Mechanical efficiency: 70%

Engine speed: 600 rpm

Calculate each of the following.
(a) Air-fuel ratio.
(b) Fuel consumption.
(c) Brake power.
(d) Thermal efficiency.

7.25. An eight-cylinder, four-stroke Diesel engine develops 900 kW of brake power at 300 rpm. The cylinder size is 37 cm by 46 cm and the engine uses 4.5 kg of fuel per minute. Upon complete combustion, the fuel releases heat energy of 45 MJ/kg. The indicated mean effective pressure is observed to be 550 kPa. Calculate each of the following.
(a) Brake thermal efficiency.
(b) Indicated thermal efficiency.
(c) Mechanical efficiency.

7.26. An air-standard Diesel cycle has a compression ratio of 17. At the beginning of the compression stroke, the temperature and pressure are 27°C and 105 kPa, respectively. The maximum temperature of the cycle is to be limited to 2000 K. Determine the outoff ratio, the thermal efficiency, the mean effective pressure, and the net-work output per kilogram.

7.27. We wish to determine the cylinder dimensions of a 200 kW (brake output), four-cylinder, 1200 rpm Diesel engine based on the data of Problem 7.26. It is anticipated that the overall thermal efficiency of the engine would be no more than 48% of the efficiency of the air-standard cycle. Calculate the cylinder dimensions and the necessary heat input per cycle. If the fuel were to have a heating value of 46 MJ/kg, what would be the fuel consumption in kilograms per hour?

7.28. Derive Equation 7.13.

7.29. Show that Equation 7.13 reduces to Equation 7.10 when the cutoff rate approaches zero and that it reduces to Equation 7.12 when the pressure ratio p_3/p_2 approaches unity.

7.30. In a dual cycle, one-half of the heat supply occurs during the constant volume process and the remaining half in the constant pressure process. The total heat supply is 600 kJ/kg per cycle. The compression ratio is 9.2. The temperature and pressure at the beginning of the compression process are 27°C and 100 kPa, respectively. Calculate the net work output and the thermal efficiency of the cycle.

7.31. In an air-standard Brayton cycle, the compressor receives air at 100 kPa and 20°C, and compresses the air to 500 kPa at a rate of 4 kg/s. The temperature of the air entering the turbine is 900°C. Determine the compressor work, turbine work, and thermal efficiency of the cycle. Assume constant specific heats and $\gamma = 1.4$.

7.32. Rework Problem 7.31 for variable specific heats. Use air tables.

7.33. A gas turbine power plant is used by a utility company for meeting peak load demands. The plant delivers 20 MW. The temperature of the gases leaving the combustion chamber is 800°C. The compressor takes in air at 15°C and 100 kPa and compresses the air isentropically to 600 kPa. Calculate each of the following.
   (a) Power requirement of the compressor.
   (b) Power output of the turbine.
   (c) Heat input rate.
   (d) Thermal efficiency.
   (e) Mass flow rate of air in kilogram per second.

7.34. Rework Problem 7.33 if the compressor and turbine isentropic efficiencies are 82% and 87%, respectively.

7.35. The maximum permissible temperature in a Brayton cycle is 500°C, while the compressor inlet temperature is 5°C. At what pressure ratio will the turbine work output exactly equal the compressor work input. What will be the net output of work if the pressure ratio is reduced by 50%?

7.36. Helium is used as the working fluid in a certain closed cycle gas turbine power plant. The compressor inlet conditions are 400 kPa and 44°C. The pressure ratio is 3 and the turbine inlet temperature is 710°C. The compressor and the turbine efficiencies are 85% and 90%, respectively. Determine the mass flow rate in kilogram per second for a power output of 50 kW.

7.37. What should be the pressure ratio for the plant in Problem 7.36 to obtain the maximum work output per cycle? What will be the mass flow rate of the helium for this output?

7.38. If an ideal regenerator is incorporated into the Brayton cycle of Problem 7.31, what would be the resulting thermal efficiency?

7.39. The following data is available for a regenerative gas turbine plant:

$$\begin{array}{lll}
\text{Compressor inlet:} & 100 \text{ kPa,} & 21°C \\
\text{Combustor inlet:} & 523 \text{ kPa,} & 280°C \\
\text{Turbine inlet:} & 523 \text{ kPa,} & 620°C
\end{array}$$

There is no temperature drop between the compressor outlet and the regenerator inlet or between the turbine outlet and the regenerator inlet. Calculate the thermal efficiency of the cycle and the regenerator effectiveness.

7.40. The isentropic efficiencies of the compressor and the turbine in a gas turbine plant are 80% and 85%, respectively. The maximum temperature in the system is 1300 K. The compressor inlet conditions are 105 kPa and 18°C. In order to obtain the maximum work from the plant, it is agreed to use $r_{p,opt}$ as given by Equation 7.17. What flow rate should be used so that the plant may produce 3000 kW of power?

7.41. If a regenerator with 92% efficiency is used for the power plant in Problem 7.40, what would be the savings in the heat supply?

7.42. The inlet conditions at the compressor of a gas turbine plant are 98 kPa and 24°C. The pressure ratio is 5.5 and the maximum temperature is 1000°C. Assume that the isentropic efficiencies of the compressor and the turbine are equal. At what efficiency value will the net work output of the system be zero?

7.43. The operating conditions of a simple gas turbine power plant are:

Compressor inlet:	100 kPa	20°C
Compressor outlet:	550 kPa	
Turbine inlet:	805°C	
Compressor efficiency:	79%	
Turbine efficiency:	83%	

Calculate the net work output per kilogram and the thermal efficiency.

7.44. If a regenerator is installed for the plant in Problem 7-43, what will be the improvement in the thermal efficiency? Compare it with the Carnot efficiency. Assume a regenerator effectiveness of 70%.

7.45. A gas turbine plant operating on the Brayton cycle receives 10 m³/s of air at 100 kPa and 25°C, and the pressure ratio is 7.5. The fuel has a heating value of 40 MJ/kg, and the air-fuel ratio is 44. Calculate each of the following.
(a) Mass of the fuel burned per hour.
(b) Power developed.
(c) Efficiency of the plant.

7.46. For the gas turbine power plant of Problem 7-45, if the compressor efficiency is 0.76, the combustion efficiency 0.94, and the turbine efficiency 0.85, what will be the (a) power developed, and (b) the efficiency?

7.47. The pressure ratio across each stage in a two-stage Brayton cycle is 2.7. The inlet conditions to the compressor are 100 kPa and 20°C. The inlet temperature to each turbine is 800°C. An ideal intercooler between the two stages is used. Calculate the compressor work and the net work output in kilojoules per kilogram. What is the thermal efficiency of the cycle?

7.48. Rework Problem 7.47 if an ideal regenerator was used.

7.49. In a gas turbine cycle there are two stages for compression, as well as for expansion. The pressure ratio for each stage is 2.5. The temperature and the pressure at the inlet to the compressor are 20°C and 100 kPa. There is intercooling at the intermediate pressure in the compression process so that the temperature of the air entering the second stage is 20°C. There is a reheat at the intermediate pressure in the expansion process and the temperature of the gases entering each turbine is 1000°C. Draw the cycle on a *T-s* diagram. Determine the compressor work, the turbine work, and the thermal efficiency with and without regeneration.

7.50. Rework Problem 7-49 if the compressor and the turbine efficiencies are 80% and 90%, respectively.

7.51. Rework Problem 7-49 if the regenerator efficiency is 78%.

7.52. A gas turbine power plant has two-stage compression and expansion, with intercooling, reheating, and regeneration. The temperature of the air leaving the second compressor is 350°C and the temperature of the air entering the combustion chamber is 740°K. The maximum temperature in the system is 950°C. The overall pressure ratio is 6. Calculate the maximum work output from the power plant and the regenerator efficiency.

7.53. A two-stage gas turbine cycle receives air at 100 kPa and 15°C. The lower stage has a pressure ratio of 3, while that for the upper stage is 4 for the compressor as well as the turbine. The temperature rise of the air compressed in the lower stage is reduced by 80% by intercooling. Also, a regenerator of 78% effectiveness is used. The upper temperature limit of the cycle is 1100°C. The turbine and the compressor efficiencies are 86%. Calculate the mass flow rate required to produce 6000 kW.

7.54. We wish to compare the performance of the following ideal gas turbine plants.

    1. Basic plant: Compression ratio 9, compressor inlet at 280 K, turbine inlet at 1200 K.

    2. Reheat plant: Equal expansion ratio for each turbine and the same inlet temperature for each turbine. Other details as in (1).

    3. Intercooled plant: Compression ratio of 3 for each compressor and the same inlet temperature for each compressor. Other details as in (1).

    4. Regenerative plant: With an ideal regenerator. Other details as in (1).

Calculate and comment on the specific work output and the thermal efficiency of each of the foregoing cycles. Assume the working fluid to be an ideal gas with $c_p = 1.01$ kJ/kg. K and $\gamma = 1.4$.

-8 ⎯⎯⎯⎯⎯⎯⎯⎯

Energy Conversion—
Vapor Cycles

8.1 INTRODUCTION

In this chapter we shall consider vapor power cycles and vapor refrigeration cycles. In a vapor power cycle, water is generally the working medium. This is not to exclude other possible media, such as sodium, potassium, and mercury. With the new interest in low-level heat recovery, organic fluids, such as tribromofluorosilane, benzene, the members of the Freon family, and Dowtherm A, are seriously being considered as the working media for a power cycle. Because of the abundance of water and its relatively high latent heat of vaporization, water is used as the working medium for vapor power cycles.

A steam power plant operates on Rankine cycle. There are many possible modifications of the basic Rankine cycle, which will be discussed in this chapter. A Rankine cycle requires a number of components for its execution: a feedwater pump, a steam generator, a turbine, and a condensor. Of these, the feed pump consumes work, while the turbine produces work, and the remaining two components exchange heat energy and accommodate a liquid-vapor change of phase. These heat exchangers have characteristics that are quite different from those of the regenerative heat exchangers discussed in relation to the Brayton cycle in the preceding chapter.

A reversed vapor cycle can be used for producing a cooling effect, or refrigeration. There are two principal types of refrigeration systems, namely, a *vapor compression system* and an *absorption system*. Like the Rankine cycle, a refrigeration cycle requires a number of components for its execution. In the case of a vapor compression system, there are a condensor, an evaporator, a throttle valve, and a vapor compressor; in the case of an absorption system, there is a liquid pump instead of the vapor compressor, and in addition, there is a source of heat.

Some of the commonly used working media for vapor compression refrigeration are Freon-11, Freon-12, Freon-22, carbon dioxide, and sulfur dioxide. The absorption systems use mixtures of water and ammonia or lithium bromide and water.

Before beginning our discussion of vapor cycles, we first consider the enthalpy-entropy chart, or the Mollier diagram, of a pure substance.

8.2 MOLLIER DIAGRAM

The Mollier, or *h-s*, diagram of a pure substance is a useful tool in analyzing a vapor cycle. An accurately constructed Mollier diagram conveys a considerable amount of information, and it is often used as a substitute for tables of properties. The Mollier diagram is a graph of the properties of a substance in the two-phase, liquid-vapor region and the superheated region. It is constructed by employing entropy values as the abscissas and enthalpy values as the ordinates. Figure 8.1 shows a sketch of the Mollier diagram for water. A complete Mollier diagram for water is given in Appendix J.

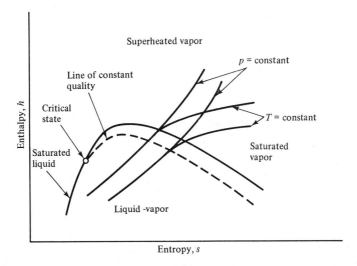

Figure 8.1   Mollier diagram.

Referring to Figure 8.1, the saturation line delineates the superheated vapor region from the liquid-vapor region. The lines of constant pressure, or isobars, on the diagram have positive slopes and the curves shift to the left of the diagram as pressure increases. In addition, the diagram shows lines of constant quality, Equations 4.1 and 4.3. These lines run somewhat parallel to the saturated vapor line and are particularly useful in determining the quality after an isentropic expansion. Also appearing on the diagram are lines of constant temperature, starting from the saturated vapor line. These constant temperature lines have positive slopes near the saturation dome and near-zero slopes for large values of entropy. A reversible adiabatic process appears as a straight vertical line on the Mollier diagram. The Mollier diagram for water shows that an isentropic expansion, if carried out to low enough pressures, will take the medium into the wet-vapor region. The throttling process, which is a constant enthalpy process (Section 5.7), appears as a horizontal line on a Mollier diagram.

8.3 THE RANKINE CYCLE

Let us first consider a Carnot cycle operating within the saturation dome of a substance such as water. Figure 8.2 shows the T-s diagram for such a cycle. In this cycle, the isothermal heat supply, path A–B, starts from the saturated liquid line and ends

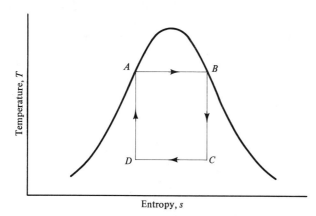

Figure 8.2 Carnot cycle operating within saturation dome.

on the saturated vapor line. This process involves evaporation from a saturated liquid state to a saturated vapor state. The isentropic expansion, path B-C, takes the substance into the low-pressure, liquid-vapor region. Process C–D, entailing condensation, takes the substance from a mostly vapor condition (a high value of quality) to a mostly liquid condition. All of these three processes are feasible in practice. The final segment of the Carnot cycle involves an isentropic compression of a low-quality, low-pressure, liquid-vapor mixture to a saturated liquid at high pressure. This is difficult to achieve in practice since compressors work reliably and efficiently when they are designed to handle either a liquid or a vapor alone, but not a liquid-vapor mixture.

Rankine proposed a modification to the cycle shown in Figure 8.2. In the Rankine cycle, the liquid-vapor mixture at the end of the isentropic expansion, state C, is condensed all the way to the saturated liquid state. The liquid is compressed to the pressure at state A and then heated at constant pressure to the temperature at state A. Figure 8.3(a) shows the Rankine cycle on T-s diagram and Figure 8.3(b) shows the physical components required to excute the various processes of the cycle. We begin with the saturated liquid state of the working medium, denoted by point 1 in Figure 8.3(a). Process 1–2 is the isentropic compression of the working fluid from a saturated liquid state to a compressed liquid state. This process is carried out in a feedwater pump. Usually, the temperature rise in this process is very small. However, the segment 1–2 in the *T-s* diagram of Figure 8.3(a) is shown disproportionately large for the sake of clarity. Process 2–2*A*–3 involves constant pressure heating from a compressed liquid state to a saturated vapor state. This operation takes place in a steam generator or in a boiler Figure 8.3(b). In process 2–2*A* the temperature increases and the process requires heat transfer from high-temperature combustion gases to the liquid, while in the heat transfer process 2*A*–3 there is no temperature change as the process involves a change of phase. The design requirements for these two processes are quite different. Therefore in large commercial plants, two separate components,

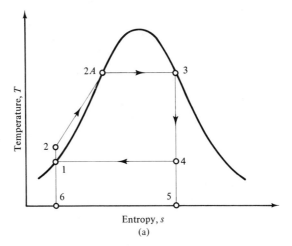

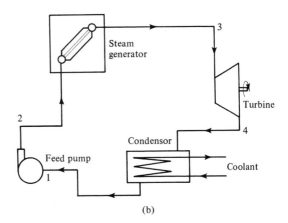

Figure 8.3 Simple Rankine cycle.
(a) *T-s* diagram.
(b) Mechanical components.

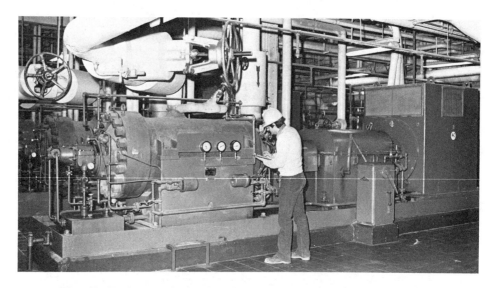

Plate 12 Feed pump of the steam power plant at the Russell Station of Rochester Gas and Electric Corporation. (Courtesy of the Rochester Gas and Electric Corporation.)

called the economizer and the steam generator, are generally used. Continuing with the Rankine cycle, the high-temperature, high-pressure saturated steam is admitted into the turbine at point 3, and it is expanded isentropically, delivering work. Finally, the wet steam at state 4 is condensed to a saturated liquid state, state 1, by using a cooling agent such as water.

In the analysis of the simple Rankine cycle, it is assumed that the changes in the kinetic energy and the potential energy of the working medium are negligible. This assumption is quite realistic. Another assumption made is that all processes are reversible, and that there are no pressure drops or heat losses in the pipes. In reality there are irreversibilities and pressure drops. The effects of these will be discussed later. For the reversible Rankine cycle, area 6–2–2*A*–3–4–5–6 in Figure 8.3(a) represents the heat energy supplied to a unit mass of the working medium, and area 4–1–6–5–4 represents the heat energy removed from the unit mass. Consequently, the difference in the two areas—namely, area 1–2–2*A*–3–4—represents the net work obtained from the cycle.

Applying the steady-flow energy equation, Equation 5.8(b), for the various processes in the simple Rankine cycle, we obtain the following for a unit mass flow rate.

Process 1–2: Isentropic compression; hence

$$q_{1\text{-}2} = 0 \quad \text{and} \quad -w_{s,\,1\text{-}2} = h_2 - h_1$$

State 2 belongs to the subcooled region. Tabulated values of the enthalpy of water in the subcooled region are not always readily available. We can resolve this difficulty in the following manner. We note that

$$h = u + pv$$

and

$$dh = du + p \, dv + v \, dp$$

Then the first law

$$\delta q = du + p \, dv$$

can be rewritten as

$$\delta q = dh - v \, dp$$

Combining this last equation with the second law

$$T \, ds = \delta q$$

gives

$$T \, ds = dh - v \, dp$$

Since $ds = 0$ for an isentropic process, the above equation beomes

$$dh = v \, dp \qquad (8.1)$$

Consequently,

$$-w_{s,1\text{-}2} = \int_1^2 dh = \int_1^2 v \, dp \qquad (8.1a)$$

It is usually a liquid that is compressed in process 1–2 of a Rankine cycle. Since the specific volume of a liquid undergoes very little change for moderate changes in pressure, we can write

$$w_{s,1\text{-}2} \simeq -v_1(p_2 - p_1) \qquad (8.1b)$$

We now continue with the rest of the processes in a Rankine cycle.

Process 2–2A–3:   constant pressure heating and

$$w_{s,2\text{-}3} = 0 \quad \text{and} \quad q_{2\text{-}3} = q_{\text{in}} = h_3 - h_2$$

Process 3–4:   Isentropic expansion; hence

$$q_{3\text{-}4} = 0 \quad \text{and} \quad w_{s,3\text{-}4} = h_3 - h_4$$

Process 4–1:   Constant pressure heat removal and

$$w_{s,4\text{-}1} = 0 \quad \text{and} \quad -q_{4\text{-}1} = h_4 - h_1$$

The net work of the cycle is

$$w_{\text{cycle}} = w_{s,3\text{-}4} + w_{s,1\text{-}2} = (h_3 - h_4) - (h_2 - h_1)$$

The thermal efficiency of the Rankine cycle is

$$\eta_{\text{Rankine}} = \frac{w_{\text{cycle}}}{q_{\text{in}}} = \frac{(h_3 - h_4) - (h_2 - h_1)}{h_3 - h_2} \qquad (8.2)$$

With the same upper and lower limits of the temperature for the Carnot and the Rankine cycles, the Rankine cycle efficiency is less than the Carnot cycle efficiency. We may reason that the average temperature at which heat is supplied in a Rankine cycle is less than the temperature at which heat is supplied in a Carnot cycle, thus leading to the lower efficiency of the Rankine cycle.

Example 8.1

A simple Rankine cycle (Figure 8.3) uses steam as the working medium and operates between 50 kPa and 2000 kPa. Determine the quality of the steam as it leaves the turbine, the thermal efficiency of the cycle, and the mass flow rate of steam to produce 10,000 kW. Compare the efficiency of the cycle with that of a Carnot cycle operating within the same temperature limits.

Solution:

Given: A simple Rankine cycle has the following characteristics.

Turbine inlet:	$p_3 = 2,000$ kPa,	state 3: saturated vapor
Turbine exhaust:	$p_4 = 50$ kPa	
Power output:	10,000 kW	
Working medium:	water	

Objective: We are to determine x_4, η_t, and $\dot{m}$ for the given cycle and η_{carnot} for the same temperature limits.

Diagram: The *T-s* diagrams for the given Rankine cycle and a Carnot cycle between the same temperature limits as those of the Rankine cycle are sketched in Figure 8.4.

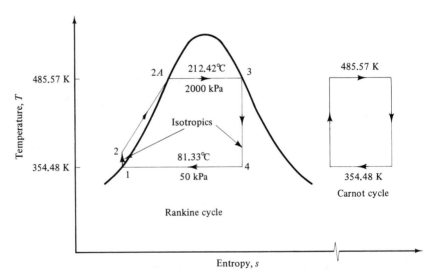

Figure 8.4

Assumptions:

1. The cycle is reversible.
2. The change in volume between states 1 and 2 is negligible.

Relevant physics: In problems of this type it is useful to set up a table of properties at all state points giving the pressure, temperature, phase, enthalpy, and entropy. Some of the properties can be read from Appendix B. Others have to be calculated.

The quality x_4 can be determined, because s_4 equals s_3—which is known— and a relationship exists among $s_{f,4}$, $s_{fg,4}$, and s_4. The thermal efficiency can be determined from Equation 8.2. The required Carnot efficiency can be determined from

$$\eta_{carnot} = 1 - \frac{T_{low}}{T_{high}}$$

Analysis: From Appendix B.2.2, we can read the values of temperature, enthalpy, and entropy for the given pressures. These are entered in the following table.

Point	t, °C	p (kPa)	Phase	h (kJ/kg)	s (kJ/kg K)
1	81.33	50	Sat. liquid	340.49	1.0910
2	81.8a	2000	Comp. liquid	342.49a	1.0910
2A	212.42	2000	Sat. liquid	908.79	2.4474
3	212.42	2000	Sat. vapor	2799.5	6.3409
4	81.33	50	Wet vapor	2201.64a	6.3409

aCalculated values.

The value of h_1 is read from the appendix. The value of h_2 can be calculated once the pump work is known. For the pump work, we can use Equation 8.1(b) to write

$$w_{s,1-2} = -v_1(p_2 - p_1)$$
$$= -0.001030(2000 - 50) \times 10^3 \text{ J/kg}$$
$$= -2 \text{ kJ/kg}$$

Thus

$$h_2 = h_1 - w_{s,1-2}$$
$$= h_{f,1} - w_{s,1-2}$$
$$= 340.49 - (-2) = 342.49 \text{ kJ/kg}$$

The temperature at point 2 is readily calculated from

$$c_p(t_2 - t_1) = (h_2 - h_1)$$

which gives

$$t_2 = t_1 + \frac{h_2 - h_1}{c_p}$$
$$= 81.33 + \frac{342.49 - 340.49}{4.18}$$
$$= 81.8°C$$

It is necessary to determine the quality of the steam leaving the turbine so that its enthalpy, h_4, can be calculated. Since process 3–4 is isentropic,

$$s_4 = s_3 = 6.3409$$

Also,
$$s_{f,4} + x_4 s_{fg,4} = s_4$$

Substituting the values of $s_{f,4}$ and $s_{fg,4}$ corresponding to the turbine exhaust pressure of 50 kPa, we find

$$1.0910 + 6.5029 x_4 = 6.3409$$

or
$$x_4 = 0.8073$$

A quality of 0.81 can also be read from the Mollier diagram for water in Appendix J. This is done by locating point 3 on the chart and coming down along a constant entropy line until the constant pressure line for 50 kPa is intersected. The value of h_4 is then given by

$$h_4 = h_{f,4} + x_4 h_{fg,4}$$
$$= 340.49 + 0.8073 \times 2305.4$$
$$= 2201.64 \text{ kJ/kg}$$

Applying the first law to a control volume around the turbine, we obtain

$$w_{s,3-4} = h_3 - h_4 = 2799.5 - 2201.64 = 597.86 \text{ kJ/kg}$$

The quantity of heat supplied is given by

$$q_{2-3} = h_3 - h_2 = 2799.5 - 342.49 = 2457.01$$

The thermal efficiency of the Rankine cycle is given by

$$\eta_t = \frac{w_{\text{cycle}}}{q_{\text{in}}} = \frac{w_{s,1-2} + w_{s,3-4}}{q_{2-3}}$$

or
$$\eta_{\text{Rankine}} = \frac{-2 + 597.86}{2457.01}$$
$$= 0.242, \text{ or } 24.2\%$$

If $\dot{m}$ is the mass flow rate of steam in kilograms per second, then

$$\dot{m}(w_{s,1-2} + w_{s,3-4}) = 10,000 \text{ kW}$$

or
$$\dot{m} = \frac{10,000}{595.86} = 16.78 \text{ kg/s}$$

If a Carnot cycle were to operate between the same temperature limits as the Rankine cycle, its efficiency would be

$$\eta_{\text{Carnot}} = 1 - \frac{81.33 + 273.15}{212.42 + 273.15}$$
$$= 0.27, \text{ or } 27\%$$

Comments: The efficiency of the Rankine cycle in this example is seen to be fairly close to that of the Carnot cycle. This is due to the fact that about 77% of the heat supply and 100% of the heat rejection occur isothermally.

Example 8.2

If the exhaust pressure of the Rankine cycle in Example 8.1 is lowered to 5 kPa, determine the net work output and the thermal efficiency of the cycle.

Solution:

Given: The simple Rankine cycle of Example 8.1 now has the following data.

Turbine inlet: $p_3 = 2000$ kPa,   state 3: saturated vapor

Turbine exhaust: $p_4 = 5$ kPa

Objective: We are to determine w_{cycle} and η_t for the cycle.

Diagram:

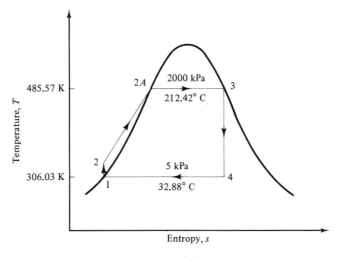

Figure 8.5

Assumptions:

1. The cycle is reversible.
2. The change in volume during the compression is negligible.

Relevant physics: Since the condenser pressure p_4 is now different, the properties at states 4 and 1 are affected. The procedure for obtaining the property values and calculating w_{cycle} and η_t is exactly the same as that for Example 8.1.

Analysis: The relevant property values at the five state points, read from Appendix B.2.2 and calculated, are entered in the following table.

Point	t (°C)	p (kPa)	Phase	h (kJ/kg)	s (kJ/kg K)
1	32.88	5	Sat. liquid	137.82	0.4764
2		2000	Comp. liquid	139.82[a]	0.4764
2A	212.42	2000	Sat. liquid	908.79	2.4474
3	212.42	2000	Sat. vapor	2799.5	6.3409
4	32.88	5	Wet vapor	1931.36[a]	6.3409

[a]Calculated values.

We can now determine the turbine exhaust quality, x_4, from

$$x_4 = \frac{s_4 - s_{f,4}}{s_{fg,4}}$$

and note that $s_4 = s_3$. We find that at 5 kPa,

$$s_{f,4} = 0.4764 \text{ kJ/kg K}, \qquad s_{fg,4} = 7.9187 \text{ kJ/kg K}$$

so that

$$x_4 = \frac{6.3409 - 0.4764}{7.9187} = 0.74$$

As in Example 8.1, this quality can be determined from the Mollier diagram, Appendix J.

Next, the enthalpy of the steam leaving the turbine is given by

$$h_4 = h_{f,4} + x_4 h_{fg,4}$$
$$= 137.82 + 0.74 \times 2423.7 = 1931.36 \text{ kJ/kg}$$

The pump work is

$$w_{s,1\text{-}2} = -v_{f,1}(p_2 - p_1)$$
$$= -0.001005(2000 - 5) = -2.0 \text{ kJ/kg}$$

and

$$h_2 = h_1 - w_{s,1\text{-}2} = 137.82 - (-2)$$
$$= 139.82 \text{ kJ/kg}$$

Also, the turbine work is

$$w_{s,3\text{-}4} = h_3 - h_4$$
$$= 2799.5 - 1931.36 = 868.14 \text{ kJ/kg}$$
$$w_{\text{cycle}} = w_{s,1\text{-}2} + w_{s,3\text{-}4} = -2.0 + 868.14$$
$$= 866.14 \text{ kJ/kg}$$

The quantity of heat supplied is given by

$$q_{2\text{-}3} = h_3 - h_2$$
$$= 2799.5 - 139.82$$
$$= 2659.68 \text{ kJ/kg}$$

The thermal efficiency is

$$\eta_t = \frac{w_{\text{cycle}}}{q_{\text{in}}} = \frac{866.14}{2659.68}$$

or

$$\eta_t = 32.56\%$$

Comment: We see that a reduction in the exhaust pressure from 50 kPa to 5 kPa results in a significant increase in the net work output, from 595.86 kJ/kg to 866.14 kJ/kg, or an increase of 45%. Furthermore, the quality of the exhaust reduces from 0.8073 to 0.74. Also, the quantity of heat supplied is larger than the one in Example 8.1 by about 8%. Thus the effect of lowering the exhaust pressure on the quantity of heat supplied is relatively small. This can be seen qualitatively from the relative sizes of areas 1–2–3–4–5–6–1 in Figure 8.4(a) and (b).

The reduction in the exhaust pressure from 50 kPa to 5 kPa improves the thermal efficiency from 24.2 percent to 32.56 percent, an improvement of 34.3%. The Carnot efficiency for the temperatures corresponding to the supply and exhaust temperatures, 212.42°C and 32.88°C, respectively—is 37%.

Example 8.3

If the inlet pressure to the turbine in Example 8.1 is raised to 10,000 kPa, determine the effect on the net work output and the thermal efficiency of the cycle.

Solution:

Given: A new value of the turbine inlet pressure is given for the simple Rankine cycle of Example 8.1.

$$\text{Turbine inlet:}\quad p_3 = 10,000 \text{ kPa},\quad \text{state 3: saturated vapor}$$

$$\text{Turbine exhaust:}\quad p_4 = 50 \text{ kPa}$$

Objective: We have to determine the new values of w_{cycle} and η_t.

Diagram:

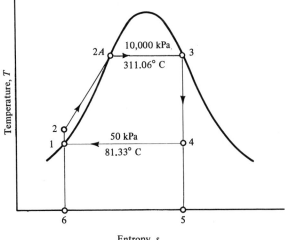

Figure 8.6

Assumptions:

1. The cycle is reversible.
2. The change in volume during the compression is negligible.
3. The specific heat of liquid water is 4.18 kJ/kg K.

Relevant physics: In view of the change, we shall have new property values at points 2A, 3, and 4. The procedure for the computation of w_{cycle} and η remains unchanged.

Analysis: The following table shows the relevant properties at the various state points in the cycle.

Point	t, (°C)	p (kPa)	Phase	h (kJ/kg)	s (kJ/kg K)
1	81.33	50	Sat. liquid	340.49	1.0910
2	83.78[a]	10,000	Comp. liquid	350.74[a]	1.0910
3	311.06	10,000	Sat. vapor	2724.7	5.6141
4	81.33	50	Wet vapor	1943.89[a]	5.6141

[a]Calculated values.

For the pump,

$$w_{s, 1\text{-}2} = -v_{f,1}(p_2 - p_1)$$
$$= -0.001030(10,000 - 50)$$
$$= -10.25 \text{ kJ/kg}$$

and

$$h_2 = h_1 - w_{s,1\text{-}2}$$
$$= 340.49 + 10.25 = 350.74 \text{ kJ/kg}$$

giving

$$t_2 = t_1 + \frac{h_2 - h_1}{c_p} = 81.33 + \frac{10.25}{4.18}$$
$$= 83.78°C$$

For the turbine,

$$x_4 = \frac{s_4 - s_{f,4}}{s_{fg,4}}$$
$$= \frac{5.6141 - 1.0910}{6.5029} \qquad (\text{since } s_4 = s_3)$$
$$= 0.6955$$

The Mollier diagram in Appendix J gives $x_4 = 0.70$

$$h_4 = h_{f,4} + x_4 h_{fg,4}$$
$$= 340.49 + 0.6955 \times 2305.4$$
$$= 1943.89 \text{ kJ/kg}$$

so that

$$w_{s, 3\text{-}4} = (h_3 - h_4) = (2724.7 - 1943.89) = 780.81 \text{ kJ/kg}$$

The net work for the cycle is

$$w_{\text{cycle}} = w_{s, 1\text{-}2} + w_{s, 3\text{-}4}$$
$$= -10.25 + 780.81$$
$$= 770.56 \text{ kJ/kg}$$

The heat supplied is

$$q_{\text{in}} = h_3 - h_2$$
$$= 2724.7 - 350.74 = 2373.96 \text{ kJ/kg}$$

The thermal efficiency is

$$\eta_t = \frac{w_{\text{cycle}}}{q_{\text{in}}} = \frac{770.56}{2373.96}$$

or
$$\eta_t = 32.45\%$$

Comment: We observe that raising the supply pressure to 10,000 kPa improves the net work output by 29%, lowers the quality to 0.6955, decreases the heat input requirement by 3%, and improves the thermal efficiency by 34%.

The two preceding examples illustrate that the efficiency of a Rankine cycle increases as the turbine inlet temperature is increased, holding the exhaust temperature constant, and as the exhaust temperature is lowered, holding the inlet temperature constant. It can be reasoned that an increase in the difference between the average temperature at which heat is supplied to the cycle and the temperature at which heat is rejected by the cycle leads to an improvement in the efficiency.

On the other hand, as the exhaust pressure is decreased, the quality of the exhaust steam can decrease to a point where the life of the turbine blades is threatened. Lower quality means greater percentage of moisture (more liquid water particles) in the steam passing through the turbine nozzles and blades. The liquid water particles, whose density is about 1000 times the density of the vapor, can cause severe damage to the turbine nozzles and blades. Thus it is industry practice to select the inlet and exhaust conditions in such a way that the quality of the exhaust is not less than 90 percent. This restriction does not really prove to be a limitation on the thermal efficiency. The exhaust quality can be maintained at an acceptable level with a simultaneous improvement in the efficiency by superheating the steam before it enters the turbine. Also, the exhaust quality can be improved by employing the reheat cycle to be discussed in Section 8.3.1.

Figure 8.7(a) shows the *T-s* diagram for a Rankine cycle with superheat, and Figure 8.7(b) shows the mechanical components used for such a cycle. The superheater is a heat exchanger transferring heat energy from the hot combustion gases to the saturated and superheated vapor. It was noted earlier that *liquid* water is heated in an economizer, and there is a temperature rise of the water in the economizer. In a steam generator, heating occurs at constant temperature since there is a *change of phase*. In a superheater, saturated vapor is superheated. The design considerations for these three components are quite different.

It can be seen from Figure 8.7(a) that superheating results in an increase in the net work output, represented by area 2B–3–4–4B–2B, an increase in the heat supply, represented by area 2B–3–5–5B–2B, and improvement (an increase) in the exhaust steam quality from x_{4B} to x_4. It also results in an improvement of the thermal efficiency because of an increase in the average temperature of heat supply.

Example 8.4

Steam at 2000 kPa and 400°C is supplied to the turbine of a power plant operating on Rankine cycle. The turbine expands the steam to 50 kPa. Determine the quality of the exhaust steam, the net work output of the cycle, the quantity of heat supplied per kilogram of steam, and the thermal efficiency of the cycle.

Solution:

Given: The data on a Rankine cycle is given.

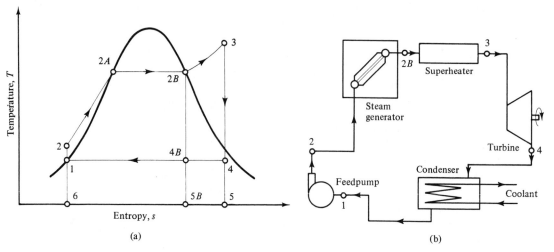

Figure 8.7 Rankine cycle with superheat.
(a) *T-s* diagram.
(b) Mechanical components.

Turbine inlet: $p_3 = 2000$ kPa, $t_3 = 400°C$

Turbine exhaust: $p_4 = 50$ kPa

Objective: We have to calculate x_4, w_{cycle}, q_{in}, and η_t.
Diagram:

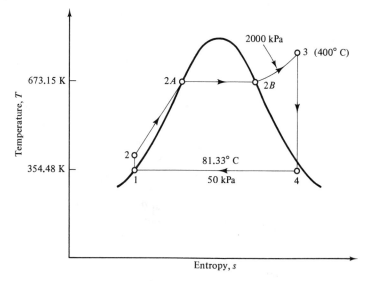

Figure 8.8 *T-s* diagram for Rankine cycle.

Assumptions:

1. The cycle is reversible.
2. The change in volume during the compression process is negligible.

Relevant physics: The pressures in this problem are the same as those in Example 8.1. The saturation temperature at 2000 kPa is 212.42°C, which is less than the given temperature. Therefore we have superheated steam and a Rankine cycle with superheat. The *T-s* diagram for the cycle is shown in Figure 8.8.

Analysis: The properties of the superheated steam at point 3 are obtained from Appendix B.3, while those at point 1 (saturated state) are obtained from

Point	t (°C)	p (kPa)	Phase	h (kJ/kg)	s (kJ/kg K)
1	81.33	50	Sat. liquid	340.49	1.0910
2	81.8[a]	2000	Comp. liquid	342.49[a]	1.0910
3	400	2000	Super. vapor	3247.6	7.1271
4	81.33	50	Wet vapor[a]	2479.9[a]	7.1271

[a]Calculated values.

Appendix B.2.2. Properties at point 2 are taken directly from Example 8.1, since the pressures are identical in the two problems at point 2.

$$x_4 = \frac{s_4 - s_{f,4}}{s_{fg,4}}$$

$$= \frac{7.1271 - 1.0910}{6.5029} \quad \text{(since } s_4 = s_3\text{)}$$

$$= 0.928$$

$$h_4 = h_{f,4} + x_4 h_{fg,4}$$

$$= 340.49 + 0.928 \times 2305.4$$

$$= 2479.9 \text{ kJ/kg}$$

$$w_{s,1-2} = -2.0 \text{ kJ/kg} \quad \text{(from Example 8.1)}$$

$$w_{s,3-4} = h_3 - h_4$$

$$= 3247.6 - 2479.9$$

$$= 767.7 \text{ kJ/kg}$$

$$w_{cycle} = w_{s,1-2} + w_{s,3-4}$$

$$= -2.0 + 767.7$$

$$= 765.7 \text{ kJ/kg}$$

$$q_{2-3} = h_3 - h_2$$

$$= 3247.6 - 342.49$$

$$= 2905.11 \text{ kJ/kg}$$

$$\eta_t = \frac{w_{s,1-2} + w_{s,3-4}}{q_{2-3}}$$

$$= \frac{-2 + 767.7}{2905.11}$$

$$= 0.264, \text{ or } 26.4\%$$

Comments: A comparison of the above results with those obtained in Example 8.1 shows that the superheating results in an improvement of the quality of turbine exhaust quality, 28% increase in the net work output, 24% increase in the quantity of heat supplied, and 3.8% increase in the thermal efficiency.

8.3.1 The Reheat Cycle

The discussion and the examples presented so far point out that the thermal efficiency of a Rankine cycle can be improved by increasing the supply pressure, decreasing the exhaust pressure, and superheating the supply steam. The maximum superheating temperature is limited by the structural strength of the mechanical components at elevated temperatures. Thus, with this limit on the turbine-inlet temperature, the thermal efficiency can be improved only by reducing the exhaust pressure. As the exhaust pressure is reduced, the quality of the exhaust steam can fall well below 90 percent. In such situations and when additional work output per unit mass of steam is desirable, the *reheat cycle* is used.

Figure 8.9 shows the *T-s* diagram and the mechanical components for a reheat cycle. There are two or more stages of expansion in a reheat cycle. Processes 1–2, 2–3, and 3–4 are similar to those in the Rankine cycle with superheat; however, the pressure at point 4—after expansion in the high-pressure stage of the turbine—is greater than the pump-inlet pressure. Point 4 is chosen in such a way that the quality at 4 is close to the minimum acceptable limit. The wet steam at point 4 is completely removed from the turbine, reheated at constant pressure (process 4–3A) to a superheated state, and then admitted to the low-pressure stage of the turbine. It is then expanded to the design exhaust pressure p_{4A}(equal to p_1), condensed, and recirculated.

We note that the steam expands in two stages, and the net work for the cycle is then given by

$$w_{\text{cycle}} = w_{s,1\text{-}2} + w_{s,3\text{-}4} + w_{s,3A\text{-}4A}$$
$$= (h_1 - h_2) + (h_3 - h_4) + (h_{3A} - h_{4A})$$

Also, the heat supply, carried out in two stages, is

$$q_{\text{in}} = (h_3 - h_2) + (h_{3A} - h_4)$$

The thermal efficiency of the reheat cycle is given by

$$\eta_t = \frac{(h_1 - h_2) + (h_3 - h_4) + (h_{3A} - h_{4A})}{(h_3 - h_2) + (h_{3A} - h_4)} \tag{8.3}$$

Example 8.5

In a Rankine cycle with reheat, steam at 10,000 kPa and 500°C enters a turbine. The steam is expanded to a pressure of 500 kPa and then reheated to 400°C. The reheated

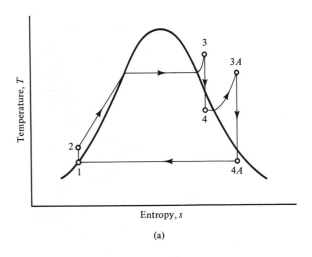

(a)

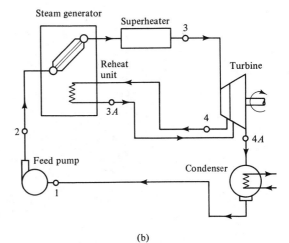

Figure 8.9 Rankine cycle with reheat.
(a) *T-s* diagram.
(b) Mechanical components.

(b)

steam at 500 kPa expands in a low-pressure stage to a pressure of 5 kPa. Determine the thermal efficiency of the cycle.

Solution:

Given: A Rankine cycle with reheat is given.

High-pressure stage: $p_3 = 10{,}000$ kPa, $t_3 = 500°C$, $p_4 = 500$ kPa

Low-pressure stage: $p_{3A} = 500$ kPa, $t_{3A} = 400°C$, $p_{4A} = 5$ kPa

Objective: We must calculate the thermal efficiency of the cycle.

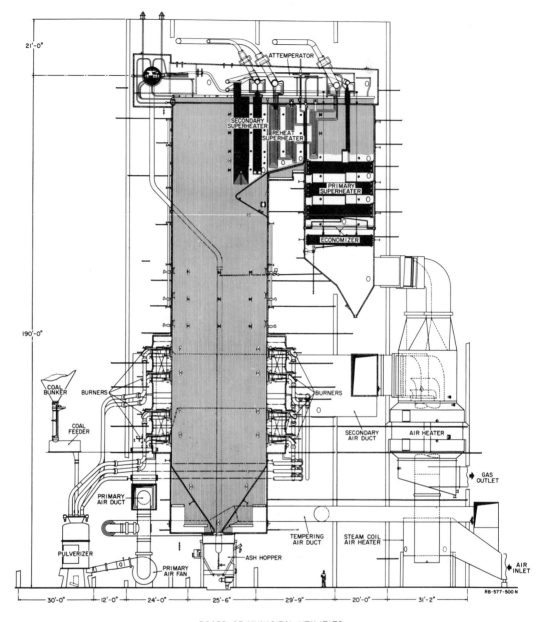

BOARD OF MUNICIPAL UTILITIES
SIKESTON POWER STATION UNIT NO. 1
SIKESTON, MISSOURI
B & W CONTRACT NO. RB-577

Plate 13 Sectional sideview of a natural circulation boiler. (Courtesy of the Babcock and Wilcox Company.)

Diagram:

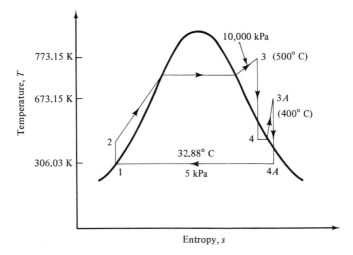

Figure 8.10

Assumptions:

1. The cyclic process is reversible.
2. The volume change during the compression process is negligible.
3. The specific heat of liquid water is 4.18 kJ/kg K.

Relevant physics: The solution to this problem involves a determination of the qualities x_4 and x_{4A}, enthalpies at all the state points, and application of Equation 8.3.

Analysis: The following table shows the relevant properties at various state points as obtained from the appendix. The calculated property values are also shown in the table.

Point	t (°C)	p (kPa)	Phase	h (kJ/kg)	s (kJ/kg K)
1	32.88	5	Sat. liquid	137.82	0.4764
2	35.28[a]	10,000	Comp. liquid	147.87[a]	0.4764
3	500	10,000	Super. vapor	3373.7	6.5966
4	151.86	500	Wet vapor	2654.48[a]	6.5966
3A	400	500	Super. vapor	3271.9	7.7938
4A	32.88	5	Wet vapor	2377.32[a]	7.7938

[a]Calculated values.

The work of the pump is

$$w_{s,1-2} = -v_{f,1}(p_2 - p_1)$$
$$= -0.001005(10,000 - 5)$$
$$= -10.05 \text{ kJ/kg}$$

and

$$h_2 = h_1 - w_{s,1-2}$$
$$= 137.82 + 10.05$$
$$= 147.87 \text{ kJ/kg}$$

This gives

$$t_2 = t_1 + \frac{h_2 - h_1}{c_p}$$
$$= 32.88 + \frac{10.05}{4.18}$$
$$= 35.28°C$$

We next calculate the qualities and enthalpies at point 4 and at point 4A.

$$x_4 = \frac{s_4 - s_{f,4}}{s_{fg,4}}$$
$$= \frac{6.5966 - 1.8607}{4.9606} \quad \text{(since } s_4 = s_3\text{)}$$
$$= 0.9547$$

$$h_4 = h_{f,4} + x_4 h_{fg,4}$$
$$= 640.23 + 0.9547 \times 2108.5$$
$$= 2653.22 \text{ kJ/kg}$$

$$x_{4A} = \frac{s_{4A} - s_{f,4A}}{s_{fg,4A}}$$
$$= \frac{7.7938 - 0.4764}{7.9187} \quad \text{(since } s_{4A} = s_{3A}\text{)}$$
$$= 0.9241$$

$$h_{4A} = h_{f,4A} + x_{4A} h_{fg,4A}$$
$$= 137.82 + 0.9241 \times 2423.7$$
$$= 2377.56 \text{ kJ/kg}$$

The net work, the heat supplied, and the thermal efficiency for the cycle can now be calculated using the enthalpy values in the table.

$$w_{\text{cycle}} = w_{s,1-2} + w_{s,3-4} + w_{s,3A-4A}$$
$$= (h_1 - h_2) + (h_3 - h_4) + (h_{3A} - h_{4A})$$
$$= (137.82 - 147.87) + (3373.7 - 2653.22) + (3271.9 - 2377.56)$$
$$= 1604.77 \text{ kJ/kg}$$

$$q_{in} = (h_3 - h_2) + (h_{3A} - h_4)$$
$$= (3373.7 - 147.87) + (3271.9 - 2653.22)$$
$$= 3844.5 \, kJ/kg$$

$$\eta_t = \frac{w_{cycle}}{q_{in}}$$

$$= \frac{1604.77}{3844.5} = 0.417, \text{ or } 41.7 \, \%$$

Comments: The carnot efficiency for the temperature limits of this reheat cycle would
be

$$\eta_{Carnot} = 1 - \frac{32.88 + 273.15}{500 + 273.15} = 0.604$$

The efficiency of the reheat cycle is well below the Carnot efficiency.

The reheat cycle in this example has an efficiency higher than the efficiency
values in the previous example problems. This is not due to the reheat feature of
the cycle, but instead to the increase in the temperature of the supply steam. The
reheat cycle in this example does succeed in maintaining the quality of the steam
in the turbine above 90 percent.

8.3.2 The Regenerative Cycle

We know that the efficiency of a Rankine cycle is less than the efficiency of a Carnot
cycle operating between the same temperature levels as the Rankine cycle. Referring
to Figure 8.11(a), we see that the Rankine cycle receives heat over a range of temperatures starting with T_2 and ending with T_3. On the other hand, the Carnot cycle would
receive heat energy only at T_3. We may inquire whether it is possible to avoid heating
process 2–2A insofar as it contributes to a lower efficiency. In theory at least, it is
possible: We do not carry out process 2–2A by external heating; instead, we use
internal heat exchange or regeneration.

Consider the mechanical configuration shown in Figure 8.8(b), where the
high-pressure liquid water from the feed pump passes through a coil wrapped around
the turbine casing and then to the boiler. We assume a reversible heat transfer from
the turbine steam to the liquid in the coil around the casing. This means that at any
given point in the coil, the temperature of the feed water is less than the temperature
of the steam, imparting heat to the water by an infintesimal amount, dT. Under
these conditions, the liquid water from the feed pump will be heated from T_2 to T_{2A}
in a reversible process by internal heat transfer. Also, the thermodynamic states of
the steam expanding in the turbine would lie on the curve 3–4A–4. Note that the
feed water experiences a temperature rise from T_1 to T_2 in the feed pump, so the
regenerative heating of the water has to begin at T_2. This means that after the temperature of the steam drops to T_{4A} (equal to T_2) due to regeneration and expansion,
the steam can be expanded isentropically along 4A–4. Since the heat exchange process
is reversible, ds for curve 2–2A will be equal and opposite of the ds for curve 3–4A;
that is, the curves 2–2A and 3–4A will be identical in slope, curvature, and length.

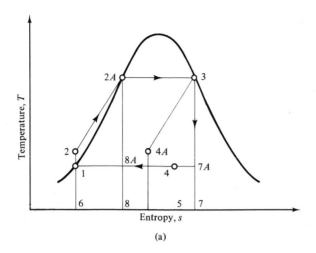

(a)

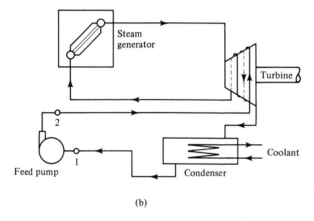

(b)

Figure 8.11 The ideal regenerative Rankine cycle.
(a) *T-s* diagram.
(b) Mechanical components.

Consequently, the areas under the two curves on the *T-s* diagram 1–2–2*A*–8*A*–8–6–1 and 3–4*A*–4–5–7–7*A*–3 are equal.

For the regenerative cycle, area 1–2–2*A*–3–4*A*–4–5–6–1 represents the heat supplied. We note that this area is given by

Area 1–2–2*A*–3–4*A*–4–5–6–1 = area 1–2–2*A*–8*A*–8–6–1

$$+ \text{ area } 2A\text{–}3\text{–}4A\text{–}4\text{–}5\text{–}8\text{–}8A\text{–}2A$$

$$= \text{area } 3\text{–}4A\text{–}4\text{–}5\text{–}7\text{–}7A\text{–}3 + \text{area } 2A\text{–}3\text{–}4A\text{–}5\text{–}8\text{–}8A\text{–}2A$$

$$= \text{area } 2A\text{–}3\text{–}7A\text{–}7\text{–}8\text{–}8A\text{–}2A$$

Also, the heat rejected by the cycle is given by area 1–4–5–6–1, which can readily be shown to be equal to area 7*A*–7–8–8*A*–7*A*. Consequently, the thermal efficiency of the ideal regenerative Rankine cycle will be given by

$$\eta_{\substack{\text{Rankine} \\ \text{regen.}}} = 1 - \frac{\text{area } 7A-7-8-8A-7A}{\text{area } 2A-3-7-8-2A}$$

$$= 1 - \frac{T_3}{T_4} = \eta_{\text{Carnot}}$$

The ideal regenerative Rankine cycle is not feasible in practice. The reversible heat transfer from the expanding steam in the turbine to the water from the feed pump cannot be achieved. Also, the quality of the exhaust from the turbine, state 4 in Figure 8.11(a), would be relatively low, the disadvantages of which have already been discussed. However, it is possible to employ the principle of regeneration to

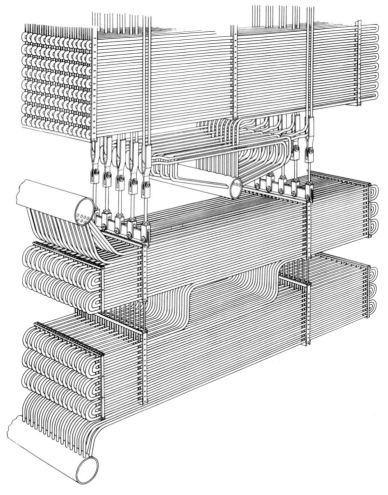

Plate 14 Economizer for a Babcock and Wilcox boiler. (Courtesy of the Babcock and Wilcox Company.)

improve the Rankine cycle efficiency. This is done by extracting a small amount of steam at an intermediate pressure from the turbine and then using the extracted steam to heat the feedwater. There are two types of feedwater heaters, open and closed. In the former, the extracted steam and the feedwater are mixed, while in the latter there is only a thermal contact between the two. Figure 8.12(a) and (b) show the *T-s* diagram and the component diagram for a single-stage extraction system with an open feedwater heater. In the open heater, the steam and the feedwater come into direct contact, resulting in good heat transfer between the extracted steam and the feedwater. We note that there are two pumps [Figure 8.12(b)], one before the heater and one after the heater.

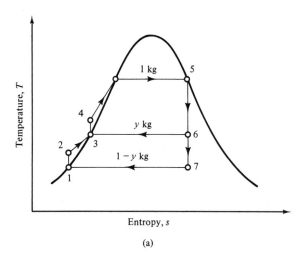

(a)

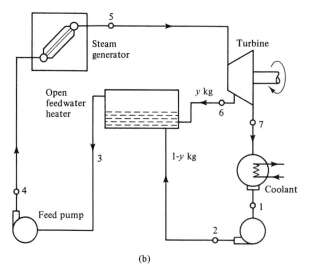

(b)

Figure 8.12 A single-stage extraction Rankine cycle with open feedwater heater.
(a) *T-s* diagram.
(b) Mechanical components.

For every 1 kg of steam entering the turbine, y kg of steam are extracted at the intermediate pressure, p_6, and the remaining $1 - y$ kg of steam are allowed to expand from p_6 to the condenser pressure, p_7, in the turbine. These $1 - y$ kg of steam are then condensed and the liquid is pressurized to pressure p_2, which is the same as the intermediate pressure p_6. This liquid water and the extracted steam, both at the same pressure, are mixed in the open feedwater heater. The fraction y of the steam extracted is chosen so that the water leaving the feedwater heater is in a saturated state at pressure p_3 [Figure 8.12(a)], which is the same as the intermediate pressure p_6. The pressure of this water is then raised to p_4 by another feed pump, and it is then admitted to the boiler.

Referring to Figure 8.12, we note that the same unit mass of water does not participate in all the processes in the cycle. This is highlighted by the labels y on the segment 6–3 and $1 - y$ on the segment 6–7–1–2–3 of the extraction cycle. Also, it is not possible to show the internal heat transfer in the feedwater heater on the T-s diagram.

The fraction y is determined by performing an energy balance on the feedwater heater.

$$\text{Energy inflow} \quad = \text{energy outflow}$$

$$\dot{m}(1 - y)h_2 + \dot{m}(y)h_6 = \dot{m}[y + (1 - y)]h_3$$

Noting that state 3 is saturated liquid, we obtain

$$(1 - y)h_2 + yh_6 = h_{f,3} \tag{8.4}$$

The pump work for 1 kg of feedwater supplied to the boiler is

$$w_{\text{pump}} = w_{s,1\text{-}2}(1 - y) + w_{s,3\text{-}4} \tag{8.4a}$$

The turbine work for one kilogram of steam entering the turbine is

$$w_{\text{turbine}} = w_{s,5\text{-}6} + w_{s,6\text{-}7}(1 - y) \tag{8.4b}$$

The heat supplied is given by

$$q_{4\text{-}5} = h_5 - h_4 \tag{8.4c}$$

A single-stage extraction cycle with a closed feedwater heater is shown in Figure 8.13. In a closed heater, the extracted steam condenses on the exterior surface of the coils carrying the feedwater. The condensate is either pumped back into the line leaving the feedwater heater [Figure 8.13(b)], or it is sent to a hot well condensor. In the latter case, only one pump is needed. In either situation, the pressure of the feedwater in the feedwater line is higher than the pressure of the extracted steam. Since the two fluids do not come into direct contact, the heat transfer characteristics of the closed feedwater heater are not as good as those of the open type.

Referring to Figure 8.13, we assume that for every 1 kg of steam entering the turbine, y kg of steam are extracted at the intermediate pressure p_6. This steam, with specific enthalpy h_6, enters the heater and raises the temperature of the high-pressure water from T_2 to T_{2A}. If the temperature T_{2A} is specified, the fraction y can be determined by performing an energy balance on the heater.

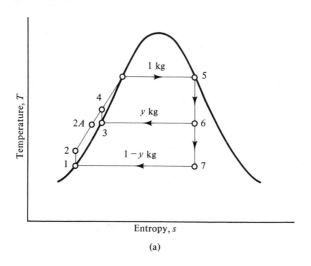

(a)

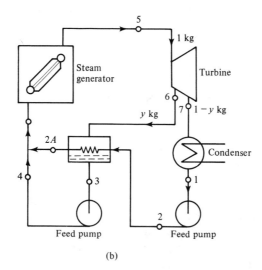

(b)

Figure 8.13 A single-stage extraction Rankine cycle with closed-feedwater heater.
(a) *T-s* diagram.
(b) Mechanical components.

Energy inflow   = energy outflow

$$(1 - y)h_2 + yh_6 = (1 - y)h_{2A} + yh_3$$

Since state 3 represents saturated liquid, we can write

$$(1 - y)h_2 + yh_6 = (1 - y)h_{2A} + yh_{f,3} \qquad (8.4d)$$

If we assume that $h_{2A} = h_3$, this equation simplifies to read

$$(1 - y)h_2 + yh_6 = h_{f,3}$$

which is the same as Equation 8.4.

The quantity of pump work for 1 kg of feedwater supplied to the boiler is

$$w_{pump} = w_{s,1-2}(1-y) + w_{s,3-4}y$$

The heat supplied is given by

$$q_{in} = (h_5 - h_{2A})(1-y) + (h_5 - h_4)y \qquad (8.4f)$$

If the difference between h_{2A} and h_4 is negligibly small, Equation 8.4(f) reduces to Equation 8.4(c).

Example 8.6 illustrates the calculations for a Rankine cycle with extraction.

Example 8.6

In an ideal regenerative Rankine cycle, the steam enters the turbine at 4000 kPa and 400°C. After isentropic expansion to an intermediate pressure of 800 kPa, a fraction y is extracted from the turbine and mixed with the feedwater to produce saturated liquid water at 800 kPa. Another feed pump raises the pressure of this liquid to 4000 kPa and delivers it to the boiler. The remainder of the steam is expanded in the turbine to 50 kPa. Determine the thermal efficiency of the cycle. If a plant operating on this cycle is to produce 10,000 kW, determine the mass flow rate of the steam.

Solution:

Given: An ideal regenerative Rankine cycle with an open feedwater heater has the following data.

High-pressure stage: $p_5 = 4000$ kPa, $t_5 = 400°C$, $p_6 = 800$ kPa

Low-pressure stage: $p_6 = 800$ kPa, $p_7 = 50$ kPa

Power output: 10,000 kW

Objective: We are to determine the thermal efficiency of the cycle, η_t, and the mass flow rate of the steam, $\dot{m}$.

Diagram:

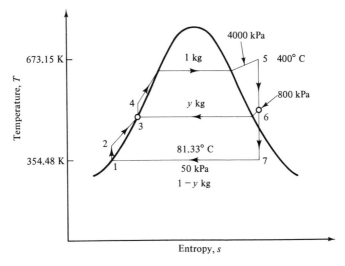

Figure 8.14

Assumptions:

1. All processes in the cycle are reversible.
2. There is no energy loss from the feedwater heater.
3. The volume change during the compression process is negligible.
4. The specific heat of liquid water is 4.18 kJ/kg.

Relevant physics: We note that there are two feed pumps with known operating pressures. This permits calculation of h_2, t_2, h_4, and t_4. We do not know the phase at point 6. The phase can be determined by comparing the value s_6, which equals the known value, s_5 with the entropy of the saturated vapor at pressure p_6. Similar considerations will help ascertain the phase at point 7. The fraction y of steam extracted can be calculated from Equation 8.4.

Once all the relevant properties at all the state points are determined, η_t can be calculated by using Equations 8.4(a), (b), and (c).

Analysis: Referring to Appendixes B.2.2 and B.3, properties at various state points are obtained and entered in the table below.

Point	t (°C)	p (kPa)	Phase	h (kJ/kg)	s (kJ/kg K)
1	81.33	50	Sat. liquid	340.49	1.0910
2	81.51[a]	800	Comp. liquid	341.26[a]	1.0910
3	170.43	800	Sat. liquid	721.11	2.0462
4	171.28[a]	4000	Comp. liquid	724.68[a]	2.0462
5	400	4000	Super. vapor	3213.6	6.7690
6	191[a]	800	Super. vapor[a]	2817.8[a]	6.7690
7	81.33	50	Wet vapor[a]	2353.1[a]	6.7690

[a]Calculated values.

In a low-pressure pump,

$$w_{s,\,1\text{-}2} = -v_{f,1}(p_2 - p_1)$$

$$= -0.001030(800 - 50)$$

$$= -0.77 \text{ kJ/kg}$$

$$h_2 = h_1 - w_{s,\,1\text{-}2}$$

$$= 340.49 + 0.77$$

$$= 341.26 \text{ kJ/kg}$$

and

$$t_2 = t_1 + \frac{h_2 - h_1}{c_p}$$

$$= 81.33 + \frac{0.77}{4.18}$$

$$= 81.51°C$$

Proceeding in a similar manner for the high-pressure pump, we get

$$w_{s,3\text{-}4} = -v_{f,3}(p_4 - p_3)$$
$$= -0.001115(4000 - 800)$$
$$= -3.57 \text{ kJ/kg}$$
$$h_4 = h_3 - w_{s,3\text{-}4}$$
$$= 721.11 + 3.57$$
$$= 724.68 \text{ kJ/kg}$$

and

$$t_4 = t_3 + \frac{h_4 - h_3}{c_p}$$
$$= 170.43 + \frac{3.57}{4.18}$$
$$= 171.28°C$$

We now consider the isentropic expansion process in the turbine and determine states 6 and 7. Examination of the entropy values for saturated vapor states from Appendix B.2 corresponding to pressure $p_6 = 800$ kPa and $p_7 = 50$ kPa shows that

$$s_{g,6} = 6.6628 \text{ kJ/kg K}$$
$$s_{g,7} = 7.5939 \text{ kJ/kg K}$$

Since

$$s_6 = s_7 = s_5 = 6.7690 \text{ kJ/kg K}$$

and $s_6 > s_{g,6}$, state 6 is superheated; since $s_7 < s_{g,7}$, state 7 is wet vapor.

To determine h_6 and t_6, we interpolate to find the h- and t-values for $p_6 = 800$ kPa and $s_6 = 6.7690$ from Appendix B.3. Thus $h_6 = 2817.8$ kJ/kg and $t_6 = 191°C$. These numbers can also be conveniently obtained, up to three or four significant digits, from the Mollier diagram for steam.

The quality at point 7 is determined from

$$x_7 = \frac{s_7 - s_{f,7}}{s_{fg,7}}$$
$$= \frac{6.7690 - 1.0910}{6.5029}$$
$$= 0.873$$

Then

$$h_7 = h_{f,7} + x_7 h_{fg,7}$$
$$= 340.49 + 0.873 \times 2305.4$$
$$= 2353.1 \text{ kJ/kg}$$

We now determine the fraction y extracted from the turbine. From an energy balance on the feedwater heater,

$$(1 - y)h_2 + yh_6 = h_{f,3} = h_3$$
$$(1 - y)341.26 + 2817.8y = 721.11$$

or

$$y = 0.153$$

The pump work for 1 kg of feedwater supplied to the boiler is

$$
\begin{aligned}
w_{pump} &= w_{s,1\text{-}2}(1 - y) + w_{s,3\text{-}4} \\
&= -0.77(1 - 0.153) - 3.57 \\
&= -4.22 \text{ kJ/kg}
\end{aligned}
$$

The turbine work for 1 kg of steam entering the turbine is

$$
\begin{aligned}
w_{turbine} &= w_{s,5\text{-}6} + (1 - y)w_{s,6\text{-}7} \\
&= (h_5 - h_6) + (1 - y)(h_6 - h_7) \\
&= (3213.6 - 2817.8) + (1 - 0.153)(2817.8 - 2353.1) \\
&= 789.4 \text{ kJ/kg}
\end{aligned}
$$

and

$$
\begin{aligned}
q_{in} &= h_5 - h_4 \\
&= 3213.6 - 724.68 \\
&= 2488.92 \text{ kJ/kg}
\end{aligned}
$$

so that

$$
\begin{aligned}
\eta_t &= \frac{w_{cycle}}{q_{in}} \\
&= \frac{-4.22 + 789.4}{2488.92} \\
&= 0.315, \text{ or } 31.5\%
\end{aligned}
$$

The mass flow rate of steam to produce 10,000 kW is

$$\dot{m} = \frac{10,000}{w_{cycle}} = \frac{10,000}{785.2} = 12.74 \text{ kg/s}$$

Comments: We note that the thermal efficiency obtained here is greater than that of Example 8.4, where the same limiting temperatures, 400°C and 81.33°C, were used. In Example 8.4, where no regeneration was involved, the efficiency was found to be 26.36%.

With an increase in the number of extraction stages, the thermal efficiency of a plant increases. As noted earlier in this section, the upper limit of the efficiency is the Carnot efficiency. However, it is not practicable to have a large number of extraction stages. The increase in the cost of the mechanical components that necessarily accompanies an increase in the number of extraction stages must be more than offset by the savings due to improved efficiency. Consequently, we seldom find more than five regenerative stages. The procedure for determining the thermal efficiency of a Rankine cycle with multiple regenerative stages is similar to the one used in Example 8.6.

8.3.3 Actual Rankine Cycle

In the earlier discussion, all the processes constituting the Rankine cycle were assumed to be ideal, or reversible, processes. In reality, there are losses due to irreversibilities due mainly to fluid friction and undesired heat losses from the mechanical components to the surroundings. Let us first consider the feed pump.

Usually, the temperature of the liquid passing through the pump is higher than the temperature of the surroundings. Consequently, there is a heat transfer from the pump to the surroundings, and the compression process is not adiabatic. However, because the flow rate through the pump is usually large, the fraction of heat transferred to the surroundings as compared to the pump work is very small; we can neglect it and still treat the compression as adiabatic. The fluid friction, however, makes the compression irreversible, requiring more work to compress the liquid through the same pressure range. In Figure 8.15(a), path 1–2s shows isentropic compression, while 1–2 denotes actual adiabatic compression. It is, therefore, customary to define the isentropic efficiency of a feedpump as

$$\eta_{i,\text{pump}} = \frac{\text{isentropic work of compression}}{\text{actual adiabatic work of compression}} \tag{8.5}$$

$$= \frac{h_{2s} - h_1}{h_2 - h_1} \tag{8.5a}$$

The effect of the irreversible process is to increase the temperature from T_{2s} to T_2. This increase in the temperature is brought about by expending high-grade mechanical work, whereas the same increase can be effected by utilizing low-grade heat energy. Consequently, it is desirable to have as high an isentropic efficiency as possible.

Furthermore, there are losses in a pump due to mechanical friction in the bearings, gears, and the transmission from the prime mover. These are usually grouped under mechanical efficiency, which is defined as

$$\eta_{m,\text{pump}} = \frac{\text{adiabatic work input}}{\text{actual work input}} \tag{8.5b}$$

There is usually a long pipe connecting the pump and the boiler. The losses due to heat transfer to the surroundings from the pipe are usually less than one-half of 1 percent of the enthalpy rise of the fluid in the boiler. The pressure drop in the pipe can be significant. The pressure drop is especially severe in the boiler, but is unavoidable since high-flow velocities in the boiler must be maintained to ensure high heat transfer rates between the combustion products of the fuel and the water. The feed pump must develop sufficient pressure head, over and above the turbine-inlet pressure, to compensate for the pressure drop in the boiler.

Within the boiler itself, fuel is burned and most of the released heat energy is imparted to the water passing through the boiler. The products of combustion leave the boiler and eventually pass through a chimney, carrying with them some of the energy released in the combustion process. It is then convenient to define a boiler

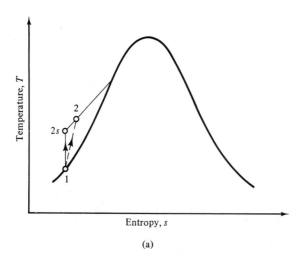

(a)

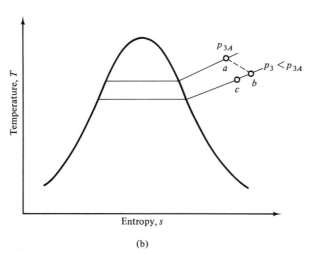

(b)

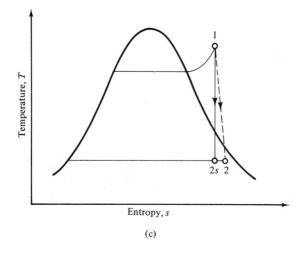

(c)

Figure 8.15 Deviations from the ideal Rankine cycle.
(a) Irreversible adiabatic compression in the feed pump.
(b) Pressure drop in the steam generator and pipe lines.
(c) Irreversible adiabatic expansion in the turbine.

efficiency to express the effectiveness of the heat transfer process in the boiler:

$$\eta_{boiler} = \frac{\text{enthalpy gain by water passing through the boiler}}{\text{energy supplied by the fuel to achieve this enthalpy gain of the water}} \quad (8.6)$$

$$= \frac{\dot{m}(h_{out} - h_{in})}{\dot{m}_{fuel}HV} \quad (8.6a)$$

where

$\dot{m}$ = mass flow rate of water through the boiler (kg/s)

h_{out} = specific enthalpy of water leaving the boiler (kJ/kg)

h_{in} = specific enthalpy of water entering the boiler (kJ/kg)

$\dot{m}_{fuel}$ = mass flow rate of the fuel into the boiler (kg/s)

HV = heating value of the fuel, that is, the amount of heat released on the complete combusion of a unit mass of fuel (kJ/kg)

(A detailed discussion of HV is presented in a later chapter.) For a well-maintained boiler, the boiler efficiency is about 80 percent. The efficiency of a superheater can can be defined in a similar manner.

As the steam is transported from the boiler, superheater, to the turbine, it experiences a pressure drop and a heat loss. The resulting change in states is shown in Figure 8.15(b). Point a in the figure represents the high-pressure superheated steam on the constant pressure line p_{34}. During the change in the state due to the pressure drop in the pipe, the enthalpy remains essentially constant (similar to a throttling process) and the new state is at b on the constant pressure line p_3. Note that the entropy at b is greater than the entropy at a. The heat transfer from the pipe to the surroundings would cause a decrease in temperature, and the state of the steam at the entrance to the turbine would be represented by point c. In practice we would know only states a and c. For convenience in discussion, the actual change from a to c has been broken into a–b and b–c.

The pressure of the steam delivered to the turbine often fluctuates. The performance of a turbine is sensitive to the changes in supply pressure. In order to maintain a steady performance of the turbine, the steam is first passed through a pressure regulating valve before it is admitted to the turbine nozzles. The condition of such steam entering the turbine nozzles is sometimes referred to as the *throttle condition* of steam. Throttling is also used for governing the turbine speed.

A turbine consists of rows of nozzles (Figure 8.16) fixed to the casing of the turbine, alternating with rows of blades attached to the rotor of the turbine. As the steam passes through a row of nozzles, it experiences a small drop in pressure and an increase in kinetic energy; then, as the steam enters a row of blades, it imparts its energy to the blades by momentum transfer and the steam further experiences a small drop in pressure. One row of nozzles and one row of moving blades constitute a *stage*. There can be as many as 40 to 50 stages in large turbines.

Due to fluid friction the expansion process in a stage is not isentropic. This means that although the desired pressure drop is achieved in a stage, the actual work

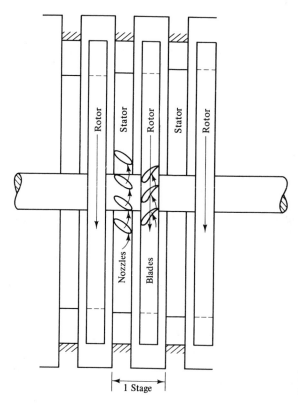

Figure 8.16 Schematic of the fixed nozzles and moving blades in a turbine.

output is less than the isentropic work output. The "loss" in the work output appears as an increase in the enthalpy of the steam leaving the stage, relative to the isentropic. A curve on an *h-s* diagram showing the actual thermodynamic states of the steam leaving each stage is called a *condition line.* Figure 8.15(c) shows a similar line on a *T-s* diagram. Had the expansion been isentropic, the turbine exhaust would have been represented by 2s in the figure. The actual expansion is irreversible adiabatic, and it is shown as the dashed curve 1–2. The heat transfer from the turbine casing to the surroundings is very small, and the process in the turbine is regarded as adiabatic. We define isentropic turbine efficiency by

$$\eta_{i,\text{turbine}} = \frac{\text{actual adiabatic work output}}{\text{isentropic work output}} \tag{8.7}$$

or

$$\eta_{i,\text{turbine}} = \frac{h_1 - h_2}{h_1 - h_{2s}} \tag{8.7a}$$

The isentropic turbine efficiency is determined by experimental measurements and is of the order of 80 to 85 percent. One beneficial side effect of the irreversible adiabatic process in a turbine is that the quality of the exhaust is increased.

In problems involving isentropic efficiencies, it is usually necessary first to compute the state properties by assuming an isentropic efficiency of 100 percent.

Plate 15 (a) The rotor of a steam turbine for a power plant. (Courtesy of the Rochester Gas and Electric Corporation.)

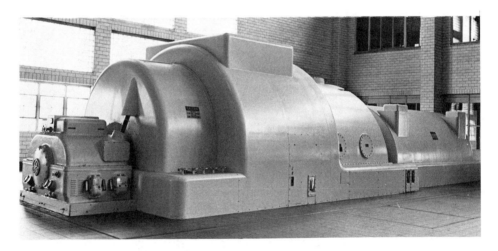

Plate 15 (b) Steam turbine-electrical generator for a power plant. (Courtesy of the Rochester Gas and Electric Corporation.)

Next, we use the equation for isentropic efficiency to calculate the actual enthalpy values at the exit. In the case of a turbine, this enthalpy value is to be used to determine the quality of the exhaust.

As in the case of the pump, we can define a mechanical efficiency of a turbine to account for the mechanical losses. This mechanical efficiency is given by

$$\eta_{m,\text{turbine}} = \frac{\text{actual work output}}{\text{adiabatic work output}} \tag{8.7b}$$

Example 8.7

If the isentropic efficiencies of the feed pump and the turbine in Example 8.4 are 0.80 and 0.90, respectively, determine thermodynamic states at the exit of the pump and the turbine, and the thermal efficiency of the cycle.

Solution:

Given: A Rankine cycle with superheat has the following data.

$$\text{Turbine inlet:} \quad p_3 = 2000 \text{ kPa}, \qquad t_3 = 400°C$$

$$\text{Condensor:} \qquad p_4 = 50 \text{ kPa}$$

$$\eta_{t,\text{pump}}: \qquad 0.80$$

$$\eta_{t,\text{turbine}}: \qquad 0.90$$

Objective: We are to determine the properties at points 2 and 4 and the thermal efficiency of the cycle.

Diagram:

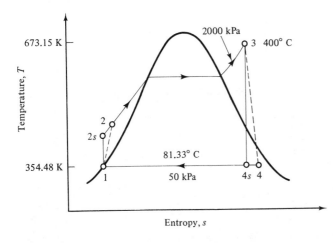

Figure 8.17   *T-s* diagram for a Rankine cycle.

Assumptions:

1. The processes in the steam generator, superheater, and condenser are reversible.
2. The volume change during the compression process is negligible.
3. The specific heat of liquid water is 4.18 kJ/kg K.

Relevant physics: Since the pump and the turbine do not operate isentropically, the actual exit enthalpies h_2 and h_4 are higher than the isentropic exit enthalpies h_{2s} and h_{4s}. The latter have been calculated for Example 8.4. The actual exit enthalpies can be calculated from Equations 8.5(a) and 8.7(a). Once the various *h*-values are known, the thermal efficiency can be calculated in the usual manner.

Analysis: We know the properties at state points 1, 2s, 3, and 4s from the solution to Example 8.4. These are entered in the table below. To determine point 2, we proceed as follows. From Equation 8.5(a)

$$\eta_{i,\text{pump}} = \frac{h_{2s} - h_1}{h_2 - h_1}$$

Point	t (°C)	p (kPa)	Phase	h (kJ/kg)	s (kJ/kg K)
1	81.33	50	Sat. liquid	340.49	1.0910
2S	81.8a	2000	Comp. liquid	342.49	1.0910
2	81.91a	2000	Comp. liquid	342.99a	
3	400	2000	Super. vapor	3247.6	7.1271
4S	81.33	50	Wet vapor	2479.9	7.1271
4	—	50	Wet vapor	2556.67a	

aCalculated values.

or

$$0.8 = \frac{342.49 - 340.49}{h_2 - 340.49}$$

Solving for h_2, we get

$$h_2 = 342.99 \text{ kJ/kg}$$

The state of the actual exhaust from the turbine can be determined by using Equation 8.7.

$$\eta_{i,\text{turbine}} = \frac{h_3 - h_4}{h_3 - h_{4s}}$$

or

$$0.9 = \frac{3247.6 - h_4}{3247.6 - 2479.9}$$

and

$$h_4 = 2556.67 \text{ kJ/kg}$$

The above value of h_4 is less than $h_{g,4}$ but greater than $h_{f,4}$ (Appendix B2.2); therefore the exhaust steam is wet vapor. Its quality is given by

$$x_4 = \frac{h_4 - h_{f,4}}{h_{fg,4}}$$

$$= \frac{2556.67 - 340.49}{2305.4}$$

$$= 0.961$$

which is indeed greater than the quality obtained in Example 8.4.

The actual work input for the pump and the actual work output of the turbine can now be calculated in two ways: (1) by using Equations 8.5 and 7.7(a), respectively, or (2) from $w_s = h_{\text{in}} - h_{\text{out}}$. Using the first approach, we have For the pump

$$0.8 = \frac{w_{i,\text{pump}}}{w_{a,\text{pump}}}$$

or

$$w_{a,\text{pump}} = \frac{-2}{0.8} = -2.5 \text{ kJ/kg}$$

For the turbine

$$0.9 = \frac{w_{a,\text{turbine}}}{w_{i,\text{turbine}}}$$

or

$$w_{a,\text{turbine}} = 0.9 \times 767.7 = 690.93 \text{ kJ/kg}$$

In the above, values of w_i for the pump and the turbine are taken from Example 8.4.

$$q_{\text{in}} = h_3 - h_1$$
$$= 3247.6 - 342.99$$
$$= 2904.61 \text{ kJ/kg}$$

The thermal efficiency is then given by

$$\eta_t = \frac{w_{a,\text{pump}} + w_{a,\text{turbine}}}{q_{\text{in}}}$$

$$= \frac{-2.5 + 690.93}{2904.61}$$

$$= 0.237, \text{ or } 23.7\%$$

Comments: The effect of the irreversible operation of the pump and the turbine is to reduce the thermal efficiency from 26.4% to 23.7%.

Example 8.8

If the mechanical efficiencies of the pump and the turbine in Example 8.7 are 0.95 and if the boiler efficiency is 0.78, determine the overall efficiency of the plant.

Solution:

Given: The data on a Rankine cycle with superheat are as follows.

$$\text{Turbine inlet:} \quad p_3 = 2000 \text{ kPa,} \qquad t_3 = 400°C$$
$$\text{Condenser:} \quad p_4 = 50 \text{ kPa}$$
$$\eta_{i,\text{pump}} = 0.80, \qquad \eta_{i,\text{turbine}} = 0.90$$
$$\eta_{m,\text{pump}} = 0.95 = \eta_{m,\text{turbine}}$$
$$\eta_{\text{boiler}} = 0.78$$

Objective: The overall efficiency of the plant is to be determined.
Diagram: See Figure 8.17.
Assumptions:

1. The condenser operates reversibly.
2. The volume change during the compression process is negligible.

Relevant physics: The less than 100% mechanical efficiencies of the pump and turbine result in an energy loss. This energy is transferred to the surroundings and has no effect on the state enthalpies.

Analysis: From Equations 8.5(b) and 8.7(b), respectively, we have

$$0.95 = \frac{w_{a,\text{pump}}}{w_{\text{actual,pump}}}$$

and

$$0.95 = \frac{W_{actual,turbine}}{W_{a,turbine}}$$

Substitution gives

$$W_{actual,pump} = \frac{-2.5}{0.95} = -2.63 \text{ kJ/kg}$$

and

$$W_{actual,turbine} = 0.95 \times 690.93 = 656.38 \text{ kJ/kg}$$

Note that the increased work input to the pump does not change the value of the exit enthalpy, since the additional work is used in overcoming mechanical losses.

From Equation 8.6,

$$0.78 = \frac{h_3 - h_2}{q_{fuel}}$$

Substitution gives

$$q_{fuel} = \frac{3247.6 - 342.99}{0.78}$$

$$= 3723.86 \text{ kJ/kg}$$

In other words, in order to obtain 1 kg of superheated steam at 2000 kPa and 400°C from liquid water at 2000 kPa and 81.91°C, 3723.86 kJ of energy of fuel is used. The overall efficiency is then given by

$$\eta_{overall} = \frac{W_{cycle}}{q_{fuel}}$$

$$= \frac{-2.63 + 656.38}{3723.86}$$

$$= 0.175, \text{ or } 17.5\%$$

Comments: This and the preceding examples illustrate that the overall efficiency of a real steam power plant can be considerably lower than the thermal efficiency of the ideal cycle.

8.3.4 Binary Vapor Cycle

The critical pressure of water is very high (22,090 kPa), but the critical temperature (374.14°C) is within the permissible structural limit of about 1000°C for boilers. On the condenser side of the Rankine cycle, very low vapor pressures are required if condensation is to occur at temperatures near ambient temperatures. Maintaining low vapor pressure requires minimizing leaks and operating vacuum pumps. For these reasons *binary* vapor cycles have been proposed and used.

A binary cycle uses two working fluids, such as mercury and water. The heat is supplied to the high-temperature fluid, in this case, mercury. Its critical temperature is 898.3°C, close to the limiting temperature of the boiler. However, its critical pressure is only about 18,000 kPa. The high-temperature fluid expands in a turbine and is then condensed in a heat exchanger. The heat energy released in the condensation process is used to heat and evaporate the low-temperature fluid, water. The high-temperature fluid is then pressurized by a pump and returned to the boiler. The low-temperature fluid, on leaving the heat exchanger, passes through a turbine, condenser, and a pump, in that order. The efficiency of such a cycle is better than that of a cycle using only water.

The toxicity of mercury and its high initial cost often stand against considering mercury-water binary vapor cycle. Also, with the modern trend towards high-temperature, high-pressure multistage regenerative cycles, the binary vapor cycle has become economically unattractive. Nonetheless, the binary cycle is an interesting cycle from the thermodynamic point of view. A binary cycle employing water and Freon-12 is currently being explored for medium temperature applications.

8.3.5 Modern Plants

In the earlier discussion of the Rankine cycle, we considered water as the working medium. Also, the pressures considered were below the critical pressure of water, namely, 22,090 kPa. As the turbine-inlet pressure is increased, the temperature also increases, and there is a corresponding improvement in the work output and efficiency. In this context, it is useful to define *heat rate*, which is the quantity of heat energy supplied to the working medium undergoing a thermodynamic cycle to produce unit net work output from the cycle. It is expressed in Btu/kWh or joules (thermal)/joule (mechanical). It can be seen from the definition that the heat rate is, in essence, the reciprocal of the thermal efficiency. Under ideal conditions as the inlet pressure is increased, the heat rate decreases, that is, the plant efficiency increases.

When we incorporate the irreversibilities due to fluid friction, pressure drops, and heat losses in calculating the heat rate, instead of monotonically decreasing with increases in pressure, the heat rate exhibits a minimum value for an optimum inlet pressure. This optimum pressure is supercritical. Consequently, many of the modern steam power plants operate with supercritical turbine-inlet pressures. A Rankine cycle with supercritical pressure is shown in Figure 8.18. Note that in no segment of the process 2–3 is the heat supply isothermal.

For power plants intended for operation in space, the working fluid for the Rankine cycle is usually a liquid metal, such as potassium, sodium, or mercury. The important considerations for space applications are the high heat transfer coefficients and the high heat capacity rates. The heat capacity rate of a fluid is the product of the mass flow rate and the specific heat of the fluid flowing through a heat exchanger. Liquid metals have these desirable characteristics, making it possible to use compact heat exchangers.

In the case of power plants employing solar energy as the heat source, the operating temperature range is rather low—from 200°C to 25°C. In such applications, Freon-11 and Freon-113 are being considered for the working medium. These fluids exhibit either a near-zero or a positive slope of the saturated vapor line on the *T-s* diagram. This means that at the end of an isentropic expansion of the vapor through a turbine, the fluid is always in a superheated vapor state. This behavior obviates the need to superheat the working medium and improves the performance of the turbine where the expansion occurs.

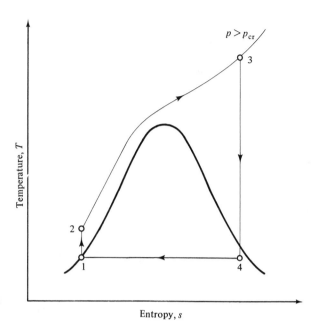

Figure 8.18 A Rankine cycle with supercritical pressure.

We now turn our attention to cycles and systems used for refrigeration, that is, for transferring heat from a low-temperature reservoir to a high-temperature reservoir.

8.4 REFRIGERATION

The Clausius statement of the second law tells us that if heat energy is to be transferred from a low-temperature body to a high-temperature body, external work must be supplied. If the Carnot cycle is traced counterclockwise, it requires work input and heat is transferred from a low-temperature (T_L) reservoir to a high-temperature (T_H) reservoir. The cycle is then said to operate as a *refrigerator* or a *heat pump*. If the objective is to extract heat from the low-temperature reservoir, then the cycle is called a *refrigeration cycle*. If the objective is to supply heat energy to the high-temperature reservoir, then the cycle is called a *heat pump*. (See Section 5.8.3).

The efficiency of a refrigeration or a heat pump cycle is measured by the *coefficient of performance*, COP. For a refrigeration cycle, it is

$$(\text{COP})_R = \frac{\text{heat extracted from the low-temperature reservoir}}{\text{work input for the refrigeration cycle}} \tag{8.8}$$

Referring to Figure 8.19,

$$(\text{COP})_{R,\text{Carnot}} = \frac{T_L(s_1 - s_4)}{(T_H - T_L)(s_1 - s_4)}$$

$$= \frac{T_L}{T_H - T_L} \tag{8.8a}$$

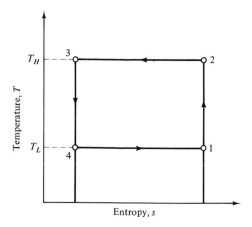

Figure 8.19 Carnot refrigeration cycle.

For a heat pump cycle, the coefficient of performance is given by

$$(COP)_{HP} = \frac{\text{heat supplied to the high-temperature reservoir}}{\text{work input for the heat pump cycle}} \qquad (8.9)$$

If the Carnot cycle in Figure 8.19 is used as a heat pump, we have

$$(COP)_{HP,Carnot} = \frac{T_H(s_2 - s_3)}{(T_H - T_L)(s_2 - s_3)}$$

$$= \frac{T_H}{T_H - T_L} \qquad (8.9a)$$

In the interest of concreteness, we shall restrict ourselves in the following discussion to a refrigeration cycle. The conclusions can be readily extended to a heat pump cycle (also, see Section 5.8.3).

Equation 8.8(a) tells us that as the temperature difference between the two reservoirs increases, area 1–2–3–4–1 in Figure 8.19 increases, and greater quantities of work input are progressively required. This results in a decrease of the coefficient of performance. For instance, for a Carnot refrigerator, if $T_L = 280$ K and $T_H = 320$ K, $(COP)_R = 7$; but if $T_L = 200$ K and $T_H = 320$ K, $(COP)_R = 1.67$.

8.4.1 Vapor Compression Refrigeration

Vapor compression refrigeration systems are used in conventional refrigeration and air-conditioning units. In a vapor compression cycle, the working medium undergoes a change of phase, and the saturated liquid, the liquid-vapor, and the vapor domains of the T-s diagram are involved. The *idealized* cycle 1–2–3–4–1 for vapor compression refrigeration is shown in Figure 8.16. Saturated vapor is compressed from state 1 to state 2 in an isentropic process. The compression is carried out in a reciprocating or a centrifugal compressor. The superheated vapor (state 2) is cooled and condensed in a condenser. The saturated liquid (state 3) leaving the condenser is expanded in a throttling valve to a low pressure. The low-pressure, saturated liquid-vapor mixture (state 4) is admitted into the evaporator, where the refrigeration of the cooling load takes place. Except for the throttling process, the vapor compression refrigeration cycle is the same as the reversed Rankine cycle. If an isentropic expansion process were used after state 3, the amount of work output in the expansion would be very small since the change in the volume of the expanding liquid is small. A throttle valve is a relatively simple mechanical component to carry out the expansion from state 3 to state 4. Since throttling is an irreversible process, the area 1–2–3–4–1 on the T-s diagram in Figure 8.20(a) does not represent the work input to the cycle.

The coefficient of performance of a vapor compression refrigeration cycle is given by

$$(COP)_R = \frac{q_{4-1}}{-w_{1-2}} = \frac{h_1 - h_4}{h_2 - h_1} \tag{8.10}$$

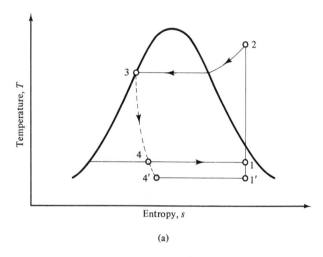

(a)

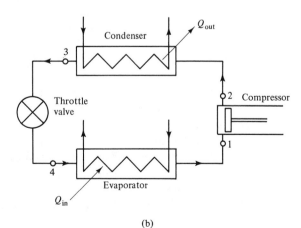

Figure 8.20 Vapor compression refrigeration cycle.
(a) T-s diagram.
(b) Mechanical components.

(b)

In commercial practice, the specification of the cooling rate is expressed in *tons of refrigeration*. One ton of refrigeration means a capacity to convert 1 ton of saturated water at 0°C from the liquid state to the solid state in a 24-hour period. This translates into 200 Btu/min or 3.517 kJ/s.

A number of different substances—such as Freon-12 (dichlorodifluoro methane), Freon-11 (trichloromonofluoro methane), Freon-13 (monochlorotrifluoro methane), Freon-113 (trichlorotrifluoro ethane), ammonia, carbon dioxide, and sulfur dioxide—can be used as refrigerants. Of these, the members of the Freon family are synthetic substances and are developed by E. I. Du Pont de Nemours and Company. A substance with a vertical saturated vapor line on the T-s diagram, or one with a positive slope, would be a desirable refrigerant. This would eliminate the need to superheat the vapor in the compression process. Some other desirable features of a refrigerant are moderately low condensing pressure, relatively high critical pressure, low freezing temperature, high latent heat of vaporization, inertness, chemical stability, nontoxicity, high heat-transfer characteristics, easy leakage detection, and low cost.

It should be noted that there is a volumetric efficiency for a given reciprocating compressor and a given pressure ratio (see Section 7.1). Furthermore, there is a pressure drop at the inlet and outlet valves, causing a reduction in the volumetric efficiency. As we lower the evaporator temperature (with a fixed condenser temperature), the pressure ratio increases. Consequently, the volumetric efficiency decreases, less mass of the refrigerant is handled per unit time, and it takes longer to establish a steady state.

Also, for a fixed condenser temperature, lowering the evaporator temperature line 4–$1'$ in Figure 8.20(a) means that the enthalpy $h_{1'}$ of the fluid entering the compressor is less than the previous enthalpy, h_1. Thus a greater work input per unit mass is required. At the same time, the cooling effect, $h_{1'} - h_{4'}$, is reduced since $h_{1'}$ is less than h_1, and $h_{4'}$ is the same as h_3 because of the throttling process. Consequently, a decrease in the evaporator temperature with a fixed condenser temperature results in a reduction in the COP. However, as noted in the preceding paragraph, with the lowering of the evaporator temperature there would also be a simulatneous decrease in the volumetric efficiency. Since the mass flow rate is proportional to the volumetric efficiency, the mass flow rate then decreases. The power input to the compressor is the product of the work input per unit mass and the mass flow rate. It turns out that as the evaporator temperature is decreased, the rate at which the work input per unit mass increases is less than the rate at which the mass flow rate decreases. Consequently, as the evaporator temperature is lowered, the power input to the compressor decreases.

Ideally, state 3 in Figure 8.20(a) should always be saturated liquid and state 1 should always be saturated vapor. With a fluctuating load, this is not always possible, and there is a possibility of admitting wet vapor in the compressor. To eliminate this possibility, an internal heat exchanger is sometimes installed between the evaporator and the condenser, as shown in Figure 8.21(a). The heat exchanger subcools the condenser discharge and superheats the evaporator output. The subcooling of the liquid in the condenser in this manner increases the cooling effect [Figure 8.21(b)], as well as the work input to the compressor. These effects are conveniently shown on a *p-h* diagram in the figure. The latter occurs because, due to the superheating of the evaporator output, a larger volume of the vapor is now involved. The coefficient of performance does not change much in such an operation. The useful cooling effect is the heat transferred to the working medium that takes place in the *evaporator*.

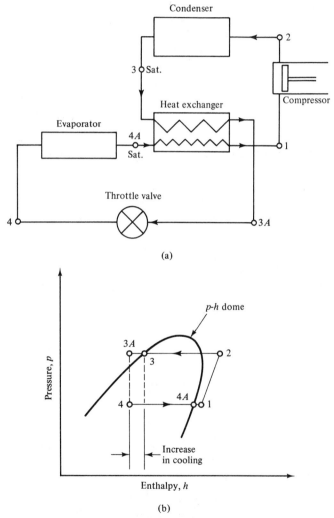

(a)

(b)

Figure 8.21   A refrigeration cycle.
(a) Mechanical components.
(b) *p-h* diagram.

Example 8.9

In a Freon-12, vapor-compression system, liquid leaves the condenser at 40°C and is then admitted into an internal heat exchanger [Figure 8.21(a)]. It then leaves the exchanger at 30°C. The evaporator temperature is −20°C. Determine the (a) cooling effect per unit mass, (b) work input per unit mass, (c) coefficient of performance, (d) volumetric efficiency, (e) power input to produce a cooling rate of 10 kJ/s, and (f) piston displacement if the compressor operates at 1200 rpm. Assume $\gamma = 1.13$ and a clearance factor of 3%.

Solution:

Given: A Freon-12 refrigeration system with an internal heat exchanger has the following data.

$$\begin{array}{ll}
\text{Condenser outlet:} & t_3 = 40°C \\
\text{Heat exchanger outlet for the warm fluid:} & t_{3A} = 30°C \\
\text{Evaporator:} & t_4 = t_{4A} = -20°C \\
\text{Cooling rate:} & 10 \text{ kJ/s.} \\
N = 1200 \text{ rpm,} \qquad \gamma = 1.13, \qquad CF = 0.03 &
\end{array}$$

Objective: We have to determine $q_{\text{evaporator}}$, W_{in}, $(COP)_R$, η_v, P, and PD.

Diagram:

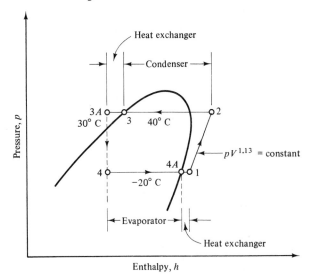

Enthalpy, h **Figure 8.22**

Assumptions:

1. The enthalpy at point $3A$ can be approximated by the enthalpy of the saturated liquid at a temperature equal to t_{3A}.
2. All processes except the throttling are reversible.

Relevant physics: The enthalpy at state 1 can be determined by performing an energy balance on the heat exchanger. Once the enthalpies at various state points are determined, the six objective items can be calculated using the following equations.

$$q_{\text{evaporator}} = h_{4A} - h_4$$

$$w_{\text{in}} = h_2 - h_1$$

$$(\text{COP})_R = \frac{h_{4A} - h_4}{w_{\text{in}}}$$

$$\eta_v = 1 + \text{CF} - \text{CF}\left(\frac{p_H}{p_L}\right)^{1/n}$$

$$\text{Power input} = \dot{m}w_{\text{in}}$$

$$\dot{m} = \frac{\text{PD}\cdot N}{v_1 \times 60}$$

where PD is the piston displacement and N is the rpm.

Analysis: The relevant properties at the various state points are given in the following table from Appendix C. The enthalpy at point $3A$(subcooled state) is determined by looking in the saturated Freon-12 table at 30°C, since in the subcooled region the enthalpy is strongly dependent on temperature and rather insensitive to small changes in pressure.

Point	t (°C)	p (kPa)	Phase	h (kJ/kg)
1	—	64.2	Super. vapor	179.467[a]
2	—	960.7	Super. vapor	232
3	40	960.7	Sat. liquid	74.527
3A	30	960.7	Subcooled liquid	64.539[a]
4	−20	64.2	Wet vapor	64.539
4A	−20	64.2	Sat. vapor	169.479

[a]Calculated or derived values.

An energy balance on the internal heat exchanger for steady flow conditions gives

$$\text{Energy in} = \text{energy out}$$

$$\dot{m}(h_3 + h_{4A}) = \dot{m}(h_{3A} + h_1)$$

or

$$h_1 = h_3 + h_{4A} - h_{3A}$$

$$= 74.527 + 169.479 - 64.539$$

$$= 179.47 \text{ kJ/kg}$$

The enthalpy at the end of the compression is obtained by referring to the *p-h* diagram for Freon-12 in Appendix J. It is

$$h_2 = 232 \text{ kJ/kg}$$

The cooling effect is the enthalpy gain in the evaporator, or

$$h_{4A} - h_4 = 169.479 - 64.539$$

$$= 104.94 \text{ kJ/kg}$$

The work input per unit mass is

$$w_{in} = h_2 - h_1$$
$$= 232 - 179.467$$
$$= 52.53 \text{ kJ/kg}$$

The coefficient of performance is

$$(COP)_R = \frac{h_{4A} - h_4}{w_{in}}$$
$$= \frac{104.94}{52.53}$$
$$= 2$$

For the evaporator and condenser temperatures considered in this example, the value of the COP is acceptable. The COP for the Carnot refrigeration cycle is 2.9.

In the absence of pressure drops at the inlet and discharge valves, the volumetric efficiency is given by Equation 7.5.

$$\eta_v = 1 + CF - CF \left(\frac{p_H}{p_L}\right)^{1/n}$$
$$= 1 + 0.03 - 0.03 \left(\frac{960.7}{64.2}\right)^{1/1.13}$$
$$= 0.70$$

The mass flow rate $\dot{m}$ of the freon can be determined from the cooling rate specified

$$\dot{m}(h_{4A} - h_4) = \text{cooling rate}$$

or

$$\dot{m} = \frac{10}{104.940} = 0.0953 \text{ kg/s}$$

The power input P is then obtained from

$$P = \dot{m}w_{in}$$
$$= 0.0953 \times 52.53$$
$$= 5 \text{ kW}$$

The piston displacement PD is determined from

$$\dot{m} = \frac{PD \cdot N}{v_1 \times 60}$$

where N is the number of revolutions per minute of the single-acting compressor. Solving for PD, we obtain

$$PD = \frac{\dot{m}v_1}{N}$$
$$= \frac{0.0953 \times 0.26 \times 60}{1200}$$
$$= 0.00124 \text{ m}^3, \text{ or } 1240 \text{ cm}^3$$

In the above, the value of v_1 is read from the graph corresponding to the known values of p_1 and h_1 in Appendix J.

8.4.2 *Multistage Vapor Compression Refrigeration*

When the evaporator temperature is low and the condenser coolant temperature is essentially fixed by the temperature of the ambient air or cooling water, the pressure rise required in the compressor is high. This results in a low volumetric efficiency. Also, the discharge temperature for a medium like ammonia can be rather high. Under such circumstances, staging is advantageously used. It not only improves the volumetric efficiency, but also the coefficient of performance.

The schematic and the *p-h* diagrams for a two-stage vapor compression refrigeration system are shown in Figure 8.23. The cooling load is applied to the evaporator in the low-pressure stage, process 4–1. In the idealized model, the saturated vapor at state 1 enters the low-pressure compressor. The superheated, intermediate pressure vapor (state 2) leaving the low-pressure compressor enters a heat exchanger, where the cooling medium is the working medium of the high pressure stage. In other words, this heat exchanger acts as a condenser for the low-pressure stage and as an evaporator for the high-pressure stage. The intermediate pressure saturated liquid (state 3) is throttled and then admitted to the evaporator. In the high-pressure stage, the working medium is compressed (1A–2A), condensed (2A–3A), throttled, and then admitted to the heat exchanger.

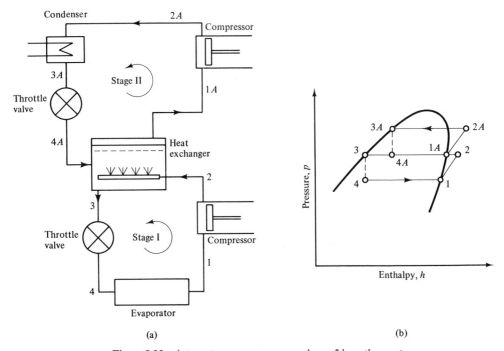

(a)   (b)

Figure 8.23   A two-stage vapor compression refrigeration system.

The heat exchanger can be open type, as shown in Figure 8.23(a), or a closed type. In a system with the open type, the same working fluid must be used in the high-pressure and low-pressure stages. This is possible if the working medium has the desirable properties discussed in Section 8.4.1 over a wide range of temperatures. Otherwise, a closed heat exchanger, in which the high-pressure stage fluid and low-pressure stage fluid come in a thermal contact but do not mix, is used. In such a case, line 4A–1A on the p-h diagram in Figure 8.23(b) will be lower than line 2–3. The temperature of the low-pressure stage fluid in the closed heat exchanger would be some 5°C to 10°C above the high-pressure stage fluid. This insures a proper heat transfer between the two fluids. The relative mass flow rates of the two fluids are determined by performing an energy balance on the heat exchanger. This gives

$$\dot{m}_I h_2 + \dot{m}_{II} h_{4A} = \dot{m}_I h_3 + \dot{m}_{II} h_{1A} \tag{8.11}$$

where $\dot{m}_I$ and $\dot{m}_{II}$ are the mass flow rates in the low-pressure stage and high-pressure stage, respectively, and the enthalpies are defined in Figure 8.23.

The cooling rate for the staged system is

$$\dot{Q}_{in} = \dot{m}_I(h_1 - h_4) \tag{8.12}$$

The work input rate to produce this cooling effect is

$$\dot{W}_{in} = \dot{m}_I w_{1-2} + \dot{m}_{II} w_{1A-2A}$$
$$= \dot{m}_I(h_2 - h_1) + \dot{m}_{II}(h_{2A} - h_{1A}) \tag{8.12a}$$

The coefficient of performance is then given by

$$(COP)_R = \frac{\dot{Q}_{in}}{\dot{W}_{in}}$$

$$= \frac{h_1 - h_4}{(h_2 - h_1) + (\dot{m}_{II}/\dot{m}_I)(h_{2A} - h_{1A})} \tag{8.13}$$

It is to be noted that to calculate $(COP)_R$, only the ratios of the mass flow rates and the enthalpies in the two stages are needed.

Example 8.10

A two-stage, Freon-12 refrigeration system operates between −40°C and 40°C. The intermediate pressure is 261 kPa. Determine the COP of the system and compare it with the COP of the system where the entire compression is carried out in a single compressor. Assume saturated conditions at the outlets of the evaporator and the condensor.

Solution:

Given: A two-stage, Freon-12 refrigeration system has the following characteristics.

HP condenser temperature:	40°C
Condenser discharge:	saturated liquid
LP evaporator temperature:	−40°C
Evaporator discharge:	saturated vapor
Intermediate pressure:	261 kPa

Objective: Here $(COP)_R$ is to be determined for the given system, and for a system where the entire compression takes place in a single compressor.

Diagram:

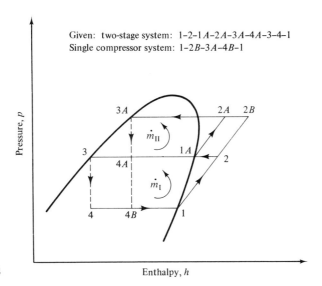

Given: two-stage system: 1–2–1A–2A–3A–4A–3–4–1
Single compressor system: 1–2B–3A–4B–1

Figure 8.24

Assumptions: An open heat exchanger is used at the intermediate pressure.

Relevant physics: The ratio of the mass flow rates in the high-pressure and low-pressure stages can be determined from Equation 8.11. The $(COP)_R$ for the staged system can then be calculated from Equation 8.13. The single compressor refrigerator will execute the cycle 1–2B–3A–4B–1 and its COP is $(h_1 - h_{4B})/(h_{2B} - h_1)$, or $(h_1 - h_{3A})/(h_{2B} - h_1)$.

Analysis: The following table of properties can be set up (Appendix G) on the basis of the given data. The enthalpy values at points 2 and 2A are determined from the p-h diagram for Freon-12 in Appendix J.

Point.	t (°C)	p (kPa)	Phase	h (kJ/kg)
1	−40	64.2	Sat. vapor	169.479
2		261	Super. vapor	192.5
3	−5	261	Sat. liquid	31.420
4	−40	64.2	Wet vapor	31.420
1A	−5	261	Sat. vapor	185.243
2A		960.7	Super. vapor	208
3A	40	960.7	Sat. liquid	74.527
4A	−5	261	Wet vapor	74.527

From Equation 8.11,

$$\dot{m}_1 h_2 + \dot{m}_{II} h_{4A} = \dot{m}_1 h_3 + \dot{m}_{II} h_{1A}$$

Substitution of the various values of h and a division by $\dot{m}_I$ gives

$$192.5 + \left(\frac{\dot{m}_{II}}{\dot{m}_I}\right)74.527 = 31.42 + \left(\frac{\dot{m}_{II}}{\dot{m}_I}\right)185.243$$

Solving for the ratio of the mass flow rates, we obtain

$$\frac{\dot{m}_{II}}{\dot{m}_I} = 1.455$$

The COP for the staged system can now be determined from Equation 8.13.

$$(COP)_R = \frac{h_1 - h_4}{(h_2 - h_1) + (\dot{m}_{II}/\dot{m}_I)(h_{2A} - h_{1A})}$$

$$= \frac{169.479 - 31.42}{(192.5 - 169.479) + 1.455(208 - 185.243)}$$

$$= 2.46$$

If the entire compression were carried out by a single compressor, the enthalpy h_{2B} of the fluid leaving such a compressor would be 216 kJ/kg (Appendix J). The compressor work would be $h_{2B} - h_1$, or 46.52 kJ/kg. The cooling effect would be $h_1 - h_{3A}$ since the enthalpy of the fluid entering the evaporator, h_{4B}, would be the same as the enthalpy of the saturated liquid leaving the condenser, h_{3A}. The COP would then be

$$(COP)_{R,\text{single stage}} = \frac{h_1 - h_{3A}}{h_{2B} - h_1} = \frac{169.479 - 74.527}{46.52}$$

$$= 2.04$$

Comments: If the entire compression is carried out in a single stage, the COP would be reduced by 17 percent.

The refrigeration cycles discussed here are ideal cycles. In practice, there are pressure drops in the pipelines connecting one component with another and in the evaporator and condenser coils. These can be accounted for in compressor analysis by suitably increasing the high-pressure value and decreasing the low-pressure value. Also, undesired heating of the refrigerent leaving the evaporator occurs due to heat transfer from the environment. This tends to superheat the vapor. With regard to evaporator, the temperature of the cooling load is several degrees higher than the temperature of the refrigerating medium. This is necessary to achieve a reasonable heat transfer rate. However, it causes irreversibilities. Likewise, irreversibilities are present in the condenser since the coolant temperature is several degrees below that of the working medium. All such deviations contribute to a reduction in the theoretical coefficient of performance, and the actual COP may be about 50 percent of the theoretical COP.

8.4.3 Reversed Brayton Cycle

In principle, all the gas cycles discussed in Chapter 7, when reversed, will operate as refrigeration cycles. However, only the reversed Brayton cycle has been found suitable for industrial applications. Figure 8.25(a) and (b) shows the schematic of a

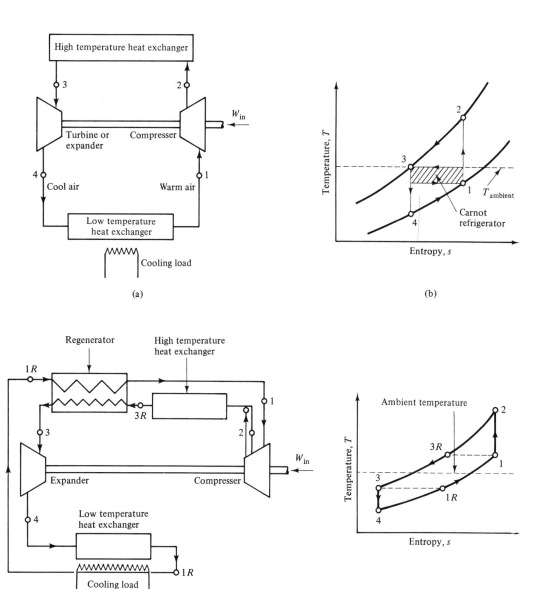

Figure 8.25 Reversed Brayton cycle for refrigeration.
(a) Mechanical components.
(b) *T-s* diagram.
(c) Mechanical components for a closed cycle with regeneration.
(d) *T-s* diagram for the closed cycle with regeneration.

reversed Brayton cycle and its *T-s* diagram. The gas is compressed isentropically from state 1 to state 2 in a compressor. The temperature at state 2 is relatively high when compared with the temperature at state 1, which is generally close to the ambient temperature. The compressed gas is then cooled in a heat exchanger, process 2–3. The high-pressure, moderate-temperature gas is expanded isentropically in a turbine to produce a low temperature. For instance, with a turbine pressure ratio of 3 and $T_3 = 300$ K, T_4 will be as low as 210 K. In process 4–1, heat energy is extracted from the medium that is to be refrigerated.

The expression for the COP of the reversed Brayton cycle is

$$(COP)_R = \frac{q_{4-1}}{-(w_{1-2} + w_{3-4})}$$

$$= \frac{c_p(T_1 - T_4)}{c_p(T_2 - T_1) - c_p(T_3 - T_4)}$$

where constant specific heats are assumed. It may be noted that the turbine work, w_{3-4}, is not always available since a throttle valve can be used to reduce the pressure from p_3 to p_4. Since processes 1–2 and 3–4 are isentropic,

$$\frac{T_2}{T_1} = \left(\frac{p_2}{p_1}\right)^{(\gamma-1)/\gamma} = r_p^{(\gamma-1)/\gamma}$$

and

$$\frac{T_3}{T_4} = \left(\frac{p_3}{p_4}\right)^{(\gamma-1)/\gamma} = r_p^{(\gamma-1)/\gamma}$$

Hence

$$\frac{T_2}{T_1} = \frac{T_3}{T_4}$$

or

$$\frac{T_1}{T_4} = \frac{T_2}{T_3}$$

After a few algebraic manipulations, we obtain

$$(COP)_R = \frac{T_4}{T_3 - T_4} \tag{8.14}$$

Since

$$\frac{T_3}{T_4} = \left(\frac{p_H}{p_L}\right)^{(\gamma-1)/\gamma} = r_p^{(\gamma-1)/\gamma}$$

we can express $(COP)_R$ in terms of the pressure ratio r_p and the exponent γ for the isentropic processes.

$$(COP)_R = [r_p^{(\gamma-1)/\gamma} - 1]^{-1} \tag{8.14a}$$

In the reversed Brayton cycle for refrigeration, there are large temperature variations in the heat exchangers (processes 2–3 and 4–1). This causes the COP of this cycle to be much less than that of a reversed Carnot cycle operating within the same temperature limits (T_3 and T_1).

The reversed Brayton cycle can successfully produce fairly low temperatures if regeneration is used. The schematic diagram of such a cycle is shown in Figure 8.25(c). Figure 8.25(d) shows the T-s diagram for the cycle. The principle of regeneration used here is similar to that for the Brayton power cycle (Figure 7.25). The working medium, a gas, enters the compressor at state 1, which is near the ambient state. On leaving the compressor (state 2), the gas is cooled to state $3R$ in a high-temperature heat exchanger by employing a coolant such as the atmospheric air. Further cooling of the gas to state 3 is accomplished in a regenerator. Then, upon leaving an expander such as a turbine or an expansion valve, the gas passes through a low-temperature heat exchanger that meets the cooling load. When the gas leaves this exchanger, it is at state $1R$. Since T_{1R} is slightly less than T_3, as the gas passes through the regenerator it is heated by the gas leaving the high-temperature heat exchanger. Finally, the gas leaves the regenerator at state 1, and the cycle is completed.

The coefficient of performance for the reversed Brayton cycle with regeneration is

$$(COP)_R = \frac{c_p(T_{1R} - T_4)}{c_p(T_2 - T_1) - c_p(T_3 - T_4)}$$

$$= \frac{T_4}{T_1 - T_4} \tag{8.15}$$

$$= \left[\frac{T_1}{T_3}(r_p)^{(\gamma-1)/\gamma} - 1\right]^{-1}$$

It is to be noted that the COP of the cycle with regeneration is less than the COP of the cycle without regeneration, since the quantity in the square bracket in Equation 8.15 is greater than that in Equation 8.14(a), and both T_1/T_3 and r_p are greater than one. The intent of using regeneration is to make it possible to attain fairly low temperatures. Referring to Figure 8.19(d), if T_1 is near ambient temperature, then the high-temperature heat exchanger can cool the working medium only to T_{3R} (equal to T_1). Further cooling of the medium to T_3 can be accomplished only by thermal contact with a cooler fluid. The cooler fluid is the working medium leaving the low-temperature heat exchanger at T_{1R}. It is the low-temperature heat exchanger which meets the cooling load requirements. Thus, with an internal heat exchanger—that is, a regenerator—it is possible to have T_3 well below the ambient temperature.

Example 8.11

A reversed Brayton cycle operates with regeneration and has a pressure ratio of 3. The compressor inlet temperature is 20°C and the lowest temperature in the cycle is −100°C. Determine the $(COP)_R$ of the cycle and the temperature range of the cooling load under the idealized conditions. Assume the working medium to be air.

Solution:

Given: A reversed Brayton cycle with regeneration operates as a refrigerator.

$$\text{Compressor inlet:} \quad T_1 = 273.15 + 20 \ = 293.15 \text{ K}$$

$$\text{Expander outlet:} \quad T_4 = 273.15 - 100 = 173.15 \text{ K}$$

$$\text{Compression ratio:} \quad r_p = 3$$

$$\text{Working medium:} \quad \text{air}$$

Objective: We have to determine $(COP)_R$ of the cycle and the lowest possible temperature of the cooling load.

Diagram:

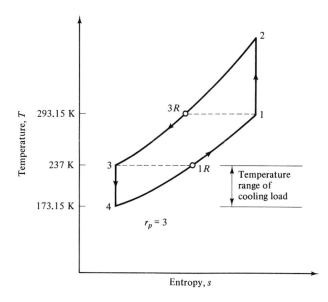

Figure 8.26

Assumptions:

1. The processes are reversible.
2. The specific heats of air are constant.
3. The ideal gas law holds for air.

Relevant physics: The temperature of the air entering the expander, T_3, can be determined using the isentropic relationships for an ideal gas. This temperature represents the upper limit of the cooling load temperature. Temperature T_4 is the lower limit. The $(COP)_R$ can be evaluated from Equation 8.15.

Analysis: For the isentropic process 3–4

$$T_3 = T_4 r_p^{(\gamma-1)/\gamma}$$
$$= 173.15(3)^{(1.4-1)/1.4}$$
$$= 237 \text{ K}$$

From Equation 8.15

$$(COP)_R = \frac{T_4}{T_1 - T_4}$$
$$= \frac{173.15}{293.15 - 173.15}$$
$$= 1.44$$

Comments: Since T_3 is less than a typical value of the ambient temperature (about 290 K), regeneration is used to cool the air in the high-temperature heat exchanger. The temperature of the coolant air entering the regenerator, T_{1R}, must be less than T_3. This means that the temperature of the cooling load must not be above T_3—that is, not above 237 K.

For this example, it is interesting to calculate the COP of a reversed Carnot cycle. Such a cycle would operate between T_1 and T_3, or between 293.15 K and 237 K. The COP would be $237/(293.15 - 237)$, or 4.22. The lowest possible temperature of the cooling load is T_4, or 173.15 K. Thus the temperature range of the cooling load is 173.15 K to 237 K.

8.4.4 Liquefaction of Gases

In today's industrial world, there is often a need to obtain temperatures below 173 K. For example, natural gas can be conveniently transported if it is in the liquid state. However, liquefaction of natural gas requires a temperature as low as 111 K. Another example is low-temperature baths of a liquefied gas such as nitrogen or oxygen used in the study of the properties of materials at low temperatures. Liquefied gases are also used for rocket propulsion. Temperatures in the range of 123 K to 173 K are often referred to as *cryogenic temperatures*. The critical temperatures of the common gases, such as air, oxygen, and nitrogen, are in this range; liquefied gases provide a satisfactory source for obtaining cryogenic temperatures. When superconductivity and superfluidity are studied, extremely low temperatures—below 30 K—are involved. In such cases liquid helium and hydrogen are used.

In liquefying a gas, use is made of the Joule-Thomson effect. In Chapter 5, we saw that a real gas undergoes a change of temperature when it is throttled. This change is measured by $(\partial T/\partial p)_h$, the Joule-Thomson coefficient. If it is positive, throttling yields cooling of the gas. In order to take advantage of this effect, the gas temperature before throttling must be low enough and the initial state must be within the inversion curve.

The schematic of a simple liquefying cycle is shown in Figure 8.27. The gas to be liquefied is compressed in the low-pressure stage, process 1–a. After intercooling (a–b), it is compressed in the high-pressure stage, process b–c. The high-pressure gas is then cooled in an aftercooler, where its temperature is brought to ambient temperature, and then it is passed through a heat exchanger. In this heat exchanger, the temperature of the gas is reduced to below its initial temperature. The high-pressure, low-temperature gas is then throttled, process 3–4, to near atmospheric pressure. This results in a low-temperature liquid-vapor mixture. The liquid is removed and the cool vapor is passed through the heat exchanger, where it gains energy from the high-pressure gas. Make-up gas is added to the low pressure gas leaving the heat exchanger, and the mixture is admitted to the compressor.

Under ideal conditions, the compression should be isothermal. In reality, it will not be isothermal; therefore line 1–2 in Figure 8.27(b) is not shown as an isothermal line. The quantity of the liquid produced is fraction y of the gas passing through the throttle valve and it is dependent on the compressor outlet pressure p_2. The higher the value of p_2, the higher is the fraction of the liquid.

The compressor work per unit mass of the gas is

$$w = (h_a - h_1) + (h_c - h_b) \tag{8.16}$$

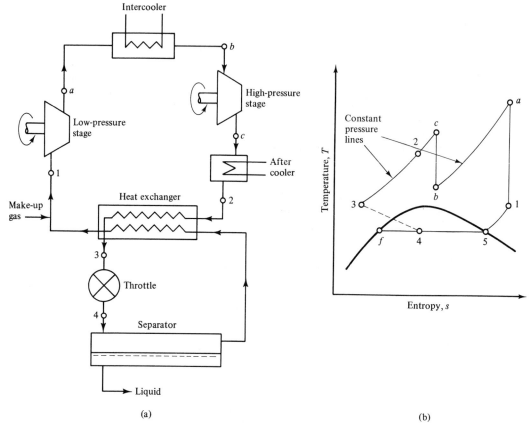

Figure 8.27 Liquefaction of gas.
(a) Mechanical components.
(b) *T-s* diagram.

The fraction y of the liquid obtained in the liquefaction process can be determined from an energy balance on the combination of the heat exchanger and the separator. For steady flow

$$\text{Energy in} = \text{energy out}$$

$$h_2 = yh_f + (1 - y)h_1$$

or

$$y = \frac{h_1 - h_2}{h_1 - h_f} \tag{8.17}$$

The cycle discussed above is the simple *Linde cycle*. For greater details on the subject of liquefaction of gases, you are referred to *Thermal Environmental Engineering* by James L. Threlkeld.*

*Threlkeld, James L., *Thermal Environmental Engineering*, 2nd ed. (Englewood Cliffs, N.J.: Prentice-Hall, Inc., 1970).

In one of the cycles used for liquefaction of natural gas, the gas is compressed to about 5 atm. It is then passed through three heat exchangers, the first one lowering the temperature to about 0°C, the second to about −100°C, and the third to −160°C. These three heat exchangers help separate heavier hydrocarbons from the natural gas. The natural gas is then throttled, resulting in a liquid-vapor mixture. The vapor is recycled and the liquid is stored.

8.4.5 Absorption Refrigeration

A close look at the vapor compression refrigeration cycle shows that high-grade mechanical energy is expended in pressurizing the vapor leaving the evaporator. An absorption refrigeration system aims at reducing the quantity of the high grade mechanical energy by employing heat energy instead. This is made possible by dissolving the refrigerant vapor (R), leaving the evaporator in a carrier liquid (C). This is accomplished in a device called an *absorber*. There are several refrigerant-carrier combinations: for example, ammonia-water, water-lithium bromide, and water-lithium chloride. Since the last two combinations use water as the refrigerant, temperatures below 0°C cannot be achieved by such combinations and these are used mainly in air conditioning applications. With the ammonia-water combination, however, temperatures well below 0°C can be produced, since ammonia is the refrigerant and water is the carrier.

Figure 8.28 shows the schematic of an absorption refrigeration system. The

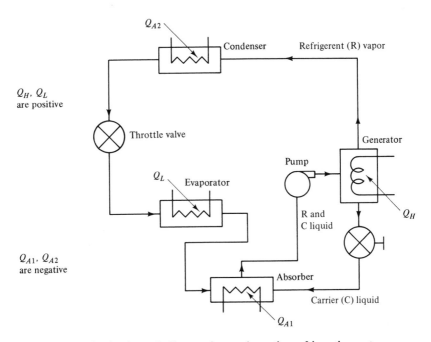

Figure 8.28   A schematic diagram for an absorption refrigeration system.

evaporator, the throttle valve, and the condenser play the same role as in a vapor compression system. After the refrigerant vapor leaving the evaporator is dissolved in a carrier, the carrier-refrigerant liquid mixture is pumped to the desired high pressure. At this point, all that is necessary is the separation of the refrigerant from the carrier. This is accomplished by heating the mixture in a device called a *generator*. The separated refrigerant vapor passes through a condenser, a throttle valve, and an evaporator, in that order.

In the vapor compression system, the vapor leaving the compressor experiences a significant increase in its pressure, as well as its temperature. The temperature of this vapor is usually several degrees above the condenser temperature. In an absorption system, the corresponding increase in the temperature is brought about by a supply of heat to the vapor and not by a supply of work, whereas the corresponding increase in pressure is brought about by a supply of work to the liquid mixture. Thus an absorption system requires a relatively small amount of work input as compared to the work of compression of the vapor in a vapor compression system.

When the carrier and the refrigerant are mixed, heat, Q_{A1}, is given off. This heat is transferred to the surroundings just as the heat, Q_{A2}, from the condenser is transferred to the surroundings. The heat supply to the system occurs in the evaporator (Q_L) and in the generator (Q_H). All these heat transfer processes are irreversible.

The coefficient of performance of an absorption system is

$$(COP)_R = \frac{Q_L}{Q_H + W_p} \tag{8.18}$$

where Q_L is the cooling effect in the evaporator, Q_H is the heat supplied to the generator, and W_p is the work input to the liquid pump. The work input is often neglected in the calculation of the COP.

It is possible to estimate an upper limit on the coefficient of performance of a given absorption refrigeration system. This is done by assuming that all the heat transfer processes are reversible isothermal and then applying the Clausius inequality. We assume that all heat transfer processes are isothermal: Q_L from the cooling load at T_L in the evaporator, Q_{A1} from the absorber at T_A to the surroundings, Q_{A2} from the condenser at T_A to the surroundings, and Q_H from the high temperature source at T_H to the generator. According to the sign convention Q_{A1} and Q_{A2} are negative, so we have

$$Q_L + Q_{A1} + Q_{A2} + Q_H = 0 \tag{8.19}$$

Application of the Clausius inequality gives

$$\Sigma \left(\frac{Q}{T} \right) \le 0$$

or

$$\frac{Q_L}{T_L} + \frac{Q_{A1}}{T_A} + \frac{Q_{A2}}{T_A} + \frac{Q_H}{T_H} \le 0$$

Substituting for $Q_{A1} + Q_{A2}$ from Equation 8.19, we have

$$\frac{Q_L}{T_L} - \frac{Q_L}{T_A} - \frac{Q_H}{T_A} + \frac{Q_H}{T_H} \leq 0$$

After some rearrangement, we obtain

$$\frac{Q_L}{Q_H} = (COP)_R \leq \frac{T_H - T_A}{T_H} \cdot \frac{T_L}{T_A - T_L} \tag{8.20}$$

Thus the maximum possible COP for an absorption system is the product of the thermal efficiency of a Carnot engine operating between T_H and T_A and the COP of a Carnot refrigerator operating between T_A and T_L. Note that Equation 8.20 does not take into account the pump work.

The coefficient of performance of practical absorption units is of order one. For the same cooling capacity (tons of refrigeration), absorption systems are bulkier than the vapor compression systems and require more maintenance. Absorption units require a longer time to attain steady-state on start-up, and these units are, therefore, suited to steady cooling loads. When waste heat, as in cogeneration systems or process steam, is readily available or when the collected solor energy is at a high enough temperature, absorption systems for air-conditioning applications are generally preferred. It should be pointed out that even in such installations, a considerable amount of electrical energy is used in pumping liquids and cooling water for the condenser and the absorber.

8.5 SUMMARY

The Rankine cycle is the basis for vapor power cycles. The work output and the thermal efficiency of a Rankine cycle can be increased by superheating, raising the supply pressure, and by lowering the condenser pressure. To insure long life of the turbine, it is desirable to keep the quality of the exhaust above 90 percent. This is accomplished by reheating of the steam after partial expansion in the turbine. Reheating also increases the work output of a given plant. The thermal efficiency of a Rankine cycle can be made to approach that of a Carnot cycle by regeneration, that is removal of partially expanded steam from the turbine and using the steam to heat the feedwater.

A vapor compression refrigeration system is similar to a reversed Rankine cycle except that throttling is used for expansion of the vapor. Members of the Freon family are the most widely used refrigerants. A reversed Brayton cycle that can be successfully used as a refrigerator uses air as the working medium and can produce temperatures of the order of $-100°C$. The simple Linde cycle is used for gas liquefaction. It involves cooling of a gas below its critical temperature and then throttling to atmospheric pressure. The absorption refrigeration system uses heat energy to produce a cooling effect and the system uses a carrier and a refrigerant.

PROBLEMS

(Draw p-V and T-s diagrams wherever applicable.)

8.1. A simple Rankine cycle uses water as the working medium and operates between 1500 kPa and 100 kPa. Determine the thermodynamic states at the end of each process in the cycle and calculate the pump work and the turbine work. Compare the thermal efficiency of the cycle with that of a Carnot cycle operating between the same temperature limits.

8.2. Steam is raised at 4 MPa and 450°C and condensed at 30°C. For a Rankine cycle operating between these limits, determine the heat absorbed, heat rejected, net work, and the cycle efficiency. Compare it with the Carnot efficiency.

8.3. In a certain power plant operating on a simple Rankine cycle, steam leaves the turbine at 40 kPa with a quality of 0.85. If the steam entering the turbine is saturated, what is the thermal efficiency of the plant? If the plant is to produce 100 kW, what should be the mass flow rate of steam in kg/s?

8.4. An engine working on Rankine cycle is supplied with saturated steam at 2 MPa, and the condenser pressure is 50 kPa. Determine each of the following.
(a) Thermal efficiency.
(b) Work output per kilogram of steam.

8.5. A simple Rankine cycle is used for a power plant. The supply pressure at the turbine inlet is 20,000 kPa and the exhaust pressure is 100 kPa. Determine the quality of the exhaust steam and the turbine work.

8.6. In the power plant of problem 8.5, it is decided to superheat the steam to 1000°C before it enters the turbine. What is the improvement in the quality and in the turbine output?

8.7. In a Rankine cycle with reheat, steam at 1200 kPa and 400°C enters the turbine. When the quality of the expanding steam becomes 0.95, the steam is removed from the turbine, reheated to 300°C and then expanded in the turbine to a pressure of 10 kPa. Determine each of the following.
(a) Pump work.
(b) Heat supplied.
(c) Work output.
(d) Thermal efficiency.
(e) If 10,000 kW are to be generated, what would be the mass flow rate of steam?

8.8. Determine the quantity of cooling water for the condenser in Problem 8.7. It is required that the temperature of the cooling water shall rise by no more than 15°C.

8.9. A Rankine cycle power plant whose steam rate is 3.34 kg/kWh receives steam at 20 MPa and 500°C. Heat losses from the turbine are 80 kJ/kg of steam. The condenser pressure is 15.1 cm of Hg. Determine each value.
(a) State of the turbine exhaust.
(b) Efficiency.

8.10. It is desired to study the effect of the inlet temperature on the performance of a Rankine cycle. Steam at 5000 kPa enters the turbine and exhausts at 20 kPa. Calculate the

thermal efficiency of the cycle for the turbine inlet temperatures of 264°C, 400°C, and 800°C.

8.11. An automobile powered by a Rankine cycle uses Freon-12 as the working medium. The engine intake is superheated vapor at 3 MPa and 180°C, while the engine exhausts at 1 MPa. Show the cycle on a *p-h* diagram. Calculate the rate of heat input for producing 40 kW and the overall thermal efficiency.

8.12. It is proposed to use focusing-type solar collectors to raise saturated steam at 200°C. The turbine exhausts at 40°C. The collectors can receive up to 25,000 kJ/m²-day. Estimate the collector area required to produce 100 kW of electrical power. Assume the following values of efficiencies:

Feed pump:	80%	Electrical generator:	97%
Turbine:	85%	Solar collection system:	25%
Boiler:	79%		

8.13. In a regenerative cycle, steam enters the turbine at 5000 kPa and 400°C and exhausts to the condenser at 10 kPa. Steam is extracted at that point in the turbine expansion where it is saturated. The extracted steam is used for feedwater heating in an open heater. Calculate the net work per kilogram of steam condensed and the thermal efficiency of the cycle.

8.14. It is decided to extract steam at 200 kPa for feedwater heating in the plant of Problem 8.13. With the extraction at two stages, what is the improvement in the thermal efficiency? Assume both feedwater heaters to be of open type.

8.15. In the plant in Problem 8.13, in addition to the extraction of steam, the remaining steam is reheated to 350°C. Determine the work output, heat supplied, and the thermal efficiency.

8.16. In a regenerative cycle, steam enters the turbine at 3500 kPa and 350°C and exhausts to the condenser at 20 kPa. Steam is extracted at 800 kPa and 200 kPa for the purpose of heating in two closed feedwater heaters. There is a single feed pump that pressurizes the condensate to 3500 kPa. Determine the thermal efficiency of the cycle and compare it with that of a Carnot cycle operating between the same temperature limits.

8.17. Rework Problem 8.13. if turbine efficiency is 85%.

8.18. Rework Problem 8.7 if turbine efficiency is 90%.

8.19. Rework Problem 8.13 if compressor efficiency is 75%.

8.20. In an ideal reheat cycle, steam enters the turbine at 4 MPa, and 350°C and expands to 100 kPa. It then passes through a reheater and emerges at 300°C, after which it expands to a condenser pressure of 7.384 kPa. For the ideal cycle, compute (a) work done in the two turbine stages, (b) heat added in the reheater, and (c) the thermal efficiency. Compare the efficiency with the Carnot efficiency.

8.21. Steam is generated at 4 MPa and 500°C in an ideal regenerative cycle. It then expands to 500 kPa where a fraction of the steam is extracted for feedwater heating. The remaining steam is then expanded in another stage to 30°C. For 1 kg of steam entering the first turbine stage, calculate each value.
(a) Fraction extracted.
(b) Total pump work.

(c) Enthalpy of the water entering the boiler.

(d) Thermal efficiency.

8.22. The following data are obtained from a steam power plant:

Turbine inlet:	$p = 5$ MPa, $T = 500°C$
Turbine outlet/condenser inlet:	5 kPa
Condenser outlet:	saturated liquid at 5 kPa
Isentropic efficiency of turbine:	0.84
Mechanical efficiency of turbine:	0.95

Determine each of the following.

(a) Quality of steam at the exit of the turbine.

(b) Mass flow rate of steam for 100 MW turbine output.

(c) Rate of heat rejection in the condenser.

8.23. A steam power plant with reheat works under the following conditions:

Boiler outlet:	5 MPa, 500°C.
Reheater pressure:	400 kPa
Reheater exit temperature:	450°C
Condenser pressure:	40 kPa
Condenser discharge:	saturated liquid
Net power developed:	250 MW

Determine each of the following.

(a) Mass flow rate.

(b) Pump power.

(c) Rate of heat addition in the reheater.

(d) Overall efficiency.

8.24. In a nuclear power plant, the primary fluid receives the energy from the nuclear reaction, and heat energy from this fluid is extracted to raise steam for a Rankine cycle. In one such system, steam is raised at 5 MPa and is then superheated to 600°C. The steam is expanded in a high-pressure turbine to 200 kPa. The efficiency of this turbine is 90%. The steam is then expanded in a low-pressure turbine, which has an efficiency of 95%. The condenser pressure is 10 kPa. An open feedwater heater is used to heat the condensate. For this purpose, a portion of the exhaust from the high-pressure turbine is used. The feed pumps have an efficiency of 65%. The power plant output is 750 MW. Calculate each of the following.

(a) Rate of the heat supply.

(b) Quantity of cooling water required in kilogram per second if its temperature rise is not to exceed 16°C.

(c) Pump power requirement.

(d) Overall thermal efficiency.

(e) Fraction of steam diverted to the open feedwater heater.

8.25. Rework Problem 8.24 if there is a reheat stage for the steam entering the low-pressure turbine. The reheat temperature is 340°C.

8.26. A steam power plant has a peak cycle pressure and temperature of 5 MPa and 600°C. There is a reheat stage at 500 kPa and to 600°C. The condenser pressure is 10 kPa. Determine the mass flow rate of steam for a power output of 500 MW. The fuel used has a heating value of 42,000 kJ/kg and 15% of the energy in the fuel is lost out the stack. Determine the overall thermal efficiency and the quantity of fuel required per hour.

8.27. Rework Problem 8.26 if the turbine and feed pump efficiencies are 75% and 65%, respectively.

8.28. A supercritical steam power plant operates at 30 MPa and 700°C. A portion of the steam, as it expands through the turbine, is extracted at 1 MPa and used in a closed feedwater heater. The liquid condensate from the heater is pressurized to the boiler pressure. The efficiency of the high-pressure turbine is 85%, that of the low pressure turbine is 90%, and that of the feed pump is 70%. The condenser pressure is 10 kPa. Calculate the fraction of steam extracted for the feedwater heater, the mass flow rate of steam to produce 8 MW, and the overall thermal efficiency.

8.29. Rework Problem 8.28 if there is a reheat stage at 1 MPa heating the steam to 500°C.

8.30. In a certain binary cycle, mercury is used for the topping cycle. The mercury cycle operates between 1800 kPa and 60 kPa, with saturated mercury vapor entering the turbine. The steam system operates between 260°C and 45.81°C on a simple Rankine cycle. Calculate the thermal efficiency of the binary cycle.

8.31. A refrigerator removes 9 kW of heat from a body when working between 245 K and 300 K. Its coefficient of performance is 75% of the COP of a Carnot refrigerator operating between the same limits. Find each of the following.
  (a) Heat rejected.
  (b) Power required.
  (c) Refrigerating effect in tons.

8.32. A Carnot heat pump is to supply 250 kW to heat rooms of a building to 22°C by extracting energy from the outside air at 0°C. Compute the COP of the system, the power required to operate the cycle, and the heat absorbed from the outside atmosphere.

8.33. Determine the maximum coefficient of performance that a refrigerator may achieve when freezing water in an environment whose temperature is 30°C. What cycle would you use to achieve this COP?

8.34. A reversed Carnot cycle has a COP of 4. What is the ratio of the high temperature to the low temperature? If the power input is 6 kW, what is the refrigerating effect in tons? What is the COP and the heat delivered if this system is used as a heat pump?

8.35. A Carnot refrigerator absorbs heat at 0°C and requires a power input of 2.0 kW/ton. Determine the COP and the temperature at which heat is rejected. If the upper temperature is changed to 40°C, what will be the input in kW/ton?

8.36. A Freon-12 refrigerator operates between 0°C and 30°C. It absorbs 3.5 kW of power. Calculate (a) the refrigerating capacity in tons, and (b) the heat rejected in kilowatts. Assume the conditions at the exits of the condenser and evaporator to be saturated.

8.37. Rework Problem 8.36 if a heat exchanger is used to superheat the evaporator discharge by 5°C and to subcool the condenser discharge.

8.38. A Carnot heat pump supplies heat energy at the rate of 24 kW to a room. The power input is 2 kW. If the environment is at $-3°C$, what is the temperature of the heated room?

8.39. In a vapor compression refrigeration system, wet vapor of Freon-12 at $-20°C$ enters a 15-cm by 15-cm twin cylinder, single-acting compressor running at 225 rpm. The compressor has a volumetric efficiency of 80% and it produces saturated vapor at $50°C$. The expansion valve receives the liquid at $40°C$. For this cycle, calculate each of the following.
 (a) Cooling capacity.
 (b) Power input in kilowatts per ton.
 (c) Input in kilowatts per ton if the actual COP is 75% of ideal value.

8.40. A Freon-12 refrigerating system has a capacity of 100 tons. The pressures in the evaporator and condenser are 0.219 MPa and 0.848 MPa, respectively. The liquid entering the expansion valve is at $30°C$, while the liquid leaving the evaporator is superheated by $5°C$. The isentropic compression efficiency is 85%, and the compression is adiabatic. Determine each of the following.
 (a) Mass flow rate.
 (b) COP.
 (c) Power per ton of refrigeration.

8.41. The following data is available for a Freon-12 refrigerator:

Evaporator temperature: $-20°C$

Condenser temperature: $30°C$

Compressor inlet state: dry saturated vapor

Condenser exit state: saturated liquid

Refrigeration load: 1 kW

Compressor: single-cylinder, single-acting, piston displacement of 100 cm³

Volumetric efficiency: 90%

Determine each of the following.
 (a) Quality of vapor entering and leaving the evaporator.
 (b) COP.
 (c) Power input to compressor and
 (d) Refrigerant flow rate.

8.42. An ideal vapor compression system uses Freon-12 as the refrigerant and operates between 900 kPa and 261 kPa. Determine the COP and the mass flow rate to produce 3 tons of refrigeration.

8.43. The Freon leaving the condenser in Problem 8.42 is subcooled by $4°C$ using an internal heat exchanger. Determine the COP and the volume displacement in cubic meters per minute to produce 3 tons of refrigeration. If the clearance factor is 4%, what is the volumetric efficiency?

8.44. An air-conditioning system uses ammonia as the working medium. The evaporator and condenser temperatures are $6°C$ and $30°C$, respectively. Due to the heat exchange with the surroundings, the liquid entering the throttle valve is subcooled by 2°C and the vapor entering the compressor is superheated by 4°C. If the compressor efficiency is 70% and $\gamma = 1.31$, determine each of the following.

(a) Power input to the compressor for a cooling load of 20 kW.

(b) System COP.

(c) Mass flow rate of the refrigerant.

8.45. A two-stage ammonia refrigeration system operates between 190.22 kPa and 1166.49 kPa. The intermediate pressure is 462.49 kPa. An open heat exchanger is used at the intermediate pressure that serves as the evaporator for the high-pressure stage and as a condenser for the low-pressure stage. Assuming saturated conditions at the inlets to the compressors and the throttle valves, determine the system COP. Also, determine the power input for a cooling effect of 15 kW.

8.46. A Freon-12 system consists of two evaporators, a compressor, a condenser, and three throttle valves. Saturated liquid at 40°C leaves the condenser. Part of this liquid is throttled to 4°C and admitted to the high-temperature evaporator and the remaining liquid is throttled to −20°C and then admitted to the low-temperature evaporator. The discharge from the high-temperature evaporator is throttled to −20°C, mixed with the discharge from the low temperature evaporator, and admitted to the compressor. The capacity of each evaporator is 5 tons. Draw schematic p-h and T-s diagrams for the cycle. Calculate the percent decrease in theoretical power required if the above system was replaced with one having a separate compressor for each evaporator. Assume saturated condition at the exit of each evaporator.

8.47. In a two-stage refrigerating system, Freon-12 is the working medium. Assume an evaporating temperature of −45°C, a condensing temperature of 40°C, and the intermediate temperature of −10°C. Determine (a) the coefficient of performance, (b) the maximum cycle temperature, and (c) the total piston displacement in cubic meters per minute per ton of refrigeration. The compressor clearance may be taken as 5%. Also, $\gamma = 1.13$.

8.48. A reversed Brayton cycle is used for refrigeration. The cycle operates between 400 kPa and 100 kPa. Air enters the expander at 35°C and enters the compressor at −15°C. Determine each of the following.

(a) Coefficient of performance.

(b) Piston displacement of the compressor in cubic meters per minute if the refrigeration capacity is 5 tons.

8.49. A reversed Brayton cycle refrigerator operates between 300 K and 250 K. Determine the COP if (a) the compression ratio is 3, (b) the compression ratio is 6. Assume the working fluid to be an ideal gas with $c_p = 1.00$ kJ/kg · K and $\gamma = 1.4$.

8.50. A reversed Brayton cycle refrigerator (see Figure 8.26) works between 290 K (T_1) and 200 K (T_{1R}) and it has a pressure ratio of 5. The working fluid is an ideal gas with $c_p = 1.04$ kJ/kg K and $\gamma = 1.3$. The system uses ideal regeneration. Determine the tonnage if the power supply is 3 kW. Also, determine the COP of the refrigerator and compare it with the COP of the corresponding Carnot refrigerator.

8.51. A reversed Brayton cycle with regeneration is to be used to produce a temperature of −100°C. The working medium is nitrogen. The ambient temperature is 20°C. If the pressure ratio is 4, what is the coefficient of performance?

8.52. Determine the theoretical work required in kilowatts per kilogram to produce liquid air at 100 kPa from dry air initially at 18°C and 100 kPa.

8.53. The simple Linde method is used to liquefy air at 18°C and 100 kPa. The air is isothermally compressed to 16,000 kPa. Determine the yield of liquid air in kilograms per kilogram of air compressed and the specific work requirement per kilogram of liquid air.

8.54. For the air in problem 8.53, determine the work input per kilogram of the air liquified.

-9 ·—·—·~·—·—·—·—

General Thermodynamic Relations

9.1 INTRODUCTION

Up to this point, we have been concerned primarily with simple systems that require only two independent properties for a complete specification of the system. The number of independent properties necessary to specify an equilibrium state of a system depends on the number of different modes of energy interactions within the system. As discussed in Chapter 2, there is only one mode of heat interaction, but there can be many different modes of work interaction. The *state postulate* specifies the number of independent properties needed to define a system. It states that the minimum number of independent properties necessary to specify the equilibrium state of a system uniquely equals the number of possible reversible work modes plus one. Heat interaction accounts for the plus one. For a simple gaseous system, the only reversible work is the $p\,dV$ work and, therefore, the number of independent properties needed to specify a simple gaseous system is two. For a simple gaseous system, we have been using pressure and volume, volume and temperature, or temperature and pressure as the independent properties.

Several additional thermodynamic properties have been introduced in the preceding chapters. These include internal energy, enthalpy, entropy, specific heats at constant pressure and at constant volume, and Joule-Thomson coefficient. For a simple gaseous system, any two of the aforementioned properties may be used for specifying the system. For instance, if absolute values of specific internal energy and entropy are known, they will uniquely define the state of a gaseous system. On the other hand, these two properties cannot be easily measured, whereas pressure, temperature, or specific volume can be measured with relative ease. Those properties

417

which cannot be measured directly are calculated from the experimental values of the measurable properties. One may wonder how these calculations are performed. The answer is provided in the thermodynamic relations. P.W. Bridgeman has shown that the maximum number of possible general thermodynamic relationships is 3.2×10^7. Fortunately, only a few of these are generally needed in practice. Since the derivatives of functions play an important role in thermodynamic relations, we shall now review partial derivatives and derivative relationships.

9.2 PARTIAL DERIVATIVES

The internal energy u of a simple gaseous system can be expressed as a function of p and v, v and T, or T and p. In each case it is a *different* function. To be specific, let the equation of state be that of an ideal gas, $pv = RT$. Consider a function given by

$$aT + bp$$

where a and b are constants. In view of the equation of state, we have

$$aT + bp = aT + \frac{bRT}{v} = \frac{apv}{R} + bp$$

The three functions would rigorously be specified as

$$F_1(T, p) = aT + bp \tag{9.1}$$

$$F_2(v, T) = aT + \frac{bRT}{v} \tag{9.1a}$$

$$F_3(p, v) = \frac{apv}{R} + bp \tag{9.1b}$$

In thermodynamics, the *same* symbol is normally used for each function. If, for example, the function represents the internal energy, we write

$$u(T, p) = aT + bp \tag{9.2}$$

$$u(v, T) = aT + \frac{bRT}{v} \tag{9.2a}$$

$$u(p, v) = \frac{apv}{R} + bp \tag{9.2b}$$

Now consider the partial derivatives with respect to v of functions F_2 and F_3 in Equations 9.1(a) and 9.1(b), respectively. These are

$$\frac{\partial F_2}{\partial v} = -\frac{bRT}{v^2} \quad \text{and} \quad \frac{\partial F_3}{\partial v} = \frac{ap}{R}$$

Since a separate symbol is used for each function in Equation 9.1, there is no ambiguity between $\partial F_2/\partial v$ and $\partial F_3/\partial v$. However, there is ambiguity inherent in the quantity $\partial u/\partial v$ since there is no indication which of the three functions in Equations 9.2 is to be considered. On the other hand, if we consider $(\partial u/\partial v)_T$, where the subscript T is used to indicate what is being held constant, we immediately conclude that the function u, involving v and T as the independent variables, is to be partially differentiated. In the example used, Equation 9.2(a) is to be differentiated partially with respect to v, holding T constant. This yields

$$\left(\frac{\partial u}{\partial v}\right)_T = -\frac{bRT}{v^2}$$

If z is a property of a simple gaseous system, there are six partial derivatives of the property when v and T, p and T, or p and v are the independent variables. These are

$$\left(\frac{\partial z}{\partial T}\right)_v, \quad \left(\frac{\partial z}{\partial v}\right)_T, \quad \left(\frac{\partial z}{\partial T}\right)_p, \quad \left(\frac{\partial z}{\partial p}\right)_T, \quad \left(\frac{\partial z}{\partial p}\right)_v, \quad \text{and} \quad \left(\frac{\partial z}{\partial v}\right)_p$$

We can derive some useful relations among these partial derivatives and the partial derivatives of the other state variables.

Let z be a function of p and v, so that

$$z = z(p, v)$$

and

$$dz = \left(\frac{\partial z}{\partial p}\right)_v dp + \left(\frac{\partial z}{\partial v}\right)_p dv \tag{9.3}$$

Since p is a function of v and T (equation of state), we can write

$$dp = \left(\frac{\partial p}{\partial v}\right)_T dv + \left(\frac{\partial p}{\partial T}\right)_v dT \tag{9.4}$$

Substituting for dp in Equation 9.3 from Equation 9.4 gives

$$dz = \left[\left(\frac{\partial z}{\partial p}\right)_v \cdot \left(\frac{\partial p}{\partial v}\right)_T + \left(\frac{\partial z}{\partial v}\right)_p\right] dv + \left(\frac{\partial z}{\partial p}\right)_v \left(\frac{\partial p}{\partial T}\right)_v dT \tag{9.5}$$

We can also consider z as a function of v and T and write

$$z = z(v, T)$$

giving

$$dz = \left(\frac{\partial z}{\partial v}\right)_T dv + \left(\frac{\partial z}{\partial T}\right)_v dT \tag{9.5a}$$

Equations 9.5 and 9.5(a) can be satisfied only if the respective coefficients of dv and dT in the two equations are equal. Thus for the coefficient of dv,

$$\left(\frac{\partial z}{\partial v}\right)_T = \left(\frac{\partial z}{\partial p}\right)_v \left(\frac{\partial p}{\partial v}\right)_T + \left(\frac{\partial z}{\partial v}\right)_p \tag{9.6}$$

and for the coefficient of dT,

$$\left(\frac{\partial z}{\partial T}\right)_v = \left(\frac{\partial z}{\partial p}\right)_v \left(\frac{\partial p}{\partial T}\right)_v \tag{9.7}$$

In general, if x, y, and z are state variables of a system and if ζ is any function (property) of the state variables, we have

$$\left(\frac{\partial \zeta}{\partial x}\right)_y = \left(\frac{\partial \zeta}{\partial z}\right)_x \left(\frac{\partial z}{\partial x}\right)_y + \left(\frac{\partial \zeta}{\partial x}\right)_z \tag{9.8}$$

and

$$\left(\frac{\partial \zeta}{\partial x}\right)_y = \left(\frac{\partial \zeta}{\partial z}\right)_y \left(\frac{\partial z}{\partial x}\right)_y \tag{9.9}$$

Equations 9.6 and 9.7, of which equations 9.8 and 9.9 are the respective generalizations, were derived by considering z as a possible function of the state variables, first of p and v and then of v and T.

To review the conditions for the exactness of the differential dz, we consider z as a function of two arbitrary variables, x and y. That is,

$$z = z(x, y)$$

Differentiation gives

$$dz = \left(\frac{\partial z}{\partial x}\right)_y dx + \left(\frac{\partial z}{\partial y}\right)_x dy$$

This differential form can also be expressed as

$$dz = M(x, y)dx + N(x, y)\, dy \tag{9.10}$$

where

$$M(x, y) = \left(\frac{\partial z}{\partial x}\right)_y$$

and

$$N(x, y) = \left(\frac{\partial z}{\partial y}\right)_x$$

If z is a point function—that is, it depends only on state and not on path (see Chapter 2)—the differential dz in Equation 9.10 is an exact differential. For z to be an exact differential, it is necessary that

$$\left(\frac{\partial M}{\partial y}\right)_x = \left(\frac{\partial N}{\partial x}\right)_y \tag{9.10a}$$

In the subsequent sections of this chapter, we shall encounter expressions for differential changes in thermodynamic properties that have a form similar to that of Equation 9.10. The property expressed by Equation 9.10(a) will be useful in working with such expressions.

When we reconsider the function $z = z(x, y)$, which was used as the basis for Equations 9.10, we find that it is possible to rewrite the function as

$$x = x(y, z) \quad \text{or} \quad y = y(z, x)$$

From the above rearrangements of the same function, we can obtain expressions for the differentials dx and dy, which lead to some useful relations. To obtain these, we differentiate the functions $x(y, z)$ and $y(z, x)$ to yield

$$dx = \left(\frac{\partial x}{\partial y}\right)_z dy + \left(\frac{\partial x}{\partial z}\right)_y dz$$

and

$$dy = \left(\frac{\partial y}{\partial z}\right)_x dz + \left(\frac{\partial y}{\partial x}\right)_z dx$$

We can eliminate dy from these two equations and obtain

$$\left[\left(\frac{\partial x}{\partial y}\right)_z \cdot \left(\frac{\partial y}{\partial z}\right)_x + \left(\frac{\partial x}{\partial z}\right)_y\right] dz - \left[1 - \left(\frac{\partial x}{\partial y}\right)_z \cdot \left(\frac{\partial y}{\partial x}\right)_z\right] dx = 0$$

Since x and z are independent, their differentials are also independent. Therefore, in order to satisfy the foregoing equation in dz and dx, their coefficients must be equal to zero. That is,

$$\left(\frac{\partial x}{\partial y}\right)_z \cdot \left(\frac{\partial y}{\partial z}\right)_x + \left(\frac{\partial x}{\partial z}\right)_y = 0 \tag{9.11}$$

and

$$1 - \left(\frac{\partial x}{\partial y}\right)_z \cdot \left(\frac{\partial y}{\partial x}\right)_z = 0 \tag{9.11a}$$

We rearrange Equation 9.11(a) to read

$$\left(\frac{\partial x}{\partial y}\right)_z = \frac{1}{(\partial y/\partial x)_z} \tag{9.12}$$

If we interchange y and z in Equation 9.12, we have

$$\left(\frac{\partial x}{\partial z}\right)_y = \frac{1}{(\partial z/\partial x)_y} \tag{9.12a}$$

Substituting the above result in Equation 9.11 gives

$$\left(\frac{\partial x}{\partial y}\right)_z \left(\frac{\partial y}{\partial z}\right)_x \left(\frac{\partial z}{\partial x}\right)_y = -1 \tag{9.13}$$

We shall find some of the equations developed in this section useful in deriving thermodynamic relations.

9.3 MAXWELL RELATIONS

For a simple compressible system, the combined first- and second-law equation from Equation 6.16 is

$$T\,ds = du + p\,dv \tag{9.14}$$

Equation 9.14 is often referred to as the *first T ds equation*, or *Gibbs equation*. Equation 9.14 is rearranged to read

$$du = T\,ds - p\,dv \tag{9.14a}$$

From the definition of enthalpy,

$$h = u + pv$$

it follows that

$$dh = du + p\,dv + v\,dp$$

or

$$dh = T\,ds + v\,dp \tag{9.15}$$

Equation 9.15 is sometimes called the *second T ds equation*.

At this point, we introduce *the specific Helmholtz function a* and *the specific Gibbs function g*. These are defined by

$$a = u - Ts \tag{9.16}$$

and

$$g = h - Ts \tag{9.17}$$

These functions are useful in the thermodynamics of chemical reactions. Differentiation of Equations 9.16 and 9.17 gives

$$da = du - T\,ds - s\,dT$$

and

$$dg = dh - T\,ds - s\,dT$$

Eliminating du and dh from the last two equations with the aid of Equations 9.14(a) and 9.15, respectively, we have

$$da = -p\,dv - s\,dT \tag{9.18}$$

and

$$dg = \quad v\,dp - s\,dT \tag{9.19}$$

The internal energy u, the enthalpy h, the Helmholtz function a, and the Gibbs function g, are properties and point functions. Their differentials are, therefore, exact. Consequently, we can apply Equation 9.10(a),

$$\left(\frac{\partial M}{\partial y}\right)_x = \left(\frac{\partial N}{\partial x}\right)_y$$

to each of the four equations, (9.14a), (9.15), (9.18), and (9.19). This gives

$$\left(\frac{\partial T}{\partial v}\right)_s = -\left(\frac{\partial p}{\partial s}\right)_v \tag{9.20}$$

$$\left(\frac{\partial T}{\partial p}\right)_s = \left(\frac{\partial v}{\partial s}\right)_p \tag{9.21}$$

$$\left(\frac{\partial p}{\partial T}\right)_v = \left(\frac{\partial s}{\partial v}\right)_T \tag{9.22}$$

$$\left(\frac{\partial v}{\partial T}\right)_p = -\left(\frac{\partial s}{\partial p}\right)_T \tag{9.23}$$

Equations 9.20 through 9.23 are known as the *Maxwell relations* for a simple compressible system. They contain partial derivatives of four properties, namely, pressure, specific volume, temperature, and entropy. Of the eight derivatives present in these relations, four derivatives involve pressure, temperature, and volume, which are all measurable properties. This means that the derivatives $(\partial T/\partial v)_s$, $(\partial T/\partial p)_s$, $(\partial p/\partial T)_v$, and $(\partial v/\partial T)_p$ can be determined experimentally and their values can be used to determine the remaining four derivatives in Equations 9.20 through 9.23, which involve changes in entropy. Thus the Maxwell relations make it possible to determine changes in entropy of a substance from its p-v-T data. For example, using the volume-temperature history of water at constant pressure, we can predict the nature of the entropy-pressure curve at constant temperature.

We know that for moderate pressures, the specific volume of the saturated water (liquid state) is minimum at 4°C. This means that $(\partial v/\partial T)_p$ for water is negative for temperatures less than 4°C, and the derivative is positive for temperatures greater than 4°C. From the Maxwell relation, Equation 9.23,

$$\left(\frac{\partial s}{\partial p}\right)_T = -\left(\frac{\partial v}{\partial T}\right)_p$$

We can then conclude that the entropy-pressure curves for liquid water at temperatures below 4°C have positive slopes. This means that the entropy of compressed liquid water at 0°C is greater than the entropy of the saturated liquid at the same temperature.

Another Maxwell relation, Equation 9.22, is helpful in deriving the Clapeyron equation, which involves the saturation pressure and temperature, and the changes in enthalpy and specific volumes due to a phase change. The Clapeyron equation is discussed in Section 9.4.

It is possible to express the four basic properties, p, v, T, and s, as partial derivatives of u, h, a, and g. Consider the following equation.

$$u = u(s, v)$$

Differentiation gives

$$du = \left(\frac{\partial u}{\partial s}\right)_v ds + \left(\frac{\partial u}{\partial v}\right)_s dv$$

A comparison of the above equation with Equation 9.14a leads us to conclude that

$$\left(\frac{\partial u}{\partial s}\right)_v = T \text{ and } \left(\frac{\partial u}{\partial v}\right)_s = -p \tag{9.24}$$

The first equation in Equation 9.24 forms the basis of a definition of the temperature of a microscopic system. For details, you are referred to any text book on statistical thermodynamics.

In a like manner, we obtain the following results from Equations 9.15, 9.18, and 9.19.

$$\left(\frac{\partial h}{\partial s}\right)_p = T \qquad \left(\frac{\partial h}{\partial p}\right)_s = v \tag{9.24a}$$

$$\left(\frac{\partial a}{\partial v}\right)_T = -p \qquad \left(\frac{\partial a}{\partial T}\right)_v = -s \tag{9.24b}$$

$$\left(\frac{\partial g}{\partial p}\right)_T = v \qquad \left(\frac{\partial g}{\partial T}\right)_p = -s \tag{9.24c}$$

When a system involves two reversible work modes, three independent state variables are needed to specify the state of the system. While taking partial derivatives with respect to one of the variables in such case, we have to hold the remaining two variables constant. We can then obtain the Maxwell relations for such a system. Obtaining the Maxwell relations becomes complicated as the number of independent state variables needed to specify a system increases.

9.4 THE CLAPEYRON EQUATION

The Clapeyron equation is a good example of how a change in a property—namely, enthalpy—can be determined from measurements of specific volume, pressure, and temperature. The equation expresses dp/dT for a substance at a saturated state in terms of the change in enthalpy, the change of specific volume, and the saturation temperature. The equation is particularly useful when phase changes—liquid-vapor, solid-liquid, and solid-vapor—are considered.

We begin by treating entropy as a function of specific volume and temperature.

$$s = s(v, T)$$

In the differential form, we have

$$ds = \left(\frac{\partial s}{\partial v}\right)_T dv + \left(\frac{\partial s}{\partial T}\right)_v dT$$

For a process involving phase change, the temperature is constant and $dT = 0$, so that

$$ds = \left(\frac{\partial s}{\partial v}\right)_T dv$$

The partial derivative in the above equation can be replaced by employing the Maxwell relation given by Equation 9.22.

$$\left(\frac{\partial s}{\partial v}\right)_T = \left(\frac{\partial p}{\partial T}\right)_v \tag{9.22}$$

We then have

$$ds = \left(\frac{\partial p}{\partial T}\right)_v dv$$

If we were to have a curve of the saturation pressure plotted against the saturation temperature for a given volume, the coefficient of dv in the above equation would then represent the slope of this curve at a given saturation state. The slope is independent of the volume change occuring during the phase change. Therefore we can replace the partial derivative by the total derivative and integrate to yield

$$s_2 - s_1 = \left(\frac{dp}{dT}\right)_{sat} (v_2 - v_1)$$

or

$$\left(\frac{dp}{dT}\right)_{sat} = \frac{s_2 - s_1}{v_2 - v_1} \tag{9.25}$$

The values of entropy and volume *before* and *after* the phase change are often designated by superscripts prime (′) and double-prime (″), respectively, so Equation 9.25 becomes

$$\left(\frac{dp}{dT}\right)_{sat} = \frac{s'' - s'}{v'' - v'}$$

Noting that $T_{sat}(s'' - s') = h'' - h'$, we have

$$\left(\frac{dp}{dT}\right)_{sat} = \frac{h'' - h'}{T_{sat}(v'' - v')} \tag{9.26}$$

In this last equation, for a liquid-vapor change the quantities $h'' - h'$ and $v'' - v'$ represent the change in enthalpy due to evaporation (h_{fg}) and the change in specific volume due to evaporation (v_{fg}), respectively.

Equation 9.26 is known as the *Clapeyron equation*, and it is valid for any phase change. It is evident from the equation that the change in enthalpy during phase change can be determined only from *p-v-T* data. On solving Equation 9.26 for $h'' - h'$, we obtain

$$h'' - h' = T_{\text{sat}}(v'' - v')\left(\frac{dp}{dT}\right)_{\text{sat}} \tag{9.26a}$$

For a liquid-vapor change, $v'' \gg v'$ and we can ignore v' in Equation 9.26. This gives

$$\left(\frac{dp}{dT}\right)_{\text{sat}} = \frac{h_{fg}}{T_{\text{sat}}v_g}$$

where the subscripts follow the nomenclature of the steam table. If ideal gas behavior is assumed, we can write RT/p for v to yield

$$\left(\frac{dp}{dT}\right)_{\text{sat}} = \frac{ph_{fg}}{RT_{\text{sat}}^2}$$

or

$$\left(\frac{dp}{p}\right)_{\text{sat}} = \frac{h_{fg}}{R}\left(\frac{dT}{T^2}\right)_{\text{sat}} \tag{9.27}$$

Equation 9.27 is called *the Clapeyron-Clausius equation*. This equation, with an appropriate change in the subscript for h, is also applicable to a solid-vapor phase change (such as for sublimation).

Example 9.1

Determine the enthalpy of vaporization of Freon-12 at 40°C using the properties listed in the appendix.

Solution:

Given: The thermodynamic properties of saturated Freon-12 are given in Appendix C2.

Objective: We have to determine h_{fg} for Freon-12 at 40°C.

Assumptions: The derivative $(dp/dT)_{\text{sat}}$ at 40°C can be approximated by

$$\frac{p_{\text{sat},45°C} - p_{\text{sat},35°C}}{45°C - 35°C}$$

Relevant physics: The required h_{fg} value can be determined from Equation 9.26(a):

$$h_{fg} = T_{\text{sat}}(v_g - v_f)\left(\frac{dp}{dT}\right)_{\text{sat}}$$

The values of the specific volumes in this equation can be read from Appendix C2. The derivative can be approximated by a finite difference expression, as noted in the assumptions.

Analysis: In order to determine the value of the derivative in the above equation, we consider values of p_{sat} at 45°C and 35°C. These are 847.7 kPa and 1084.3 kPa, respectively. We then have

$$\left(\frac{dp}{dT}\right)_{sat} \simeq \frac{(1084.3 - 847.7) \times 10^3}{45 - 35}$$

$$= 23.66 \times 10^3 \text{ Pa/K}$$

Also, from Appendix B4, at 40°C

$$v_g - v_f = 0.017373 \text{ m}^3/\text{kg}$$

so that

$$h_{fg} = (273.15 + 40) \times 0.017373 \times 23.66 \times 10^3$$

$$= 128.7 \text{ kJ/kg}$$

Comments: This value of h_{fg} is within 0.2% of the tabulated value of 128.525 kJ/kg. If v_f were ignored in this calculation, we would obtain a value of 134.6 kJ/kg for h_{fg}, resulting in an error of about 5%.

Example 9.2

Estimate the saturation pressure of water vapor at -50°C using the data in the steam tables.

Solution:

Given: The properties of saturated water at low temperatures are given in Appendix B4.

Objective: We have to determine p_{sat} at -50°C.

Assumptions: The lowest temperature for which the properties of water are listed in Appendix B4 is -40°C, but we have a temperature of -50°C. Therefore we assume the following:

1. The properties at -40°C can be used.

2. The h_{ig} value does not change appreciably between -40°C and -50°C.

Relevant physics: Equation 9.27 relates the change in the saturation pressure due to a change in the saturation temperature. Integrating this equation between two saturation states, we obtain

$$\int_1^2 \frac{dp}{p} = \frac{h_{ig}}{R} \int_1^2 \frac{dT}{T^2}$$

or

$$\ln\left(\frac{p_2}{p_1}\right) = \left(\frac{h_{ig}}{R}\right)\frac{T_2 - T_1}{T_1 T_2}$$

The subscript 1 refers to conditions at -40°C, while the subscript 2 refers to conditions at -50°C.

Analysis: We know that

$$p_1 = 0.0129 \text{ kPa} \qquad T_1 = 273.15 - 40 = 233.15 \text{ K}$$

$$p_2 = ? \qquad T_2 = 273.15 - 50 = 233.15 \text{ K}$$

$$h_{ig} = 2838.9 \text{ kJ/kg} \qquad R = 461.52 \text{ J/kg K}$$

Substitution gives

$$\ln\left(\frac{p_2}{0.0129}\right) = \frac{2838.9}{0.46152}\left(\frac{223.15 - 233.15}{233.15 \times 223.15}\right)$$

so
$$p_2 = 0.00395 \text{ kPa, or } 3.95 \text{ Pa}$$
The saturation pressure of water vapor at $-50°C$ is estimated at 3.95 kPa.

9.5 THERMODYNAMIC RELATIONS FOR INTERNAL ENERGY, ENTHALPY, AND ENTROPY

We begin by considering the internal energy fo a pure substance as a function of temperature and volume,
$$u = u(T, v)$$
In the differential form, the above becomes
$$du = \left(\frac{\partial u}{\partial T}\right)_v dT + \left(\frac{\partial u}{\partial v}\right)_T dv$$
or
$$du = c_v \, dT + \left(\frac{\partial u}{\partial v}\right)_T dv \tag{9.28}$$
We now seek a suitable expression for the partial derivative $(\partial u/\partial v)_T$ in this equation. The first $T \, ds$ equation
$$T \, ds = du + p \, dv$$
can be divided by dv while holding the temperature constant, which gives
$$T\left(\frac{\partial s}{\partial v}\right)_T = \left(\frac{\partial u}{\partial v}\right)_T + p$$
Using the Maxwell relation, Equation 9.22, and rearranging, we get
$$\left(\frac{\partial u}{\partial v}\right)_T = T\left(\frac{\partial p}{\partial T}\right)_v - p$$
Substitution of this result in Equation 9.28 gives
$$du = c_v \, dT + \left[T\left(\frac{\partial p}{\partial T}\right)_v - p\right] dv \tag{9.29}$$
For a constant volume process, Equation 9.29 reduces to the definition of the constant volume specific heat.

For a constant temperature process, Equation 9.29 becomes
$$du_T = \left[T\left(\frac{\partial p}{\partial T}\right)_v - p\right] dv_T \tag{9.30}$$

For an ideal gas $(\partial p/\partial T)_v = R/v$, and the quantity in the square bracket in Equation 9.29 becomes identically zero. This proves that the internal energy of an ideal gas does not change with a change in volume; rather, it is a function of temperature alone. For any substance other than an ideal gas, the change in its internal energy can be calculated if we know the value of constant volume specific heat and the equation of state. A knowledge of the equation of state permits evaluation of the derivative $(\partial p/\partial T)_v$.

Next, we consider the enthalpy of a pure substance as a function of temperature and pressure:

$$h = h(T, p)$$

In the differential form, this becomes

$$dh = \left(\frac{\partial h}{\partial T}\right)_p dT + \left(\frac{\partial h}{\partial p}\right)_T dp$$

or

$$dh = c_p\, dT + \left(\frac{\partial h}{\partial p}\right)_T dp \tag{9.31}$$

We now proceed to obtain a suitable expression for the partial derivative $(\partial h/\partial p)_T$ in the above equation. From the second $T\, ds$ equation

$$T\, ds = dh - v\, dp$$

it follows that

$$\left(\frac{\partial h}{\partial p}\right)_T = v + T\left(\frac{\partial s}{\partial p}\right)_T$$

The derivative on the right side of the above equation can be replaced by the derivative $(\partial v/\partial T)_p$ using the Maxwell relation, Equation 9.23. This gives

$$\left(\frac{\partial h}{\partial p}\right)_T = v - T\left(\frac{\partial v}{\partial T}\right)_p \tag{9.31a}$$

Inserting the expression for $(\partial h/\partial p)_T$ from Equation 9.31(a) in Equation 9.31 gives

$$dh = c_p\, dT + \left[v - T\left(\frac{\partial v}{\partial T}\right)_p\right] dp \tag{9.32}$$

We can substitute for the derivative in the above equation by using the definition of the coefficient of isobaric expansion, β (defined in Chapter 4).

$$\beta = \frac{1}{v}\left(\frac{\partial v}{\partial T}\right)_p \tag{4.5b}$$

Then Equation 9.32 becomes

$$dh = c_p\, dT + (1 - \beta T)v\, dp \tag{9.32a}$$

For a constant pressure process, Equation 9.32 becomes

$$dh_p = c_p\, dT_p$$

while for a constant temperature process,

$$dh_T = \left[v - T\left(\frac{\partial v}{\partial T}\right)_p\right] dp_T \tag{9.33}$$

It can be shown that for an ideal gas, $dh_T = 0$.

The equations for a change in internal energy and a change in enthalpy, Equations 9.29 and 9.32, respectively, can be readily integrated if the equation of state and the change of state are known. It is not necessary to know the actual process bringing about the change in state.

To develop general expressions for the change of entropy, we begin by considering temperature and volume as the independent variables.

$$s = s(T, v)$$

$$ds = \left(\frac{\partial s}{\partial T}\right)_v dT + \left(\frac{\partial s}{\partial v}\right)_T dv \tag{9.34}$$

The first $T\,ds$ equation,

$$T\,ds = du + p\,dv$$

can be divided by dT, holding v constant; this gives

$$T\left(\frac{\partial s}{\partial T}\right)_v = \left(\frac{\partial u}{\partial T}\right)_v = c_v \tag{9.35}$$

Also, from the Maxwell relation, Equation 9.22,

$$\left(\frac{\partial s}{\partial v}\right)_T = \left(\frac{\partial p}{\partial T}\right)_v \tag{9.22}$$

Substituting for the partial derivatives in Equation 9.34 from Equations 9.35 and 9.22, we have

$$ds = c_v \frac{dT}{T} + \left(\frac{\partial p}{\partial T}\right)_v dv \tag{9.36}$$

For a finite change in entropy,

$$s_2 - s_1 = \int_1^2 c_v \frac{dT}{T} + \int_1^2 \left(\frac{\partial p}{\partial T}\right)_v dv \tag{9.36a}$$

We now obtain an expression for the change in entropy in terms of the changes in temperature and pressure by considering the entropy as a function of temperature and pressure.

$$s = s(T, p)$$

$$ds = \left(\frac{\partial s}{\partial T}\right)_p dT + \left(\frac{\partial s}{\partial p}\right)_T dp \tag{9.37}$$

The second $T\,ds$ equation,

$$T\,ds = dh - v\,dp$$

when divided by dT, holding p constant, gives

$$T\left(\frac{\partial s}{\partial T}\right)_p = \left(\frac{\partial h}{\partial T}\right)_p = c_p \tag{9.38}$$

From the Maxwell relation, Equation 9.23,

$$\left(\frac{\partial s}{\partial p}\right)_T = -\left(\frac{\partial v}{\partial T}\right)_p \tag{9.23}$$

Substituting Equations 9.38 and 9.23 in Equation 9.37, we obtain

$$ds = \frac{c_p}{T} dT - \left(\frac{\partial v}{\partial T}\right)_p dp \tag{9.39}$$

In view of Equation 4.5(b), a finite change in entropy is given by

$$s_2 - s_1 = \int_1^2 \frac{c_p}{T} \, dT - \int_1^2 \beta v \, dp \qquad (9.39a)$$

Equations 9.36 and 9.39 can be used to determine the change of entropy of a pure substance. In order to use these equations, we need to know only the equation of state and the variation of c_p or c_v with respect to temperature. If one of the three properties p, v, or T is constant during the process in question, then the expression for change of entropy becomes relatively simple. It can be readily shown that Equations 9.36 and 9.39 are reduced to Equations 6.18 and 6.21, respectively, for an ideal gas.

Example 9.3

The van der Waals equation of state is

$$p = \frac{RT}{v - b} - \frac{a}{v^2}$$

Derive an expression for the change of entropy of a van der Waals gas undergoing a constant temperature process.

Solution: We can use either Equation 9.36(a) or 9.39(a). We prefer the former, since it is easy to evaluate $(\partial p / \partial T)_v$. For a van der Waals gas,

$$\left(\frac{\partial p}{\partial T} \right)_v = \frac{R}{v - b}$$

For a constant temperature process, Equation 9.36(a) then yields

$$s_2 - s_1 = \int_1^2 \frac{R}{v - b} \, dv = R \ln \left(\frac{v_2 - b}{v_1 - b} \right)$$

9.6 RELATIONS FOR c_p AND c_v

Equation 9.35 for c_v and Equation 9.38 for c_p involve derivatives of entropy and enthalpy at constant volume and constant pressure, respectively. In dealing with real gases, it is often necessary to know the variation in specific heats with respect to pressure or specific volume. Mathematically speaking, we seek expressions for $(\partial c_v / \partial v)_T$ and $(\partial c_p / \partial p)_T$. The desired expressions are obtained by starting with Equations 9.36 and 9.39,

$$ds = c_v \frac{dT}{T} + \left(\frac{\partial p}{\partial T} \right)_v dv \qquad (9.36)$$

$$ds = c_p \frac{dT}{T} - \left(\frac{\partial v}{\partial T} \right)_p dp \qquad (9.39)$$

Because these equations contain exact differentials, we can apply the test for exactness to obtain

$$\left[\frac{\partial}{\partial v} \left(\frac{c_v}{T} \right) \right]_T = \left[\frac{\partial}{\partial T} \left(\frac{\partial p}{\partial T} \right)_v \right]_v$$

and

$$\left[\frac{\partial}{\partial p}\left(\frac{c_p}{T}\right)\right]_T = -\left[\frac{\partial}{\partial T}\left(\frac{\partial v}{\partial T}\right)_p\right]_p$$

Carrying out the differentiation, we obtain

$$\left(\frac{\partial c_v}{\partial v}\right)_T = T\left(\frac{\partial^2 p}{\partial T^2}\right)_v \tag{9.40}$$

and

$$\left(\frac{\partial c_p}{\partial v}\right)_T = -T\left(\frac{\partial^2 v}{\partial T^2}\right)_p \tag{9.41}$$

The usefulness of these two equations lies in the fact that the changes in c_v and c_p can be found once the equation of state is known.

It is relatively easy to measure c_p, as compared to c_v. If c_p is known and if we have an expression for the difference between the two specific heats, we can determine c_v. Such an expression can be obtained by equating Equations 9.36 and 9.39.

$$ds = c_v\frac{dT}{T} + \left(\frac{\partial p}{\partial T}\right)_v dv = c_p\frac{dT}{T} - \left(\frac{\partial v}{\partial T}\right)_p dp$$

or

$$\frac{c_p - c_v}{T}dT = \left(\frac{\partial p}{\partial T}\right)_p dp + \left(\frac{\partial v}{\partial T}\right)_p dp$$

For a constant pressure process, $dp = 0$ and we obtain

$$\frac{c_p - c_v}{T}dT = \left(\frac{\partial p}{\partial T}\right)_v dv \qquad \text{(for } p = \text{constant)}$$

Division by dT gives

$$c_p - c_v = T\left(\frac{\partial v}{\partial T}\right)_p\left(\frac{\partial p}{\partial T}\right)_v \tag{9.42}$$

An alternate form can be obtained by using the relation

$$\left(\frac{\partial p}{\partial T}\right)_v = -\left(\frac{\partial v}{\partial T}\right)_p\left(\frac{\partial p}{\partial v}\right)_T$$

which is based on Equations 9.12(a) and 9.13. Substituting the above relation in Equation 9.42, we get

$$c_p - c_v = -T\left(\frac{\partial v}{\partial T}\right)_p^2\left(\frac{\partial p}{\partial v}\right)_T \tag{9.42a}$$

On the basis of experimental data it is known that $(\partial p/\partial v)_T$ is always negative. With the presence of the squared term $(\partial v/\partial T)_p^2$, which is always positive, we then conclude that $(c_p - c_v)$ is always positive, except at absolute-zero temperature. At absolute-zero temperature, the two specific heats are equal. They are also equal when $(\partial v/\partial T)_p$ is identically zero. For water this occurs at 4°C.

For most solids and liquids, the quantity $(\partial v/\partial T)_p$ is very small, which means that $c_p - c_v \simeq 0$. Table 9.1 gives the values of c_p/c_v and $c_p - c_v$ for liquid water at 1 atm. As the temperature increases, these values also increase. For temperatures well below the boiling point of water, the value of $c_p - c_v$ is less than 2 percent of the value of c_p. Even at the boiling point, the difference between c_p and c_v is only about 10 percent. This is true of most solids and liquids. Therefore, tables of properties for solids and liquids often give value of specific heat without qualifying it as constant volume or constant pressure. The values tabulated are usually values of c_p. Since $c_p \geq c_v$, the constant volume lines on a T-s diagram have steeper slopes than the constant pressure lines.

TABLE 9.1 VALUES OF c_p/c_v AND $c_p - c_v$
FOR LIQUID WATER AT 1 ATM

t (°C)	$(c_p/c_v) \times 10^4$	$c_p - c_v$(J/kg K)
0	1.5	0.6
10	13.4	5.6
20	63.3	26.3
30	143.5	59.1
40	246.7	100.5
50	369	148.5
60	506	201.5
70	658	258.7
80	822	318.8
90	999	381.4
100	1184	446.1

From: N. Ernest, Dorsey, *Properties of Ordinary Water-Substance* (New York: Reinhold Publishing Co., 1940), p. 263.

The difference between the specific heats can be expressed in terms of the coefficient of isothermal compressibility κ and the coefficient of isobaric expansion β. Both κ and β were introduced in Section 4.5 and defined by Equations 4.5 and 4.5(a).

$$\kappa = -\frac{1}{v}\left(\frac{\partial v}{\partial p}\right)_T \quad \text{or} \quad \left(\frac{\partial p}{\partial v}\right)_T = -\frac{1}{\kappa v}$$

$$\beta = \frac{1}{v}\left(\frac{\partial v}{\partial T}\right)_p \quad \text{or} \quad \left(\frac{\partial v}{\partial T}\right)_p = \beta v$$

Substituting the above definitions into Equation 9.42(a) yields

$$c_p - c_v = -T(\beta v)^2\left(-\frac{1}{\kappa v}\right)$$

or

$$c_p - c_v = \frac{\beta^2 v T}{\kappa} \tag{9.42b}$$

Example 9.4

Calculate the specific heat at constant volume for liquid water at 300 K and 1 atm using the values of β, v, and κ from a handbook.

Solution:

Given: The properties of liquid water are to be taken from a handbook.

Objective: We are to calculate the value of c_v for liquid water at 300 K and 1 atm.

Assumptions: None.

Relevant physics: Equation 9.42(b) permits a calculation of the quantity $c_p - c_v$

$$c_p - c_v = \frac{\beta^2 v T}{\kappa} \tag{9.42b}$$

Since the relevant properties in the above equation are known from the handbook data, we can compute c_v.

Analysis: We can use Equation 9.42(b) to determine the value of $c_p - c_v$. To do this, we need values of β, v, T, and κ. From the CRC Handbook of Tables for Applied Engineering Science, 2nd ed. pp. 88–95*, we find that for $T = 77°F$ and $p = 1$ atm,

$$c_p = 0.998 \text{ Btu/lb}_m \text{ °R}$$

$$\beta = 0.00011 \text{ °R}^{-1}$$

$$v = \frac{1}{62.23} \text{ ft}^3/\text{lb}_m$$

and

$$\kappa = 0.0046\%, \text{ atm}^{-1}$$

We first convert these values into SI units.

$$c_p = 0.998(\text{Btu/lb}_m \text{ °R}) \times [4180 \text{ (J/kg K) per (Btu/lb}_m \text{ °R)}]$$

$$= 4172 \text{ J/kg K}$$

$$\beta = 0.00011 °R^{-1} \times (1.8°R/K)$$

$$= 1.98 \times 10^{-4} \text{ K}^{-1}$$

$$v = \frac{1}{62.23} (\text{ft}^3/\text{lb}_m) \times (0.02832 \text{ m}^3/\text{ft}^3) \times (1 \text{ lb}_m/0.4535 \text{ kg})$$

$$= 1.0034 \times 10^{-3} \text{ m}^3/\text{kg}$$

$$\kappa = 0.0046 \times 10^{-2} \text{ atm}^{-1} \times \left(\frac{1 \text{ atm}}{101,325 \text{ Pa}}\right)$$

$$= 4.541 \times 10^{-10} \text{ Pa}^{-1}$$

From Equation 9.42(b)

$$c_p - c_v = \frac{\beta^2 v T}{\kappa}$$

$$= \frac{(1.98 \times 10^{-4})^2 \times 1.0034 \times 10^{-3} \times 300}{4.541 \times 10^{-10}}$$

$$= 26 \text{ J/kg K}$$

Since the value of c_p is 4172 J/kg K,

$$c_v = 4172 - 26 = 4146 \text{ J/kg K}$$

Comments: The calculated difference between c_p and c_v agrees with the value listed in Table 9.1.

*CRC Handbook of Tables for Applied Engineering Science, 2nd ed.

At this point we define a few additional properties. The *isothermal bulk modulus*, B_T, is defined as

$$B_T = -v\left(\frac{\partial p}{\partial v}\right)_T \qquad (9.43)$$

It can readily be seen that B_T is the reciprocal of the isothermal compressibility κ.

The *adiabatic compressibility* κ_s is a measure of change in volume per unit volume due to a change in pressure in the absence of any heat interaction. It is given by

$$\kappa_s = -\frac{1}{v}\left(\frac{\partial v}{\partial p}\right)_s \qquad (9.44)$$

The *adiabatic bulk modulus*, B_s, is defined by

$$B_s = -v\left(\frac{\partial p}{\partial v}\right)_s \qquad (9.45)$$

We can readily see that B_s is the reciprocal of the adiabatic compressibility κ_s.

The properties $\kappa, \beta, \kappa_s, B_T$, and B_s are thermodynamic properties and, for a simple substance, they are functions of two variables. These properties are listed in standard handbooks of physical properties.

9.7 JOULE-THOMSON COEFFICIENT

The Joule-Thomson coefficient was introduced in the discussion of the throttling process in Chapter 5. The coefficient is equal to $(\partial T/\partial p)_h$ and is denoted by μ_J (Equation 5.9). We can now obtain a general relation for the Joule-Thomson coefficient. We have

$$dh = c_p\, dT + \left[v - T\left(\frac{\partial v}{\partial T}\right)_p\right] dp \qquad (9.32)$$

For a constant enthalpy process, $dh = 0$. We then obtain

$$\mu_J = \left(\frac{\partial T}{\partial p}\right)_h = \frac{1}{c_p}\left[T\left(\frac{\partial v}{\partial T}\right)_p - v\right] \qquad (9.46)$$

Since the Joule-Thomson coefficient can be measured with relative ease, Equation 9.46 proves useful in determining c_p.

For an ideal gas $(\partial v/\partial T)_p$ is R/p and $\mu_J = 0$. This means that for an ideal gas undergoing a throttling process, the temperature change is zero. If we consider a generalized equation of state based on the compressibility factor z,

$$pv = Z\,RT$$

Equation 9.46 becomes

$$\mu_J = \frac{RT^2}{pc_p}\left(\frac{\partial Z}{\partial T}\right)_p \qquad (9.47)$$

A close examination of Equation 9.47 reveals that the sign of μ_J, which determines whether cooling will occur in a throttling process or not, is governed by the sign of the derivative $\partial Z/\partial T$. It is possible to draw some conclusions about the sign of this derivative by inspecting carefully the generalized compressibility chart in Figure 4.14. Consider the low-pressure region of the chart. For a low value of pressure in this region, as the value of T_r increases, the Z-value increases until T_r is approximately equal to 5. That is, for $T_r < 5$, $(\partial Z/\partial T)_p$ is positive in the low-pressure region. However, as T_r exceeds 5, Z decreases with an increase in T_r. That is, for $T_r > 5$, $(\partial Z/\partial T)_p$ is negative. This means that for a very low pressure, $\partial Z/\partial T$ changes sign, so μ_J also changes sign from positive to negative at $T_r = 5$. As the behavior of $(\partial Z/\partial T)_p$ is examined for increasing values of pressure, we find that there are two values of T_r at which the change in the sign of the derivative occurs. The dome-shaped curve in Figure 9.1 is the locus of such values of T_r, and the curve is called *Joule-Thomson inversion curve*. Points on this curve signify a zero value of μ_J. For points within the dome, μ_J is positive, signifying a cooling effect as a result of the throttling. Points outside the dome mean $\mu_J < 0$, signifying a warming effect.

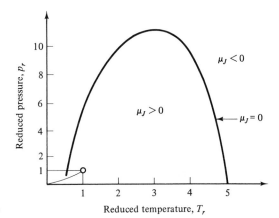

Figure 9.1   Inversion curve.

9.8 SUMMARY

The topics covered in this chapter were related to thermodynamic properties and relationships among them. Partial derivatives and differential expressions were used extensively in arriving at the relationships. It was noted that a simple system requires a specification of only two independent properties.

The Maxwell relations among pressure, specific volume, temperature, and entropy were obtained for a simple system. The Clapeyron equation was then derived, which relates $(dp/dT)_{sat}$ in a change of phase with changes in enthalpy and volume and with the saturation temperature. Generalized relations for changes in internal energy, enthalpy, and entropy were obtained. Also obtained were expressions for the difference between c_p and c_v, from which it was concluded that $c_p \geq c_v$. Finally a general expression for the Joule-Thomson coefficient was obtained.

PROBLEMS

9.1. Starting with Equation 9.13 and the Maxwell relation, Equation 9.23, derive the remaining three Maxwell relations.

9.2. Prove that the constant pressure lines in the wet region on an h-s diagram are straight lines but not parallel lines. Also, prove that the slope of a constant pressure line on an h-s diagram increases with temperature.

9.3. Replace x, y, and z in Equation 9.13 by p, v, and T. Evaluate the three derivatives in the equation by employing (a) the Van der Waal equation of state and (b) the Redlich-Kwong equation of state (Chapter 4), and verify Equation 9.13.

9.4. Demonstrate that $(\partial g/\partial p)_T = v$ for steam at 200°C. You may consider a number of superheated states in the vicinity of 200°C, plot curves of Gibbs function against pressure with temperature held constant, and differentiate numerically.

9.5. Derive the following partial derivatives involving the internal energy.

(a) $\left(\dfrac{\partial u}{\partial T}\right)_v = c_p - \dfrac{vT}{\kappa}\beta$ 　　(b) $\left(\dfrac{\partial u}{\partial T}\right)_p = c_p - pv\beta$

(c) $\left(\dfrac{\partial u}{\partial v}\right)_p = \dfrac{c_p}{v\beta} - p$ 　　(d) $\left(\dfrac{\partial u}{\partial p}\right)_p = pv\kappa - vT\beta$

9.6. Derive the following relations involving the enthalpy.

(a) $\left(\dfrac{\partial h}{\partial p}\right)_T = -T\left(\dfrac{\partial v}{\partial T}\right)_p + v$ 　　(b) $\left(\dfrac{\partial h}{\partial s}\right)_v = T - v\left(\dfrac{\partial T}{\partial v}\right)_s$

(c) $\left(\dfrac{\partial h}{\partial v}\right)_s = v\left(\dfrac{\partial p}{\partial v}\right)_s$ 　　(d) $\left(\dfrac{\partial h}{\partial T}\right)_s = v\left(\dfrac{\partial p}{\partial T}\right)_s$

9.7. Derive the following relations for the Helmholtz and Gibbs functions.

(a) $a - T\left(\dfrac{\partial a}{\partial T}\right)_v = u$ 　　(b) $\left(\dfrac{\partial a}{\partial T}\right)_v = -s$

(c) $c_v = -T\left(\dfrac{\partial^2 a}{\partial T^2}\right)_v$ 　　(d) $g - T\left(\dfrac{\partial g}{\partial T}\right)_p = h$

(e) $\left(\dfrac{\partial g}{\partial T}\right)_p = -s$ 　　(f) $c_p = -T\left(\dfrac{\partial^2 g}{\partial T^2}\right)_p$

9.8. When a force F acts on an elastic bar causing it to stretch by dL, the work done in a reversible process is $\delta W = F\,dL$. Show that for the bar,

$$\left(\dfrac{\partial u}{\partial L}\right)_T = F - T\left(\dfrac{\partial F}{\partial T}\right)_L$$

9.9. Based on the generalized relations, derive the following equations for the van der Waals equation of state.

(a) $(h_2 - h_1)_T = (p_2 v_2 - p_1 v_1) + a\left(\dfrac{1}{v_1} - \dfrac{1}{v_2}\right)$

(b) $(s_2 - s_1)_T = R \ln\left[\dfrac{(v_2 - b)}{(v_1 - b)}\right]$

9.10. Using the van der Waals gas model, evaluate the change in enthalpy at constant temperature for nitrogen at 27°C when it is compressed from 100 kPa to 10 MPa. How does your answer compare with the ideal gas solution?

9.11. Show that in the vapor region of a T-s diagram, the value of $(\partial T/\partial s)_p$ increases with increasing temperature.

9.12. Extrapolate the sublimation pressure for water from the triple point to $-20°C$ using the triple-point data in the appendix. Determine the percentage of error in this calculated value by comparing with the pressure in the appendix.

9.13. Estimate the melting temperature of ice at 5000 kPa. Refer to the *Handbook of Chemistry and Physics* for properties of ice at 1 atm.

9.14. A woman weighing 55 kg is ice skating. The area of the skate blades in contact with ice is 18 mm². The temperature of the ice is $-1°C$. Use appropriate calculations to determine whether the ice under the blades will melt. Triple-point data for water is $p = 0.6113$ kPa, $T = 0.01°C$, $h_f = 0.01$ kJ/kg, $h_i = -333.4$ kJ/kg, $v_f = 0.001$ m³/kg, and $v_i = 0.001091$ m³/kg.

9.15. Estimate the pressure at which helium 4 will boil corresponding to a temperature of 3°K. It is known that it boils at 4.22 K and 100 kPa and has an h_{fg} value of 83.3 J/gm-mol.

9.16. Calculate the value of h_{fg} for ammonia at 20°C using the Clapeyron equation and verify it with the value in the appendix.

9.17. One of the equations of state of gas is $p(v - b) = RT$, where b is a constant. Derive an expression for the Joule-Thomson coefficient of this gas.

9.18. Show that the equation of state for a pure substance can be written as

$$\frac{dv}{v} = \beta \, dT - \kappa \, dp$$

9.19. Show that the isothermal compressibility is always greater than or equal to the adiabatic compressibility.

9.20. Show that

$$\frac{c_p}{c_v} = \frac{\kappa}{\kappa_s}$$

9.21. Prove that $(\partial c_v / \partial v)_T = 0$ for a van der Waals gas.

9.22. If the specific heat, c_v for a van der Waals gas is given by $c_v = a + bT$, calculate the entropy change for the gas when its state changes from point 1 to point 2.

9.23. Calculate the c_p value of oxygen at (a) 400 K and 4 MPa, and (b) 400 K and 2 MPa. Use Equation 9.42 or 9.42(a).

9.24. Determine the Joule-Thomson coefficient for water in each case.
  (a) 3000 kPa and 300°C.
  (b) 9000 kPa and 300°C.

9.25. Show that the inversion temperature for a Dieterici gas (Chapter 4) is $2a(v - b)/Rbv$.

9.26. The boiling point of helium 4 at 1 atm is 4.22 K and its enthalpy of vaporization is 83.3 kJ/kg-mol. A reduction in pressure causes the boiling point to drop. Estimate the pressure necessary to produce a temperature of 1 K.

-10

Nonreacting Mixtures

10.1 INTRODUCTION

Up to now, only single-component systems have been considered. Many engineering situations involve mixtures of gases, mixtures of gases and vapors, and mixtures of liquids in which the constituents of the mixtures do not react chemically. In our earlier discussions, we treated air as if it were a single substance. We know that air consists of a mixture of nitrogen gas, oxygen gas, water vapor, and traces of other gases. We have assumed that it is permissible to treat air, a mixture of several gases, as a single component gas. Such a model was satisfactory for most of the applications presented in earlier chapters. The fraction of water vapor in air is usually small and it can often be ignored, as in the eariler chapters.

Now, consider the compression of atmospheric air in a compressor. Typically, as the pressure of the air is raised, its temperature also increases. In most cases, the air is eventually cooled and its temperature is allowed to return to the ambient conditions. If the fraction of the water vapor in the air is sufficiently large, there is the possibility that some of the water vapor will condense upon cooling of the compressed air. Water in the liquid state can reduce the life of the equipment and, therefore, the water must be removed. In order to predict the quantity of the condensate, the presence of the water vapor in air must be taken into consideration in the thermodynamic analysis. In air-conditioning applications, the fraction of the water vapor in air may have to be altered for human comfort. The extent of the alteration can be determined only if air is treated as a two-component mixture. Solutions of different liquids, such as water and lithium bromide, are used in absorption air-conditioning

438

systems. When a change of phase of such a solution takes place, we find that the change does not occur at a constant temperature.

In this chapter we shall consider the different types of mixtures discussed in the preceding paragraph and examine thermodynamic properties of mixtures.

10.2 MIXTURES OF IDEAL GASES

A mixture of N ideal gases is the simplest type of mixture. Let m_i and M_i be the mass and the molar mass (molecular weight), respectively, of the ith component. Then, the number of moles n_i of the ith component is given by

$$n_i = \frac{m_i}{M_i} \tag{10.1}$$

The total mass m of the mixture is given by

$$m = \sum_{i=1}^{N} m_i \tag{10.1a}$$

and the total number of moles n in the mixture is given by

$$n = \sum_{i=1}^{N} n_i \tag{10.1b}$$

We now define the *mole fraction* x_i of the ith component as the ratio of the number of moles of the ith component in a mixture to the total number of moles in the mixture. That is,

$$x_i = \frac{n_i}{n} \tag{10.2}$$

The *mass fraction* y_i of the ith component is defined by

$$y_i = \frac{m_i}{m} \tag{10.3}$$

We can determine the pressure, temperature, volume, mass, and the composition of a mixture experimentally. With a given set of constituents, we can form mixtures having an almost infinite variety of compositions. It would then seem that for every composition, we have to have a table of properties. Such, however, is not the case. The Gibbs-Dalton law permits calculation of internal energy, enthalpy, and entropy of a mixture, once the specific properties of the components are known. Before discussing the law, we shall introduce the concepts of partial pressure and partial volume.

Consider a mixture of N components occuppying a volume V. The pressure of a mixture of N components is called the *total pressure*, p. If the ith component alone is allowed to occupy volume V and is at the same temperature T as that of the mixture, then the ith component will exist at a *partial pressure*, p_i, which is the pressure experienced by the walls of the container due to the presence of the ith component alone. In a mixture of ideal gases, because of the large intermolecular spaces compared to the molecular dimensions, there is hardly any interference in the movement of the molecules of the other species present in the mixture. Consequently, the molecules of the ith component behave as if the other components were absent, and the partial pressure of the ith component in a mixture is p_i.

Now consider the same ith component in the mixture again. Let the ith component exist at total pressure p and temperature T of the mixture. Then the volume occupied by the ith component is known as the *partial volume*, V_i. The partial pressure and the partial volume are illustrated in Figure 10.1.

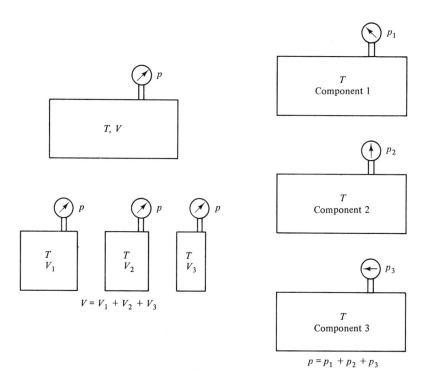

Figure 10.1   Schematic to illustrate partial pressures and partial volumes.

10.2.1 Gibbs-Dalton Law

The Gibbs-Dalton law has two parts. The first part, known as Dalton's law of partial pressure, relates to pressure, an intensive property of the mixture. It states that the total pressure of a mixture of perfect gases is the sum of the partial pressures of the components of the mixture. That is, for N components

$$p = p_1 + p_2 + \cdots + p_N = \sum_{i=1}^{N} p_i \tag{10.4}$$

Since the ith component is assumed to obey the ideal gas law,

$$p_i V = n_i \bar{R} T \tag{10.5}$$

where $\bar{R}$ is the universal gas constant and n_i is the number of moles of the ith component. For the n moles of mixture at a total pressure p, we can write

$$pV = n\bar{R}T \tag{10.5a}$$

Dividing Equation 10.5 by Equation 10.5(a) gives

$$\frac{p_i}{p} = \frac{n_i}{n} = x_i = \text{mole fraction} \tag{10.6}$$

Thus for a mixture of ideal gases, we have shown that the ratio of the partial pressure of a component to the total pressure of the mixture equals the mole fraction of the component.

The second part of the Gibbs-Dalton law relates to the extensive properties, internal energy, enthalpy, and entropy. The law states that the extensive property (U, H, or S) of a mixture is the sum of the extensive properties of the components of the constituents of the mixture. Thus, for example, the internal energy U of a mixture is given by

$$U = U_1 + U_2 + \cdots + U_N = \sum_{i=1}^{N} U_i \tag{10.7}$$

or

$$U = \sum_{i=1}^{N} m_i u_i = \sum_{i=1}^{N} n_i \bar{u}_i \tag{10.7a}$$

where $\bar{u}_i$ is the *molal internal energy* of the ith component.

The gas constant R of a mixture of ideal gases can be determined by using the Gibbs-Dalton law. For a mixture behaving like an ideal gas, we can write

$$pV = mRT$$

or

$$R = \frac{V}{mT}(p_1 + p_2 + \cdots + p_N)$$

where the law of partial pressures is used. Noting that the ith component behaves like an ideal gas, $p_i V = m_i R_i T$, the above equation becomes

$$R = \frac{1}{m}(m_1 R_1 + m_2 R_2 + \cdots + m_N R_N)$$

or in view of Equation (10.3)

$$R = \frac{1}{m} \sum_{i=1}^{N} m_i R_i = \sum_{i=1}^{N} y_i R_i \tag{10.8}$$

We can define the mixture molar mass M of a mixture of ideal gases. We begin with

$$nM = m = n_1 M_1 + n_2 M_2 + \cdots + n_N M_N$$

Then

$$M = \frac{n_1}{n} M_1 + \frac{n_2}{n} M_2 + \cdots + \frac{n_N}{n} M_N = \sum_{i=1}^{N} x_i M_i \tag{10.9}$$

10.2.2 Amagat-Leduc Law

We have already defined the partial volume of a component of a gaseous mixture. The Amagat-Leduc law states that the volume of a mixture of ideal gases is given by the sum of the partial volumes of the components of the mixture. If $V_1, V_2, \ldots$ are the partial volumes of the components, then the volume V of the mixture is given by

$$V = V_1 + V_2 + \cdots + V_N = \sum_{i=1}^{N} V_i \qquad (10.10)$$

Also, from the definition of partial volume V_i, we have

$$pV_i = n_i \bar{R} T \qquad (10.10a)$$

Dividing Equation 10.10(a) by Equation 10.5(a), we obtain

$$x_i = \frac{V_i}{V} = \frac{n_i}{n} \qquad (10.11)$$

Thus we have shown that the ratio of the partial volume of a component to the total volume of the mixture equals the mole fraction of that component.

Example 10.1

A tank of volume 0.2 m^3 contains 4 kg nitrogen, 1 kg oxygen, and 0.5 kg carbon dioxide. If the temperature of the mixture is 20°C, determine (a) the total pressure of mixture, (b) the gas constant of the mixture and the mixture molar mass, and (c) the specific enthalpy of the mixture.

Solution:

Given: A tank contains a mixture of nitrogen, oxygen, and carbon dioxide.

$$N_2: \quad m_1 = 4 \text{ kg}, \qquad t_1 = t = 20°C, \qquad V_1 = V = 0.2 \text{ m}^3$$
$$O_2: \quad m_2 = 1 \text{ kg}, \qquad t_2 = t = 20°C, \qquad V_2 = V = 0.2 \text{ m}^3$$
$$CO_2: \quad m_3 = 0.5 \text{ kg}, \qquad t_3 = t = 20°C, \qquad V_3 = V = 0.2 \text{ m}^3$$

Objective: We have to calculate p, R, M, and h for the gaseous mixture.

Assumptions:

1. The component gases and the mixture behave like an ideal gas.
2. The mixture obeys the Gibbs-Dalton law.
3. Specific heats of the component gases are constant.
4. Specific enthalpies of the component gases at 0°C are zero.

Relevant physics: The partial pressures of the component gases can be determined, since their volumes, temperatures, and masses are known. The required quantities can be calculated using the equations developed in Section 10.2.

Analysis: (a) The partial pressures of the components can be readily determined since the volume, temperature, and mass are given and the values of R for each component can be read from the appendix.

$$N_2:\ p_1 = \frac{m_1 R_1 T}{V} = \frac{4 \times 296.80(273.15 + 20)}{0.2}\ \text{Pa}$$

$$= 1740.2\ \text{kPa}$$

$$O_2:\ p_2 = \frac{m_2 R_2 T}{V} = \frac{1 \times 259.83(273.15 + 20)}{0.2}\ \text{Pa}$$

$$= 380.8\ \text{kPa}$$

$$CO_2:\ p_3 = \frac{m_3 R_3 T}{V} = \frac{0.5 \times 188.92(273.15 + 20)}{0.2}\ \text{Pa}$$

$$= 138.4\ \text{kPa}$$

Then

$$p = p_1 + p_2 + p_3 = 2259.4\ \text{kPa}$$

(b) From Equation 10.8, the gas constant of the mixture is

$$R = \frac{1}{m}(m_1 R_1 + m_2 R_2 + m_3 R_3)$$

$$= \frac{4 \times 296.8 + 1 \times 259.83 + 0.5 \times 188.92}{4 + 1 + 0.5}$$

$$= 280.27\ \text{J/kg K}$$

From Equation 10.9, the molar mass of the mixture is

$$M = x_1 M_1 + x_2 M_2 + x_3 M_3$$

or

$$M = \frac{p_1}{p} M_1 + \frac{p_2}{p} M_2 + \frac{p_3}{p} M_3$$

where Equation 10.6 is used. Substitution of the various values for pressure and molar mass gives

$$M = \frac{1}{2259.4}(1740.2 \times 28.013 + 380.8 \times 32 + 138.4 \times 44.01) = 29.66\ \text{kg/kg-mol}$$

(c) We assume constant specific heats of the components and zero enthalpy at 0°C. Reading the values of c_p from the appendix, we obtain

$$N_2:\ h_1 = c_{p1}t_1 = 1.0416 \times 20 = 20.832\ \text{kJ/kg}$$

$$O_2:\ h_2 = c_{p2}t_2 = 0.9216 \times 20 = 18.432\ \text{kJ/kg}$$

$$CO_2:\ h_3 = c_{p3}t_3 = 0.8418 \times 20 = 16.836\ \text{kJ/kg}$$

Then

$$H = m_1 h_1 + m_2 h_2 + m_3 h_3$$

$$= 4 \times 20.832 + 1 \times 18.432 + 0.5 \times 16.836$$

$$= 110.2\ \text{kJ}$$

One may calculate the equivalent specific heat c_p at constant pressure from $H = mc_p t$.

Example 10.2

The volumetric analysis of a gaseous mixture is H_2, 15%; CO, 20%; and N_2, 65%. Determine the mass fraction of each component and the R-value for the mixture.

Solution:

Given: A gaseous mixture has the following composition.

$$H_2: \quad 15\% \text{ by volume}$$

$$CO: \quad 20\% \text{ by volume}$$

$$N_2: \quad 65\% \text{ by volume}$$

Objective: The mass fraction y for each constituent and the R-value for the mixture are to be calculated.

Assumptions: Assume ideal gas behavior.

Relevant physics: We are given the volumetric analysis of the mixture. For convenience, we may assume that we have 1 kg-mol of the mixture. Then the number of moles n_i of the ith constituent exactly equals the mole fraction x_i of that constituent. From Equation 10.11

$$\frac{V_i}{V} = \frac{n_i}{n} = x_i$$

Since

$$n = 1 \text{ kg-mol}$$

$$n_i = x_i \text{ kg-mol}$$

Once the mole fraction of a component is determined, the relative mass m_i of the ith component is determined from Equation 10.1. It is

$$m_i = n_i M_i$$

where M is the molar mass.

The mass fraction y_i is given by

$$y_i = \frac{m_i}{\sum m_i}$$

Analysis: The calculations for problems of this type are best presented in a tabular form. The table below lists the percent by volume, the mole fraction, the relative mass, and the mass fraction for each constituent.

Component	% by volume	x_i	$x_i \times M_i$ = relative mass, m_i	Mass fraction, y_i
H_2	15	0.15	$0.15 \times 2.016 = 0.302$	0.0125
CO	20	0.20	$0.20 \times 28.01 = 5.602$	0.2323
N_2	65	0.65	$0.65 \times 28.013 = 18.208$	0.7552
Sum:	100	1	$\sum m_i$ 24.112	1.0000

The molar mass of the mixture is the sum of the relative masses, or 24.112 kg/kg-mol, and the gas constant is

$$R = \frac{\bar{R}}{M} = \frac{8314.34}{24.112} = 344.82 \text{ J/kg K}$$

Example 10.3

Suppose n_1 moles of gas 1 are mixed adiabatically with n_2 moles of gas 2. Both the gases are at the same temperature and pressure before mixing. What is the change in the entropy as a result of the mixing?

Solution:

Given: Two gases, initially at the following conditions, are mixed adiabatically.

$$\text{Gas 1:}\quad n_1, p, T$$

$$\text{Gas 2:}\quad n_2, p, T$$

Objective: The entropy change due to mixing is to be determined.

Assumptions:

1. The gases obey the ideal gas law.
2. The specific heats of the gases are constant.

Relevant physics: The change of entropy of an ideal gas can be expressed as (Chapter 6)

$$\Delta \bar{s} = \bar{c}_p \ln \frac{T_f}{T_i} - \bar{R} \ln \frac{p_f}{p_i}$$

where the subscripts i and f denote the initial and final states, respectively, and the overbars on s and c_p denote values per kg-mole of the gas.

Analysis: The given mixing process does not involve work or heat interaction. Therefore the internal energy remains constant and the temperature of the mixture is the same as that of the original components. Thus, $T_f = T_i$. Also $p_i = p$ for both gases and p_f equals the partial pressure in the mixture. Since the pressure of the mixture also equals p, the quantity (p_f/p_i) equals the mole fraction x. Consequently,

$$\Delta S_1 = n_1 \Delta \bar{s}_1 = -n_1 \bar{R} \ln (p_f/p_i)_1 = -n \bar{R} \ln x_1$$

and

$$\Delta S_2 = n_2 \Delta \bar{s}_2 = -n_2 \bar{R} \ln (p_f/p_i)_2 = -n_2 \bar{R} \ln x_2$$

where x_1 and x_2 are the mole fractions of the two gases. The total change in the entropy due to mixing is

$$\Delta S = \Delta S_1 + \Delta S_2$$
$$= -\bar{R}(n_1 \ln x_1 + n_2 \ln x_2)$$

Note that since the mole fractions x's are less than 1, values of $\ln x$ are negative and ΔS is positive. In general, when N gases are mixed, the change in entropy is given by

$$\Delta S = -\bar{R} \sum_{i=1}^{N} n_i \ln x_i \tag{10.12}$$

The above result shows that the change of entropy due to mixing depends only on the number of moles of the constituent gases and not on the nature of the gases. Thus if 1 kg-mol of hydrogen is mixed with 1 kg-mol of nitrogen, the resulting change of entropy is the same as in the case of mixing of 1 kg-mol of methane and 1 kg-mol of oxygen. However, when 1 kg-mol of nitrogen is mixed with 1 kg-mol of nitrogen, there is no change in the entropy. This is due to the fact that there is no way to distinguish between 1 kg-mol of nitrogen and another kg-mol of nitrogen. In other words, for entropy to increase due to mixing of gases, the gases must be dissimilar and distinguishable.

10.3 PROPERTIES OF MOIST AIR

Atmospheric air contains water vapor. The mole fraction of the vapor is rather small and so the partial pressure of the vapor is also small. Under these conditions the water vapor in air can be treated as an ideal gas. The water vapor content of air can increase if it is brought in contact with a body of liquid water. Also, if the air is sufficiently cooled, some of the water vapor can condense. In the engineering approach for working with moist air, it is assumed that the mixture of dry air and water vapor behaves as an ideal gas. It is further assumed that when the vapor condenses, there are no dissolved gases in the condensate. Another assumption is that when the water vapor in moist air is in thermal equilibrium with the condensate, the vapor pressure is the saturation pressure corresponding to the temperature of the mixture. These assumptions are commonly used when considering other mixtures of gases and condensable vapors.

Moist air means a mixture of *dry air* and *water vapor*. In air-conditioning applications, *standard dry air* is *defined* as consisting of 0.2095 mole fraction of oxygen, 0.7809 mole fraction of nitrogen, 0.0093 mole fraction of argon, and 0.0003 mole fraction of carbon dioxide. The following definitions relate to moist air.

The *humidity ratio W* is defined as the mass of water vapor associated with a unit mass of *dry air* in a given sample of moist air. Note that W is not the mass fraction, since the former is based on a unit mass of *dry air* while the latter is based on a unit mass of the *mixture*. Using w and a as the subscripts to denote water vapor and dry air, respectively, we can write

$$p_w V = m_w R_w T \quad \text{for water vapor} \tag{10.13}$$

and

$$p_a V = m_a R_a T \quad \text{for dry air} \tag{10.13a}$$

Dividing Equation 10.13 by Equation 10.13(a), we obtain

$$\frac{p_w}{p_a} = \frac{p_w}{p - p_w} = \frac{m_w}{m_a} \frac{R_w}{R_a}$$

Inserting the numerical values of R_w and R_a and noting that p_w/p equals the mole fraction x_w, we have

$$W = \frac{0.622 \, x_w}{1 - x_w} \tag{10.14}$$

Consider a sample of moist air. If the partial pressure of the water vapor is equal to the saturation pressure corresponding to the temperature of the given sample of the moist air, then the given sample is said to be *saturated moist air*. This is illustrated by the T-s diagram in Figure 10.2 for the moisture in the air. If the moisture in the air can be represented by a point on the saturated vapor line on a T-s diagram, such as points a or b, the air is saturated moist air. Next, consider states of the moisture in moist air represented by points such as c or d. These are in the superheated region; the moist air, then, is not saturated. If such air is cooled at constant pressure, its temperature will drop along dcb, and it will ultimately reach the temperature T_b. Up to this point the mole fraction, and hence the partial pressure of the water vapor, are constant. Once point b is reached, further cooling results in condensation of the water vapor. Point b is the dew point of all states of the moisture in moist air on line bcd and the temperature at point b is the *dew point temperature*, t_d, of the given sample of air of temperature t.

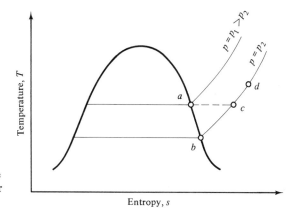

Figure 10.2 *T-S* diagram for the condensable vapor in a gas-vapor mixture.

When a given sample of moist air is saturated, the corresponding humidity ratio is designated by W_s. The quantity W_s represents the maximum quantity of water vapor that can be associated with a unit mass of dry air at a given temperature. Quite often the actual moisture content W is less than this maximum. We define *degree of saturation* μ to express the relative moisture content by

$$\mu = \frac{W}{W_s} \qquad (10.15)$$

where W and W_s are determined at the same temperature.

We have seen that W and x are related by Equation 10.14. The same equation can be used for the saturation conditions by using the subscript s on x and W. The relative moisture can also be defined in terms of the mole fractions, x_w and $x_{w,s}$. It is then known as *relative humidity*, ϕ, and it is the ratio of the volumetric fraction of the water vapor in a given sample of moist air to the fraction necessary to saturate the sample of the moist air at constant temperature and constant total pressure. Note that the degree of saturation involves mass fraction, while the relative humidity involves the volumetric fractions. Since the mole fraction of a component in a mixture of ideal gases equals the volumetric fraction of the component, the relative humidity can be expressed as

$$\phi = \frac{x_w}{x_{w,s}} \qquad (10.16)$$

where x_w and $x_{w,s}$ are determined at the same temperature. Before introducing additional definitions, we show a method of determining W, μ, and ϕ for a given sample of air.

Example 10.4

A sample of moist air at 1 atm and 25°C has a moisture content of 0.01 % by volume. Determine the humidity ratio, the partial pressure of the water vapor, the degree of saturation, the relative humidity, and the dew-point temperature.

Solution:

Given: Certain moist air has the following characteristics.

$$p = 1 \text{ atm}, \qquad t = 25°C$$

$$x_w = 0.01$$

Objective: We have to calculate W, p_w, ϕ, and t_d.

Assumptions: The mixture of dry air and water vapor behaves like an ideal gas.

Relevant physics: This problem involves a straightforward application of the material in Section 10.3. Equation 10.14 can be used to determine W and W_s. The partial pressure p_w can be obtained from the relationship for x_w, p_w, and p. The mole fraction $x_{w,s}$ of saturated water vapor at 25°C can be calculated once $p_{w,s}$ is known. The value of $p_{w,s}$ is simply the saturation pressure corresponding to 25°C. We can represent the sequence of computations diagrammatically as shown below.

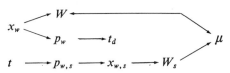

Analysis: We are given the value of x_w. From Equation 10.14, W can be determined.

$$W = 0.622 \frac{x_w}{1 - x_w}$$

$$= 0.622 \cdot \frac{0.01}{1 - 0.01}$$

$$= 0.00628 \text{ kgw/kga}$$

The partial pressure p_w of the water vapor can be determined from

$$x_w = \frac{p_w}{p}$$

A rearrangement gives

$$p_w = px_w$$

$$= 101.3 \times 0.01$$

$$= 1.013 \text{ kPa}$$

We need to determine W_s so that the degree of saturation, μ, can be evaluated. On checking the steam tables in Appendix B2.1, we find that the saturation

pressure, $p_{w,s}$, corresponding to 25°C is 3.169 kPa. This means that when sufficient moisture is added to our moist air, its partial pressure will increase to $p_{w,s}$
and, correspondingly, W will increase to W_s. This process is represented by the
horizontal line ca in Figure 10.2. Rewriting Equation 10.9 for the saturation
conditions, we have

$$x_{w,s} = \frac{p_{w,s}}{p} = \frac{3.169}{101.3} = 0.0313$$

From Equation 10.14

$$W_s = 0.622 \cdot \frac{x_{w,s}}{1 - x_{w,s}}$$

$$= 0.622 \cdot \frac{0.0313}{1 - 0.0313}$$

$$= 0.0201$$

The degree of saturation is

$$\mu = \frac{W}{W_s} = \frac{0.00628}{0.0201}$$

$$= 0.313$$

The relative humidity is

$$\phi = \frac{x_w}{x_{w,s}} = \frac{0.01}{0.0313}$$

$$= 0.319$$

The dew point can be determined by reading the saturation temperature
of water corresponding to the partial pressure p_w. From interpolation of the
data in the steam tables, we find that the saturation temperature for 1.013 kPa
is 7.1°C. Thus the dew-point temperature is 7.1°C.

In air-conditioning calculations, it is often necessary to know the value of the
enthalpy of the moist air. It is expressed in joules per kilogram of *dry air* present in
a given sample of moist air. If we assign a zero value to the enthalpy of dry air, as
well as to the enthalpy of liquid water at 0°C, then we can write the following
equation for the enthalpy of moist air in joules per kilogram of air.

$$h = h_a + W h_w$$

$$= c_{p,a} t + W h_g \tag{10.17}$$

where

$c_{p,a}$ is the constant pressure specific heat of the dry air in joules per kilogram-degree
celsius;

t is the temperature of the moist air in degrees Celsius;

W is the mass of the water vapor per kilogram of dry air;

h_g is the enthalpy of the water vapor at the partial pressure of the water vapor
and at the temperature of the moist air in joules per kilogram of the water
vapor.

The following approximate formula may be used to calculate the enthalpy of
the moist air in kilojoules per kilogram of air.

$$h = 1.003t + (1.88t + 2468)W \tag{10.18}$$

10.3.1 The Adiabatic Saturation Process

The adiabatic saturation process involves flow of moist air over a body of liquid water whose temperature is lower than the temperature of the moist air. This results in a heat transfer from the moist air to the cooler liquid water, which brings about evaporation of the water and a reduction in the temperature of the moist air. The vapor generated escapes into the moist air thus increasing its humidity ratio. This is shown in Figure 10.3. It is assumed that there is long enough residence time for

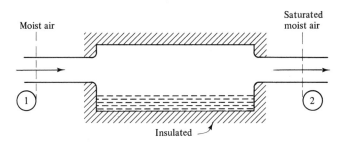

Figure 10.3 Adiabatic saturation of moist air.

the air in the chamber shown in the figure. As this process continues, the moist air ultimately becomes saturated. In this process, there is only internal heat transfer between the body of water and the moist air, and *no external heat transfer* is involved. Therefore the process is regarded as an adiabatic one. Since the process results in the saturation of the moist air, the process is called an *adiabatic saturation process*.

In a thermodynamic analysis of the process it is usually assumed that the temperature of the body of water is exactly the same as that of the moist air at section 2. We can write an energy balance equation for the adiabatic saturation process. We consider a unit mass of dry air associated with W kg of moist air entering at section 1. Since the humidity ratio changes from W_1 to $W_{2,s}$, the mass of water picked up by the air between sections 1 and 2 is $W_{2,s} - W_1$. The specific enthalpy of this mass before vaporization is $h_{f,2}$. Thus an energy balance yields

$$h_1 + (W_{2,s} - W_1)h_{f,2} = h_2 \tag{10.19}$$

where h_1 and h_2 are the specific enthalpies of the moist air per kilogram of dry air at sections 1 and 2, respectively. Since state 2 is a saturation state obtained in an adiabatic manner, it is customary to denote it by the superscript $*$. By using Equation 10.17, Equation 10.19 may be rewritten as

$$(c_{p,a}t_1 + W_1 h_{g,1}) + (W_{2,s}^* - W_1)h_{f,2}^* = (c_{p,a}t_2^* + W_{2,s}^* h_{g,2}^*) \tag{10.20}$$

A feature of the adiabatic saturation process that is not readily evident is that for a given inlet condition at section 1, condition 2* is automatically fixed if the total pressure of the moist air does not change in the process. Consider Equation 10.20. The state of moist air is completely determined once we know its temperature, total pressure, and one of the several parameters defined in the previous section (for instance, W, μ, or ϕ). Suppose we use relative humidity as the parameter. The quantity in the first pair of parentheses in Equation 10.20 can be evaluated for a given inlet condition. For an assumed value of t_2^*, $p_{w,2}^*$ can be determined from the steam tables; then $W_{2,s}^*$ can be determined using Equations 10.9 and 10.14. The values $h_{f,2}^*$ and $h_{g,2}^*$ can be read from the steam tables for the assumed value of t_2^*. Thus all the remaining quantities in Equation 10.20 can be evaluated. If the assumed value of t_2^* is the right value, then Equation 10.20 will be satisfied; otherwise the calculations should be repeated with a different value of t^* value. Thus we see that there is a unique value of t^* corresponding to the temperature t of a given sample of moist air. Therefore we drop the subscripts 1 and 2 in Equations 10.19 and 10.20 and rewrite them to read

$$h + (W_s^* - W)h_f^* = h^* \tag{10.21}$$

$$(c_{p,a}t + Wh_g) + (W_s^* - W)h_f^* = (c_{p,a}t^* + W_s^*h_g^*) \tag{10.21a}$$

If Equation 10.18 is used for h, Equation 10.21 becomes

$$[1.003t + (1.88t + 2468W)] + (W_s^* - W)h_f^*$$
$$= [1.003t^* + (1.88t^* + 2468)W_s^*] \tag{10.21b}$$

Among the properties of moist air defined so far (x_w, W, μ, ϕ, t_d, and t^*), it happens that the adiabatic saturation temperature, t^*, is relatively easy to measure. A practical way to measure t^* is to use a wet-bulb thermometer. A wet-bulb thermometer consists of an ordinary thermometer with its bulb covered with cotton wick. One end of the wick is immersed in a water reservoir. Air at a velocity of greater than 3 m/s is passed over the wet bulb to ensure that the convective heat transfer is greater than the radiative heat transfer. From considerations of the heat and mass transfer, the temperature recorded by the wet bulb thermometer is correlated with the adiabatic saturation temperature t^*. Fortunately, for moist air at atmospheric pressure and temperature, the difference between the wet-bulb temperature and the adiabatic saturation temperature is negligible and the former is taken to be t^*.

Example 10.5

Determine the adiabatic saturation temperature of the air in Example 10.4.

Solution:

Given: From the data and solution to Example 10.4, we have

$$t = 25°C, \qquad W = 0.00628 \text{ kgw/kga}$$
$$p = 1 \text{ atm}, \qquad c_{p,a} = 1.003 \text{ kJ/kg·K}$$

Objective: We have to determine t^*.
Assumptions: Ideal gas behavior for the moist air is assumed.

Relevant physics: We know that for the correct value of t^* for the given sample of air, Equation 10.21 for adiabatic saturation must be satisfied. Equation 10.21(a) is the same as Equation 10.21 but contains the expressions for h and h^*.

   Once a value of t^* is assumed, the following computation leads to the results required for Equation 10.21.

$$t^*\binom{\text{steam}}{\text{tables}} \underset{\searrow}{\overset{\nearrow}{\longrightarrow}} \begin{array}{l} p_{w,s} \longrightarrow x^*_{w,s} \longrightarrow W_s \\ h^*_f \\ h^*_g \end{array}$$

$$t\binom{\text{steam}}{\text{table}} \underset{\searrow}{\overset{\nearrow}{\longrightarrow}} \begin{array}{l} p_w \longrightarrow x_w \longrightarrow W \\ h_f \\ h_g \end{array}$$

   With known values of $c_{p,a}$, t, W, h_g, W^*_s, h^*_f, t^*, and h^*_g, we can then verify if Equation 10.21(a) or 10.21(c) is satisfied.

Analysis: We first determine the enthalpy using Equation 10.18.

$$\cdot h = 1.003t + (1.88t + 2468)W$$
$$= 1.003 \times 25 + (1.88 \times 25 + 2468)0.00628$$
$$= 40.87 \text{ kJ/kg}a$$

Assume $t^* = 15°C$. Then from Appendix B2.1, $p^*_{w,s} = 1.7051$ kPa. From Equation 10.9,

$$x^*_{w,s} = \frac{p^*_{w,s}}{p} = \frac{1.7051}{103.25} = 0.0165$$

From Equations 10.14 and 10.21

$$W^*_s = 0.622 \cdot \frac{x^*_{w,s}}{1 - x^*_{w,s}}$$
$$= 0.622 \cdot \frac{0.0165}{1 - 0.0165} = 0.01044 \text{ kgw/kg}a$$

$$h^* = 1.033t^* + (1.88t^* + 2468)W^*_s$$
$$= 1.003 \times 15 + (1.88 \times 15 + 2468)0.01044$$
$$= 41.11 \text{ kJ/kg}a$$

$$(W^*_s - W)h^*_f = (0.01044 - 0.00628) \times 63 = 0.262 \text{ kJ/kg}a$$

The equation to be satisfied by the assumed value of t^* is

$$h + (W^*_s - W)h^*_f = h^*$$ (10.21)
$$40.87 + 0.262 \overset{?}{=} 41.11$$

Our assumed value of $t^* = 11.6°C$ essentially satisfies Equation 10.21.

10.3.2 *Psychrometric Chart and Thermodynamic Processes*

A computation of all the properties of moist air, especially in air-conditioning calculations, can take a fairly long time. A psychrometric chart obviates the need to
calculate a number of these properties. The chart is constructed using the humidity
ratio, W, and the enthalpy, h. The humidity ratio appears as the ordinate and the
enthalpy is used as an oblique coordinate. The constant enthalpy lines are shown
as straight lines with negative slopes. On a chart constructed in this manner (see
Figure 10.4), the constant dry-bulb temperature lines appear as vertical lines, especially at low temperatures. These lines are not parallel. If we move up on a constant
temperature line on a psychrometric chart, we soon reach the saturation line. The
higher the temperature, the larger the value of W_s. There are constant wet-bulb
temperature lines on the chart, and these are almost parallel to the constant enthalpy
lines. The chart also shows constant volume lines. A given state can be shown on the
chart if any two properties are known.

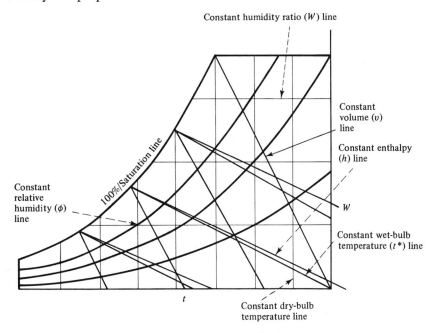

Figure 10.4 Psychrometric chart.

The dew point for a given state can be located by moving to the left of the state
on a constant W-line until the saturation line is reached and then by reading the
dry-bulb temperature corresponding to this point. A given psychrometric chart is
meant for a specified total pressure. A psychrometric chart for 1 atm is given in
Appendix K.

Now, let us consider some thermodynamic processes on a psychrometric chart.
Let a given state be represented by point 1 on Figure 10.5.

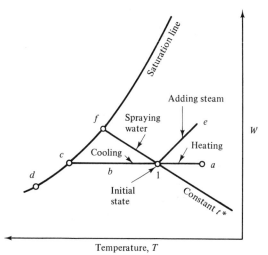

Figure 10.5 Processes on a psychrometric chart.

Heating. In a heating process the moisture content is not affected, so W must remain constant. Also, the heating would bring about a rise in temperature. Thus heating will change the state along a constant W-line along line 1–a.

Cooling. A cooling process will take the moist air on a path exactly opposite to 1–a, that is, along 1–b. If cooling is continued, the air will be eventually saturated and will be at state c. Further cooling will initiate condensation and the state would begin to change along c–d. This is one way of accomplishing dehumidification.

Addition of steam. An addition of steam will bring about an increase in humidity, as well as in enthalpy. This will also raise the temperature. The state of the moist air will change along line 1–e in Figure 10.5. The exact slope of this line will depend upon the quantity of the steam added per kilogram of dry air and on the enthalpy of the steam.

Water spraying. Water spraying is essentially an adiabatic saturation process. Therefore the state of the moist air will change along the constant wet-bulb temperature line passing through the given state. Only if enough water is sprayed to make the air saturated will the final state be at point f in Figure 10.5; otherwise, it will lie somewhere on line 1–f. Note that the process results in a reduction of the dry-bulb temperature. Because of this, the process is effectively used in cooling air in dry, warm climates.

Mixing of two streams. Let the states of the two streams of moist air be represented by points 1 and 2, as in Figure 10.6. Since mass and energy are conserved, we can write

Dry air: $\qquad \dot{m}_{a,1} + \dot{m}_{a,2} = \dot{m}_{a,3}$ $\qquad\qquad$ (10.22)

Water: $\quad \dot{m}_{a,1} W_1 + \dot{m}_{a,2} W_2 = \dot{m}_{a,3} W_3$ $\qquad\qquad$ (10.22a)

Energy: $\quad \dot{m}_{a,1} h_1 + \dot{m}_{a,2} h_2 = \dot{m}_{a,3} h_3$ $\qquad\qquad$ (10.23)

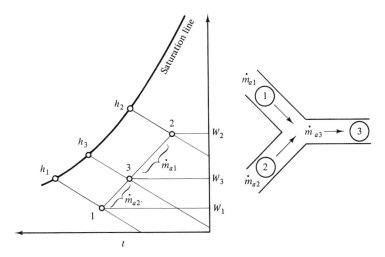

Figure 10.6 Adiabatic mixing of two streams of moist air.

The last two equations enable us to determine h_3 and W_3 and thus to locate state 3 after mixing. The equations can be manipulated to yield

$$\frac{\dot{m}_{a,1}}{\dot{m}_{a,2}} = \frac{h_2 - h_3}{h_3 - h_1} = \frac{W_2 - W_3}{W_3 - W_1} \tag{10.24}$$

Equation 10.24 mean that point 3 (Figure 10.6) divides segment 1–2 in such a manner that

$$\frac{\dot{m}_{a,1}}{\dot{m}_{a,2}} = \frac{\text{segment } 3\text{–}2}{\text{segment } 1\text{–}3} \tag{10.25}$$

Example 10.6

Saturated moist air is heated by steam condensing inside the tubes of a heating coil, as shown in Figure 10.7. Some of the air passes over the coil (path 1–2–3) and the remain-

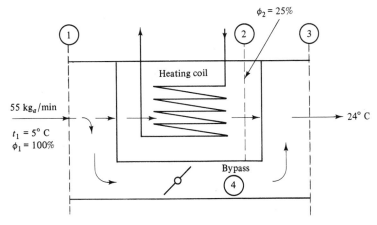

Figure 10.7

455

ing is by-passed (path 1–4–3). Assuming a total pressure of 1 atm, determine the quantity of air bypassing the coil and the heat added by the coil.

Solution:

Given: A portion of a cold air stream passes over a heating coil, and the remainder is bypassed. These two streams are then mixed.

$$\text{Sections 1 and 4:}\quad t_1 = 5°C = t_4;\qquad \phi_1 = 100\% = \phi_4;$$

$$\dot{m}_1 = 55 \text{ kga/min}$$

$$\text{Section 2:}\qquad \phi_2 = 25\%$$

$$\text{Section 3:}\qquad t_3 = 24°C$$

Objective: We have to determine $\dot{m}_4$ and the heat added by the heating coil, $\dot{Q}_{2\text{-}3}$.

Assumptions:

1. The moist air behaves like an ideal gas.
2. The moist air obeys Dalton's law.

Relevant physics: Since the air at section 1 is saturated, ($\phi_1 = 100\%$), the conditions of moisture at this section are represented by the subscript s. The value of $p_{w,s1}$ for 5°C is read from Appendix B2.1. With this pressure known, we can calculate $x_{w,s1}$ and then $W_{s,1}$, which equals W_2, W_3, and W_4 since no moisture is added or removed. Thus two properties each at section 2 and section 3 are known. Then, an energy balance and a mass balance give the desired results.

Analysis: We can write the following equations.

$$\dot{m}_{a,1} = \dot{m}_{a,2} + \dot{m}_{a,4} = \dot{m}_{a,3} = 55 \text{ kga/min}$$

$$W_1 = W_{s,1} = W_4 = W_3 = W_2$$

The saturation pressure for 5°C is 0.8721 kPa. The mole fraction can be calculated from

$$x_{w,s1} = \frac{p_{w,s1}}{p_{a,1} + p_{w,s1}} = \frac{0.8721}{101.3} = 0.0086$$

Then

$$W_{s,1} = 0.622 \left(\frac{x_{w,s1}}{1 - x_{w,s1}} \right)$$

$$= 0.622 \left(\frac{0.00866}{1 - 0.00866} \right)$$

$$= 0.0054 \text{ kgw/kga} = W_3 = W_4$$

We are given that $\phi_2 = 0.25$, so that

$$\phi_2 = \frac{x_{w,2}}{x_{w,s2}} = 0.25$$

Since the moisture content of the air passing over the heating coil does not change, the humidity ratio is constant, as is the mole fraction of the water vapor (Equation 10.14). Thus $x_{w,2} = x_{w,s1} = 0.0086$. Then we have

$$x_{w,s2} = \frac{x_{w,2}}{0.25} = \frac{0.0086}{0.25}$$

$$= 0.0344$$

Hence

$$p_{w,s2} = px_{w,s2}$$
$$= 101.3 \times 0.0344 \text{ kPa}$$
$$= 3.484 \text{ kPa}$$

The saturation temperature corresponding to this pressure is interpolated from Appendix B2.1 to be 26.44°C. Therefore

$$t_2 = 26.44°C$$

The enthalpies at sections 1, 2, and 3 are calculated from Equation 10.18.

$$h = 1.003t + (1.88t + 2468)W$$

The enthalpy values are

$$h_1 = 1.003 \times 5 + (1.88 \times 5 + 2468) \times 0.0054$$
$$= 18.39 \text{ kJ/kg}$$
$$h_2 = 1.003 \times 26.44 + (1.88 \times 26.44 + 2468) \times 0.0054$$
$$= 40.11 \text{ kJ/kg}$$
$$h_3 = 1.002 \times 24 + (1.88 \times 24 + 2468) \times 0.0054$$
$$= 37.64 \text{ kJ/kg}$$

The amount of air bypassed can be calculated from the energy balance equation

$$\dot{m}_{a,2}h_2 + \dot{m}_{a,4}h_4 = \dot{m}_{a,3}h_3$$

We note that $h_4 = h_1 = 18.39$ J/kg$a$, and $\dot{m}_{a,2} + \dot{m}_{a,4} = 55$ kg$a$/min. The energy balance equation then becomes

$$40.11 (55 - \dot{m}_{a,4}) + 18.39 \times \dot{m}_{a,4} = 55 \times 37.64$$

or

$$\dot{m}_{a,4} = 6.26 \text{ kg}a/\text{min}$$

The heat added by the coil can be calculated from

$$\dot{m}_{a,2}h_1 + \dot{Q}_{2-3} = \dot{m}_{a,2}h_2$$

or

$$\dot{Q}_{2-3} = (55 - 6.26)(40.11 - 18.39)$$
$$= 1058.85 \text{ kJ/min}$$
$$= 17.6 \text{ kW}$$

The problem can also be readily solved using a psychrometric chart, such as the one in Appendix K.

10.4 AIR CONDITIONING

The objective of air conditioning is to make appropriate modifications to the atmospheric air, mix it with air drawn from the space to be conditioned, if necessary, and supply this mixture of air at a predetermined temperature and relative humidity. Air conditioners for residental applications do not mix the conditioned air with the air drawn from the space, while commercial systems do mix the two streams of air. The environmental conditions to be maintained in a space depend upon the purpose for which the space is used. The requirements for human occupancy are quite different from those for warehousing tobacco or for livestock. In this introductory discussion, we shall assume that a given space is to be conditioned for human occupancy. This means that the temperature in the conditioned space should be maintained between 18°C and 26°C with the relative humidity between 45 percent and 20 percent.

There are a number of mechanisms contributing to heat gains and losses in a space. We shall shortly examine these contributions. Let $\dot{Q}$ be the net rate of heat gain in kilowatts. Note that $\dot{Q}$ can be positive or negative. Let $\dot{m}_a$ be the mass flow rate of the moist air entering the space and h_{in} and h_{out} be the specific enthalpies of the air entering and leaving the space, respectively. When the space is conditioned, under steady flow,

$$\dot{m}_a h_{in} + \dot{Q} = \dot{m}_a h_{out} \qquad (10.26)$$

The quantity h_{out} corresponds to the temperature and the relative humidity to be maintained in the conditioned space. The net heat gain $\dot{Q}$ will help determine the mass flow rate since, for human occupancy, the conditions corresponding to h_{in} cannot be arbitrarily chosen. In summer months, the temperature of the conditioned air entering a space, t_{in}, is generally not less than 10°C to avoid near-freezing conditions that would otherwise prevail in the space adjacent to the air inlet. Also, in winter months the temperature of the entering air is seldom allowed to exceed 50°C. Keeping such considerations in mind, the temperature and the relative humidity of the incoming air are specified by the designer. This means that h_{in} is known. Thus, with the net heat gain $\dot{Q}$ and the supply and discharge conditions known, $\dot{m}_a$ can be calculated.

In conditioning air, we also have to keep track of the moisture gain. If $\dot{W}$ is the rate of the moisture gain in a space, then

$$\dot{m}_a W_{in} + \dot{W} = \dot{m}_a W_{out} \qquad (10.26a)$$

This equation permits a determination of the humidity ratio at inlet.

The major factors affecting the net heat gain $\dot{Q}$ in a space are the following.

1. People.
2. Heat transfer through walls, roof, and glass areas.
3. Energy-consuming devices such as electrical motors, furnaces, and lights.
4. Infiltration of air.

Let us consider these factors in some detail.

The human body, as a result of its metabolic processes, continuously releases heat energy, as well as moisture by perspiration. The heat-release rate can be anywhere from 75 W to 700 W, depending on the nature of the activity in which the body is engaged; the lower figure corresponds to sleeping, while the higher figure corresponds to sustained hard work. The corresponding figures for moisture release are 0.05 kg/h and 0.45 kg/h, respectively. For light and moderate work, the heat-release rates are about 160 W and 220 W, respectively. The corresponding figures for moisture release are 0.23 kg/h and 0.30 kg/h, respectively.

The heat released by the body must be carried away by convection, radiation, and evaporation of perspiration. If the environment temperature is higher than the body temperature, then the only mechanism for removing body heat is evaporation. Under such conditions, a person feels highly uncomfortable. On the other hand, if the environmental temperature is below 18°C, then there is excessive heat loss from a human body by convection and radiation. Then a person has to increase body activity, thus releasing more heat energy. Evidently, there is a limit to this.

The heat transfer through walls and roof is proportional to the temperature difference between the inside and outside environments and to the overall heat conductance of the walls and roof. The overall heat conductance is a function of the thermal conductivities of the materials used in the construction of the walls and roof, the geometry of the roof and wall, and the wind velocity. In summer months, the heat stored in the walls and roof strongly enhances the heat gain into a space. The heat transfer through glass depends upon the incidence angle of solar rays, the total solar radiation incidence on the glass surface, the thermal properties of glass, and the air space between thermopane windows. The incidence angle and the solar radiation depend on the local solar time, date, latitude, and orientation of the glass surface.

The energy released by a machine in the space to be conditioned can be readily calculated from the rating of the machine and the load factor of the machine. In case of incandescent lights, their wattage rating directly gives the heat-release rate. In case of fluorescent lights, the energy consumed by the ballast must be accounted for.

The heat gain due to infiltration is very difficult to calculate accurately. Infiltration occurs through cracks in walls, joints, window and door frames and whenever windows or doors are not shut tightly. Infiltration increases as the wind velocity increases. The ASHRAE handbook* gives empirical equations for calculating heat gain or loss due to infiltration.

10.5 CHEMICAL POTENTIAL AND GIBBS FUNCTION

When earlier we considered a single-component system, any property of the system was expressed as a function of any two other properties. In deriving Maxwell's relations, we saw that these two properties, used as independent variables, are usually chosen from the four properties p, T, V, and S. Of these, pressure and temperature are intensive, while volume and entropy are extensive. When we consider a mixture of two species, we have to specify two additional properties, namely, the number of moles of each species. For example, we may express the internal energy U of a mixture of two species A and B as

$$U = U(S, V, n_A, n_B)$$

*ASHRAE Handbook & Product Directory—1977 Fundamentals, American Society of Heating, Refrigerating and Air-Conditioning Engineers, Inc, (New York: 1977).

A differential change in the internal energy is then given by

$$dU = \left(\frac{\partial U}{\partial S}\right)_{V,n_A,n_B} dS + \left(\frac{\partial U}{\partial V}\right)_{S,n_A,n_B} dV + \left(\frac{\partial U}{\partial n_A}\right)_{S,V,n_B} dn_A + \left(\frac{\partial U}{\partial n_B}\right)_{S,V,n_A} dn_B$$

On the right side of this equation, the first two partial derivatives require that the composition of the mixture be held constant. Under these conditions, we can use Equation 9.24 to write

$$\left(\frac{\partial U}{\partial S}\right)_{V,n_A n_B,} = T \quad \text{and} \quad \left(\frac{\partial U}{\partial V}\right)_{S,n_A,n_B} = -p$$

Also, we let $(\partial U/\partial n_A)_{S,V,n_B}$, which is the rate of change of the internal energy with respect to the change in the number of moles of species A, equal μ_A. In a similar manner, we let $(\partial U/\partial n_B)_{S,V,n_A}$ equal μ_B. Thus

$$\mu_A = \left(\frac{\partial U}{\partial n_A}\right)_{S,V,n_B} \quad \text{and} \quad \mu_B = \left(\frac{\partial U}{\partial n_B}\right)_{S,V,n_A} \tag{10.27}$$

The equation for a differential change in internal energy then becomes

$$dU = T\,dS - p\,dV + \mu_A dn_A + \mu_B dn_B \tag{10.28}$$

The quantities μ_A and μ_B are intensive properties of the mixture. They are called *chemical potentials* associated with n_A and n_B, respectively.

It is possible to relate μ_A and μ_B to the enthalpy, to the Gibbs function, and to the Helmholtz function (Section 9.3). This is done as follows. Consider the Gibbs function of a mixture in the form

$$G = G(T, p, n_A, n_B)$$

For a differential change,

$$dG = \left(\frac{\partial G}{\partial T}\right)_{p,n_A,n_B} dT + \left(\frac{\partial G}{\partial p}\right)_{T,n_A,n_B} dp + \left(\frac{\partial G}{\partial n_A}\right)_{T,p,n_B} dn_A + \left(\frac{\partial G}{\partial n_B}\right)_{T,p,n_A} dn_B$$

or

$$dG = -S\,dT + V\,dp + \left(\frac{\partial G}{\partial n_A}\right)_{T,p,n_B} dn_A + \left(\frac{\partial G}{\partial n_B}\right)_{T,p,n_A} dn_B \tag{10.29}$$

We also know that

$$G = U + pV - TS$$

or

$$dG = dU + p\,dV + V\,dp - T\,dS - S\,dT$$

Inserting the expression for dU from Equation 10.28 gives

$$dG = -S\,dT + V\,dp + \mu_A dn_A + \mu_B dn_B \tag{10.29a}$$

On comparing Equations 10.29 and 10.29(a), we conclude that

$$\mu_A = \left(\frac{\partial G}{\partial n_A}\right)_{T,p,n_B} \quad \text{and} \quad \mu_B = \left(\frac{\partial G}{\partial n_B}\right)_{T,p,n_A} \tag{10.30}$$

Next, consider the enthalpy function

$$H = H(S, p, n_A, n_B)$$

Differentiation yields

$$dH = \left(\frac{\partial H}{\partial S}\right)_{p,n_A,n_B} dS + \left(\frac{\partial H}{\partial p}\right)_{S,n_A,n_B} dp + \left(\frac{\partial H}{\partial n_A}\right)_{S,p,n_B} dn_A + \left(\frac{\partial H}{\partial n_B}\right)_{S,p,n_A} dn_B$$

$$= T \, dS + V \, dp + \left(\frac{\partial H}{\partial n_A}\right)_{S,p,n_B} dn_A + \left(\frac{\partial H}{\partial n_B}\right)_{S,p,n_A} dn_B \tag{10.31}$$

Also

$$H = U + pV$$

so that

$$dH = dU + p \, dV + V \, dp$$

Inserting the expression for dU from Equation 10.28, we obtain

$$dH = T \, dS + V \, dp + \mu_A dn_A + \mu_B dn_B \tag{10.31a}$$

A comparison of Equations 10.31 and 10.31(a) leads to the conclusion

$$\mu_A = \left(\frac{\partial H}{\partial n_A}\right)_{S,p,n_B} \quad \text{and} \quad \mu_B = \left(\frac{\partial H}{\partial n_B}\right)_{S,p,n_A} \tag{10.32}$$

In a similar manner, by considering the Helmholtz function, A, to be a function of T, V, n_A, and n_B, we can show that

$$\mu_A = \left(\frac{\partial A}{\partial n_A}\right)_{T,V,n_B} \quad \text{and} \quad \mu_B = \left(\frac{\partial B}{\partial n_B}\right)_{T,V,n_A} \tag{10.33}$$

In Equations 10.27, 10.30, 10.32, and 10.33, we have four different expressions for the chemical potential, μ. If we examine these expressions carefully, we find that Equation 10.30 requires holding the intensive properties T and p constant, while in the remaining expressions one intensive property and one extensive property are to be held constant while evaluating the derivatives. A derivative such as $(\partial G/\partial n_A)_{T,p,n_B}$ is called a *partial molal property*. In particular this derivative is called the *partial molal Gibbs function*. Since a large number of chemical reactions take place at constant temperature and pressure, the partial molal Gibbs function is a very important property in the thermodynamic analysis of chemical reactions. This potential is a measure of the driving force for a chemical reaction. We shall use the partial molal Gibbs function in Chapter 11.

10.6 SOLUTIONS

A solution may be defined as a phase (solid, liquid, or gas) containing more than one species. According to this definition, when one gas is mixed with another a solution is formed. Generally speaking, gases can be mixed in any proportion to form a solution. The word *solution*, as it is used in practice, more often denotes a solution of liquids or solids than a solution of gases.

Typically, when two species A and B are mixed to form a solution, there is a change in the volume, enthalpy, and entropy. This means that these properties, when measured after the solution is formed, have different values than the sum of the values of the properties for the two species before mixing. The change in volume is due to the cohesive force between molecules in A and molecules in B. The change in enthalpy is due to the heat of solution, that is, the heat energy absorbed or released during a mixing process. The change in entropy is due to the irreversible nature of mixing.

In an *ideal solution*, the cohesive forces are assumed to be uniform. That is, the forces among the molecules of species A, the forces among the molecules of species B, and the forces between molecules of A and molecules of B are identical. This means that it does not matter which molecules in the mixture surround molecules of species A, and the change in volume in forming an ideal solution is zero. For such a solution, the heat of solution is also zero.

The partial vapor pressure of a component situated above a liquid solution is an important property in the study of solutions. If the vapor pressure is rather low, it would mean that few molecules have escaped from the liquid to form the vapor phase. Thus the partial pressure is indicative of the tendency of liquid molecules to escape into vapor phase. *Raoult's law* states that in an ideal solution, the partial vapor pressure of each component is equal to the product of the mole fraction of that component in the liquid solution and the vapor pressure of the component in the pure liquid state at the temperature of the solution. Consider an ideal solution of species A and B, with mole fractions x_A and x_B, respectively. The total vapor pressure, p, is given by Dalton's law.

$$p = p_A + p_B$$

where p_A and p_B are the partial pressures of the two species. If p_A^* is the vapor pressure of component A in the pure liquid state at a temperature equal to that of the solution, then according to Raoult's law,

$$p_A = x_A p_A^* \tag{10.34}$$

Likewise

$$p_B = x_B p_B^* \tag{10.34a}$$

Equation 10.34 can be rearranged to read

$$\frac{p_A^* - p_A}{p_A^*} = 1 - x_A \tag{10.35}$$

This last form expresses the fractional reduction in the partial pressure of a species in terms of the mole fraction of the species.

Raoult's law applies reasonably well to solutions of hydrocarbons. The straight lines in Figure 10.8 show Raoult's law graphically for a two-component solution. The ordinate and the abscissa for the figure are partial pressures p_A and p_B and the mole fraction x_A. When the mole fraction of species A is zero, that is, only species B is present, the partial pressure equals p_B^*. In a like manner, for $x_A = 1$, the ordinate has a value of p_A^*. As x_A increases from zero to 1, according to Raoult's law, p_A increases linearly from zero to p_A^*. When this happens, x_B simultaneously decreases from 1 to zero, and p_B decreases from p_B^* to zero. Note that the figure relates to a specific temperature. Mixtures of fluorobenzene and chlorobenzene and of benzene and toluene obey Raoult's law.

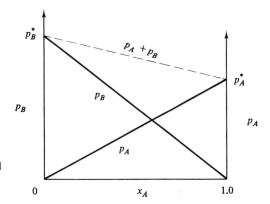

Figure 10.8 Raoult's law and partial pressures of the constituents of a two-component gaseous mixture.

In general, Raoult's law does not hold over the entire range, $0 < x_A < 1$, for many solutions. Figure 10.9 shows the deviation from Raoult's law. The dashed lines show the behavior of an ideal solution that obeys Raoult's law, while the solid lines represent the behavior of real solutions. The curved lines in Figure 10.9(a) show the partial-pressure behavior when the deviation from Raoult's law is positive. A positive deviation means that the partial pressure at any mole fraction is greater than that predicted by Raoult's law. Some of the solutions exhibiting a positive deviation from Raoult's law are acetone and n-heptane, ethyl alcohol and water, phenol and mesitylene, and resorcinol and fluorene. The positive deviation is attributed to relatively weak cohesive forces between molecules of the two species (A and B) of the solution in relation to the cohesive forces among the molecules of species A or of species B.

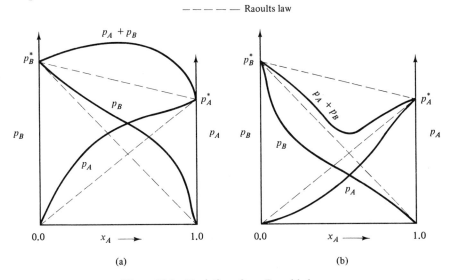

Figure 10.9 Deviations from Raoult's law.

We see from Figure 10.9(a) that the total-pressure $(p_A + p_B)$ curve exhibits a maximum. If a temperature-mole fraction curve with constant pressure were plotted for a solution with positive deviation from Raoult's law, the curve would exhibit a minimum. When the mole fraction of such a solution corresponds to the maximum total pressure in Figure 10.9(a), (that is, to the minimum boiling point on the T-x curve), the vapor solution and the liquid solution in equilibrium have the same composition. Substances exhibiting the constant-boiling behavior are known as *azeotropes*.

A negative deviation from Raoult's law is shown in Figure 10.9(b). Here the partial pressure at any mole fraction is less than that predicted by Raoult's law. Some of the solutions exhibiting such behavior are chloroform and isopropyl bromide and 2,2,3-trimethylbutane and 2,4-dimethylpropane. The negative deviation from Raoult's law is due to strong cohesive forces between molecules of the two species (A and B) of the solution in relation to the cohesive forces among the molecules of species A or of species B.

The total-pressure curve for a solution with negative deviation exhibits a minimum [Figure 10.9(b)]. A temperature-mole fraction curve for such a solution exhibits a maximum. When the mole fraction of such a solution corresponds to the minimum total pressure $(p_A + p_B)$ or the maximum temperature, the vapor and the liquid in equilibrium have the same composition. The components of the solutions are then azeotropes. For a thorough discussion of such substances, see *Azeotropy and Polyazeotropy*, edited by K. Ridgway.*

We can see from Figure 10.9(a) and (b) that as p_A approaches p_A^*, the curve for p_A becomes tangential to the straight line of Raoult's law. This behavior is expressed by *Henry's law*, which states that the partial pressure of the dilute component is proportional to its mole fraction:

$$p_{\text{dilute}} = kx_{\text{dilute}} \tag{10.35}$$

where k is a constant.

In dealing with binary solutions, the quantity x_A is often referred to as the *concentration* of species A. When a two-component ideal solution undergoes a liquid-vapor phase change, the T-x diagram apears as in Figure 10.10. The upper curve in the figure is the saturated vapor curve, while the lower curve is the saturated liquid curve. Let us assume species B has a lower boiling point, T_B^*, than species A. Let the concentration of species A in a given solution be x_0. As the temperature of the solution is raised to the saturation temperature, T_0, the solution is ready for a phase change The solution is in a saturated liquid state at T_0, and this state is denoted by a in Figure 10.10. A supply of heat to the solution results in the vaporization of the more volatile component, species B. The first formed vapor has a concentration represented by point b on the diagram. The vapor mixture is saturated at temperature T_0 and the saturated vapor state is represented by point b. The vapor thus formed is rich in species B but weak in species A. The remaining liquid becomes richer in A, and its concentration increases to x_1. The boiling point for this state is higher than T_0. Further

*K. Ridgway, ed., *Azeotropy and Polyazeotropy* (New York: The Macmillan Company, 1963).

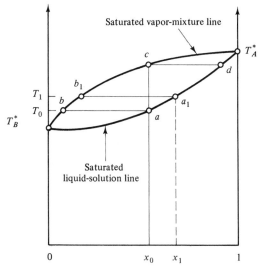

Figure 10.10 Change of phase (liquid-vapor) of a two-component mixture.

x is the concentration of species A in the binary solution.

supply of heat raises the temperature to T_1, bringing about additional vaporization. The new vapor formed has a state b_1, and the concentration of the remaining liquid further increases. The horizontal lines a–b and a_1–b_1 denote the liquid-vapor phase change at temperatures T_0 and T_1, respectively. The vaporization continues as additional heat is supplied. The liquid solution moves along a–a_1–d and the vapor solution moves along the saturated vapor mixture line b–b_1–c. When the vaporization process is complete, the vapor is at state C with exactly the same concentration as existed at point a.

This behavior is utilized in distilling a component from a binary solution in chemical industries. The process involves alternate heating and cooling and is carried out in fractionating or distillating columns. This behavior of binary solutions is also important in absorption air conditioning.

10.7 SUMMARY

In this chapter we dealt with ideal gas mixtures, mixtures of gases and vapors, and solutions of liquid. The Gibbs-Dalton law enabled us to evaluate the pressure, internal energy, enthalpy, and entropy of a mixture of ideal gases. The Amagat-Leduc law related the partial volumes of the components with the total volume of a mixture. Several new properties were introduced in the discussion of moist air: humidity ratio, degree of saturation, relative humidity, dew-point temperature, and adiabatic saturation temperature. The psychrometric chart was introduced and used to explain the change in state of moist air due to a number of thermodynamic processes. The basics of air conditioning were also introduced. The concept of the chemical potential was presented and partial molal Gibbs function was defined. Raoult's law as related to an ideal solution was stated and deviations from Raoult's law leading to Henry's law were discussed. Finally, the change of phase of a liquid solution with reference to a temperature-concentration diagram was discussed.

PROBLEMS

10.1. A 0.2-m³ tank contains a mixture of 2 kg of CO_2 and 4 kg of N_2. The temperature of the mixture is 25°C. Determine (a) the partial pressures, partial volumes, and the mole fractions of CO_2 and N_2, and (b) the gas constant, the molecular weight, the enthalpy, and the entropy of the mixture. Use gas tables.

10.2. Determine the amount of O_2 at 25°C that should be added to the tank of Problem 10.1 to raise its pressure by a factor of 2.

10.3. The specific heat ratio for a mixture is defined by

$$\gamma_m = \frac{m_1 c_{p1} + m_2 c_{p2} + \cdots}{m_1 c_{v1} + m_2 c_{v2} + \cdots}$$

Show that the mixture obeys $pv^{\gamma_m} = $ constant during a reversible adiabatic process.

10.4. A rigid insulated tank of volume 1 m³ is divided into two compartments by a partition. One compartment has a volume of 0.3 m³ and contains O_2 at 200 kPa and 25°C. The other compartment contains N_2 at 1000 kPa and 100°C. If the partition is removed, determine the partial pressures of O_2 and N_2 in the resulting mixture.

10.5. A tank contains 6 kg of CO. How much nitrogen should be added to the tank to produce a mixture to contain equal parts by volume of the constituents? If the pressure of the mixture is 500 kPa, determine the partial pressures of the constituents. Also, determine the molar mass, the gas constant, and the specific heat at constant pressure of the mixture.

10.6. A 0.1-m³ tank contains a gaseous mixture of O_2 and methane (CH_4) at 700 kPa and 38°C. The composition of the mixture is 25% O_2 and 75% CH_4 by volume. It is desired to alter the composition to 75% O_2 and 25% CH_4 by volume at the original pressure and temperature. How many kilograms of mixture must be bled and what mass of O_2 should be added?

10.7. The volumetric analysis of a gaseous mixture is CO; 3%; O_2, 8%; CO_2, 35%; and N_2, 54%. Determine the proportions of the constituents by mass and the partial pressures of the constituents if the total pressure is 100 kPa.

10.8. The analysis by mass of a certain gaseous mixture is Ar, 5%; He, 11%; O_2, 17%; and N_2, 67%. Determine each of the following.
(a) Volumetric composition of the mixture.
(b) Gas constant of the mixture.
(c) Mass of the mixture of volume 8 m³ at 100 kPa and 27°C.

10.9. For the mixture in Problem 10.8, determine each of the following.
(a) Partial pressures and volumes of the constituents.
(b) Specific heat at constant pressure.
(c) Enthalpy and the internal energy of the mixture.

10.10. For the mixture in Problem 10.7, determine each of the following.
(a) Partial pressures of the constituents.

(b) Mixture molar mass and gas constant.

(c) Specific enthalpy at 400 K.

10.11. If the temperature of the mixture in Problem 10.4 is to be raised by 100°C, determine the amount of heat that must be supplied.

10.12. A mixture of gases contains 25% CO_2, 30% H_2, and 45% N_2 on a volumetric basis. The temperature and pressure of the mixture are 350 K and 200 kPa. How much heat energy must be removed from the mixture to cool it to 300 K?

10.13. Show that the entropy change for a mixture of two perfect gases is given by

$$dS = (m_1 c_{p1} + m_2 c_{p2}) \frac{dT}{T} - (m_1 R_1 + m_2 R_2) \frac{dp}{p}$$

where the subscripts 1 and 2 refer to the two gases.

10.14. A mixture of H_2, and O_2 in equal amounts is cooled from 60°C to 30°C at a constant pressure of 300 kPa. Determine the change in internal energy, enthalpy, and entropy per kilogram of the mixture. These properties may be taken as zero at 0°C.

10.15. Determine the maximum work and the irreversibility for the process of Problem 10.4.

10.16. In a mixture of water vapor and nitrogen at 500 kPa and 25°C, the water vapor is saturated. What is the humidity ratio?

10.17. A sample of moist air at 27°C and 100 kPa has a humidity ratio of 0.012 kg w/kga. Calculate each value.

(a) Partial pressure of water vapor.

(b) Dew point.

(c) Volumetric and gravimetric (by mass) proportions of the water vapor and dry air.

10.18. A sample of moist air at 100 kPa and 25°C has a humidity ratio of 0.016. Calculate each value.

(a) Mole fraction of the water vapor.

(b) Dew point.

(c) Relative humidity.

10.19. A steam condenser operates at 24.08°C, and within the condenser there are 0.08 kg of dry air for every kilogram of steam. The steam can be assumed to be saturated vapor. Calculate each value.

(a) Humidity ratio.

(b) Condenser vacuum.

(c) Volumetric composition.

10.20. The dew point of a sample of moist air at 25°C is 10°C. Determine (a) the humidity ratio, (b) the relative humidity, and (c) the volumetric composition. Assume the total pressure to be 101 kPa.

10.21. Calculate the adiabatic saturation temperature of the air in Problem 10.20.

10.22. A given sample of moist air has a relative humidity of 60% and a dew point of 8°C. Determine each of the following.

(a) Dry-bulb temperature.

(b) Humidity ratio.

(c) Adiabatic saturation temperature.

10.23. Determine the adiabatic saturation temperature of the moist air of Problem 10.18.

10.24. Show that the dew point of a vapor-gas mixture when cooled at constant volume is of slightly lower temperature than the dew point obtained by cooling at constant pressure.

10.25. The air in a room has a dry-bulb temperature of 23°C. The wall temperature is found to be 16°C. Determine the highest relative humidity of the air without causing condensation on the wall.

10.26. A compressor adiabatically compresses air from 1 atm and 18°C, to 4 atm. Determine the temperature after compression and the amount of work to be supplied. If the compressed air is cooled to 18°C, what is the amount of condensate formed in the compressed air?

10.27. Consider moist air undergoing an isentropic process. The total change in the entropy of the moist air for such a process is zero. Show that the change in the entropy (non-zero) of the moisture component is equal and opposite to the change in the entropy of the dry air component.

10.28. A gaseous mixture leaving a stack at 110 kPa and 80°C has the following volumetric analysis: H_2O, 16%; CO_2, 32%; N_2, 52%. Calculate (a) the humidity ratio, (b) the relative humidity, and (c) the degree of saturation. What should be the temperature of the stack walls so as to prevent the condensation of the water vapor?

10.29. It is desired to reduce the relative humidity of saturated moist air at 25°C and 100 kPa by 25% by heating the air. What should be the final temperature of the air to meet this objective? Verify your answer by consulting a psychrometric chart.

10.30. A stream of moist air, 6°C and 50% relative humidity, enters a heating unit at a rate of 15 m³/s. Determine the rate of heat input necessary to raise the temperature to 30°C and the relative humidity after heating. How much water should be added to this heated air to raise its relative humidity to 40%?

10.31. Moist air from a room at 20°C and 40% relative humidity is mixed with outside air at −10°C and 80% relative humidity. The mass flow rates are 50 kg/s and 20 kg/s, respectively. Determine the dry-bulb temperature, the relative humidity, and the enthalpy of the mixture.

10.32. If the state of the mixture in Problem 10.31 is to be changed to 50°C and 70% relative humidity, determine (a) the rate of heat and moisture supply to the mixture, and (b) the rate of heat and moisture loss from the room. Assume that the air-exhaust rate from the room is 20 kg/s.

10.33. In a dry, warm climate area, the outside air is at 40°C and 20% relative humidity. The heat and moisture gain for a space is 4 kW and 40 kg/h. It is proposed to maintain the temperature in the conditioned space at 20°C and 60% relative humidity by spraying water at 20°C in the stream of the outside air to be admitted to the space. The space has 100% exhaust. Determine the temperature, relative humidity, and the mass flow rate of the air that will be admitted into the space. Also, determine the quantity of water that will be used in the spraying chamber.

10.34. When hydrogen gas is produced in a certain process, it is found to contain 1 part by volume of water vapor for every 10 parts of hydrogen. The temperature and pressure of this mixture are 70°C and 150 kPa. In order to remove the water vapor, the mixture is cooled at constant pressure to 10°C at a rate of 30 m³/min. Calculate the rate at which condensation occurs and the final composition of the mixture.

10.35. In an air-conditioning system for a precision gauge laboratory, atmospheric air at 32°C and 80% relative humidity is first chilled to 10°C to remove moisture. The air is then heated until its relative humidity reaches 40%. If the air flow rate is 800 m³/min, determine each of the following.

(a) Heat removed in the cooling section in kilojoules per minutes.

(b) Heat added in the heating section in kilojoules per minute.

(c) Rate of condensate removal.

10.36. In order to save energy in the system in Problem 10.35, it is decided to replace the heating section by a mixing chamber in which a quantity of atmospheric air is mixed with the chilled air at 10°C. How much atmospheric air should be added, in kilogram per minute, to result in a humidity of 40%? What is the dry-bulb temperature of the mixture?

10.37. It is proposed to use evaporative cooling for a building where the outside air is at 34°C and 12% relative humidity. If the final relative humidity is to be 26%, what would be the dry bulb temperature of the air leaving the cooler? If the quantity of the air handled is 860 m³/min, determine the amount of water used by the cooler.

10.38. The conditioned space in Problem 10.37 is to be maintained at 22°C and 40% relative humidity. Calculate the cooling load that can be handled by the system.

10.39. The heat losses from a structure are 500 kW when the outside air is saturated at 2°C. The space is to be maintained at 20°C and 48% relative humidity. In order to provide heat to the structure to compensate the losses, the outside air is first heated and then saturated steam at 200 kPa is added to produce air at 48°C. Determine each of the following.

(a) Humidity ratio and the relative humidity of the air entering the space.

(b) Quantity of air to be heated in kilogram per minute.

(c) Amount of steam used in kilogram per minute.

10.40. Air at 20°C and 35% relative humidity is heated to 55°C. The heated air is used to produce a dry product containing no more than 4.5% moisture. The air leaves the dryer saturated at 42°C, while the product enters the dryer with 31% moisture. If the product leaves the dryer at a rate of 1 metric ton per hour, calculate the air-flow rate and the heat required in the preheater.

10.41. It is a common experience that foods dry out when exposed to very cold air, such as air at −10°C. Discuss this phenomenon and make appropriate calculations.

10.42. Consider a large vertical tank; water enters at the top at a rate of 4.2 kg/s at 40°C and leaves at the bottom at 18°C. Moist air of a dry-bulb temperature 23°C and 100 kPa and a wet-bulb temperature of 14°C is admitted at the bottom of the tank and bubbled through the water in the tank. The moist air leaves the tank at 32°C, 92 kPa, and 75% relative humidity. Determine the mass-flow rate of dry air and the rate at which water leaves the tank bottom. This method is proposed as a solution to the thermal pollution of rivers due to power plants.

10.43. An insulated vessel of volume 50 m³ contains liquid water occupying 1.2 m³. The space above the water is filled with moist air. The temperature and pressure of the tank are 30°C and 100 kPa. If 75 kg of steam at 900 kPa and a quality of 50% is admitted into the vessel, what are the final temperature and pressure in the vessel? This method is used for an emergency blowdown of a nuclear reactor.

10.44. A *cooling tower* is a device that uses evaporative cooling to cool warm water below the dry bulb temperature of the air. The cooled water is then returned to the condenser for condensing the turbine exhaust. Water to be cooled is introduced at the top of the tower. Atmospheric air, introduced at the bottom of the tower, comes in contact with water falling over baffles and leaves the tower as saturated air. In one cooling tower, water enters at 43°C and leaves at 18°C. The state of the entering air is 15°C dry-bulb temperature and 35% relative humidity. What should be the flow rate of atmospheric air to cool 50 kg/s of water? What is the loss of water?

10.45. Prove that the chemical potential of a perfect gas is given by

$$\mu = \bar{R}T \ln p + f(T)$$

where $f(T)$ is a function of temperature alone. What is the effect of p on μ when T is constant?

10.46. Show that

$$U = G - T\left(\frac{\partial G}{\partial T}\right)_p - p\left(\frac{\partial G}{\partial p}\right)_T$$

10.47. Calculate the phase compositions in the liquid-vapor system of methane and carbon monoxide at −100°C and 40 atms.

10.48. Standard air is cooled to 100 K at 1 atm pressure. Determine the liquid-vapor composition.

-11-

Chemical Reactions
and Equilibrium

11.1 INTRODUCTION

A study of chemical reactions generally belongs to chemistry so you may wonder about the reasons for including the topic in a text on thermodynamics. Chemical reactions can be *exothermic*, where heat is released, or *endothermic*, where heat energy must be supplied to bring about a reaction. The heat released in an exothermic reaction can be used, for example, to produce steam for a power plant. We are interested in the study of exothermic reactions involving those substances that can be used as fuels. In other words, we should like to inquire into chemical reactions to the extent that these reactions are a source of heat energy for heat engines.

A *chemical reaction* is sometimes defined as a rearrangement of the atoms of the reactants due to a redistribution of electrons to form products that are different from the reactants. The implication of this definition is that the number of atoms of each element participating in a reaction remains unchanged. This principle is used in writing the equation for a chemical reaction. In considering a reaction, there are three points of considerable interest: (1) the direction and the extent of the reaction; (2) the quantity of heat released or absorbed; and (3) the rate of reaction. Chemical thermodynamics deals with the first two points, which will be examined in this chapter. Chemical kinetics deals with the rate of reaction, which is beyond the scope of this text. We shall consider (1) the principle of conservation of mass, which is known as *stoichiometry* when applied to chemical reactions, (2) the principle of conservation of energy involving the application of the first law to reactants and products, (3) the second law of thermodynamics for a reaction, providing us with information on the direction of a reaction, and (4) conditions for the equilibrium of a system with particular reference to reacting systems. Since we are mainly interested in the chemical reactions of fuels resulting in a release of heat energy, we begin our discussion with fuels.

11.2 FUELS

Fuels exist in nature as a solid (for example, coal), as a liquid (for example, petroleum), and as a gas (for example, natural gas). The combustible elements in a fuel of any form are carbon, hydrogen, and traces of sulfur. The composition of solid and liquid fuels is usually given as *ultimate* analysis, that is, the percentages by mass of the constituents of the fuel are specified. On the other hand, the composition of a gaseous fuel is given in terms of a *volumetric* analysis.

Among the solid fuels, coal figures prominently. It has been used for more than a century in its natural form or as a raw material for synthetic fuels. Products of coal gasification, such as synthetic natural gas and liquid fuels and coke, are derived from coal. Although coal is an important source of fuel, there is an increasing interest in wood, peat, and solid waste. Wood contains, on an average, 50 percent carbon and 43 percent oxygen. For the same mass, wood releases about 50 percent of the heat released by coal on combustion. Peat and lignite have about 58 and 70 percent carbon, respectively. While the heat released by peat is slightly more than that from wood, the heat released by lignite is about 50 percent greater. All three of these fuels have about 5 to 6 percent hydrogen. Solid waste, which generally has a large moisture content, is about 50 percent as effective as wood as a heat source. At present, there is some interest in the so-called refuse-derived fuel (RDF).

The carbon and hydrogen in liquid and gaseous fuels are chemically bonded, while this is not generally the case with coal. The liquid and gaseous fuels are often referred to as *hydrocarbons*. The hydrocarbons, chemically bonded hydrogen and carbon, are classified as *paraffins*, $C_n H_{2n+2}$, *olefins* and *napthenes*, $C_n H_{2n}$, and *aromatics*, $C_n H_{2n-2}$. The paraffins and olefins have chain structures for the carbon atoms, while the other two have ring structures. The hydrocarbons are classified as *saturated* and *unsaturated*. In a saturated hydrocarbon, the maximum possible number of hydrogen atoms consistent with the structure of the hydrocarbon molecule is present. An unsaturated hydrocarbon molecule has the capability of absorbing at least one hydrogen atom to produce yet another hydrocarbon compound.

Liquid fuels are mixtures of several hydrocarbons. The different liquid fuels that are commercially available—for example, gasoline, kerosene, and diesel oil—are obtained by cracking crude oil and distilling it. Depending upon the temperature of distillation, different types of fuels are obtained. Figure 11.1 shows some typical distillation curves. A distillation curve for a fuel is a plot of the distillation temperature versus the fraction of the fuel recovered at that temperature. In accordance with the discussions in Section 10.7, the curve is not a horizontal straight line. As the crude oil is heated, the most volatile component evaporates first. The portion evaporated is removed and cooled. Commercial gasoline is actually a mixture of many hydrocarbons, all evaporating in a relatively narrow temperature range.

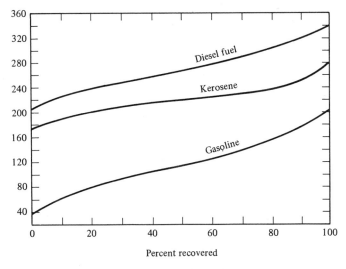

Gaseous fuels are found in natural gas wells and they are also produced in certain chemical processes and in coal gasification. The principle products of coal gasification attracting considerable interest are methane, alcohol, and certain liquid fuels. Table 11.1 lists the constituents of several gaseous fuels, both natural and derived. A major percentage of natural gas consists of methane.

TABLE 11.1 VOLUMETRIC ANALYSES OF SOME GASEOUS FUELS*

Constituent	Various Natural Gases				Producer Gas from Bituminous Coal	Carbureted Water Gas	Coke-Oven Gas
	A	B	C	D			
Methane	93.9	60.1	67.4	54.3	3.0	10.2	32.1
Ethane	3.6	14.8	16.8	16.3			
Propane	1.2	13.4	15.8	16.2			
Butanes plus[a]	1.3	4.2		7.4			
Ethene						6.1	3.5
Benzene						2.8	0.5
Hydrogen					14.0	40.5	46.5
Nitrogen		7.5		5.8	50.9	2.9	8.1
Oxygen					0.6	0.5	0.8
Carbon monoxide					27.0	34.0	6.3
Carbon dioxide					4.5	3.0	2.2

[a]This includes butane and all heavier hydrocarbons.

Most fuels contain several different hydrocarbons. In such cases, percentages by mass of carbon, hydrogen, and other elements are noted and the composition is expressed as that of a single complex hydrocarbon or organic compound.

11.3 THE COMBUSTION PROCESS

The combustion process involves oxidation of the carbon and hydrogen present in a fuel. Because combustion is a chemical reaction, the mass of each of the chemical elements present in the reactants remains constant. This makes it possible to write an equation of reaction. For example, consider the combustion of hydrogen, where hydrogen atoms combine with oxygen atoms to form water molecules.

$$\begin{array}{ccc} \textit{Reactants} & & \textit{Products} \\ 2H_2 + O_2 & \longrightarrow & 2\,H_2O \\ 4 \quad 32 & & 36 \end{array} \tag{11.1}$$

This equation indicates that 2 kg-mol of hydrogen react with 1 kg-mol of oxygen to produce 2 kg-mol of water. The equation thus gives molal proportions of the reactants and products. The phase of the water in the combustion product can be liquid or vapor. Since at constant pressure and temperature conditions, a mole of a gaseous substance is a measure of its volume, the equation tells us the volumetric proportions of the reactants and products. Equation 11.1 also tells us the proportions by mass, called *gravimetric proportions*, of the reactants and products. We note that the molar mass (molecular weight) of H_2 is 2 kg/kg-mol, that of O_2 is 32 kg/kg-mol, and that of H_2O is 18 kg/kg-mol. Then the equation indicates that 4 parts by mass of H_2 combine with 32 parts by mass of O_2 to produce 36 parts by mass of H_2O. These numbers appear below the respective symbols in Equation 11.1. We observe from these numbers that the total mass of the reactants equals the total mass of the products. Thus the equation contains enough information to determine the molal and gravimetric proportions of the reactants and products. Equations of this type are called *stoichiometric equations*. In a stoichiometric equation, the number of atoms of each element participating in the reaction is the same before and after the reaction takes place.

Another example is combustion of methane (CH_4). In this case both carbon and hydrogen are oxidized. The reaction eaquation is

$$\begin{array}{cccc} CH_4 + 2O_2 & \longrightarrow & CO_2 + 2H_2O \\ 16 \quad 64 & & 44 \quad 36 \end{array} \tag{11.1a}$$

In this instance, the products are carbon dioxide and water. We note that one atom of carbon in one molecule of CH_4, when oxidized, will produce one molecule of CO_2 requiring one molecule of O_2. Also, four atoms of hydrogen in one molecule of CH_4 will produce two molecules of H_2O requiring one molecule of O_2. Thus complete oxidation of 1 kg-mol of CH_4 requires 2 kg-mol of O_2, resulting in 1 kg-mol of CO_2 and 1 kg-mol of H_2O. This is expressed by Equation 11.1(a), which is also a stoichiometric equation. Observe that in an actual combustion process there can be a number of intermediate products of combustion. Here we are concerned only with the end products.

In most of the combustion processes encountered in practice, air—which contains oxygen—is used for oxidation. On a molal basis, air consists of 78 percent nitrogen, 21 percent oxygen, and 1 percent argon. The argon in air is usually neglected in combustion calculations, and it is assumed that the molal analysis of the combustion air is 79 percent nitrogen and 21 percent oxygen. This means that for every kilogram-mole of oxygen in air there are 79/21, or 3.76, kg-mol of nitrogen, and for every kilogram-mole of oxygen in air there are $1 + 3.76$, or 4.76, kg-mol of air.

As noted above, we consider combustion air to consist of nitrogen (79 percent) and oxygen (21 percent) on a molal basis. Since ideal gas behavior is assumed, in accordance with the principle of partial volumes (Section 10.2) these proportions also represent the volumetric proportions of nitrogen and oxygen, respectively. The nitrogen in such air is called *atmospheric nitrogen* and includes other constituents. The molar mass of the atmospheric nitrogen, after taking into consideration 1 percent argon in air, is taken to be 28.16 kg/kg-mol, compared with 28.013 kg/kg-mol for pure nitrogen. It is assumed that the nitrogen present does not participate in the chemical reactions of the combustion process. In reality, if the temperature of the end products is high, some of the nitrogen does become oxidized. Nitrogen oxides produced in this manner contribute to pollution of the atmosphere.

Let us rewrite Equation 11.1(a) to reflect the presence of nitrogen in air.

$$CH_4 + 2\,O_2 + 2(3.76)\,N_2 \longrightarrow CO_2 + 2\,H_2O + 7.52\,N_2 \qquad (11.2)$$

The quantity 3.76 appearing in the coefficient of N_2 in Equation 11.2 is the ratio of the number of kg-moles of the nitrogen in air to the number of kg-moles of the oxygen in air $(79 \div 21 = 3.76)$. The minimum quantity of air that supplies just enough oxygen for complete oxidation of the fuel is called the *theoretical air*. In practice, it is often found that the theoretical air is not adequate for complete combustion. Therefore the actual amount of air supplied is somewhat greater than the theoretical air. The amount of air actually supplied is expressed as a percentage of the theoretical air. Thus 200 percent theoretical air means that the actual air supplied is twice the theoretical air. This can also be expressed by the statement that the *excess air* supplied is 100 percent. Whenever excess air is supplied, some oxygen will appear in the products of combustion. With 100 percent excess air, the equation of reaction for methane is

$$CH_4 + 2(2.0)\,O_2 + 2(3.76)(2.0)\,N_2 \longrightarrow CO_2 + 2\,H_2O$$
$$+ 2\,O_2 + 15.04\,N_2 \qquad (11.2a)$$

Consider Equation 11.2(a). It indicates that the reactants consist of 1 kg-mol of CH_4, 4 kg-mol of O_2, and 4×3.76 kg-mol of N_2. The molar mass of methane in the equation is 16 kg/kg-mol. Therefore we consider 16 kg of methane, that is, the mass of 1 kg-mol of methane, for the reaction in Equation 11.2(a). Since 4 kg-mol of O_2 are to be supplied, and the molar mass of oxygen is 32 kg/kg-mol, we need 4×32, or 128, kg of O_2. In air, there are 4×3.76, or 15.04, kg-mol of N_2 associated with 4 kg-mol of O_2. The molar mass of nitrogen is 28.16 kg/kg-mol, and therefore the 4×3.76 kg-mol of N_2 in the reaction translate into $4 \times 3.76 \times 28.16$, or 423.5, kg of N_2. Consequently, for combusting 16 kg of CH_4 with 100 percent excess air, according to Equation 11.2(a) we need 128 kg of O_2 and 423.5 kg of N_2, that is $128 + 423.5$ kg of air.

Another way of referring to the quantity of air used in a combustion reaction is to specify the *air-fuel* ratio, AF. It is defined as the mass of the air used for the combustion of a unit mass of fuel, or

$$AF = \frac{\text{mass of the air supplied}}{\text{mass of the fuel burned}} \tag{11.3}$$

The air-fuel ratio for the reaction in Equation 11.2(a) is $(128 + 423.5)/16$, or 34.5.

Example 11.1

Butane, C_4H_{10}, is burned with 150% theoretical air. Determine the air-fuel ratio and the volumetric composition of the products of combustion.

Solution:

Given: Butane is burned with 50% more air than that required for the stoichiometric combustion.

Objective: The air-fuel ratio and the volumetric composition of the products of combustion are to be calculated.

Assumptions: The pressure and temperature of the air are the same as those of the fuel, so that the molal ratio of the air and the fuel also equals the volume ratio of the air and the fuel.

Relevant physics: The reaction equation will enable us to calculate the air-fuel ratio. The volumetric composition is computed using the method discussed in Chapter 10.

Analysis: The equation for stochimoetric reaction is

$$2\,C_4H_{10} + 13\,O_2 + 13(3.76)\,N_2 \longrightarrow 8\,CO_2 + 10\,H_2O + 13(3.76)\,N_2$$

With 150% theoretical air, the equation becomes

$$2\,C_4H_{10} + 19.5\,O_2 + 19.5(3.76)\,N_2 \longrightarrow$$
$$8\,CO_2 + 10\,H_2O + 6.5\,O_2 + 73.3\,N_2$$

The air-fuel ratio, AF, is (Equation 11.3)

$$AF = \frac{\text{mass of the air supplied}}{\text{mass of butane}}$$

$$= \frac{\overbrace{19.5 \times 32}^{O_2} + \overbrace{19.5 \times 3.76 \times 28.16}^{N_2}}{2(4 \times 12 + 10 \times 1)}$$

$$= 23.18$$

From the reaction equation, we see that $19.5(1 + 3.76)$ kg-mol of air react with 2 kg-mol of the fuel. Therefore the air-fuel ratio on a volume basis is $19.5 \times (1 + 3.76)/2$, or 46.43.

The number of moles of a constituent in the products of combustion is representative of its parts by volume in the products. Therefore the percentage by volume of a constituent equals the ratio of the number of moles of the constituent to the total number of moles of the product, multiplied by 100.

The calculations for the volumetric analysis of the combustion products are tabulated below.

Product	Number of moles in the products, n_i	Percentage by volume $= \dfrac{n_i}{\Sigma n_i} \times 100$
CO_2	8	8.18
H_2O	10	10.22
O_2	6.5	6.65
N_2	73.3	74.95
Sum	97.8	100.00

Example 11.2

A natural gas has the following composition by volume: 54.3% methane, 16.2% propane, 16.3% ethane, 7.4% butane, 2% O_2, and 3.8% N_2. This natural gas is burned with 50% excess air. Determine the air-fuel ratio, molal analysis of the products of combustion, and the dew point of the products if the pressure is 100 kPa.

Solution:

Given: Natural gas with the following composition is burned.

Component 1: methane,   54.3% by volume

Component 2: propane,   16.2% by volume

Component 3: ethane,   16.3% by volume

Component 4: butane,   7.4% by volume

Component 5: oxygen,   2.0% by volume

Component 6: nitrogen,   3.8% by volume

Excess air for combustion: 50% by mass

Pressure of the products: 100 kPa

Objective: We have to determine the air-fuel ratio, the molal analysis of the products of combustion, and the dew point of the products.

Assumptions: None.

Relevant physics: We need to know the chemical formula for each constituent so that the reaction equation for each constituent can be written. While writing such an equation, we write it for the mole fraction of the constituent, which equals the volume fraction in 1 kg-mol of the natural gas. Once this is done, the stoichiometric oxygen requirement for the combustion of 1 kg-mol of natural gas is simply the sum of the number of kilogram-moles of oxygen in the reaction equation of each constituent *minus* the oxygen content of the fuel. With this information, we can readily calculate the air-fuel ratio.

At constant pressure and temperature, the number of moles of a gaseous substance is a measure of its volume. Therefore the molal analysis of the products of combustion also represents the volumetric analysis of the products.

We recall from Chapter 10 that the dew point of a mixture of gases containing a condensable vapor is that temperature at which the vapor would condense if the mixture were cooled. The dew-point temperature is the saturation temperature corresponding to the partial pressure of the vapor.

Analysis: We shall first calculate the stoichiometric air requirement. For the various constituents present in one mole of natural gas, we can write the following.

Methane:

$$0.543 \text{ CH}_4 + 0.543(2 \text{ O}_2) \longrightarrow 0.543(\text{CO}_2 + 2 \text{ H}_2\text{O})$$

Propane:

$$0.162 \text{ C}_3\text{H}_8 + 0.162(5 \text{ O}_2) \longrightarrow 0.162(3 \text{ CO}_2 + 4 \text{ H}_2\text{O})$$

Ethane:

$$0.163 \text{ C}_2\text{H}_6 + 0.163(3.5 \text{ O}_2) \longrightarrow 0.163(2 \text{ CO}_2 + 3 \text{ H}_2\text{O})$$

Butane:

$$0.074 \text{ C}_4\text{H}_{10} + 0.074(6.5 \text{ O}_2) \longrightarrow 0.074(4 \text{ CO}_2 + 5 \text{ H}_2\text{O})$$

Thus the oxygen requirement for 1 kg-mol of the natural gas is

$$0.543 \times 2 + 0.162 \times 5 + 0.163 \times 3.5 + 0.074 \times 6.5 = 2.95 \text{ kg-mol}$$

But there are 0.02 kg-mol of O_2 present in the natural gas. Therefore the actual stoichiometric requirement of oxygen is $2.95 - 0.02 = 2.93$ kg-mol. The number of moles of air actually supplied is

$$2.93(1 + 3.76) \times 1.50, \text{ or } 20.91 \text{ kg-mol}$$

The mass of this air is

$$20.91 \times 28.97, \text{ or } 605.76 \text{ kg}$$

The mass of 1 kg-mol of the fuel is

$$0.543(12 + 4) + 0.162(12 \times 3 + 8) + 0.163(12 \times 2 + 6)$$
$$+ 0.074(12 \times 4 + 10) + 0.02(32) + 0.038(28.01) = 26.70$$

The air-fuel ratio by mass is then 605.76/26.7, or 22.68

In order to determine the dew point, we have to know the partial pressure, p_w, of the water vapor in the products of combustion. The partial pressure can be determined if we know the mole fraction of the water vapor.

The products of combustion consist of O_2, N_2, CO_2, and H_2O. In calculating the number of kilogram-moles of oxygen present in the products of combustion, we note that there are 20.91×0.21 kg-mol of O_2 in 20.91 kg-mol of the air supplied, of which 2.93 kg-mol of O_2 are used up in the combustion process. The number of moles of the products of combustion are obtained from the reaction equations as shown below:

$$O_2: \quad (20.91 \times 0.21 - 2.93) \qquad\qquad\qquad\qquad = 1.46$$

$$N_2: \quad (0.038 + 20.91 \times 0.79) \qquad\qquad\qquad\quad = 16.56$$

$$CO_2: \quad \overbrace{(0.543}^{\text{Methane}} + \overbrace{0.162 \times 3}^{\text{Propane}} + \overbrace{0.163 \times 2}^{\text{Ethane}} + \overbrace{0.074 \times 4)}^{\text{Butane}} = 1.65$$

$$H_2O: \quad \overbrace{(0.543 \times 2}^{\text{Methane}} + \overbrace{0.162 \times 4}^{\text{Propane}}$$

$$+ \overbrace{0.163 \times 3}^{\text{Ethane}} + \overbrace{0.074 \times 5)}^{\text{Butane}} \qquad\qquad = 2.59$$

$$\text{Total number of kilogram moles} = 22.26$$

Thus the molal composition of the products of combustion is as follows.

O_2	1.46 kg-mol, or 6.7% by volume
N_2	16.56 kg-mol, or 7.4% by volume
CO_2	1.65 kg-mol, or 7.4% by volume
H_2O	2.59 kg-mol, or 11.6% by volume
Total	22.26 kg-mol, or 100% by volume

The mole fraction of the water vapor, x_w, is

$$x_w = \frac{2.59}{22.26} = 0.1163$$

Then

$$p_w = x_w p = 0.1163 \times 100 = 11.63 \text{ kPa}$$

The saturation temperature corresponding to the above value of p_w is 48.5°C. The dew point is then 48.5°C.

11.3.1 Orsat Apparatus

It is possible to determine the quantity of air supplied for the combustion of a given fuel from the volumetric analysis of the products of combustion. The volumetric composition can be determined by using the Orsat apparatus, shown in Figure 11.2. It consists of three pipettes, each one capable of receiving a flue gas sample. The Orsat apparatus utilizes the concept of partial volume discussed in Section 10.2. A measured volume of the flue gases, usually 100 cc, is admitted into the apparatus burette by lowering the leveling cup. The gas is then forced into the first pipette, which contains potassium hydroxide. This chemical absorbs the carbon dioxide in the flue gas. The volume of the remaining gas is measured and the gas is forced into the second pipette by raising the leveling cup. This pipette contains pyrogallic acid, which absorbs the oxygen. After the oxygen is absorbed, the remaining gas is passed through the third pipette containing cuprous chloride, which absorbs carbon monoxide. Thus of the major components of the flue gas, namely, CO_2, CO, O_2, and N_2, the first three are absorbed. The remainder gas is, therefore, only nitrogen. The water

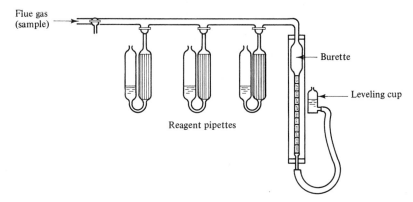

Figure 11.2 Orsat's apparatus.

in the flue gas usually condenses out, since the test is conducted at room temperature and the dew point of the flue gases is usually above room temperature. Therefore the Orsat analysis gives results on a dry basis.

Example 11.3

The Orsat analysis of the products of combustion of methane is as follows: 8% CO_2, 5.85% O_2, 0.6% CO, and 85.55% N_2. Determine the air-fuel ratio.

Solution:

Given: The Orsat analysis of the products of combustion of methane is

$$CO_2: \quad 8\%$$
$$O_2: \quad 5.85\%$$
$$CO: \quad 0.6\%$$
$$N_2: \quad 85.55\%$$

Objective: The air-fuel ratio is to be determined.

Assumptions: None.

Relevant physics: In problems of this type we do not know beforehand how much oxygen is used for the combustion process. Therefore an equation of reaction is written, where the number of kg-moles of O_2 and H_2O are the unknowns, and the number of kg-moles of CO_2, CO, and N_2 are known. Then a mass balance on each of the basic elements, namely, C, H_2, N_2, and O_2, gives a set of equations in the unknowns.

Analysis: We shall write the equation for reaction by assuming 100 kg-mol of dry products of combustion. Since we do not know the proportions of the different reactants, we shall use symbols $a, b, c, \ldots$ as the coefficients of the reactants.

$$a\,CH_4 + b\,O_2 + b(3.76)\,N_2 \longrightarrow$$
$$8\,CO_2 + 3\,O_2 + 0.6\,CO + c\,H_2O + 85.55\,N_2$$

We now perform a mass balance on each element, that is, equate the number of atoms of an element on the left side and the right side of the reaction equation.

480

$$3.76b = 85.55 \quad \text{or} \quad b = 22.75 \quad \text{(nitrogen)}$$
$$a = 8 + 0.6 \quad \text{or} \quad a = 8.6 \quad \text{(carbon)}$$
$$4a = 2c \quad \text{or} \quad c = 17.2 \quad \text{(hydrogen)}$$

The equation for the reaction is then

$$8.6\,CH_4 + 22.75\,O_2 + 85.55\,N_2 \longrightarrow$$
$$8\,CO_2 + 5.85\,O_2 + 0.6\,CO + 17.2\,H_2O + 85.55\,N_2$$

The actual air-fuel ratio on a mass basis is

$$\frac{22.75 \times 32 + 85.55 \times 28.16}{8.6(12 + 4)} = 22.8$$

Comments: It is to be noted that the Orsat's apparatus does not yield the percentages of moisture, argon, and pollutants (such as oxides of nitrogen) present in combustion gases.

11.4 CHEMICAL REACTION AND THE FIRST LAW

Let us assume that a chemical reaction takes place in a closed rigid container of fixed volume. Let the subscripts *react* and *prod* denote the reactants and the products, respectively. The first law for a closed system is written as

$$Q - W = \Delta U$$

Since there is no change in the volume of our system, $W = 0$ and the equation becomes

$$Q = \Delta U$$

The change of internal energy due to the chemical reaction is given by

$$\Delta U = \sum_i (n_i \bar{u}_i)_{\text{prod}} - \sum_j (n_j \bar{u}_j)_{\text{react}} = U_{\text{prod}} - U_{\text{react}} \qquad (11.4)$$

when n represents the number of kg-moles and $\bar{u}$ is the specific internal energy in kilojoules per kg-mole. Since there are at least two reactants and often there is more than one product of reaction, we have used summation signs in Equation 11.4.

We can measure the quantity Q by using a constant volume bomb calorimeter* and thus find the difference in the internal energies of the products and the reactants. Since the species and the composition of the species in the system before and after the reaction are not necessarily the same, some care must be exercised in determining the internal energy of the reactants and the products of reaction.

Let us rewrite Equation 11.4 to read

$$\Delta U = (U_{\text{prod}} - U_{\text{prod},b}) + (U_{\text{prod},b} - U_{\text{react},a}) + (U_{\text{react},a} - U_{\text{react}}) \qquad (11.4a)$$

The various terms in this equation are depicted in Figure 11.3. The quantity in the first pair of parentheses in Equation 11.4(a) is the increase in the internal energy of the products as the temperature is raised from T_0 to T_2, T_2 being the temperature

*H. A. Skinner, *Experimental Thermochemistry*, vol. 1 and 2 (New York: Interscience Publishers, New York, 1962).

at the end of the reaction. This increase can be calculated once the number of kg-moles and the molal specific heat of each of the products are known. The quantity in the third pair of parentheses is the decrease in the internal energy of the reactants as the temperature is lowered from T_1 to T_0, T_1 being the temperature of the reactants just before the reaction. This decrease can also be calculated. The quantity in the second pair of parentheses in Equation 11.4(a) is the change in the internal energy when the temperatures before and after the reaction are identical. This change can be determined only by experiment.

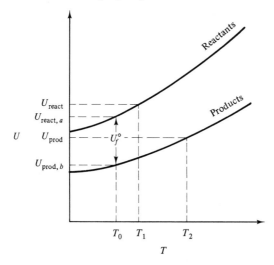

Figure 11.3 Internal energies of reactants and products of a reaction.

If the temperature and pressure before and after a reaction are T_0 and p_0 and if the reactants are *basic elements*, then the internal energy of the product formed equals the heat transferred during the reaction. This quantity is called the *internal energy of formation*. It is designated by $\bar{u}_f^0$ on a molal basis. The values of T_0 and p_0 used in the determination of u_f^0 are 25°C and 100 kPa. Thus if 1 kg-mol of carbon and 1 kg-mol of oxygen react and if the initial and final states are both at 25°C and 100 kPa, then the internal energy of formation of the 1 kg-mol of CO_2 produced equals the quantity of heat supplied during the reaction. This quantity is also the internal energy of 1 kg-mol of CO_2 at 25°C and 100 kPa.

The molal internal energy $\bar{u}$ of a substance at temperature T and pressure p is the sum of the internal energy of formation, $\bar{u}_f^0$, at 25°C and 100 kPa and the change in the internal energy, $\Delta\bar{u}$, as the state of the substance is changed from 25°C and 100 kPa to T and p. That is,

$$\bar{u} = \bar{u}_f^0 + \Delta\bar{u} \tag{11.4b}$$

As noted before, $\bar{u}_f^0$ is zero for basic elements, while it is nonzero for compounds.

Referring to Figure 11.3, the internal energy-temperature curves for the reactants and the products must be displaced vertically in such a manner that, at 25°C, the quantity $U_{prod,b} - U_{react,a}$ exactly equals the internal energy of formation.

Now, let us consider a steady-flow process where the reactants enter a reaction vessel such as a combustion chamber and the products leave the chamber. We assume that the changes in the kinetic and potential energies are negligible. The first law for the complete reaction becomes

$$\dot{Q} - \dot{W}_s = (\dot{m}h)_{\text{prod}} - (\dot{m}h)_{\text{react}}$$

Since there is no shaft work in a combustion chamber, $\dot{W}_s = 0$.

We are interested in obtaining an expression for the quantity of heat transferred to the control volume when *a given mass* of reactants enters the control volume, the reaction is completed within the control volume, and the products whose total mass equals that of the reactants leave the control volume. In other words, we wish to determine Q for a complete reaction of a *given mass* of reactants occurring in a flow process. Therefore we drop the dots in $\dot{Q}$ and $\dot{m}$ in the last equation and write

$$Q = (mh)_{\text{prod}} - (mh)_{\text{react}}$$

We can write $n\bar{h}$ for mh, where $\bar{h}$ is the specific molal enthalpy and n is the number of kilogram-moles. Then the first law can be rewritten as

$$Q = \sum_i (n_i\bar{h}_i)_{\text{prod}} - \sum_j (n_j\bar{h}_j)_{\text{react}} \tag{11.5}$$

where the two summations encompass all the products or all the reactants, respectively. It is assumed in this equation that the reactants enter and the products leave the control volume.

Let us suppose that the reactants are basic elements, such as carbon and oxygen or hydrogen and oxygen. Earlier, we had arbitrarily assigned a zero value of internal energy for water at $0.01°C$ (triple point). In an analogous manner, we assign a zero value of enthalpy to *all the elements* at standard temperature and pressure, namely $25°C$ and 100 kPa. This assumes that an element is in its natural phase at $25°C$ and 100 kPa. Once this is done, the enthalpy of an element at any other temperature and pressure can be determined. A zero value cannot be *assigned* to the enthalpy of a product, such as carbon dioxide or water, at an arbitrary temperature and pressure. This issue did not arise when we applied the first law to nonreacting systems. Since the heat transferred, Q, can now be measured, we are able to determine the enthalpy of the products. For example, if hydrogen and oxygen react to produce water and if the temperature and pressure of the reactants and the product are $25°C$ and 100 kPa, the heat transferred is found to be $-285,838$ kJ/kg-mol of liquid water formed. The negative sign tells us that heat is rejected during the reaction. The quantity of heat transferred is known as the *enthalpy of formation* and is designated by $\bar{h}_f^0$. To be precise, the enthalpy of formation of a chemical compound is the enthalpy of the compound produced in a reaction where the temperatures just before and after the reaction are identical, the pressures just before and after the reaction are identical, and the compound is the only product and the reactants are the basic chemical elements in their state of zero enthalpy. The superscript 0 in $\bar{h}_f^0$ denotes standard conditions of $25°C$ and 100 kPa. All hydrocarbons are combinations of hydrogen and carbon elements. These combinations are brought about in chemical reactions.

Consequently all hydrocarbons have enthalpies of formation. Table 11.2 lists the enthalpies of formation for a variety of compounds. The enthalpy of formation for the basic elements is zero, while the enthaply of a compound at 25°C and 100 kPa equals its enthalpy of formation.

TABLE 11.2 ENTHALPY OF FORMATION, GIBBS FUNCTION OF FORMATION
AND ABSOLUTE ENTROPY OF VARIOUS SUBSTANCES AT 25°C, 100 kPa PRESSURE*

Substance	Formula	M kg/kg-mol	State	$\bar{h}_f^0$ kJ/kg-mol	$\bar{g}_f^0$ kJ/kg-mol	$\bar{s}^0$ kJ/kg-mol K
Carbon monoxide[a]	CO	28.011	gas	−110,529	−137,150	197.653
Carbon dioxide[a]	CO_2	44.011	gas	−393,522	−394,374	213.795
Water[a,b]	H_2O	18.015	gas	−241,827	−228,583	188.833
Water[b]	H_2O	18.015	liquid	−285,838	−237,178	70.049
Methane[a]	CH_4	16.043	gas	−74,873	−50,751	186.256
Acetylene[a]	C_2H_2	26.038	gas	+226,731	+209,234	200.958
Ethene[a]	C_2H_4	28.054	gas	+52,283	+68,207	219.548
Ethane[c]	C_2H_6	30.070	gas	−84,667	−32,777	229.602
Propane[c]	C_3H_8	44.097	gas	−103,847	−23,316	270.019
Butane[c]	C_4H_{10}	58.124	gas	−126,148	−16,914	310.227
Octane[c]	C_8H_{18}	114.23	gas	−208,447	+16,859	466.835
Octane[c]	C_8H_{18}	114.23	liquid	−249,952	+6,940	360.896
Carbon[a] (graphite)	C	12.011	solid	0	0	5.795

[a]From JANAF *Thermochemical Data*, The Dow Chemical Company, Thermal Laboratory, Midland, Mich.
[b]From *Circular 500*, National Bureau of Standards.
[c]From F. D. Rossini et al., *API Research Project 44*.

*(G. J. Van Wylen and R. E. Sonntag, *Fundamentals of Classical Thermodynamics*, SI Version. New York: John Wiley & Sons, Inc. © 1976. Used with permission.)

The enthalpy of a compound such as CO_2 or H_2O at temperature T and pressure p is then determined from

$$\bar{h} = \bar{h}_f^0 + \Delta\bar{h} \qquad (11.6)$$

The quantity $\Delta\bar{h}$ is the change in the enthalpy of a compound as its temperature and pressure are changed from 298 K and 100 kPa to T and p. The values of $\Delta\bar{h}$ for a number of gases are given in Appendix I1, where 298 K is used as the reference temperature at which $\Delta\bar{h}$ is taken as zero.

It was stated that the enthalpy of formation can be determined from the heat transfer measurements during a reaction. In practice, the enthalpy of formation is determined by the application of statistical thermodynamics. This application involves use of spectroscopic data. In some instances, an element or compound can exist in more than one state at 25°C and 100 kPa. For example, carbon can be in the form of graphite, diamond, or plain powder. It is important that the state of a substance be clearly specified while tabulating the value of $\bar{h}_f^0$.

11.4.1 Enthalpy and Internal Energy of Reaction

When a reaction goes to completion, the resulting products have an enthalpy value which is different from that for the reactants. The ratio of the difference between these two values and the number of moles of the combustible reactant, n_{CR}, is called the *enthalpy of reaction*, $\bar{h}_R$.

$$\bar{h}_R = \frac{[(mh)_{\text{prod}} - (mh)_{\text{react}}]}{n_{CR}}$$

or

$$\bar{h}_R = \frac{[\sum_i (n_i \bar{h}_i)_{\text{prod}} - \sum_j (n_j \bar{h}_j)_{\text{react}}]}{n_{CR}} \tag{11.7}$$

where $\bar{h}$ is given by Equation 11.6. The enthalpy of reaction is usually expressed in kilojoules per kilogram or kilojoules per kilogram-mole. Also, the values of enthalpies of reaction are generally tabulated for reaction temperature and pressure of 25°C and 100 kPa, respectively. Table 11.3 lists the values of h_R for several substances for complete combustion with 100 percent theoretical air. Note that as the temperature of the products increases, the enthalpy of reaction decreases.

The definition of the internal energy of reaction $\bar{u}_R$ is similar to that of $\bar{h}_R$.

$$\bar{u}_R = \frac{[\sum_i (n_i \bar{u}_i)_{\text{prod}} - \sum_j (n_j \bar{u}_j)_{\text{react}}]}{n_{CR}} \tag{11.7a}$$

where

$$\bar{u} = (\bar{u}_f^0 + \Delta\bar{u}) = (\bar{h}_f^0 + \Delta\bar{h} - p\bar{v})$$

The heating value of a fuel is closely related to the enthalpy of formation. The *heating value* of a fuel is generally defined as the quantity of heat released in a reaction at 25°C and 100 kPa due to combustion of a unit mass of fuel. It is expressed in kilojoules per kilogram of fuel, kilojoules per kilogram-mole of fuel or sometimes as kilojoules per cubic meter of fuel. From this definition it is evident that the heating value of a fuel is the negative enthalpy of reaction.

Water is present in most products of combustion. Depending upon the temperature of the products of combustion, the water can be in the form of liquid or vapor. The difference in the enthalpies of water in liquid and vapor states is quite significant. This differential enthalpy is not available for use if the water formed in the reaction escapes as vapor. Therefore it is customary to distinguish between the *lower heating value*, LHV, and *higher heating value*, HHV. The two are related by

$$\text{HHV} = \text{LHV} + mh_{fg} \tag{11.8}$$

where m is the mass of the water vapor formed on combustion of a unit mass of the fuel.

The higher heating value gives the maximum heat obtainable from the combustion of a fuel, but it is seldom realized in practice. This is because, from practical

TABLE 11.3 ENTHALPY OF REACTION, h_R, OF SOME HYDROCARBONS AT 25°C AND 100 kPa*

Hyrocarbon	Formula	Liquid H₂O in products (Negative of higher heating value)		Vapor H₂O in products (Negative of lower heating value)	
		Liquid hydro-carbon, kJ/kg fuel	Gaseous hydro-carbon, kJ/kg fuel	Liquid hydro-carbon, kJ/kg fuel	Gaseous hydro-carbon, kJ/kg fuel
Paraffin family					
Methane	CH_4		−55,496		−50,010
Ethane	C_2H_6		−51,875		−47,484
Propane	C_3H_8	−49,975	−50,345	−45,983	−46,353
Butane	C_4H_{10}	−49,130	−49,500	−45,344	−45,714
Pentane	C_5H_{12}	−48,643	−49,011	−44,983	−45,351
Hexane	C_6H_{14}	−48,308	−48,676	−44,733	−45,101
Heptane	C_7H_{16}	−48,071	−48,436	−44,557	−44,922
Octane	C_8H_{18}	−47,893	−48,256	−44,425	−44,788
Decane	$C_{10}H_{22}$	−47,641	−48,000	−44,239	−44,598
Dodecane	$C_{12}H_{26}$	−47,470	−47,828	−44,109	−44,467
Olefin family					
Ethene	C_2H_4		−50,296		−47,158
Propene	C_3H_6		−48,917		−45,780
Butene	C_4H_8		−48,453		−45,316
Pentene	C_5H_{10}		−48,134		−44,996
Hexene	C_6H_{12}		−47,937		−44,800
Heptene	C_7H_{14}		−47,800		−44,662
Octene	C_8H_{16}		−47,693		−44,556
Nonene	C_9H_{18}		−47,612		−44,475
Decene	$C_{10}H_{20}$		−47,547		−44,410
Alkylbenzene family					
Benzene	C_6H_6	−41,831	−42,266	−40,141	−40,576
Methylbenzene	C_7H_8	−42,437	−42,847	−40,527	−40,937
Ethylbenzene	C_8H_{10}	−42,997	−43,395	−40,924	−41,322
Propylbenzene	C_9H_{12}	−43,416	−43,800	−41,219	−41,603
Butylbenzene	$C_{10}H_{14}$	−43,748	−44,123	−41,453	−41,828

*(G. J. Van Wylen and R. E. Sonntag, *Fundamentals of Classical Thermodynamics, SI Version*. New York: John Wiley & Sons, Inc. © 1976. Used with permission.)

considerations, the temperature of the flue gases must be above the dew point, that is, the water must leave as vapor.

Example 11.4

Calculate the enthalpy of CO_2 at 500 K and 3000 kPa on a kg-mol basis.

Solution:

Given: One kilogram-mole of CO_2 at 500 K and 3000 kPa is given.

Objective: The specific enthalpy of the CO_2 is to be determined.

Assumptions: The effect of the 3000 kPa pressure on the enthalpy value of CO_2 is negligible and the enthalpy is essentially dependent on temperature.

Relevant physics: We note from Equation 11.6 that the enthalpy of a gas like CO_2 consists of two parts: the enthalpy of formation, $\bar{h}_f^0$, and the temperature-dependent part, $\Delta\bar{h}$. The quantity $\Delta\bar{h}$ can be obtained in two ways, (1) from the enthalpy table for CO_2 in Appendix I1, and (2) from the specific heat data in Appendix H.

Analysis: From Equation 11.6 we have

$$\bar{h} = \bar{h}_f^0 + \Delta\bar{h}$$

From Table 11.2, for CO_2,

$$\bar{h}_f^0 = -393{,}522 \text{ kJ/kg-mol}$$

1. From Appendix I1, corresponding to 500 K and 3000 kPa,

$$\Delta\bar{h} = 8314 \text{ kJ/kg-mol}$$

The required $\bar{h}$ is then

$$\bar{h}_{T,p} = -393{,}522 + 8314$$
$$= -385{,}208 \text{ kJ/kg-mol}$$

2. The specific heat at constant pressure for CO_2 is given by Appendix H.

$$\bar{c}_p = -3.7357 + 30.529\theta^{0.5} - 4.1034\theta + 0.024198\theta^2$$

where $\bar{c}_p$ is in kJ/kg-mol and $\theta = \dfrac{T}{100}$ with T in degrees Kelvin. Then

$$\Delta\bar{h} = \int_{298.15}^{500} \bar{c}_p(T)\, dT = 100 \int_{2.9815}^{5.00} \bar{c}_p(\theta)\, d\theta$$

$$= 100\left[-3.7357\theta + \frac{30.529}{1.5}\theta^{1.5} - \frac{4.1034}{2}\theta^2 + \frac{0.024198}{3}\theta^3 \right]_{2.9815}^{5.00}$$

$$= 8297 \text{ kJ/kg-mol}$$

Therefore

$$\bar{h}_{T,p} = -393{,}522 + 8297$$
$$= -385{,}225 \text{ kJ/kg-mol}$$

We note that the difference in the enthalpy values calculated by the two methods is less than 0.005 %.

Example 11.5

Determine the internal energy of reaction $\bar{u}_f^0$ when hydrogen is completely oxidized into H_2O at 25°C and 100 kPa.

Solution:

Given: Hydrogen is completely oxidized at 25°C and 100 kPa.

Objective: The internal energy of the reaction, $\bar{u}_f^0$, is to be determined.

Assumptions: The ideal gas law holds.

Relevant physics: Inasmuch as the reaction occurs at the standard temperature and pressure (25°C and 100 kPa), the change in enthalpy due to the reaction is exactly equal to the enthalpy of formation, $\bar{h}_f^0$, of H_2O. This quantity can be read from Table 11.2 and also expressed in terms of the internal energy of formation and the product pV.

Analysis: The equation for the reaction is

$$2\,H_2 + O_2 \longrightarrow 2\,H_2O$$

From Table 11.2, the enthalpy of formation of H_2O is $-241,827$ kJ/kg-mol. Since there are 2 kg-mol of water in the reaction equation, we can write the following equation for the enthalpy change in the reaction.

$$2\bar{h}_f^0 = H_{H_2O} - (H_{H_2} + H_{O_2})$$

$$= (U_{H_2O} - U_{H_2} - U_{O_2}) + [(pV)_{H_2O} - (pV)_{H_2} - (pV)_{O_2}]$$

or

$$2\bar{h}_f^0 = U_{f,H_2O}^0 + (n_{H_2O} - n_{H_2} - n_{O_2})\bar{R}T_0$$

where the ideal gas law has been used to replace pV by $n\bar{R}T$. Rearrangement gives

$$U_{f,H_2O}^0 = 2\bar{h}_f^0 - \bar{R}T_0(n_{H_2O} - n_{H_2} - n_{O_2})$$

$$= 2 \times (-241,827) - 8.314 \times 298.15(2 - 2 - 1)$$

$$= -481,175 \text{ kJ}$$

or

$$\bar{u}_f^0 = -240,588 \text{ kJ/kg-mol}$$

Example 11.6

Gaseous octane, C_8H_{18}, is burned with 50% excess air. The combustion air and octane enter the combustion chamber at 400 K and the products of combustion leave at 1200 K. Neglecting changes in the potential and kinetic energies, determine the heat transferred per kilogram of octane.

Solution:

Given: Gaseous octane is combusted.

Excess air:	50%
Inlet temperature of the reactants:	400 K
Outlet temperature of the products:	1200 K

Objective: We have to determine the heat transferred, q, per kilogram of octane.

Diagram:

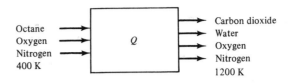

Octane → | | → Carbon dioxide
Oxygen → | Q | → Water
Nitrogen → | | → Oxygen
400 K | | → Nitrogen
1200 K

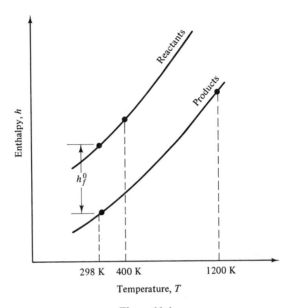

Figure 11.4

Assumptions: The changes in the potential and kinetic energies are negligible.
Relevant physics: The quantity of heat transferred can be calculated by applying the
first law to the steady-flow reaction and by employing Equation 11.5.

$$Q = H_{prod} - H_{react}$$

where

$$H_{prod} = \sum_i (n_i \bar{h}_i)_{prod}$$

and

$$H_{react} = \sum_j (n_j \bar{h}_j)_{react}$$

From the reaction equation for octane with 50% excess air,

$$C_8H_{18} + 1.5(12.5)\,O_2 + 1.5(12.5)(3.76)\,N_2 \longrightarrow$$
$$8\,CO_2 + 9\,H_2O + 6.25\,O_2 + 70.54\,N_2$$

we note that the reactants are C_8H_{18}, O_2, and N_2 and the products are CO_2,
H_2O, O_2, and N_2.

In determining the enthalpies of the reactants and the products, the enthalpies of formation for C_8H_{18}, CO_2, and H_2O have to be accounted in conjunction with Equation 11.6. Also, we note that the temperature of the reactants is 400 K, while that of the products is 1200 K.

Analysis: We first proceed to calculate the enthalpy of the reactants. To determine the enthalpy of octane, we integrate $c_p\, dT$ in the desired temperature range by assuming c_p to be constant. The value of $\bar{c}_p$ is obtained from Appendix A, which—when multiplied by the molar mass of octane, 114 kg/kg-mol—gives c_p. The enthalpy values for O_2 and N_2 are obtained from Appendix I1.

Octane:
$$H_1 = n_1\bar{h}_1 = 1\int_{298}^{400} \bar{c}_p\, dT + \bar{h}_f^0$$
$$= (1.7113 \times 114)(400 - 298) - 208,447$$
$$= -188,548 \text{ kJ}$$

Oxygen:
$$H_2 = n_2(\bar{h}_2 \text{ at } 400 \text{ K})$$
$$= (1.5 \times 12.5)(3029)$$
$$= 56,793 \text{ kJ}$$

Nitrogen:
$$H_3 = n_3(\bar{h}_3 \text{ at } 400 \text{ K})$$
$$= (1.5 \times 12.5 \times 3.76)(2971)$$
$$= 209,455 \text{ kJ}$$

$$H_{\text{react}} = H_1 + H_2 + H_3$$
$$= -188,548 + 56,793 + 209,455$$
$$= 77,700 \text{ kJ}$$

The enthalpies of the products are obtained in the following manner:

Carbon Dioxide:
$$H_4 = n_4[(\bar{h}_f^0 + \Delta h)_4 \text{ at } 1200 \text{ K}]$$
$$= 8(-393,522 + 44,484)$$
$$= -2,792,304 \text{ kJ}$$

Water:
$$H_5 = n_5[(\bar{h}_f^0 + \Delta h)_5 \text{ at } 1200 \text{ K}]$$
$$= 9(-241,827 + 34,476)$$
$$= -1,866,159 \text{ kJ}$$

Oxygen:
$$H_6 = n_6(\bar{h}_6 \text{ at } 1200 \text{ K})$$
$$= 6.25 \times 29,765$$
$$= 186,031 \text{ kJ}$$

Nitrogen:
$$H_7 = n_7(\bar{h}_7 \text{ at } 1200 \text{ K})$$
$$= 70.54 \times 28,108$$
$$= 1,982,738 \text{ kJ}$$

$$H_{\text{prod}} = H_4 + H_5 + H_6 + H_7$$
$$= -2,792,304 - 1,866,159 + 186,031 + 1,982,738$$
$$= -2,489,694 \text{ kJ}$$

The heat transferred, Q, is given by

$$Q = H_{prod} - H_{react}$$
$$= -2,489,694 - 77,700$$
$$= -2,567,394 \text{ kJ per kg-mole of octane}$$

Since these computations were based on 1 kg-mol of octane, the calculated value of Q is the heat transferred in kiloJoules per kg-mole of octane. The negative sign tells us that heat is rejected. The heat released per kilogram of octane is Q/M, or 22,522 kJ/kg.

Comments: In this example, the enthalpies of the products and the reactants are calculated with reference to 298 K. The enthalpies of formation at 298 K assist us in calculating the change in ehtalpy from reactants to products.

Example 11.7

Determine the higher and lower heating values of methane in kilogram Joules per kilogram.

Solution:

Assumptions: The combustion of metane occurs at the standard temperature and temperature.

Relevant physics: The equation of the reaction is

$$CH_4 + 2O_2 = CO_2 + 2H_2O$$

If the water in the products is in a vapor phase, the quantity $-(H_{prod} - H_{react})$ represents the lower heating value; if the water is in a liquid phase, the quantity represents the higher heating value.

Analysis: We first calculate the higher heating value, HHV. The following enthalpy values are obtained from Table 11.2.

Methane: $H_1 = n_1 \bar{h}^0_{f1} = 1 \times (-74,873) = -74,873 \text{ kJ}$

Oxygen: $H_2 = n_2 \bar{h}^0_{f2} = 2 \times 0 = 0$

Carbon Dioxide: $H_3 = n_3 \bar{h}^0_{f3} = 1 \times (-395,522) = -393,522 \text{ kJ}$

Water (liquid): $H_4 = n_4 \bar{h}^0_{f4} = 2 \times (-285,838) = -571,676 \text{ kJ}$

Then

$$HHV = -[(H_3 + H_4) - (H_1 + H_2)]$$
$$= -[(-393,522) + (-571,676)] - [(-74,873)]$$
$$= -890,325 \text{ kJ/kg-mole}$$
$$= -55,496 \text{ kJ/kg}$$

which is exactly the same value as in Table 11-3.

The lower heating value is calculated by considering the water in the products to be in a vapor state. The enthalpy of water vapor, $\bar{h}_4$, is $-241,827$ kJ/kg-mol. Then the lower heating value is

$$LHV = -[(H_3 + H_4) - (H_1 + H_2)]$$
$$= -[(-393,522) + 2 \times (-241,827)] - [(-74,873)]$$
$$= 802,303 \text{ kJ/kg-mol}$$
$$= 50,010 \text{ kJ/kg}$$

Comments: Note that the heating value depends on the reaction temperature. If we were to calculate the heating value of the methane in this problem at a reaction temperature of 200°C, the result would be quite different from the one obtained here.

11.4.2 Adiabatic Flame Temperature

In the combustion reactions considered so far, we referred to the heat transferred during a reaction. If we consider a reaction to occur under conditions where no heat transfer is permitted, then all the energy released during the reaction will cause a significant increase in the enthalpies of the end products. This, in turn, causes an increase in the temperature of the end products. When a combustion reaction proceeds adiabatically, the temperature attained by the end products is called *adiabatic flame temperature*. With the assumptions of no work and no heat transfer, this is the maximum temperature that can be reached for a given reaction. Since the air supplied for combustion can be in excess of the stoichiometric quantity, which is the minimum to ensure combustion, it follows that the adiabatic flame temperature is the highest when no excess air for combustion is supplied.

A knowledge of the adiabatic flame temperature for a given combustion process is essential since the internal structure of the combustion chamber has to withstand the high temperatures produced. By using excess air it is possible to control the adiabatic flame temperature.

The calculation for the adiabatic flame temperature involves a trial-and-error procedure. We assume a value of the adiabatic flame temperature, determine the enthalpies of the end products corresponding to the assumed temperature, and perform an energy balance according to the first law. If the energy equation is not satisfied, we repeat the process by assuming another value of the temperature.

Example 11.8

Gaseous propane is burned with 400% excess air. Determine the adiabatic flame temperature. The temperature of the propane and the supply air is 25°C.

Solution:

Given: Gaseous propane is burned adiabatically.

$$\text{Excess air:} \qquad 400\%$$

$$\text{Initial temperature:} \quad 25°C$$

Objective: We are to determine the adiabatic flame temperature.

Relevant physics: Since no heat transfer occurs, Equation 11.5 becomes

$$H_{prod} = H_{react}$$

or

$$\sum_i (n_i \bar{h}_i)_{prod} = \sum_j (n_j \bar{h}_j)_{react}$$

We have to determine the temperature of the products that satisfies the last equation.

Analysis: The equation for the reaction is

$$C_3H_8 + 5(5\,O_2 + 5 \times 3.76\,N_2) \longrightarrow 3\,CO_2 + 4\,H_2O + 20.0\,O_2 + 94\,N_2$$

Since the enthalpies of oxygen and nitrogen are zero at 25°C, the enthalpy of the reactants is equal to $\bar{h}_f^0$ for C_3H_8, that is, $-103,847$ kJ/kg mol. Since the process is adiabatic, this quantity also equals the enthalpy of the products, H_{prod}.

$$H_{prod} = -103,847 \text{ kJ}$$

We can write the following equation for H_{prod}.

$$H_{prod} = 3 \times \bar{h}_{f,CO_2}^0 + 4\bar{h}_{f,H_2O}^0 + 3 \int_{298}^{T^*} \bar{c}_{p,CO_2} \, dT + 4 \int_{298}^{T^*} \bar{c}_{p,H_2O} \, dT$$

$$+ 20 \int_{298}^{T^*} \bar{c}_{p,O_2} \, dT + 94 \int_{298}^{T^*} \bar{c}_{p,N_2} \, dT$$

where T^* is the unknown adiabatic flame temperature. As a first approximation, we assume constant specific heats to obtain

$$H_{prod} = 3 \times (-393,522) + 4 \times (-241,827) + 3 \times 44.01 \times 0.8418$$
$$\times (T^* - 298) + 4 \times 18.015 \times 1.8723 \times (T^* - 298) + 20 \times 32$$
$$\times 0.9216 \times (T^* - 298) + 94 \times 28.16 \times 1.0416(T^* - 298)$$

Simplification gives

$$H_{prod} = -3,214,312 + 3578T^* = -103,847$$

or

$$T^* = 869 \text{ K}$$

We can assume 900 K as the adiabatic flame temperature and read the enthalpy values from Appendix Il for CO_2, H_2O, O_2, and N_2. Then, at 900 K,

$$H_{prod} = 3 \times (-393,522) + 4(-241,827) + 3(28,041)$$
$$+ 4(21,924) + 20(19,246) + 94(18,221)$$
$$= -2,147,874 + 2,269,513$$
$$= 121,639 \text{ kJ}$$

which is greater than H_{react}. A value of $T^* = 890$ K gives H_{prod} that is close enough to H_{react}. Thus the adiabatic flame temperature of combustion of propane with 400% excess air is 890 K.

11.4.3 Explosion Pressure

If the rate of reaction is much faster than the rate at which the volume increases, then we essentially have a constant volume reaction. Under these conditions, there is not enough time for heat transfer to occur and the process is close to adiabatic. In such a situation the pressure can rise very rapidly. The peak pressure generated in this manner is called *explosion pressure*. The explosion pressure can be determined by applying the first law to a constant volume reaction with no heat transfer.

$$\cancel{Q}^{\,0} - \cancel{W}^{\,0} = \Delta U$$

or

$$U_{prod} - U_{react} = 0$$

or

$$\sum_i \int_{T_0}^{T_e} (n_i \bar{c}_{v,i}\, dT)_{\text{prod}} + \sum_i (n_i \bar{u}_{f,i}^0)_{\text{prod}} - \sum_j (n_j \bar{u}_{f,j}^0)_{\text{react}} = 0 \qquad (11.9)$$

where the summation is carried out for all the products, and T_0 and T_e are the reference temperature (25°C) and the flame temperature corresponding to the explosion pressure, respectively. In Equation 11.9 it is assumed that the reactants are at the reference temperature. Equation 11.9 then lets us calculate T_e, which is different from the adiabatic flame temperature, T^*. The explosion pressure p_e is then determined by assuming an ideal gas behavior.

Example 11.9

Liquid octane at 25°C is burned with 200% theoretical air at constant volume. The flame temperature is calculated to be 2900 K. Determine the explosion pressure.

Solution:

Given: Liquid octane undergoes constant volume, adiabatic combustion.

Initial temperature: 25°C

Flame temperature, T_e: 2900 K

Objective: The explosion pressure is to be determined.

Assumptions:

1. The ideal gas law holds.
2. The initial pressure is 100 kPa.

Relevant physics: The initial molal volume can be determined since the initial temperature and pressure are known. The initial number of moles of the reactants can be ascertained from the reaction equation. This, then, permits a computation of the volume of the reactants, which equals the volume of the products for the given process. Since T_e is known, an application of the ideal gas law yields the explosion pressure.

Analysis: The equation for the reaction is

$$\text{C}_8\text{H}_{18} + 2(12.5)\,\text{O}_2 + 2(12.5)(3.76)\,\text{N}_2 \longrightarrow$$

$$8\,\text{CO}_2 + 9\,\text{H}_2\text{O} + 12.5\,\text{O}_2 + 94.0\,\text{N}_2$$

or

$$120 \text{ moles of reactants} \longrightarrow 123.5 \text{ moles of products}$$

Volume of 1 kg-mol at 25°C and 100 kPa is

$$\bar{v} = \frac{\bar{R}T}{p}$$

$$= \frac{8{,}314(273.15 + 25)}{100 \times 10^3}$$

$$= 24.79 \text{ m}^3/\text{kg-mol}$$

so that

$$V_{\text{react}} = (\bar{v} \times n)_{\text{react}}$$

$$= 24.79 \times 120$$

$$= 2974.8 \text{ m}^3$$

This volume is now occupied by 123.5 moles of reactants at 2900 K. The explosion pressure is

$$p_e = \frac{n_{\text{react}} \bar{R} T_e}{V_{\text{react}}}$$

$$= \frac{123.5 \times 8{,}314 \times 2900}{2974.8} \, \text{Pa}$$

$$= 1000 \, \text{kPa}$$

The above value of explosion pressure is only moderately high due to the fact that 100 percent excess air is used.

11.5 THE THIRD LAW OF THERMODYNAMICS AND ABSOLUTE ENTROPY

When we applied the first law of thermodynamics to chemical reactions, we found that we had to choose a common reference state, namely 25°C at 100 kPa, and to assign reference values arbitrarily to the enthalpy of stable elements. Only in this manner could we calculate the enthalpies of products in a chemical reaction. When we consider the entropy change in a chemical reaction, the same problem of selecting a reference state arises, because the equations developed in Chapter 6 allow us to calculate only the change in entropy of a pure substance. In a chemical reaction the composition and the constituents change. While it is possible to assign arbitrary values to entropies of elements for a predetermined reference state, the third law of thermodynamics makes this unnecessary.

The third law of thermodynamics is a result of the pioneering work by W. H. Nernst. The Nernst heat theorem states that if a chemical change takes place between pure crystalline solids at absolute-zero temperature, then there is no change in entropy. This theorem was later generalized by Simon, Planck, and others. One well-known generalization states that in any isothermal reversible process, the entropy change of a solid or a liquid system approaches zero as the absolute temperature approaches zero. This statement is known as the *Nernst-Simon theorem*. An alternate form of the third law states that it is impossible to lower the temperature of a system to absolute zero in a finite number of steps. This form of the third law is sometimes called the *unattainability statement*. It is possible to demonstrate that the Nernst-Simon theorem and the unattainability statement are equivalent in the sense that they both produce the same consequences. This is done by considering a magnetic cooling process for a paramagnetic system.

Max Planck generalized the third law with the following statement: The entropy of a substance in a perfect crystalline structure at absolute-zero temperature is zero. The law is based on experimental data. A perfect crystalline structure implies perfect order and absolute-zero temperature implies zero energy. The slightest disorder in the form or some randomness in the arrangement of the molecules of a substance at absolute zero would mean a nonzero entropy for the substance. The experimental evidence supporting the third law pertains primarily to the data on chemical reactions.

Using this data, it can be also shown that the entropy of a crystalline substance at absolute zero is independent of pressure, that is, $(\partial s/\partial p)_T = 0$. At temperatures above absolute zero, entropy does depend on pressure.

Consider a substance undergoing a constant pressure process and experiencing a change in temperature from absolute zero to T degrees Kelvin. For this process we can write the following equation for the change in entropy of the substance.

$$\bar{s}(T) - \bar{s}(0) = \int \frac{\delta q}{T} = \int_0^T \frac{\bar{c}_p\, dT}{T}$$

In this equation, $\bar{s}(T)$ and $\bar{s}(0)$ are the entropies of the substance at temperatures T and zero degrees, respectively. In accordance with the third law, the entropy of a substance at absolute-zero temperature is zero, so that $\bar{s}(0)$ is zero. Also, we note that the quantity $\bar{c}_p/T$ approaches a finite limit as the temperature approaches zero. Consequently, we can calculate the absolute entropy of a substance along an isobar by using the above equation.

The absolute values of entropies for a number of substances have been calculated for the isobar corresponding to a pressure of 100 kPa. These values are designated by $\bar{s}^0$ and these are dependent on temperature. Table 11.2 and Appendix I1 list values of $\bar{s}^0$ for a number of substances.

Now let us suppose we wish to determine the entropy of a substance at a temperature T and a pressure p, both chosen arbitrarily. We can visualize two processes, the first taking the substance from absolute zero to T along the isobar for $p = 100$ kPa, and the second taking the substance from 100 kPa and T to p and T along the isotherm for T. The entropy change for the first process is $\bar{s}^0(T)$, while that for the second process is $\Delta \bar{s}(100 \text{ kPa} \rightarrow p)$. Then the absolute entropy of the substance at T and p is given by

$$\bar{s}(T, p) = \bar{s}^0(T) + \Delta \bar{s}(100 \text{ kPa} \longrightarrow p) \qquad (11.10)$$

If the substance is an ideal gas, then $\Delta \bar{s}(100 \text{ kPa} \rightarrow p)$ can be calculated from

$$\Delta \bar{s}(100 \text{ kPa} \longrightarrow p) = -\bar{R} \int_{100}^p \frac{dp}{p} = -\bar{R} \ln \frac{p}{100} \qquad (11.10a)$$

where p is in kilopascals.

If tabulated entropy values are not available and the absolute entropy is known at only one such state, as 25°C and 100 kPa, then we need the equation of state and relevant specific heat data to calculate the entropy for other states.

When we have a mixture of gases, as is often the case with products of combustion, we must calculate the partial pressure p_i of each component and then calculate the entropy $\bar{s}_i$ of each component by Equations 11.10. Then the entropy S of the mixture is

$$S(T, p) = \sum_i n_i \bar{s}_i(T, p_i) \qquad (11.10b)$$

The change in entropy due to a chemical reaction is then given by

$$\Delta s = \sum_i [n_i \bar{s}_i(T, p_i)]_{\text{prod}} - \sum_j [n_j \bar{s}_j(T, p_j)]_{\text{react}} \qquad (11.10c)$$

When we use Equations 11.10 in the rest of this chapter, we shall drop the parenthetical quantities after $\bar{s}$, $\bar{s}^0$, $\Delta\bar{s}$, $\bar{s}_i$, and $\bar{s}_j$.

Example 11.10

Methane and oxygen at 25°C and 100 kPa react to form water vapor and carbon dioxide at the same temperature. Assuming that the reaction is completed, determine the change in entropy per kilogram mole of methane.

Solution:
Given: CH_4 and O_2 react at 25°C and 100 kPa.
Objective: The change in entropy due to the reaction is to be determined.
Assumptions: None.
Relevant physics: From Equation 11.10(c) the change in entropy is

$$\Delta S = \sum_i (n_i \bar{s}_i)_{\text{prod}} - \sum_j (n_j \bar{s}_j)_{\text{react}}$$

Since the reaction equation is

$$CH_4 + 2\,O_2 \quad \longrightarrow \quad CO_2 + 2\,H_2O,$$
$$\Delta S = (n_3 \bar{s}_3 + n_4 \bar{s}_4) - (n_1 \bar{s}_1 + n_2 \bar{s}_2)$$

where subscripts 1, 2, 3, and 4 refer to CH_4, O_2, CO_2, and H_2O, respectively. The values of $\bar{s}$ are determined from Equations 11.10 and 11.10a,

$$\bar{s}(T) = \bar{s}^0(T) - \bar{R} \ln\left(\frac{p}{100}\right)$$

where p is in kiloPascals and values of $\bar{s}^0$ are listed in Appendix I1.
Analysis: From the reaction equation

$$CH_4 + 2\,O_2 \quad \longrightarrow \quad CO_2 + 2\,H_2O$$

we observe that there are 3 kg-mol before and after the reaction. Therefore the total pressure p before and after the reaction is the same. Since the partial pressure of a component equals the product of the mole fraction of the component and the total pressure, we have

$$CH_4: \quad p_1 = 0.333p \quad n_1 = 1 \text{ kg-mol}$$
$$O_2 \; : \quad p_2 = 0.667p \quad n_2 = 2 \text{ kg-mol}$$
$$CO_2: \quad p_3 = 0.333p \quad n_3 = 1 \text{ kg-mol}$$
$$H_2O: \quad p_4 = 0.667p \quad n_4 = 2 \text{ kg-mol}$$

The value of $\bar{s}^0$ for CH_4 is read from Table 11.2, and those for O_2, CO_2, and H_2O are read from Appendix I1. The values of entropies of the products and the reactants and their number of moles are given below.

$$CH_4: \quad \bar{s}_1 = 186.256 - 8.314 \ln(0.333) = 195.390 \text{ kJ/kg-mol/K,}$$
$$O_2: \quad \bar{s}_2 = 205.142 - 8.314 \ln(0.667) = 208.513 \text{ kJ/kg-mol/K,}$$
$$CO_2: \quad \bar{s}_3 = 213.795 - 8.314 \ln(0.333) = 222.929 \text{ kJ/kg-mol/K,}$$
$$H_2O: \quad \bar{s}_4 = 188.833 - 8.314 \ln(0.667) = 192.204 \text{ kJ/kg-mol/K,}$$

The change in entropy per kilogram-mole of CH_4 for the reaction is

$$\Delta S = (n_3 \bar{s}_3 + n_4 \bar{s}_4) - (n_1 \bar{s}_1 + n_2 \bar{s}_2)$$
$$= (1 \times 222.929 + 2 \times 192.204) - (1 \times 195.390 + 2 \times 208.513)$$
$$= -5.079 \text{ kJ/K per kg-mol of } CH_4.$$

Comments: The negative value of ΔS is not a violation of the second law. In order to return the products of combustion to 25°C and 100 kPa, 74,873 kJ/kg-mol $(-\bar{h}_f^0)$ of CH_4 are rejected to the surroundings. We can calculate the change of entropy of the surroundings using this quantity of heat rejected if we assume the surroundings to be at 25°C, or 298 K.

$$(\Delta S)_{\text{surr}} = \frac{Q}{T}$$
$$= \frac{-(-74,873)}{298}$$
$$= +251.25 \text{ kJ/K}$$
$$(\Delta S)_{\text{uni}} = 251.25 - 5.079 = 246.17 \text{ kJ/K}$$

Thus the entropy of the universe increases in this chemical reaction.

11.6 CHEMICAL REACTIONS AND THE SECOND LAW

It was shown in Chapter 6 that we can determine the available work, availability, and irreversibility for a process. When a chemical reaction takes place, we are interested in determining the available work that can be obtained by employing a reversible engine to harness the heat energy released. We are also interested in knowing the maximum available work—that is, the availability—and the irreversibility of the reaction. Another matter of interest is whether a set of given reactants mixed in stoichiometric proportions will indeed react, and if so, what portion of the reactants will be converted into the reaction products. The answer to this query is provided by the second law.

We know from Chapter 6 that for negligible changes in kinetic and potential energies, the rate of available work for a steady flow process is given by Equation 6.35,

$$A\dot{W}_s = \sum [\dot{m}_{\text{in}}(h_{\text{in}} - T_0 s_{\text{in}})] - \sum \dot{m}_{\text{out}}(h_{\text{out}} - T_0 s_{\text{out}})$$

where the subscripts *in* and *out* denote conditions at the inlet and outlet, respectively, and T_0 denotes the temperature of the surroundings. When we consider a reaction occuring within a control volume, this equation takes the form

$$AW_s = \sum [n(\bar{h}_f^0 + \Delta \bar{h} - T_0 \bar{s})]_{\text{in, react}} - \sum [n(\bar{h}_f^0 + \Delta \bar{h} - T_0 \bar{s})]_{\text{out, prod}} \quad (11.11)$$

In Equation 11.11, the first summation is performed for all the incoming reactants and the second summation is performed for all the outgoing products. Let us consider the steady-flow reaction to occur in such a manner that both the reactants and the products are at the standard conditions, namely, 25°C and 100 kPa. Under these conditions the $\Delta\bar{h}$ terms in Equation 11.11 drop out. Also, we recall the definition of the Gibbs function, $g = h - Ts$. Equation 11.11 then becomes

$$\text{AW}_s = \sum (n\bar{g}_f^0)_{\text{in, react}} - \sum (n\bar{g}_f^0)_{\text{out, prod}} \tag{11.12}$$

Equation 11.12 confirms our earlier assertion that the Gibbs function plays an important role in chemical reactions. Consequently, the Gibbs function of formation, $\bar{g}_f^0$, has been defined in a manner analogous to that for the enthalpy of formation. To be consistent with the assumption that $\bar{h}_f^0$ is zero at 25°C and 100 kPa, the value of $\bar{g}^0$ at this state should equal the negative value of the product of the absolute entropy at 25°C and 100 kPa, and 298.15 K. Unfortunately, this is not the case. The Gibbs function for each basic element at 25°C and 100 kPa is taken as zero, and the Gibbs function of other substances are found relative to this base. The values the Gibbs function of formation are listed in Table 11.2. Note that if the temperature of the products is higher than 25°C, the reversible work would be less than that given by Equation 11.12.

The equation (Equation 6.44(a)) for the availability ψ developed in Chapter 6 was

$$\psi = (h - T_0 s) - (h_0 - T_0 s_0)$$

where the changes in potential and kinetic energies are neglected. The subscript zero in the equation denotes the ground state. The equation can be rewritten for the products emerging from a reaction.

$$\psi_{\text{prod}} = \sum \{n[(\bar{h} - \bar{h}_0) - T_0(\bar{s} - \bar{s}_0)]\}_{\text{out, prod}} \tag{11.13}$$

where $\bar{h}$ is the molal enthalpy of a product at temperature T_{out} and $\bar{s}$ is the absolute molal entropy of a product at T_{out}. The quantities $\bar{h}_0$ and $\bar{s}_0$ are molal enthalpy and absolute molal entropy of the product at the temperature of the surroundings. Note that the calculation of $\bar{s}$ for a product may involve partial pressure of the product and use of Equations 11.10, 11.10(a), and 11.10(b).

Extending the equation for irreversibility, I, developed in Chapter 6 to a steady-flow process involving a chemical reaction, we obtain

$$I = T_0 [\sum_i (n_i \bar{s}_i)_{\text{out, prod}} - \sum_j (n_j \bar{s}_j)_{\text{in, react}}] \tag{11.13a}$$

Example 11.11

Methane at 25°C and 100 kPa is burned with 300% theoretical air at 25°C and 100 kPa. Assume that the reaction occurs reversibly and that each constituent of the products and the reactants is at 25°C and 100 kPa. Calculate the available work for the process.

Solution:

Given: Methane is burned at 25°C and 100 kPa.

$$\text{Excess air: } 200\%$$

$$\text{Process: } \quad \text{reversible}$$

Objective: The available work AW_s for the process is to be determined.

Assumptions: None.

Relevant physics: The equation for the reaction is

$$\text{CH}_4 + 3(2)\,\text{O}_2 + 3(2)(3.76)\,\text{N}_2 \longrightarrow \text{CO}_2 + 2\,\text{H}_2\text{O} + 4\,\text{O}_2 + 22.56\,\text{N}_2$$

Since the reaction occurs reversibly at 25°C, AW_s can be calculated from Equation 11.12 by using the Gibbs function of formation from Table 11.2.

$$AW_s = \sum (n g_f^0)_{in, react} - \sum (n \bar{g}_f^0)_{out, prod}$$

Analysis: The Gibbs function of formation for O_2 and N_2 at 25°C and 100 kPa are zero, so that AW_s is given by

$$AW_s = (\bar{g}_f^0)_{CH_4} - (\bar{g}_f^0)_{CO_2} - 2(g_f^0)_{H_2O}$$
$$= -50,751 - (-394,374) - 2(-228,583)$$
$$= 800,789 \text{ kJ/kg-mol } (CH_4)$$
$$= 49,915 \text{ kJ/kg } (CH_4)$$

Comments: Since the state at the end of the reaction process is ground state—that is, the temperature and the pressure of the products are the same as those of the commonly available sink, the atmosphere—the available work calculated is also the availability of 1 kg of methane.

The next example deals with the irreversibility of a combustion process.

Example 11.12

Methane at 25°C and 100 kPa is burned with 300% theoretical air in an adiabatic manner. The temperature of the surroundings is 25°C. The partial pressure of each of the reactant product gases is 100 kPa. Determine the increase in entropy during combustion and the availability of the products of combustion.

Solution:

Given: Methane at 25°C and 100 kPa is burned.

Excess air: 200%

Process: adiabatic

Objective: The change of entropy ΔS for the process and the availability of the products Ψ_{prod} are to be determined.

Assumptions: The temperature of the surroundings is 298 K.

Relevant physics: The change of entropy is

$$\Delta S = S_{prod} - S_{react}$$

Since the partial pressure of each of the gases in reactants and in products is specified as 100 kPa, the entropy of the gases can be read from the appendix once the relevant temperature is known. The temperature of the product gases is the adiabatic flame temperature, which can be determined by using Equation 11.5.

The availability of the products can be calculated by employing Equation 11.13 or, more simply, by subtracting the irreversibility I of the reaction from the available work for the process, calculated in Example 11.11.

Analysis: The equation for the reaction is

$$CH_4 + 3(2) O_2 + 3(2)(3.76) N_2 \longrightarrow CO_2 + 2 H_2O + 4 O_2 + 22.56 N_2$$

We have to determine the adiabatic flame temperature first. To that end, Equation 11.5 becomes

$$\sum n(\bar{h}_f^0 + \Delta \bar{h})_{in, react} = \sum [n(\bar{h}_f^0 + \Delta \bar{h})]_{out, prod}$$

The values of $\bar{h}_f^0$ for the reactants and products are obtained from Table 11.2. The values of $\Delta\bar{h}$ for the reactants are identically zero at $T = 298$ K. Inserting these values in the foregoing equation, we have

$$-74{,}873 = [-393{,}522 + \Delta\bar{h}(\mathrm{CO_2})] + 2[-241{,}827 + \Delta\bar{h}(\mathrm{H_2O})]$$
$$+ 4\,\Delta\bar{h}(\mathrm{O_2}) + 22.56\,\Delta\bar{h}(\mathrm{N_2})$$

The adiabatic flame temperature is found by trial and error (see Example 11.8) to be 1130 K.

To find the entropy change for the combustion process, we note that the entropy of the reactants at 298 K is

$$S_{\mathrm{react}} = \bar{s}^0(\mathrm{C_2H_4}) + 6\bar{s}^0(\mathrm{O_2}) + 22.56\bar{s}^0(\mathrm{N_2})$$
$$= 186.256 + 6 \times 205.142 + 22.56 \times 191.611$$
$$= 5739.85 \text{ kJ/kg-mol (CH}_4\text{) K}$$

The entropy of the products at 1130 K is

$$S_{\mathrm{prod}} = \bar{s}(\mathrm{CO_2}) + 2\bar{s}(\mathrm{H_2O}) + 4\bar{s}(\mathrm{O_2}) + 22.56\bar{s}(\mathrm{N_2})$$
$$= 276.014 + 2 \times 233.902 + 4 \times 244.588 + 22.56 \times 232.184$$
$$= 6960.236 \text{ kJ/kg-mole (CH}_4\text{) K}$$

$$\Delta S = S_{\mathrm{prod}} - S_{\mathrm{react}}$$
$$= 6960.236 - 5739.85$$
$$= 1220.386 \text{ kJ/kg-mol (CH}_4\text{) K}$$

The irreversibility is then determined from

$$I = T_0\,\Delta S$$
$$= 298.15 \times 1220.386$$
$$= 363{,}858 \text{ kJ/kg-mol (CH}_4\text{)}$$
$$= 22{,}680 \text{ kJ/kg (CH}_4\text{)}$$

We had determined the available work, AW_s, of the isothermal reversible reaction in Example 11.11 to be 49,915 kJ/kg (CH$_4$). Therefore the availability of the products of combustion of the adiabatic process is

$$\psi_{\mathrm{prod}} = \mathrm{AW}_s - I = 49{,}915 - 22{,}680 = 27{,}235 \text{ kJ/kg (CH}_4\text{)}$$

We can be verify that the availability can also be determined from Equation 11.13 as

$$\psi_{\mathrm{prod}} = \sum [n(\bar{h} - T_0\bar{s}) - n(\bar{h}_0 - T_0\bar{s}_0)]_{\mathrm{out,\,prod}}$$

where the first parenthetical quantity in the summation is evaluated at 1130 K and the second parenthetical quantity is evaluated at $T_0 = 298$ K.

Comments: In this example, if the total pressure for the products of combustion is specified as 100 kPa, then the molal entropy values ($\bar{s}^0$) for each constituent will have to be augmented by ($-\bar{R}\ln x$), where x is the mole fraction of the constituent.

In Example 11.12, the change of entropy was positive in accordance with the second law. However, this does not mean that a proposed reaction will always go to completion. As a reaction proceeds, the entropy increases. It may turn out that once a reaction reaches a certain point, the entropy change reaches a maximum value. A continuation of the reaction beyond this point would mean a decrease in the entropy of the products. In such a case, the reaction will not go to completion and the end result is a mixture of the products of combustion and a portion of the reactants.

Consider a reaction where carbon monoxide reacts in an adiabatic reaction vessel with steam to produce carbon dioxide and hydrogen. The equation of the reaction is

$$CO + H_2O \longrightarrow CO_2 + H_2$$

If the initial temperature and pressure are 25°C and 100 kPa and if the reaction is adiabatic, the entropy of the products would be 439.862 kJ/kg K, provided the reaction goes to completion. This entropy value can be calculated by first determining the adiabatic flame temperature and then applying Equations 11.10, 11.10(a), 11.10(b), and 11.10(c) to each of the product constituents. If it is assumed that the reaction goes only halfway; that is, only 1/2 mol of CO_2 is formed for 1 mol of CO in the reactants, and 1/2 mol of H_2 is formed for 1 mol of H_2O in the reactants, the entropy of the products is found to be 446.77 kJ/kg K. This value is greater than the entropy value of the products for complete combustion. This means that if the reaction were to continue from the halfway stage to completion, the entropy of the chemical system under consideration, which is undergoing the adiabatic process, would decrease and thus the second law would be violated. If a plot of entropy of the products against the moles of CO_2 or of H_2 formed were to be made, we would obtain a curve exhibiting a maximum corresponding to 0.7 moles of CO_2. The calculations involved in the above procedure are lengthy and tedious. We then naturally ask this question: Is it possible to predict if a reaction is possible, and, if so, how far will it proceed? Considerations of chemical equilibrium make it possible to answer this question.

11.7 EQUILIBRIUM

Earlier in the text we required homogeneity of all properties of a system to insure equilibrium of a system. When we consider a chemical system this requirement proves insufficient. What we would like is a quantitative criterion valid for any system. We begin with the second-law restriction on an isolated system.

$$(dS)_{\substack{\text{isolated} \\ \text{system}}} \geq 0$$

In an isolated system, since both work and heat interactions are not permissible, the internal energy is constant. Also, since work is zero the boundaries of the system cannot expand or contract and the system volume must remain constant. The fact that the mass of an isolated system is constant is designated by the subscript n, where n is the number of moles in the system. The second law restriction can then be rewritten as

$$(dS)_{U,V,n} \geq 0 \tag{11.14}$$

The symbol greater than corresponds to a spontaneous change in the system, with an accompanying entropy increase. Once all spontaneous changes are over, the entropy reaches its maximum and the system is in equilibrium.

Criteria for equilibrium in terms of the Helmholtz and Gibbs functions can also be obtained. Since the Helmholtz function is defined as

$$A = U - TS$$

$$dA = dU - T\,dS - S\,dT$$

For an isothermal system, $dT = 0$. Using the first law, we then obtain

$$dA_T = \delta Q - \delta W - T\,dS \qquad (11.15)$$

From the second law, we know that

$$T\,dS - \delta Q \geq 0$$

so that Equation 11.15 becomes

$$-dA_T - \delta W = T\,dS - \delta Q \geq 0 \qquad (11.15a)$$

When a chemical reaction occurs in a constant volume vessel, the work interaction is absent and the last equation becomes

$$-dA_{T,V,n} \geq 0$$

or

$$dA_{T\,V,n} \leq 0 \qquad (11.16)$$

Equation 11.16 means that in a spontaneous reaction, the Helmoltz function of an isolated system—that is, of a system of constant temperature, volume, and mass—must decrease in the absence of all forms of work. When the Helmholtz function attains a minimum value, the system has attained a state of equilibrium.

We now establish a criterion for equilibrium in terms of the Gibbs function.

$$G = H - TS = U + pV - TS$$

or

$$G = A + pV$$

so that

$$dG = dA + d(pV)$$

For a constant pressure, constant temperature chemical reaction,

$$dG_{T,p} = dA_T + p\,dV$$

or

$$dG_{T,p} - p\,dV = dA_T \leq -\delta W \qquad (11.17)$$

where Equation 11.15(a) is used. In Equation 11.17, $p\,dV$ represents the boundary work and δW represents the total work. We can write

$$\delta W_v = \delta W - p\,dV$$

where δW_v represents the sum of all types of work interactions except those due to a volume change. Equation 11.17 then becomes

$$dG_{T,p} \leq -\delta W_v \qquad (11.18)$$

In a simple closed system, only boundary work is present and other mechanisms of work are absent. In such a case, Equation 11.18 reduces to

$$dG_{T,p,n} \leq 0 \tag{11.18a}$$

Equation 11.18(a) implies that the Gibbs function always decreases for a spontaneous change in a closed system under conditions of constant temperature and pressure and in the absence of all work effects. Equation 11.18(a) is the criterion for chemical equilibrium. It is to be emphasized that Equation 11.18(a) as the criterion for equilibrium is equivalent to the criterion in Equation 11.14. However, the former involves properties that are convenient to use—namely, temperature and pressure—in its application to reactive systems.

In the $CO + H_2O$ reaction discussed at the end of Section 11.6, we can calculate the Gibbs function of the remaining reactants and the products for various degrees of reaction. The number of moles of CO_2 produced corresponding to the minimum value of the Gibbs function of the remaining reactants and products represents the equilibrium composition. We should note that the Gibbs function is temperature-dependent and, therefore, for a given mixture of reactants and products, the equilibrium composition is a function of temperature.

Equation 11.18(a) is also applicable to a mixture of two phases. That is, for a two-phase mixture to be in equilibrium, the Gibbs function of the mixture must have a minimum value.

11.7.1 Chemical Equilibrium and the Reaction Constant

We now consider the chemical equilibrium of a reaction involving only one phase. Such a reaction is called a *homogeneous chemical reaction*. For simplicity, let the reactants A and B and the products C and D, be in a gaseous phase, although the procedure that will be followed in the analysis is equally applicable to other phases. Consider the species A, B, C, and D to be in chemical equilibrium at temperature T and pressure p. We can write the following reaction equation for the four species

$$v_A A + v_B B \rightleftharpoons v_C C + v_D D \tag{11.19}$$

where the v's are the stoichiometric coefficients.

We saw in the preceding section that a chemical reaction would proceed in such a way that the Gibbs function of the system would decrease (Equation 11.18(a) until a minimum value is attained. This means that at equilibrium

$$dG_{T,p} = 0 \tag{11.20}$$

The Gibbs function for our system is given by

$$G = G(n_A, n_B, n_C, n_D, T, p)$$

where n_A, n_B, n_C, and n_D are the number of moles of species A, B, C, and D, respectively. The infinitesimal change in the Gibbs function at constant temperature and pressure is given by

$$dG_{T,p} = \left(\frac{\partial G}{\partial n_A}\right) dn_A + \left(\frac{\partial G}{\partial n_B}\right) dn_B + \left(\frac{\partial G}{\partial n_C}\right) dn_C + \left(\frac{\partial G}{\partial n_D}\right) dn_D$$

The derivatives of G in the foregoing equation were defined in Equation 10.30 to be the chemical potentials, μ's, so that we have

$$dG_{T,p} = \mu_A dn_A + \mu_B dn_B + \mu_C dn_C + \mu_D dn_D \qquad (11.20a)$$

We observe that dn_A, dn_B, dn_C, and dn_D in the above equation are infinitesimal changes in the number of moles of species A, B, C, and D, respectively. In our reacting system, these changes cannot be arbitrary, but they must be proportional to the respective stoichiometric coefficients, ν_A, ν_B, ν_C, and ν_D. Also, on the basis of Equation 11.19, an increase in n_C and n_D must necessarily mean a decrease in n_A and n_B. Consequently, we find that the following equations hold.

$$-\frac{dn_A}{\nu_A} = -\frac{dn_B}{\nu_B} = \frac{dn_C}{\nu_C} = \frac{dn_D}{\nu_D} \qquad (11.20b)$$

Combining Equations 11.20 and 11.20(a) and then substituting for dn_B, dn_C, and dn_D from Equation 11.20(b), we obtain

$$dG_{T,p} = \left(\frac{-dn_A}{\nu_A}\right)(\mu_C \nu_C + \mu_D \nu_D - \mu_A \nu_A - \mu_B \nu_B) = 0$$

Since ν_A is a positive number and dn_A is nonzero, we conclude that

$$\nu_C \mu_C + \nu_D \mu_D - \nu_A \mu_A - \nu_B \mu_B = 0 \qquad (11.21)$$

Equation 11.21 is called the *equation of reaction equilibrium*. The equation is valid for any chemical reaction. In order to be able to use the equation, we must have an expression for the chemical potential μ. The chemical potential is given by

$$\mu = \left(\frac{\partial G}{\partial n_i}\right)_{T,p,n_j} = \bar{g}_i$$

where $\bar{g}_i$ is the molal specific Gibbs function at a pressure p and temperature T. The specific Gibbs function is, by definition,

$$\bar{g} = \bar{h} - T\bar{s}$$

The right side of the above equation can be determined with reference to the standard state of 100 kPa and 25°C. Then $\bar{h}(T) = \bar{h}^0(T)$, where $\bar{h}^0(T)$ incorporates the enthalpy of formation, as well as the enthalpy change due to temperature change from 298 K to T. Also, the absolute entropy is given by Equations 11.10 and 11.10(a):

$$\bar{s}(T) = \bar{s}^0(T) - \bar{R}\ln\left(\frac{p}{p_{\text{total}}}\right)$$

where p is the partial pressure of the component and p_{total} is the total pressure of the mixture. The partial molal Gibbs function or the chemical potential is then given by the above expressions for $\bar{h}(T)$ and $\bar{s}(T)$.

$$\mu(T) = \bar{g}(T) = \bar{h}^0(T) - T\bar{s}^0(T) + \bar{R}T\ln\left(\frac{p}{p_{\text{total}}}\right)$$

The quantity $\bar{h}^0(T) - T\bar{s}^0(T)$ is the standard state Gibbs function, $\bar{g}^0(T)$. Using this function, we obtain

$$\mu(T) = \bar{g}(T) = \bar{g}^0(T) + \bar{R}T\ln\left(\frac{p}{p_{\text{total}}}\right) \qquad (11.22)$$

We now substitute Equation 11.22 into Equation 11.21 and drop the functional notation (T) after g and the subscript total on p. We then obtain

$$v_C\left[\bar{g}_C^0 + \bar{R}T\ln\left(\frac{p_C}{p}\right)\right] + v_D\left[\bar{g}_D^0 + \bar{R}T\ln\left(\frac{p_D}{p}\right)\right]$$

$$- v_A\left[\bar{g}_A^0 + \bar{R}T\ln\left(\frac{p_A}{p}\right)\right] - v_B\left[\bar{g}_B^0 + \bar{R}T\ln\left(\frac{p_B}{p}\right)\right] = 0$$

A rearrangement of this equation leads to

$$(v_C\bar{g}_C^0 + v_D\bar{g}_D^0 - v_A\bar{g}_A^0 - v_B\bar{g}_B^0) + v_C\bar{R}T\ln\left(\frac{p_C}{p}\right)$$

$$+ v_D\bar{R}T\ln\left(\frac{p_D}{p}\right) - v_A\bar{R}T\ln\left(\frac{p_A}{p}\right) - v_B\bar{R}T\ln\left(\frac{p_B}{p}\right) = 0 \quad (11.23)$$

We denote the quantity in the first pair of parentheses by $\Delta G^0(T)$; it is called the *standard-state, Gibbs-function change*. It represents the change in the Gibbs function that would occur if the stoichiometric reaction were to go to completion. The quantity ΔG^0 is evaluated by assuming that each of the reactants and the products is at the standard pressure of 100 kPa and a temperature T. Since the standard pressure is used, the superscript zero is appended. Once the stoichiometric equation and the temperature for a reaction are known, $\Delta G^0(T)$ can be calculated. We now recast Equation 11.23 by using $\Delta G^0(T)$, to read

$$-\Delta G^0(T) = v_C\bar{R}T\ln\left(\frac{p_c}{p}\right) + v_D\bar{R}T\ln\left(\frac{p_D}{p}\right) - v_A\bar{R}T\ln\left(\frac{p_A}{p}\right) - v_B\bar{R}T\ln\left(\frac{p_B}{p}\right)$$

Simplification results if we assume the total pressure p to be 1 atm and if we express the partial pressures p_A, p_B, p_C, and p_D in atmospheres. We then have

$$-\Delta G^0(T) = \bar{R}T[v_C\ln p_C + v_D\ln p_D - v_A\ln p_A - v_B\ln p_B]$$

or

$$-\Delta G^0(T) = \bar{R}T\ln\left[\frac{p_C^{v_C}p_D^{v_D}}{p_A^{v_A}p_B^{v_B}}\right]$$

The quantity in the square bracket is called the *equilibrium constant, K_p*. The equation then simplifies to read

$$-\Delta G^0(T) = \bar{R}T\ln K_p \qquad (11.24)$$

where

$$K_p = \frac{p_C^{v_C}\cdot p_D^{v_D}}{p_A^{v_A}\cdot p_B^{v_B}} \qquad (11.24a)$$

On the basis of Equation 11.24, the equilibrium constant can be also expressed as

$$K_p = e^{-(\Delta G^0(T)/\bar{R}T)} \qquad (11.25)$$

It is common to find values of K_p or of $\log K_p$ tabulated instead of $\Delta G^0(T)$. These values are temperature-dependent and are listed for some of the common reactions, along with the reaction equations, in Appendix M.

Note that the equation of reaction used in the determination of the K_p value must be known. The reaction

$$2\,CO + O_2 \;\rightleftharpoons\; 2\,CO_2$$

can be also written as

$$CO + \tfrac{1}{2}O_2 \;\rightleftharpoons\; CO_2$$

The stoichiometric coefficients in the first equation are twice the corresponding coefficients in the second equation. Consequently, in view of Equation 11.24(a), the value of K_p in the first equation equals the square of the value of the K_p in the second equation. It is a general practice to assume that the terms in the numerator of Equation 11.24(a), used for determining K_p, represent the products, while those in the denominator represent the reactants.

If there are a number of reactions going on simultaneously, we can determine the equilibrium constant for each reaction and then determine the composition for equilibrium. The reactions must be independent of one another.

11.7.2 The Gibbs-Helmholtz and van't Hoff Equations

In the foregoing discussions, it was assumed that the temperature and pressure before and after a reaction remained the same. Equation 11.24 then permits us to calculate the standard-state, Gibbs-function change, $\Delta G^0(T)$. If the temperature of the reaction is changed, the value of $\Delta G^0(T)$ will change. We shall now determine the change in the values of $\Delta G^0(T)$ of a chemical reaction as the reaction temperature is changed, bearing in mind that during each reaction at a given temperature, the pressure is constant. From the definition of the Gibbs function

$$G = H - TS = U + pV - TS$$

so that

$$dG = dU + p\,dV + V\,dp - T\,dS - S\,dT$$

In view of the first $T\,dS$ equation, we can write

$$dG = V\,dp - S\,dT$$

For a constant pressure reaction, we obtain

$$\left(\frac{\partial G}{\partial T}\right)_p = -S \tag{11.26}$$

The change in the Gibbs function for a chemical reaction is

$$\Delta G = G_{\text{prod}} - G_{\text{react}}$$

$$= (H - TS)_{\text{prod}} - (H - TS)_{\text{react}} \tag{11.27}$$

or

$$\Delta G = \Delta H - T\Delta S \tag{11.27a}$$

since the temperature before and after the reaction is the same. To determine the effect of change in the temperature at which a reaction occurs on the change in the Gibbs function, we partially differentiate Equation 11.27 with respect to T, yielding

$$\left(\frac{\partial \Delta G}{\partial T}\right)_p = \left(\frac{\partial G_{\text{prod}}}{\partial T}\right)_p - \left(\frac{\partial G_{\text{react}}}{\partial T}\right)_p$$

In view of Equation 11.26, this last equation becomes

$$\left(\frac{\partial \Delta G}{\partial T}\right)_p = -S_{\text{prod}} + S_{\text{react}} = -\Delta S$$

Solving Equation 11.27(a) for ΔS and substituting the result in the above equation, we obtain

$$\left(\frac{\partial \Delta G}{\partial T}\right)_p = \frac{\Delta G - \Delta H}{T} \tag{11.28}$$

Equation 11.28 is known as the *Gibbs-Helmholtz equation*. Rewriting Equation 11.28 for a reaction at standard-state pressure, we have

$$\left(\frac{\partial \Delta G^0}{\partial T}\right)_p = \frac{\Delta G^0 - \Delta H^0}{T} \tag{11.29}$$

where ΔG^0 and ΔH^0 are the standard-state Gibbs-function change and enthalpy change, respectively.

The variation of the equilibrium constant K_p with respect to temperature can now be determined. We note that in the equation for K_p, Equation 11.25, the quantity $\Delta G^0(T)/T$ appears as a part of the exponent. If we partially differentiate this quantity with respect to temperature, holding p constant, the following result is obtained.

$$\left[\frac{\partial}{\partial T}\left(\frac{\Delta G^0}{T}\right)\right]_p = \frac{1}{T}\left(\frac{\partial \Delta G^0}{\partial T}\right)_p - \frac{\Delta G^0}{T^2}$$

Substitution for $(\partial \Delta G^0/\partial T)_p$ from Equation 11.29 yields

$$\left[\frac{\partial}{\partial T}\left(\frac{\Delta G^0}{T}\right)\right]_p = -\frac{\Delta H^0}{T^2} \tag{11.30}$$

We note from Equation 11.24 that

$$\bar{R} \ln K_p = -\frac{\Delta G^0}{T}$$

so that

$$\left[\frac{\partial(\ln K_p)}{\partial T}\right]_p = \frac{1}{\bar{R}}\frac{\partial}{\partial T}\left(\frac{\Delta G^0}{T}\right)_p = \frac{\Delta H^0}{\bar{R}T^2} \tag{11.31}$$

where Equation 11.30 is used. Equation 11.31 is known as the *van't Hoff isobar equation*. If the equation is rewritten for a reaction, the term ΔH^0 then represents the enthalpy of reaction, $\bar{h}_R^0$ (Equation 11.7).

If we assume that the enthalpy of reaction is constant over a small range of temperature T_1 to T_2, which is a reasonable assumption, we can readily integrate Equation 11.31 to yield

$$\ln\left(\frac{K_{p2}}{K_{p1}}\right) = -\frac{h_R^0}{\bar{R}}\left(\frac{1}{T_2} - \frac{1}{T_1}\right) \tag{11.32}$$

This result means that for an exothermic reaction (h_R^0 negative), the value of K_p decreases as the temperature of the reaction is raised. This means that the extent to which the reaction proceeds also decreases. This aspect is discussed in Section 11.7.3. The energy released by combustion in adiabatic conditions is then diminished, as the reaction does not go to completion.

Example 11.13

What are the values of the standard-state Gibbs function of change and the equilibrium constant for the reaction

$$H_2O \;\rightleftharpoons\; H_2 + \tfrac{1}{2}O_2$$

Assume the reaction temperature to be 3000 K.

Solution:

Given: The dissociation of a water molecule into hydrogen and oxygen and the reaction of the latter two to give a water molecule occurs at 3000 K.

Objective: The values of ΔG° and K_p for the reaction are to be determined.

Assumptions: None.

Relevant physics: The standard-state Gibbs-function change, ΔG^0, can be determined once the standard-state Gibbs functions, $\bar{g}^0$, for hydrogen, oxygen, and water are known.

$$\Delta G^0 = \bar{g}^0_{H_2} + \tfrac{1}{2}\bar{g}^0_{O_2} - \bar{g}^0_{H_2O}$$

Alternatively, ΔG^0 can be determined from

$$\Delta G^0 = \Delta H^0 - T\Delta S^0$$

where

$$\Delta H^0 = H^0_{prod} - H^0_{react}$$

and

$$\Delta S^\circ = S^0_{prod} - S^0_{react}$$

The equilibrium constant can be determined from Equation 11.25.

Analysis:

Enthalpy of the products at 3000 K:

$$\text{Hydrogen:}\quad H_1 = n_1\bar{h}_1 = 1 \times 88{,}743 \quad = \quad 88{,}743 \text{ kJ}$$

$$\text{Oxygen:}\quad H_2 = n_2\bar{h}_2 = 0.5 \times 98{,}098 = \quad 49{,}049 \text{ kJ}$$

$$H^0_{prod} = 137{,}792 \text{ kJ}$$

Enthalpy of the reactant at 3000 K:

$$\text{Water vapor:}\quad H_3 = n_3(\bar{h}^0_{f3} + \Delta\bar{h}_3) = 1(-241{,}827 + 126{,}361) = -115{,}466 \text{ kJ}$$

Entropy of the products at 3000 K:

$$\text{Hydrogen:}\quad S^0_1 = n_1\bar{s}_1 = 1 \times (202.887) \quad = 202.887 \text{ kJ/K}$$

$$\text{Oxygen:}\quad S^0_2 = n_2\bar{s}_2 = 0.5 \times (284.508) = 142.254 \text{ kJ/K}$$

$$S^0_{prod} = 345.141 \text{ kJ/K}$$

Entropy of the reactant at 3000 K:

$$\text{Water vapor:}\quad S_3 = n_3\bar{s}_3 = 1 \times 286.383 = 286.383 \text{ kJ/K}$$

For the reaction temperature of 3000 K,

$$\Delta G^0 = (H^0_{prod} - H^0_{react}) - T(S^0_{prod} - S^0_{react})$$

$$= (137{,}792 + 115{,}466) - 3000(345.141 - 286.383)$$

$$= 76{,}984 \text{ kJ}$$

From Equation 11.25,

$$K_p = \exp\left(\frac{-\Delta G^0}{\bar{R}T}\right)$$

Substituting the values of $\Delta G°$, $\bar{R}$, and T, we obtain

$$K_p = \exp \frac{-76,984}{8.314 \times 3000}$$

$$= 0.0458$$

The value of K_p obtained above agrees well with the value in Appendix M.

11.7.3 Extent of Reaction

As a reaction proceeds, there is a decrease in the mass of the reactants and an increase in the mass of the products. In accordance with *the law of definite proportions*, the increase in the mass of a reaction component is proportional to its molar mass M and to its stoichiometric coefficient v. If m_i is the initial mass of a component and m_f is the final mass of a component after the reaction, we define the *extent of reaction**, λ, by

$$\lambda = \frac{m_f - m_i}{vM} \qquad \text{(for product)} \qquad (11.33)$$

and

$$\lambda = -\frac{m_f - m_i}{vM} \qquad \text{(for reactant)} \qquad (11.33a)$$

The above definition ensures a positive value for the extent of reaction λ, even though the mass of a reaction product increases while that of a reactant decreases.

The extent of a reaction is zero if λ equals zero, that is, no change in mass has occurred. On the other hand, if λ equals unity, then we say that a complete conversion of v moles of the component has occurred.

Since the number of moles n of a component is the ratio of the mass m to the molar mass M, Equations 11.33 and 11.33(a) can be recast to read

$$\lambda = \frac{n_f - n_i}{v} \qquad \text{(for product)} \qquad (11.33b)$$

and

$$\lambda = -\frac{n_f - n_i}{v} \qquad \text{(for reactant)} \qquad (11.33c)$$

In view of the above definition of the extent of reaction, for the reaction equation

$$v_A A + v_B B \longrightarrow v_C C + v_D D \qquad (11.34)$$

we can write the following for the extent of reaction:

$$\lambda = -\frac{(n_f - n_i)_A}{v_A} = -\frac{(n_f - n_i)_B}{v_B} = \frac{(n_f - n_i)_C}{v_C} = \frac{(n_f - n_i)_D}{v_D} \qquad (11.34a)$$

where v_A, v_B, v_C, and v_D are the stoichiometric coefficients of components A, B, C, and D respectively, and $(n_f - n_i)_j$ is the change in the number of moles of the jth component due to the reaction.

*I. Prigogine and R. Defay, *Chemical Thermodynamics* (Longman Group Limited, 1973), p. 10.

Let us assume that the reaction in Equation 11.34 does not go to completion and the extent of reaction is λ. Then there will be some masses of the reactants A and B left over. The number of moles corresponding to these masses will be given by Equation 11.33(c).

$$n_{f,A} = -v_A\lambda + n_{i,A} \quad \text{and} \quad n_{f,B} = -v_B\lambda + n_{i,B}$$

Noting that the initial masses of compounds C and D are zero, the number of moles of these products will be given by Equation 11.33(b).

$$n_{f,C} = v_C\lambda \quad \text{and} \quad n_{f,D} = v_D\lambda$$

Then the equation for the partially completed reaction can be written as

$$n_{i,A}A + n_{i,B}B \longrightarrow (n_{i,A} - v_A\lambda)A + (n_{i,B} - v_B\lambda)B + v_C\lambda C + v_D\lambda D \quad (11.35)$$

If the initial number of moles of A and B are v_A and v_B, respectively, the last equation becomes

$$v_AA + v_BB \longrightarrow v_A(1 - \lambda)A + v_B(1 - \lambda)B + v_C\lambda C + v_D\lambda D \quad (11.35a)$$

Example 11.14

One mole of CO is burned with 200% theoretical air in a steady flow process at 1 atm. The carbon monoxide and air are both supplied at 25°C and the products leave at 2400 K. Determine the extent of reaction and the amount of heat transferred. Assume that the initial number of moles of the reactants numerically equal the respective stoichiometric coefficients.

Solution:

Given: Carbon monoxide is burned in a steady flow process at 1 atmosphere.

Excess air:	100%
Reactant temperature:	25°C
Product temperature:	2400 K
Number of moles of CO:	1
Number of moles of O_2:	1
Number of moles of N_2:	3.76

Objective: The extent of reaction λ and the amount of heat transferred are to be determined.

Assumptions: None.

Relevant physics: The reaction equation can be written as

$$CO + O_2 + 3.76\,N_2 \rightleftarrows CO_2 + \tfrac{1}{2}O_2 + 3.76\,N_2$$

Since the reaction may not go to completion, we represent the extent of reaction by λ and rewrite the reaction equation following Equation 11.35(a).

$$CO + O_2 + 3.76\,N_2 \longrightarrow \overbrace{(1 - \lambda)\,CO + (1 - \lambda)O_2 + 3.76\,N_2}^{\text{Left over reactants}} + \overbrace{\lambda\,CO_2 + \tfrac{1}{2}\lambda\,O_2}^{\text{Products}}$$

From the number of moles of a given constituent in the equilibrium mixture and the total number of moles of the mixture, the partial pressure of each constituent can be calculated in terms of λ. The value of K_p can be determined from the data in Appendix M. Then, substituting the values of the various partial pressures and the equilibrium constant in Equation 11.24(a), we can solve for λ. The heat transferred, Q, is the difference between H_{prod} and H_{react}.

Analysis: At equilibrium, the total number of moles, n_T, is

$$n_T = n_1 + n_2 + n_3 + n_4$$

where

$$n_1 = \text{number of moles of } CO_2 = \lambda$$

$$n_2 = \text{number of moles of } CO = 1 - \lambda$$

$$n_3 = \text{number of moles of } O_2 = 1 - \lambda + \frac{\lambda}{2} = 1 - \frac{\lambda}{2}$$

$$n_4 = \text{number of moles of } N_2 = 3.76$$

Substitution gives

$$n_T = 5.26 - \frac{\lambda}{2}$$

Expressions for the partial pressures of the gases in the products are determined below.

$$CO_2: \quad p_1 = \frac{n_1}{n_T} = \frac{\lambda}{5.26 - \lambda/2}$$

$$CO: \quad p_2 = \frac{n_2}{n_T} = \frac{1 - \lambda}{5.26 - \lambda/2}$$

$$O_2: \quad p_3 = \frac{n_3}{n_T} = \frac{1 - \lambda/2}{5.26 - \lambda/2}$$

The basic reaction in this example is $CO + \frac{1}{2}O_2 \rightarrow CO_2$. From Appendix M, the value of $\ln K_p$ for the reverse of this reaction at 2400 K is -3.860. The value of K_p for the present reaction is then given by

$$K_p = \frac{1}{e^{-3.86}} = 47.46$$

Also from Equation 11.24(a)

$$K_p = \frac{p_1}{p_2 p_3^{1/2}}$$

Substituting the expressions for p_1, p_2, and p_3, we obtain

$$\frac{\lambda/(5.26 - \lambda/2)}{[(1 - \lambda)/(5.26 - \lambda/2)][(1 - \lambda/2)/(5.26 - \lambda/2)]^{1/2}} = 47.46$$

Simplification gives

$$\frac{\lambda(10.52 - \lambda)^{1/2}}{(1 - \lambda)(2 - \lambda)^{1/2}} = 47.46$$

A trial-and-error procedure gives

$$\lambda = 0.94$$

We can now calculate the number of moles of each constituent in the products.

$$CO_2: \quad n_1 = \lambda \quad\quad = 0.94$$

$$CO: \quad n_2 = 1 - \lambda \ = 0.06$$

$$O_2: \quad n_3 = 1 - \frac{\lambda}{2} = 0.53$$

$$N_2: \quad n_4 = 3.76 \quad\quad = 3.76$$

This means that 0.06 moles of CO remain unburned. Now we can determine the amount of heat transferred. The reaction can be written as

$$CO + O_2 + 3.76\,N_2 \longrightarrow$$

$$0.94\,CO_2 + 0.06\,CO + \left(1 - \frac{0.94}{2}\right)O_2 + 3.76\,N_2$$

The heat transferred is given by

$$Q = H_{\text{prod}} - H_{\text{react}}$$

Enthalpy of the products:

$$CO_2: \quad H_1 = n_1(\bar{h}^0_{f1} + \bar{h}_1) = 0.94(-393,522 + 115,788) = -261,070$$

$$CO: \quad H_2 = n_2(\bar{h}^0_{f2} + \bar{h}_2) = 0.06(-110,529 + 71,346) = -2,351$$

$$O_2: \quad H_3 = n_3(\bar{h}^0_{f3} + \bar{h}_3) = 0.53(0 + 74,492) = 39,481$$

$$N_2: \quad H_4 = n_4(\bar{h}^0_{f4} + \bar{h}_4) = 3.76(0 + 70,651) = 265,648$$

$$H_{\text{prod}} = 41,708$$

Enthalpy of the reactants at 298 K:

$$CO: \quad H_5 = H_{\text{react}} = -110,529 + 0$$

Then

$$Q = H_{\text{prod}} - H_{\text{react}}$$

$$= 41,708 - (-110,529)$$

$$= 152,237 \text{ kJ/kg-mol (CO)}$$

A compound that is otherwise stable at ordinary temperatures, often partially dissociates into its constituent elements and radicals at high temperatures. For instance, at high temperature water molecules may dissociate to form hydrogen and oxygen molecules. However, only some of the molecules in a given sample may dissociate. The degree of dissociation can be determined in exactly the same manner as the extent of reaction, λ, is determined in Example 11.14. If a given sample of gaseous mixture contains two species that can dissociate, then we determine λ_1 and λ_2, the degrees of dissociation for the two species, by using two equilibrium constants simultaneously.

11.8 SUMMARY

Chemical reactions and equilibrium were the main topics considered in this chapter. After a brief discussion of the solid, liquid, and gaseous fuels, chemical reactions and stiochiometric equations were examined. Concepts of theoretical air and excess air were presented. The concepts of internal energy and enthalpy of formation were introduced. Also introduced were the heat of reaction, and higher and lower heating values of fuels. For combustion processes where no heat transfer takes place, the adiabatic flame temperature and the explosion pressure were defined.

The third law of thermodynamics, which states that the entropy of a pure crystalline substance is zero at absolute-zero temperature, was introduced in this chapter. The third law enabled us to calculate the absolute entropy of a substance.

As a result of the application of the second law to a chemical reaction, we could determine how far a given reaction can proceed. Conditions for the equilibrium of an isolated system were expressed in terms of the Helmholtz function and the Gibbs function. The Gibbs function of a system is at its minimum when the system is in equilibrium. This leads to the definition of the equilibrium constant, K_p. Once the value of K_p is known for a reaction, we can calculate the equilibrium composition. The Gibbs-Helmholtz equation and van't Hoff isobar equation, relating the effect of change of the reaction temperature on the standard-state, Gibbs-function change and the equilibrium constant, respectively, were derived. Finally, the extent of a reaction was defined and a method was presented to determine the extent of a given reaction.

PROBLEMS

11.1. A certain fuel, the chemical formula for which can be written as $C_6H_{17}O$, is burned in a stoichiometric reaction. Write the equation for reaction and determine the air-fuel ratio on mass and volume bases.

11.2. Octane is burned with air, and an Orsat's analysis gives the following composition of the products of combustion on a dry basis: CO_2, 9%; CO, 1.3%; O_2, 7.8%; N_2, 81.9%. Determine the excess air used in this combustion process.

11.3. A certain coal has the following composition by mass: CO, 85%, H_2 4.3%, O_2 1.5%, S 0.8%, and N_2, 1.1%, the balance being ash. Find the equivalent formula for the coal and write the stoichiometric equation of reaction. Also, determine the air-fuel ratio.

11.4. Calculate the stoichiometric air-fuel ratio by mass and the corresponding volumetric analysis of the products of combustion of ethyl alcohol, C_2H_6O.

11.5. In practice, the carbon dioxide content of exhaust gases is used to determine the excess air used for combustion. If z denotes the fraction of excess air in the reaction

$$(CH_2)_n + 1.5n(1 + z) O_2 + 5.67n(1 + z) N_2 \longrightarrow$$
$$n\, CO_2 + n\, H_2O + 1.5nz\, O_2 + 5.67n(1 + z) N_2$$

show that the mole fraction of CO_2 after the condensation of the water vapor in the exhaust is given by $1/(6.67 + 7.17z)$.

11.6. For the coal in Problem 11.3, convert the analysis to a dry basis and then determine the air-fuel ratio.

11.7. Octane is burned with 75% excess air. Assuming the pressure of the products of combustion to be 100 kPa, determine the dew-point temperature.

11.8. One kilogram of sugar, $C_{12}H_{22}O_{11}$, is completely burned by using 100% excess air. Determine the volumetric composition of the products of combustion.

11.9. One kg-mole of heptane (C_7H_{16}) is burned with 40% excess air. Assume that the combustion is complete. Find (a) stoichiometric (theoretical) air-fuel ratio, (b) percentage of oxygen present in products by volume and by weight, and (c) change in the number of moles between reactants and products. Molar masses of atmospheric nitrogen and of air are 28.2 kg/kg-mol and 29.0 kg/kg-mol, respectively.

11.10. A certain type of coal with the gravimetric composition of C, 74.4%, H_2, 12%, O_2, 2%, and the rest ash, is burned with 160% theoretical air. Compute each value.
  (a) Stoichiometric and actual air-fuel ratios.
  (b) Exhaust gas composition on a dry basis assuming complete combustion.

11.11. A flue gas sample analyzed by Orsat's apparatus gave the following percentage values: CO_2, 7.2%; CO, 1%; O_2, 10%; N_2, 81.8%. Express this volumetric composition on mass basis. Assuming that the fuel is hydrocarbon, find its approximate formula.

11.12. A certain fuel has the following gravimetric composition: C, 88%; H_2, 4%; ash, 8%. Find each of the following.
  (a) Stoichiometric air-fuel ratio.
  (b) Gravimetric composition of the products.
  (c) Volumetric composition of the products.

11.13. During the combustion of the fuel in Problem 11.12, the dry flue gas composition determined by Orsat's apparatus was as follows: CO_2, 12.6%; O_2, 7.051%; N_2, 80.349%. Find each value.
  (a) Mass of dry products (flue gases) per kilogram of fuel burned.
  (b) Excess air.

11.14. A producer gas having the volumetric composition of 10% H_2, 3% CH_4, 18% CO, 6% CO_2, 2% O_2, and 60.3% N_2, is burned in a process boiler. The volumetric composition of the dry flue gases determined by Orsat's apparatus is 15% CO_2, 4.7% O_2, and 80.3% N_2. Determine the excess air used for combustion.

11.15. Propane is burned with 50% excess air. Determine the amount of heat transferred if the exhaust is at 1100 K. Assume the propane and supply air to be at 25°C.

11.16. Liquid octane at 25°C is burned with 10% excess air. The supply air is at 400 K and the products of combustion are at 1400 K. Determine the amount of heat that must be removed from the combustion chamber.

11.17. If the temperature of the combustion gases of Problem 11.16 is to be restricted to 1000 K and if no heat is removed from the combustion chamber, how much excess air should be used?

11.18. Determine the heat of reaction of gaseous propane at 25°F when burned with 25% excess air, for an exhaust temperature of (a) 1000 K, and (b) 2000 K.

11.19. Hydrogen peroxide, H_2O_2, at 25°C and 800 kPa is admitted into a gas generator at a rate of 0.2 kg/s and decomposed into oxygen and steam. The products are expanded from 800 kPa and 1000 K to 100 kPa in a gas turbine. The enthalpy of formation of liquid H_2O_2 is $-187,583$ kJ/kg-mol. Determine the heat transfer to the gas generator and the power output of the turbine.

11.20. One mole of acetylene gas and 2 mol of oxygen react to produce 1 mol of carbon monoxide, 1 mol of carbon doixide, and some water vapor. The products of combustion leave at 600 K. Calculate the quantity of heat transferred in kiloJoules per kilogram mole of acetylene gas. Assume that the pressure of the reactants and products is one atmosphere.

11.21. A certain type of coal has the following composition by weight: carbon, 77%; sulfur, 1.2%; the remainder is ash. The coal has a heating value of 11,000 kJ/kg and is used as the fuel for a 800-MW plant which has a thermal efficiency of 39%. Calculate the quantity of SO_2 given off in kilograms per hour.

11.22. Ethane at standard temperature and pressure is burned with 22% excess air of 60% relative humidity in a steady-flow system. If the products of combustion leave at 500°K, determine the quantity of heat energy released in kiloJoules per kilogram of ethane.

11.23. Rework Problem 11.22 if the fuel is butane.

11.24. Estimate the adiabatic flame temperature when the coal of Problem 11.21 is burned with 50% excess air. State your assumptions.

11.25. What is the adiabatic flame temperature of ethane in a stoichiometric reaction?

11.26. If the adiabatic flame temperature of Problem 11.25 were to be reduced by 25%, how much excess air should be used?

11.27. Sulfur at 25°C is burned with 25% excess air. The supply air temperature is 500 K. The $\bar{h}_f^0$ value for SO_2 is $-296,831$ kJ/kg-mol. Determine the adiabatic flame temperature. The specific heat of sulfur is given by

$$\bar{c}_p = 32.238 + 0.0219T - 3.527 \times 10^{-6} \, T^2$$

where $\bar{c}_p$ is in kJ/kg-mol K and T is in degrees Kelvin.

11.28. The combustion in the cylinder of an automobile engine when the piston is at its dead center can be treated as a constant volume process. The air-fuel ratio for a certain engine is 16 to 1 by mass, and the fuel used is octane. The temperature and pressure of the air-fuel mixture just before ignition are 300°C and 900 kPa, respectively. Neglecting dissociation, determine the temperature and pressure immediately after combustion.

11.29. Two grams of liquid benzene are placed in a container of fixed volume of 0.025 m². The initial temperature of the benzene and the air is 25°C. If the benzene is burned with no excess air, determine the pressure in the container.

11.30. Evaluate the change in entropy when CO and O_2 react under stoichiometric conditions at 25°C.

11.31. Determine the heat released and the change of entropy when 1 kg-mol of CO reacts with 1 kg-mol of water vapor at 500 K and 100 kPa.

11.32. One mole of carbon and 2 mol of oxygen at 25°C enter a combustion chamber and the products leave at 1600 K and 100 kPa. Determine the heat transferred to the surroundings and the net entropy change of the process. The surroundings are at 25°C.

11.33. One kilogram-mole of carbon monoxide reacts with 1 kg-mol of water vapor in a steady-flow adiabatic process under a constant pressure condition. Assuming that the reaction goes to completion, determine the change in the entropy in kJ/kg-mol K of CO.

11.34. Propane gas at 25°C and 100 kPa is burned with 200% theoretical air, also at 25°C and 100 kPa. The temperature and pressure of the products of combustion are 1000 K and 100 kPa. The air enters the combustion chamber at 50 m/s and the products which are fed to a gas turbine leave at 500 m/s. Determine each of the following.

(a) Net heat transfer.

(b) Increase in entropy.

(c) Irreversibility of the process.

11.35. Calculate the maximum work for an isothermal reaction of methane with 200% theoretical air at standard conditions for the reactants and products.

11.36. One mole of CO and 1 mol of water vapor can react to produce CO_2 and H_2. Plot a graph of the Gibbs function of the products versus the number of moles of CO_2 formed. Determine equilibrium composition.

11.37. Calculate ΔG^0 for combustion of butane at 1800 K.

11.38. Determine the equilibrium constant for the reaction in Problem 11.37.

11.39. Calculate the Gibbs function of formation for water vapor.

11.40. The products of combustion of a hydrocarbon fuel and air at 1 atm pressure are

$$7.6\ CO_2 + 4.35\ CO + 5.45\ H_2O + 0.35\ H_2 + 1.98\ O_2 + 54.45\ N_2$$

Determine the equilibrium constant.

11.41. Determine the equilibrium constant for methane when it is completely oxidized at standard pressure.

11.42. At high temperatures, oxygen molecules dissociate. What is the dissociation of oxygen molecules, expressed as a percent, in each case?

(a) 3000 K and 1 atm

(b) 3000 K and 3 atm

11.43. When fuel is burned in the combustion chamber of a boiler in a power plant, the nitrogen from the combustion air dissociates and then forms oxides of nitrogen because of the high temperatures. We may expect that the nitrogen from such oxides should be released as the combustion gases are cooled. However, this does not happen since sufficient time is not available for cooling of the combustion gases as these gases come in contact with 1200 K surface of the boiler tubes. Consequently, oxides of nitrogen escape into the atmosphere. Calculate the percentage of NO present in 2500 K combustion gases that will be retained after sudden cooling to 1200 K.

11.44. In a steady-flow process, air is heated at 1 atm. Determine the mole fraction of NO in the hot air for each final temperature.

(a) 2000 K

(b) 3000 K

(c) 4000 K

(d) 5000 K

11.45. A mixture of 1 mol of CO and 1 mol of H_2O vapor is heated from 298 K to (a) 1000 K, (b) 2000 K, and (c) 3000 K. For each case, determine the quantity of heat energy to be supplied and the extent of the reaction.

11.46. Isooctane, C_8H_{18}, is burned with 20% excess air. Calculate the equilibrium composition if the products of combustion are at 4 atm and 2500 K.

11.47. If 200% excess O_2 is supplied for the combustion of CO and if the products are at 2500 K and 1 atm, determine the extent of reaction.

Bibliography

A. Books With Broad Coverage of Various Topics in Thermodynamics

ABBOTT, MICHAEL M., *Schaum's Outline of Theory and Problems of Thermodynamics*. New York: McGraw-Hill, 1972.

BACON, DENNIS HENRY, *Engineering Thermodynamics*. London: Butterworths, 1972.

BALZHISER, RICHARD E., *Chemical Engineering Thermodynamics, The Study of Energy, Entropy, and Equilibrium*. Englewood Cliffs, N.J.: Prentice-Hall, 1972.

BOSNJAKOVIC, FRAN, *Technical Thermodynamics*. trans. Perry L. Blackshear Jr. New York: Holt, Rinehart and Winston, 1965.

BRZUSTOWSHI, THOMAS A., *Introduction to the Principles of Engineering Thermodynamics*. Reading, Mass: Addison-Wesley, Pub. Co. 1969.

CALLEN, H.B., *Thermodynamics*, New York: John Willey, 1960.

FAIRES, VIRGIL MORING, *Thermodynamics*, 5th ed. New York: Macmillan, 1970.

———, and CLIFFORD MAX SIMMANG, *Thermodynamics*, 6th ed. New York: Macmillan, 1978.

FINCK, JOSEPH LOUIS, *Generalized Thermodynamics: Its Philosophy and Rationale*. Jerusalem: Jerusalem Academic Press, 1974.

GIBBINGS, J.C., *Thermodynamics, an Introduction to the Governing Equations of Thermodynamics, and of the Mechanics of Fluids*, 1st ed. Elmsford, N.Y.: Pergamon Press, 1970.

HABERMAN, WILLIAM L., and JAMES E. A. JOHN. *Engineering Thermodynamics*. Boston: Allyn & Bacon 1980.

HOFFMAN, JACK PHILLIP, *Thermodynamics*, 2nd ed. New York: McGraw-Hill, 1974.

HOLMAN, JACK, *Thermodynamics*. New York, McGraw Hill, 1977.

518

Huang, Francis F., *Engineering Thermodynamics*. New York: Macmillan, 1976.

Keenan, Joseph Henry, *Thermodynamics*. Cambridge, Mass: M.I.T. Press, 1970.

Mark, Melvin, and Arthur R. Foster. *Thermodynamics Principles and Applications*. Boston: Allyn & Bacon, 1979.

Reynolds, William Craig, *Engineering Thermodynamics*, 2nd ed. New York: McGraw-Hill, 1977.

Rogers, G.F.C., and Y. R. Mayhew. *Engineering Thermodynamics Work and Heat Transfer*. London: Longmans, Green and Co. Ltd., 1967.

Silver, Howard F., and John E. Nydahl. *Introduction to Engineering Thermodynamics*. New York: West Pub. Co., 1977.

Silver, Robert Simpson, *An Introduction to Thermodynamics With Some New Derivations Based on Real Irreversible Processes*. Cambridge, England: University Press, 1971.

Swalin, Richard A., *Thermodynamics of Solids*, 2nd ed. SI version New York: John Wiley, 1972.

Truesdell, C. and S. Bharatha, *Classical Thermodynamics as a Theory of Heat Engines*, Springer-Verlag, New York Inc. Federal Republic of Germany, 1977.

VanNess, H. C., and M. W. Zemanski. *Basic Engineering Thermodynamics*. New York: McGraw-Hill, 1966.

Van Wylen, Gordon John, and Richard E. Sonntag, *Fundamentals of Classical Thermodynamics*, 2nd ed. New York: John Wiley, 1978.

Wales, Charles, E., *Programmed Thermodynamics*, Vol. 1. New York: McGraw-Hill, 1970.

Wall, Frederick Theodore, *Chemical Thermodynamics: A Course of Study*, 3rd ed. San Francisco Calif.: W. H. Freeman, 1974.

Wallace, Duane C., *Thermodynamics of Chrystals*. New York: John Wiley, 1972.

Wark, Kenneth, *Thermodynamics*, 2nd ed. New York: McGraw-Hill, 1977.

Ziegler, Hans, *Intro to Thermodynamics*. New York: Elsevier-North-Holland, 1976.

B. Internal and External Combustion Engines

Anderson, John Wallace, *Diesel Engines*, 1st ed. New York: McGraw-Hill, 1935.

Bailey, Neil Phillips, *Principles of Heat Engineering*. New York: John Wiley, (n.d.)

Gillispie, Charles Coulston, *Lazare Carnot Savant, A Monograph Treating Carnot's Scientific Work*. Princeton, NJ: Princeton University Press, 1971.

Kates, Edgar Jesse, *Diesel and High Compression Gas Engines: Fundamentals*. Chicago: American Tech. Soc., 1954.

Levine, David Allan, *Internal Combustion: The Race in Detroit*. Westport, Conn: Greenwood Press, 1976.

Maleev, V. L., *Internal Combustion Engines Theory and Design*, 2nd. ed. New York: McGraw-Hill, 1945.

Orbert, Edward F., *Internal Combustion Engines and Air Pollution*. New York: Harper & Row Pub., 1973.

Purday, Herbert Frank Percy, *Diesel Engine Design*, 4th ed. New York: D. Van Nostrand, 1937.

WALKER, GRAHAM, *Stirling Cycle Machine*, Oxford: Clarington Press, 1973.

ZARINCHANG, JAFAR, *The Stirling Engine*. London: Intermediate Technology Group, 1972.

C. Steam Power Plants, and Steam and Turbines

COHEN, H. ROGERS, *Gas Turbine Theory*. New York: John Wiley, 1973.

Combustion in Fossil Power Systems, 3rd ed. Windsor, Conn: Combustion Engineering, 1981.

CROFT, TERREL WILLIAMS, *Steam Engine Principles and Practices*, 2nd ed. New York: Mc-Graw-Hill, 1939.

DELAVAL STEAM TURBINE CO., *DeLaval Handbook; An Engineering Data Book for Users of Pumps, Turbines Compressors and Gears:* Trenton, NJ: 1947.

DIXON, SYDNEY LAWRENCE, *Fluid Mechanics, Thermodynamics of Turbomachinery*, 2nd ed. Pergamon Press, N.Y.: Oxford, 1975.

GAFFERT, GUSTAF ADOLF, *Steam Power Stations*, 2nd and 3rd eds. New York: McGraw-Hill, 1946.

KEARTON, W. J., *Steam Turbine Theory and Practice*, 7th ed. Liverpool, U.K., London U.K.: The English Language Book Society and Pitman Publishing, 1973.

NORBYE, JAN P., *The Gas Turbine Engine: Design Development, Applications*. Radnor, PA: Chilton, 1975.

RIMBERG, DAVID, *Utilization of Waste Heat from Power Plants*. Park Ridge, NJ: Noyes Data Corp., 1974.

SPANNHAKE, WILHELM, *Centrifugal Pumps, Turbines and Propellers: Basic Theory and Characteristics*. Cambridge; Mass: The Technology Press, M.I.T., 1934.

STODOLA, AUREL, *Steam & Gas Turbines, With a Supplement on the Prospects of the Thermal Prime Mover*. Glouster, Mass: P. Smith, 1945.

STORER, J. D., *A Simple Story of the Steam Engine*. London: J. Baker, 1969.

WOODRUFF, EVERETT BOWMAN, *Steam Plant Operation*, 3rd ed. New York: McGraw-Hill, 1967.

VAVRA, MICHAEL H., *Aero-Thermodynamics and Flow in Turbomachines*. New York: R. E. Krieger Pub. Co., 1974.

D. Fuels and Combustion

DELORENZI, OTTO, *Combustion Engineering a Reference Book on Fuel Burning and Steam Generating*. New York: New York Combustion Engineering Co., 1947.

Fluid Mechanics of Combustion. New York: A.S.M.E., 1974.

HILL, PHILLIP GRAHAM, *Power Generation Resources, Hazards Technology, and Cost*. Cambridge, Mass: M.I.T. Press, 1977.

PEARSON, JERRY DEAN, *Thermodynamic Properties of Combustion Gasses*, 1st ed. Ames, Iowa: Iowa State University Press, 1965.

The Efficient Use of Energy/Guilford Engineering. IPC Science and Technology Press in Collaboration with The Institute of Fuel Acting on Behalf of the U.K. Dept. of Energy, 1975.

WILLIAMS, FORMAN A., *Combustion Theory. The Fundamental Theory of Chemical Reacting Flow Systems*. Reading, Mass: Addison-Wesley, 1965.

E. Miscellaneous

KREIDER, JAN F. and FRANK KREITH, *Solar Heating and Cooling*. New York: McGraw-Hill, 1975.

TRINKS, W. and M. H. MAWHINNEY, *Industrial Furnaces*, 5th ed, Vol. I. New York: John Wiley, 1961.

Answers to Selected Problems

Chapter 1

1.2 1062.7 kPa   **1.3(a)** 798.7 kPa absolute   **1.6(a)** 730.8 mm of Hg
1.9(a) 1,438 Pascals **(b)** 11 Torr   **1.11(a)** 7.18 cm **(b)** 30 cm   **1.16(a)** 0.638 **(b)** 31.9 units
1.20.(a) 77.2 kg **(b)** 2.413 kg·mol   **1.21.** 12.85 kg   **1.23** 451.6 kPa

Chapter 2

2.2 17.6 J/s   **2.8** −101.25 N·cm   **2.9.** −89.75 N·cm   **2.12** 1.806×10^{-3} N·m
2.17.(a) 0.138 MJ/kg **(b)** zero   **2.18(a)** 0.65 MJ **(b)** 2863 K
2.19.(a) −1.229 MJ **(b)** 4.25 MPa   **2.21** 45.17 kJ   **2.24.(a)** 2.55 kJ **(b)** 0.995 kJ
2.25.(a) −17.98 MJ **(b)** −14.74 MJ
2.27.(a) $V_2 = 0.0335$ m³, $W = 3.604$ kJ, $W_{\text{nonusable}} = 1.35$ kJ **(b)** $\eta = 1.216$, $W_{\text{quasi}} = 4.89$ kJ,
$W_{\text{lost}} = 767$ J   **2.29.** zero

Chapter 3

3.1. 11.0 kN   **3.2.** 0.707 N   **3.4.(a)** 373 W/m^2 **(b)** 500 W/m^2 **3.5.** −56 kW
3.10.(a) 2.29 MJ **(b)** zero   **3.12.(a)** −2.0kJ **(b)** zero **(c)** 2.0 MJ   **3.15** 3.451 MJ/kg·mol
3.16 1.126   **3.32** −4948 MJ   **3.34.(a)** 0.618 **(b)** 275 J
3.35 204 kJ/cycle; 0.51   **3.36.(a)** 221.6 kJ **(b)** 0.554

Chapter 4

4.1.(a) Superheated vapor **(b)** Compressed liquid
4.3.(a) $v_g = 16.265$ m³/kg **(b)** $v = 0.001029$ m³/kg to 3.407 m³/kg
4.4.(a) $u_f = 503.5$ kJ/kg; $h_f = 504.03$ kJ/kg **(c)** $u_g = 2572.2$ kJ/kg; $h_g = 2763.5$ kJ/kg
4.6.(a) 0.0906 m³/kg **(b)** 2084.5 kJ/kg   **4.7.** 912.76 kJ/kg   **4.17.(a)** .968 **(b)** 2.55 MJ/kg
4.19.(a) 0.63 **(b)** .1301 m³/kg   **4.20.** .288   **4.21.(a)** 0.0 **(c)** 104.95 kJ/kg **(e)** 231.2 kJ/kg

4.22.(a) 6318.3 kJ **(b)** 6818 kJ   **4.26.** −2141.9 kJ/kg   **4.27.** 28,218 kJ
4.28(a) 320.46 mJ **(b)** 25.23% vapor; 74.77% liquid   **4.29.** 17,034 kJ
4.32.(a) 0.08643 m³/kg **(b)** 0.8921 m³/kg **(c)** 0.0874 m³/kg   **4.34.(a)** for ideal gas, error 1.0%

Chapter 5

5.1.(a) 1.146 kg/m³ **(b)** 46 cm   **5.2.(a)** 7.42 kg/s **(b)** 46 cm   **5.4.** 2.08 cm
5.6. 13.1 MJ/s   **5.9.** −915.8 J/s   **5.14.** 10.4 kg/s   **5.17.** 1.691 kg/s   **5.18.** 17.6 g/s
5.23.(a) 16.2 m/s **(b)** 370.4 kPa   **5.27.(a)** 36.2 m/s **(b)** 233.04 K **(c)** 111.4 kPa
5.28.(a) 13.77 m/s **(b)** 8.101 kw   **5.32.** 249 K   **5.34.** 95.14%   **5.35.** 12.74 kg/hr
5.38. 10.25 kg/hr   **5.39.(a)** 17.7 m/s **(b)** 94.3 m/s **(c)** 200.6 kg/s   **5.41.(a)** −30°C
(b) 0.46   **5.43.(a)** −21.1 kJ/kg **(b)** 640.7 kJ/kg **(c)** −2173.28 **(d)** 2830.1 kJ/kg **(e)** 22.46%
5.51. 417.4 K

Chapter 6

6.3. 0.067   **6.4.** impossible   **6.7.(a)** −2904 J **(b)** 1096 J **(c)** 0.267
6.9.(a) 33.85 m³ **(b)** 0.298 kJ/K **(c)** 159.0 kJ, −87.73 kJ **(d)** 71.27 kW **(e)** 44.8%
6.11. −16.65 J/K, 473 + dT, where dT is positive   **6.12.(a)** 9.13 kJ/K **(b)** 51.47 kJ/K
6.15.(a) irreversible **(b)** 12.7 kJ   **6.16.** 0.512 kJ/K   **6.18.** −0.146 kJ/K   **6.19.** 4.45 J/K
6.21.(a) −8.06 J/K **(b)** 8.33 J/K **(c)** 0.27 J/K   **6.23.(a)** 1.2025 kJ/K **(b)** −1.045 kJ/K
(c) 0.157 kJ/K   **6.24.** 0.0829 kJ/K   **6.27.(a)** 2108.4 kJ/kg **(b)** 4.961 kJ/kg·K
6.29 3.92 kJ/K   **6.33.(a)** 18.7 m³, 371.54 K, 10.33 kPa **(b)** 0.1616 kJ/K   **6.34.** 0.144 kJ/K
6.38.(a) 0.0105 kJ/kg·K **(b)** 1.07   **6.41(a)** 0.3121 kJ/K **(b)** 353.15 K   **6.43.** 552.7 kJ/kg
6.46. 168.85 kJ/kg   **6.50.(a)** −19.47 kJ **(b)** 6.46 kJ

Chapter 7

7.1.(a) 0.81 **(b)** 0.74   **7.3.(a)** 3.28 kW **(b)** 735 cc **(c)** 0.808
7.6.(a) 0.797 **(b)** 393.5 K **(c)** 17.35 kW   **7.11.(a)** −229.3 kJ/kg **(b)** −274.0 kJ/kg
7.13.(a) 1176.9 kJ/kg **(b)** 601.1 kJ/kg **(c)** 0.51   **7.14.** 342.6 kPa
7.16.(a) 0.553, 1646 kPa, 638 K **(b)** 0.454, 1345 kPa, 521K
7.18.(a) 7.67 kW **(b)** 323.63 kPa **(c)** 0.51   **7.23.** 0.6037
7.25.(a) 0.267 **(b)** 0.322 **(c)** 0.827   **7.30.(a)** 1446 kJ/kg **(b)** 0.578
7.31.(a) 171.75 kJ/kg **(b)** 434.0 kJ/kg **(c)** 0.369
7.33.(a) 193.3 kJ/kg **(b)** 431.5 kJ/kg **(c)** 49.9 MW **(d)** 0.40 **(e)** 83.9 kg/s
7.35.(a) 35.8 **(b)** 78.3 kJ/kg   **7.36.** 0.456 kg/s   **7.38.** 0.604   **7.39.(a)** 0.467
(b) 0.958   **7.41.** 338.9 kW   **7.43.(a)** 112.7 kJ/kg **(b)** 0.203   **7.46.(a)** 2772.9 kW
(b) 0.261   **7.49.(a)** 176.1 kJ/kg **(b)** 588.6 kJ/kg **(c)** $\eta_t = 0.347$, $(\eta_{th})_{reg} = 0.70$ **7.51.** 0.467

Chapter 8

8.1.(a) $w_p = -1.46$ kJ/kg, $w_t = 457.7$ kJ/kg **(b)** $\eta_t = 0.192$, $\eta_c = 0.209$
8.3.(a) 0.195 **(b)** 0.210 kg/s   **8.5.(a)** 0.598 **(b)** 641 kJ/kg
8.7.(a) −1.2 kJ/kg **(b)** 3628.7 kJ/kg **(c)** 980.5 kJ/kg **(d)** 0.2702 **(e)** 10.20 kg/s
8.9.(a) 1077.8 kJ/kg **(b)** 0.363   **8.12.** 7072 m²   **8.13.(a)** 1194.1 kJ/kg **(b)** 0.384
8.17.(a) 814.1 kJ/kg **(b)** 0.327   **8.20.(a)** 1221.1 kJ/kg **(b)** 689.9 kJ/kg **(c)** 0.338
8.22.(a) 0.906 **(b)** 96.23 kg/s **(c)** 2196.8 kJ/kg
8.24.(a) 1946 MW **(b)** 17,882 kg/s **(c)** 5.0 MW **(d)** 0.385 **(e)** 0.1176 kg
8.27.(a) 321.85 kg/s **(b)** 0.381 **(c)** 112,536 kg/hr **8.31.(a)** 11.7 kW **(b)** 2.69 kW **(c)** 2.57 tons
8.32.(a) COP = 13.41 **(b)** 18.64 kW **(c)** 231.36 kW   **8.33.** 9.1
8.35.(a) COP = 1.75, $T_H = 429$ K **(b)** 0.513 kW/ton   **8.38.** 21.4°C
8.40.(a) 2.92 kg/s **(b)** 4.27 **(c)** 0.82 kW/ton   **8.42.(a)** 5.93 **(b)** 0.0930 kg/s
8.45.(a) 4.52 **(b)** 3.32 kW/ton   **8.46.** 23.34% decrease   **8.48.(a)** 0.387
(b) 14.92 m³/min·ton   **8.50.(a)** 0.774 tons **(b)** 0.907 **(c)** 2.22   **8.51.** 0.590

Chapter 9

9.10.(a) $-14{,}431$ J/kg **(b)** -19358 J/kg   **9.12.** 0.82% **9.15.** 38.1 kPa   **9.16.** 1162.2 kJ/kg
9.26. 48.5 Pa

Chapter 10

10.2. 0.1882 kg-mol   **10.4.(a)** 221.6 kPa **(b)** 517.1 kPa   **10.6.(a)** 0.361 kg **(b)** 0.577 kg
10.7.(a) $m_{CO} = 0.84$ kg, $m_{O_2} = 2.56$ kg, $m_{CO_2} = 15.40$ kg, $m_{N_2} = 15.13$ kg
(b) $p_{CO} = 3$ kPa, $p_{O_2} = 8$ kPa, $p_{CO_2} = 35$ kPa, $p_{N_2} = 54$ kPa
10.9.(a) $p_{Ar} = 2159$ Pa, $V_{Ar} = 0.173$ m³, $p_{He} = 47410$ Pa, $V_{He} = 3.793$ m³, $p_{O_2} = 9166$ Pa,
$V_{O_2} = 0.733$ m³, $p_{N_2} = 41264$ Pa, $V_{N_2} = 3.301$ m³
(b) $c_p = 1.452$ kJ/kg·K **(c)** $h = 435.8$ kJ/kg, $u = 291.2$ kJ/kg   **10.12.** -1133.6 kJ
10.14.(a) -36.61 kJ/kg **(b)** -51.27 kJ/kg **(c)** 35.54 kJ/kg·k
10.17(a) 98.11 Pa **(b)** $99.15°C$ **(c)** 24.49 m³, 0.472 m³, $m_a = 28.42$ kg, $m_w = 0.340$ kg
10.18.(a) 0.0251 **(b)** $98.84°C$ **(c)** 79.2% **10.20.(a)** $\phi = 9\%$ **(b)** $V_w = 0.00191$ m³,
$V_a = 0.8462$ m³   **10.21.** $10.5°C$ **10.22.(a)** $8°C$ **(b)** 0.005 kg_w/kg_a **10.25.** 65%
10.26.(a) $T_2 = 24.4°C$, $w = -4568.8$ J/kg **(b)** $W = 0.00966$ kg_w/kg_a
10.30. 473.5 kJ/s, 12%, 0.140 kg_w/s
10.35.(a) -1014.68 kJ/min **(b)** 214.43 kJ/min **(c)** 0.261 kg_w/min
10.37. $21.67°C$, 4.12 kg_w/min
10.39.(a) 0.007 kg_w/kg_a, 10% **(b)** 17.06 kg_a/s **(c)** 0.0452 kg_w/ks, 0.007 kg_w/kg_a
10.42. 6.30 kg_a/s, 4.07 kg_w/s   **10.44.** 100.32 kg/s, 1.655 kg_w/s

Chapter 11

11.2. 18.58% **11.3.(a)** $C_{7083}H_{2150}O_{469}S_{25}N_{39}$ **(b)** 10.509
11.4.(a) 8.99 **(b)** $V_{H_2O} = 18.4\%$, $V_{N_2} = 69.29\%$, $V_{CO_2} = 12.29\%$
11.8. $V_{CO_2} = 9.58\%$, $V_{H_2O} = 8.78\%$, $V_{O_2} = 9.58\%$, $V_{N_2} = 72.05\%$
11.11. $C_{205}H_{406}$ **11.15.** $-1{,}042{,}987$ kJ/kg·mol
11.20. $-873{,}394$ kJ/kg·mol   **11.25.** 2400 K   **11.26.** 85.1%
11.27. 1765.75 K   **11.28.** 2502.7 K, 4.145 MPa
11.31.(a) $-39{,}818$ kJ/kg·mol **(b)** 40.95 kJ/K
11.34.(a) 604.791 kJ/kg·mol **(b)** 2334 kJ/kg-K·mol **(c)** $15{,}783$ kJ/kg
11.37. -1.513 GJ/kg·mol   **11.45.(a)** 280.3 kJ/kg·mol; 52.6% complete
(b) 2242.6 kJ; 10.85% complete **(c)** 4355.0 kJ, 4.24% complete   **11.47.** 91% complete

Appendices

Gas/Chemical Formula		Molar Mass	$R \dfrac{kJ}{kg\ K}$	$c_{po} \dfrac{kJ}{kg\ K}$	$c_{vo} \dfrac{kJ}{kg\ K}$	γ
Acetylene	C_2H_2	26.038	0.319 30	1.711	1.393	1.23
Air	—	28.97	0.287 00	1.0035	0.7165	1.400
Ammonia	NH_3	17.03	0.488 20	2.092	1.607	1.30
Argon	Ar	39.948	0.208 13	0.5203	0.3122	1.667
Butane	C_4H_{10}	58.124	0.143 04	1.7164	1.5734	1.091
Carbon Dioxide	CO_2	44.01	0.188 92	0.8418	0.6529	1.289
Carbon Monoxide	CO	28.01	0.296 83	1.0413	0.7445	1.400
Ethane	C_2H_6	30.07	0.276 50	1.7662	1.4897	1.186
Ethylene	C_2H_4	28.054	0.296 37	1.5482	1.2518	1.237
Freon-12	CCl_2F_2	120.92	0.068 80	0.586	0.519	1.13
Freon-22	$CHClF_2$	86.476	0.096 1	0.628	0.536	1.17
Helium	He	4.003	2.077 03	5.1926	3.1156	1.667
Hydrogen	H_2	2.016	4.124 18	14.2091	10.0849	1.409
Methane	CH_4	16.04	0.518 35	2.2537	1.7354	1.299
Neon	Ne	20.183	0.411 95	1.0299	0.6179	1.667
Nitrogen	N_2	28.013	0.296 80	1.0416	0.7448	1.400
Octane	C_8H_{18}	114.23	0.072 79	1.7113	1.6385	1.044
Oxygen	O_2	31.999	0.259 83	0.9216	0.6618	1.393
Propane	C_3H_8	44.097	0.188 55	1.6794	1.4909	1.126
Steam	H_2O	18.015	0.461 52	1.8723	1.4108	1.327

APPENDIX B1   PROPERTIES OF COMPRESSED LIQUID WATER

(v in m^3/kg, u in kJ/kg, h in kJ/kg, and s in kJ/kgK) (G. J. Van Wylen and R. E. Sonntag. *Fundamentals of Classical Thermodynamics, SI Version*, 2nd Edition. New York: John Wiley & Son, Inc. © 1976. Reprinted by permission.)

Properties of Compressed Liquid Water

T	p = 5,000 kPa (263.99)				p = 10,000 kPa (311.06)				p = 15,000 kPa (342.24)			
	v	u	h	s	v	u	h	s	v	u	h	s
Sat.	.001 285 9	1147.8	1154.2	2.9202	.001 452 4	1393.0	1407.6	3.3596	.001 658 1	1585.6	1610.5	3.6848
0	.000 997 7	.04	5.04	.0001	.000 995 2	.09	10.04	.0002	.000 992 8	.15	15.05	.0004
20	.000 999 5	83.65	88.65	.2956	.000 997 2	83.36	93.33	.2945	.000 995 0	83.06	97.99	.2934
40	.001 005 6	166.95	171.97	.5705	.001 003 4	166.35	176.38	.5686	.001 001 3	165.76	180.78	.5666
60	.001 014 9	250.23	255.30	.8285	.001 012 7	249.36	259.49	.8258	.001 010 5	248.51	263.67	.8232
80	.001 026 8	333.72	338.85	1.0720	.001 024 5	332.59	342.83	1.0688	.001 022 2	331.48	346.81	1.0656
100	.001 041 0	417.52	422.72	1.3030	.001 038 5	416.12	426.50	1.2992	.001 036 1	414.74	430.28	1.2955
120	.001 057 6	501.80	507.09	1.5233	.001 054 9	500.08	510.64	1.5189	.001 052 2	498.40	514.19	1.5145
140	.001 076 8	586.76	592.15	1.7343	.001 073 7	584.68	595.42	1.7292	.001 070 7	582.66	598.72	1.7242
160	.001 098 8	672.62	678.12	1.9375	.001 095 3	670.13	681.08	1.9317	.001 091 8	667.71	684.09	1.9260
180	.001 124 0	759.63	765.25	2.1341	.001 119 9	756.65	767.84	2.1275	.001 115 9	753.76	770.50	2.1210
200	.001 153 0	848.1	853.9	2.3255	.001 148 0	844.5	856.0	2.3178	.001 143 3	841.0	858.2	2.3104
220	.001 186 6	938.4	944.4	2.5128	.001 180 5	934.1	945.9	2.5039	.001 174 8	929.9	947.5	2.4953
240	.001 226 4	1031.4	1037.5	2.6979	.001 218 7	1026.0	1038.1	2.6872	.001 211 4	1020.8	1039.0	2.6771
260	.001 274 9	1127.9	1134.3	2.8830	.001 264 5	1121.1	1133.7	2.8699	.001 255 0	1114.6	1133.4	2.8576
280					.001 321 6	1220.9	1234.1	3.0548	.001 308 4	1212.5	1232.1	3.0393
300					.001 397 2	1328.4	1342.3	3.2469	.001 377 0	1316.6	1337.3	3.2260
320									.001 472 4	1431.1	1453.2	3.4247
340									.001 631 1	1567.5	1591.9	3.6546

APPENDIX B1 (Continued)

T	$p = 20,000$ kPa (365.81) v	u	h	s	$p = 30,000$ kPa v	u	h	s	$p = 50,000$ kPa v	u	h	s
Sat.	.002 036	1785.6	1826.3	4.0139								
0	.000 990 4	.19	20.01	.0004	.000 985 6	.25	29.82	.0001	.000 976 6	.20	49.03	.0014
20	.000 992 8	82.77	102.62	.2923	.000 988 6	82.17	111.84	.2899	.000 980 4	81.00	130.02	.2848
40	.000 999 2	165.17	185.16	.5646	.000 995 1	164.04	193.89	.5607	.000 987 2	161.86	211.21	.5527
60	.001 008 4	247.68	267.85	.8206	.001 004 2	246.06	276.19	.8154	.000 996 2	242.98	292.79	.8052
80	.001 019 9	330.40	350.80	1.0624	.001 015 6	328.30	358.77	1.0561	.001 007 3	324.34	374.70	1.0440
100	.001 033 7	413.39	434.06	1.2917	.001 029 0	410.78	441.66	1.2844	.001 020 1	405.88	456.89	1.2703
120	.001 049 6	496.76	517.76	1.5102	.001 044 5	493.59	524.93	1.5018	.001 034 8	487.65	539.39	1.4857
140	.001 067 8	580.69	602.04	1.7193	.001 062 1	576.88	608.75	1.7098	.001 051 5	569.77	622.35	1.6915
160	.001 088 5	665.35	687.12	1.9204	.001 082 1	660.82	693.28	1.9096	.001 070 3	652.41	705.92	1.8891
180	.001 112 0	750.95	773.20	2.1147	.001 104 7	745.59	778.73	2.1024	.001 091 2	735.69	790.25	2.0794
200	.001 138 8	837.7	860.5	2.3031	.001 130 2	831.4	865.3	2.2893	.001 114 6	819.7	875.5	2.2634
220	.001 169 3	925.9	949.3	2.4870	.001 159 0	918.3	953.1	2.4711	.001 140 8	904.7	961.7	2.4419
240	.001 204 6	1016.0	1040.0	2.6674	.001 192 0	1006.9	1042.6	2.6490	.001 170 2	990.7	1049.2	2.6158
260	.001 246 2	1108.6	1133.5	2.8459	.001 230 3	1097.4	1134.3	2.8243	.001 203 4	1078.1	1138.2	2.7860
280	.001 296 5	1204.7	1230.6	3.0248	.001 275 5	1190.7	1229.0	2.9986	.001 241 5	1167.2	1229.3	2.9537
300	.001 359 6	1306.1	1333.3	3.2071	.001 330 4	1287.9	1327.8	3.1741	.001 286 0	1258.7	1323.0	3.1200
320	.001 443 7	1415.7	1444.6	3.3979	.001 399 7	1390.7	1432.7	3.3539	.001 338 8	1353.3	1420.2	3.2868
340	.001 568 4	1539.7	1571.0	3.6075	.001 492 0	1501.7	1546.5	3.5426	.001 403 2	1452.0	1522.1	3.4557
360	.001 822 6	1702.8	1739.3	3.8772	.001 626 5	1626.6	1675.4	3.7494	.001 483 8	1556.0	1630.2	3.6291
380					.001 869 1	1781.4	1837.5	4.0012	.001 588 4	1667.2	1746.6	3.8101

528

(p in kPa or MPa; v_f and v_g in m³/kg; u_f, u_{fg}, and u_g in kJ/kg; h_f, h_{fg}, and h_g in kJ/kg; and s_f, s_{fg}, and s_g in kJ/kgK) (G. J. Van Wylen and R. E. Sonntag, *Fundamentals of Classical Thermodynamics, SI Version*, 2nd Edition. New York: John Wiley & Sons, Inc. © 1976. Reprinted by permission.)

Saturated Steam (temperature table)

Temp. °C T	Press. kPa	Specific Volume		Internal Energy			Enthalpy			Entropy		
		Sat. Liquid v_f	Sat. Vapor v_g	Sat. Liquid u_f	Evap. u_{fg}	Sat. Vapor u_g	Sat. Liquid h_f	Evap. h_{fg}	Sat. Vapor h_g	Sat. Liquid s_f	Evap. s_{fg}	Sat. Vapor s_g
0.01	0.6113	0.001 000	206.14	.00	2375.3	2375.3	.01	2501.3	2501.4	.0000	9.1562	9.1562
5	0.8721	0.001 000	147.12	20.97	2361.3	2382.3	20.98	2489.6	2510.6	.0761	8.9496	9.0257
10	1.2276	0.001 000	106.38	42.00	2347.2	2389.2	42.01	2477.7	2519.8	.1510	8.7498	8.9008
15	1.7051	0.001 001	77.93	62.99	2333.1	2396.1	62.99	2465.9	2528.9	.2245	8.5569	8.7814
20	2.339	0.001 002	57.79	83.95	2319.0	2402.9	83.96	2454.1	2538.1	.2966	8.3706	8.6672
25	3.169	0.001 003	43.36	104.88	2304.9	2409.8	104.89	2442.3	2547.2	.3674	8.1905	8.5580
30	4.246	0.001 004	32.89	125.78	2290.8	2416.6	125.79	2430.5	2556.3	.4369	8.0164	8.4533
35	5.628	0.001 006	25.22	146.67	2276.7	2423.4	146.68	2418.6	2565.3	.5053	7.8478	8.3531
40	7.384	0.001 008	19.52	167.56	2262.6	2430.1	167.57	2406.7	2574.3	.5725	7.6845	8.2570
45	9.593	0.001 010	15.26	188.44	2248.4	2436.8	188.45	2394.8	2583.2	.6387	7.5261	8.1648
50	12.349	0.001 012	12.03	209.32	2234.2	2443.5	209.33	2382.7	2592.1	.7038	7.3725	8.0763
55	15.758	0.001 015	9.568	230.21	2219.9	2450.1	230.23	2370.7	2600.9	.7679	7.2234	7.9913
60	19.940	0.001 017	7.671	251.11	2205.5	2456.6	251.13	2358.5	2609.6	.8312	7.0784	7.9096
65	25.03	0.001 020	6.197	272.02	2191.1	2463.1	272.06	2346.2	2618.3	.8935	6.9375	7.8310
70	31.19	0.001 023	5.042	292.95	2176.6	2469.6	292.98	2333.8	2626.8	.9549	6.8004	7.7553
75	38.58	0.001 026	4.131	313.90	2162.0	2475.9	313.93	2321.4	2635.3	1.0155	6.6669	7.6824
80	47.39	0.001 029	3.407	334.86	2147.4	2482.2	334.91	2308.8	2643.7	1.0753	6.5369	7.6122
85	57.83	0.001 033	2.828	355.84	2132.6	2488.4	355.90	2296.0	2651.9	1.1343	6.4102	7.5445
90	70.14	0.001 036	2.361	376.85	2117.7	2494.5	376.92	2283.2	2660.1	1.1925	6.2866	7.4791
95	84.55	0.001 040	1.982	397.88	2102.7	2500.6	397.96	2270.2	2668.1	1.2500	6.1659	7.4159

[a] Adapted from Joseph H. Keenan, Frederick G. Keyes, Philip G. Hill, and Joan G. Moore, *Steam Tables*, (New York: John Wiley & Sons, Inc., 1969).

APPENDIX B2-1 (Continued)

Saturated Steam (Temperature Table)

Temp. °C T	Press. MPa p	Specific Volume		Internal Energy			Enthalpy			Entropy		
		Sat. Liquid v_f	Sat. Vapor v_g	Sat. Liquid u_f	Evap. u_{fg}	Sat. Vapor u_g	Sat. Liquid h_f	Evap. h_{fg}	Sat. Vapor h_g	Sat. Liquid s_f	Evap. s_{fg}	Sat. Vapor s_g
100	0.101 35	0.001 044	1.6729	418.94	2087.6	2506.5	419.04	2257.0	2676.1	1.3069	6.0480	7.3549
105	0.120 82	0.001 048	1.4194	440.02	2072.3	2512.4	440.15	2243.7	2683.8	1.3630	5.9328	7.2958
110	0.143 27	0.001 052	1.2102	461.14	2057.0	2518.1	461.30	2230.2	2691.5	1.4185	5.8202	7.2387
115	0.169 06	0.001 056	1.0366	482.30	2041.4	2523.7	482.48	2216.5	2699.0	1.4734	5.7100	7.1833
120	0.198 53	0.001 060	0.8919	503.50	2025.8	2529.3	503.71	2202.6	2706.3	1.5276	5.6020	7.1296
125	0.2321	0.001 065	0.7706	524.74	2009.9	2534.6	524.99	2188.5	2713.5	1.5813	5.4962	7.0775
130	0.2701	0.001 070	0.6685	546.02	1993.9	2539.9	546.31	2174.2	2720.5	1.6344	5.3925	7.0269
135	0.3130	0.001 075	0.5822	567.35	1977.7	2545.0	567.69	2159.6	2727.3	1.6870	5.2907	6.9777
140	0.3613	0.001 080	0.5089	588.74	1961.3	2550.0	589.13	2144.7	2733.9	1.7391	5.1908	6.9299
145	0.4154	0.001 085	0.4463	610.18	1944.7	2554.9	610.63	2129.6	2740.3	1.7907	5.0926	6.8833
150	0.4758	0.001 091	0.3928	631.68	1927.9	2559.5	632.20	2114.3	2746.5	1.8418	4.9960	6.8379
155	0.5431	0.001 096	0.3468	653.24	1910.8	2564.1	653.84	2098.6	2752.4	1.8925	4.9010	6.7935
160	0.6178	0.001 102	0.3071	674.87	1893.5	2568.4	675.55	2082.6	2758.1	1.9427	4.8075	6.7502
165	0.7005	0.001 108	0.2727	696.56	1876.0	2572.5	697.34	2066.2	2763.5	1.9925	4.7153	6.7078
170	0.7917	0.001 114	0.2428	718.33	1858.1	2576.5	719.21	2049.5	2768.7	2.0419	4.6244	6.6663
175	0.8920	0.001 121	0.2168	740.17	1840.0	2580.2	741.17	2032.4	2773.6	2.0909	4.5347	6.6256
180	1.0021	0.001 127	0.194 05	762.09	1821.6	2583.7	763.22	2015.0	2778.2	2.1396	4.4461	6.5857
185	1.1227	0.001 134	0.174 09	784.10	1802.9	2587.0	785.37	1997.1	2782.4	2.1879	4.3586	6.5465
190	1.2544	0.001 141	0.156 54	806.19	1783.8	2590.0	807.62	1978.8	2786.4	2.2359	4.2720	6.5079
195	1.3978	0.001 149	0.141 05	828.37	1764.4	2592.8	829.98	1960.0	2790.0	2.2835	4.1863	6.4698
200	1.5538	0.001 157	0.127 36	850.65	1744.7	2595.3	852.45	1940.7	2793.2	2.3309	4.1014	6.4323
205	1.7230	0.001 164	0.115 21	873.04	1724.5	2597.5	875.04	1921.0	2796.0	2.3780	4.0172	6.3952
210	1.9062	0.001 173	0.104 41	895.53	1703.9	2599.5	897.76	1900.7	2798.5	2.4248	3.9337	6.3585

215	2.104	0.001 181	0.094 79	918.14	1682.9	2601.1	920.62	1879.9	2800.5	2.4714	3.8507	6.3221
220	2.318	0.001 190	0.086 19	940.87	1661.5	2602.4	943.62	1858.5	2802.1	2.5178	3.7683	6.2861
225	2.548	0.001 199	0.078 49	963.73	1639.6	2603.3	966.78	1836.5	2803.3	2.5639	3.6863	6.2503
230	2.795	0.001 209	0.071 58	986.74	1617.2	2603.9	990.12	1813.8	2804.0	2.6099	3.6047	6.2146
235	3.060	0.001 219	0.065 37	1009.89	1594.2	2604.1	1013.62	1790.5	2804.2	2.6558	3.5233	6.1791
240	3.344	0.001 229	0.059 76	1033.21	1570.8	2604.0	1037.32	1766.5	2803.8	2.7015	3.4422	6.1437
245	3.648	0.001 240	0.054 71	1056.71	1546.7	2603.4	1061.23	1741.7	2803.0	2.7472	3.3612	6.1083
250	3.973	0.001 251	0.050 13	1080.39	1522.0	2602.4	1085.36	1716.2	2801.5	2.7927	3.2802	6.0730
255	4.319	0.001 263	0.045 98	1104.28	1496.7	2600.9	1109.73	1689.8	2799.5	2.8383	3.1992	6.0375
260	4.688	0.001 276	0.042 21	1128.39	1470.6	2599.0	1134.37	1662.5	2796.9	2.8838	3.1181	6.0019
265	5.081	0.001 289	0.038 77	1152.74	1443.9	2596.6	1159.28	1634.4	2793.6	2.9294	3.0368	5.9662
270	5.499	0.001 302	0.035 64	1177.36	1416.3	2593.7	1184.51	1605.2	2789.7	2.9751	2.9551	5.9301
275	5.942	0.001 317	0.032 79	1202.25	1387.9	2590.2	1210.07	1574.9	2785.0	3.0208	2.8730	5.8938
280	6.412	0.001 332	0.030 17	1227.46	1358.7	2586.1	1235.99	1543.6	2779.6	3.0668	2.7903	5.8571
285	6.909	0.001 348	0.027 77	1253.00	1328.4	2581.4	1262.31	1511.0	2773.3	3.1130	2.7070	5.8199
290	7.436	0.001 366	0.025 57	1278.92	1297.1	2576.0	1289.07	1477.1	2766.2	3.1594	2.6227	5.7821
295	7.993	0.001 384	0.023 54	1305.2	1264.7	2569.9	1316.3	1441.8	2758.1	3.2062	2.5375	5.7437
300	8.581	0.001 404	0.021 67	1332.0	1231.0	2563.0	1344.0	1404.9	2749.0	3.2534	2.4511	5.7045
305	9.202	0.001 425	0.019 948	1359.3	1195.9	2555.2	1372.4	1366.4	2738.7	3.3010	2.3633	5.6643
310	9.856	0.001 447	0.018 350	1387.1	1159.4	2546.4	1401.3	1326.0	2727.3	3.3493	2.2737	5.6230
315	10.547	0.001 472	0.016 867	1415.5	1121.1	2536.6	1431.0	1283.5	2714.5	3.3982	2.1821	5.5804
320	11.274	0.001 499	0.015 488	1444.6	1080.9	2525.5	1461.5	1238.6	2700.1	3.4480	2.0882	5.5362
330	12.845	0.001 561	0.012 996	1505.3	993.7	2498.9	1525.3	1140.6	2665.9	3.5507	1.8909	5.4417
340	14.586	0.001 638	0.010 797	1570.3	894.3	2464.6	1594.2	1027.9	2622.0	3.6594	1.6763	5.3357
350	16.513	0.001 740	0.008 813	1641.9	776.6	2418.4	1670.6	893.4	2563.9	3.7777	1.4335	5.2112
360	18.651	0.001 893	0.006 945	1725.2	626.3	2351.5	1760.5	720.5	2481.0	3.9147	1.1379	5.0526
370	21.03	0.002 213	0.004 925	1844.0	384.5	2228.5	1890.5	441.6	2332.1	4.1106	.6865	4.7971
374.14	22.09	0.003 155	0.003 155	2029.6	0	2029.6	2099.3	0	2099.3	4.4298	0	4.4298

APPENDIX B2-2 SATURATED STEAM (PRESSURE TABLE)

(p in kPa or MPa; v_f and v_g in m³/kg; u_f, u_{fg}, and u_g in kJ/kg; h_f, h_{fg}, and h_g in kJ/kg; and s_f, s_{fg}, and s_g in kJ/kgK) (G. J. Van Wylen and R. E. Sonntag, *Fundamentals of Classical Thermodynamics, SI Version*, 2nd Edition. New York: John Wiley & Sons, Inc. © 1976. Reprinted by permission.)

Saturated Steam (pressure table)

Press. kPa p	Temp. °C T	Specific Volume		Internal Energy			Enthalpy			Entropy		
		Sat. Liquid v_f	Sat. Vapor v_g	Sat. Liquid u_f	Evap. u_{fg}	Sat. Vapor u_g	Sat. Liquid h_f	Evap. h_{fg}	Sat. Vapor h_g	Sat. Liquid s_f	Evap. s_{fg}	Sat. Vapor s_g
0.6113	0.01	0.001 000	206.14	.00	2375.3	2375.3	.01	2501.3	2501.4	.0000	9.1562	9.1562
1.0	6.98	0.001 000	129.21	29.30	2355.7	2385.0	29.30	2484.9	2514.2	.1059	8.8697	8.9756
1.5	13.03	0.001 001	87.98	54.71	2338.6	2393.3	54.71	2470.6	2525.3	.1957	8.6322	8.8279
2.0	17.50	0.001 001	67.00	73.48	2326.0	2399.5	73.48	2460.0	2533.5	.2607	8.4629	8.7237
2.5	21.08	0.001 002	54.25	88.48	2315.9	2404.4	88.49	2451.6	2540.0	.3120	8.3311	8.6432
3.0	24.08	0.001 003	45.67	101.04	2307.5	2408.5	101.05	2444.5	2545.5	.3545	8.2231	8.5776
4.0	28.96	0.001 004	34.80	121.45	2293.7	2415.2	121.46	2432.9	2554.4	.4226	8.0520	8.4746
5.0	32.88	0.001 005	28.19	137.81	2282.7	2420.5	137.82	2423.7	2561.5	.4764	7.9187	8.3951
7.5	40.29	0.001 008	19.24	168.78	2261.7	2430.5	168.79	2406.0	2574.8	.5764	7.6750	8.2515
10	45.81	0.001 010	14.67	191.82	2246.1	2437.9	191.83	2392.8	2584.7	.6493	7.5009	8.1502
15	53.97	0.001 014	10.02	225.92	2222.8	2448.7	225.94	2373.1	2599.1	.7549	7.2536	8.0085
20	60.06	0.001 017	7.649	251.38	2205.4	2456.7	251.40	2358.3	2609.7	.8320	7.0766	7.9085
25	64.97	0.001 020	6.204	271.90	2191.2	2463.1	271.93	2346.3	2618.2	.8931	6.9383	7.8314
30	69.10	0.001 022	5.229	289.20	2179.2	2468.4	289.23	2336.1	2625.3	.9439	6.8247	7.7686
40	75.87	0.001 027	3.993	317.53	2159.5	2477.0	317.58	2319.2	2636.8	1.0259	6.6441	7.6700
50	81.33	0.001 030	3.240	340.44	2143.4	2483.9	340.49	2305.4	2645.9	1.0910	6.5029	7.5939
75	91.78	0.001 037	2.217	384.31	2112.4	2496.7	384.39	2278.6	2663.0	1.2130	6.2434	7.4564
MPa												
0.100	99.63	0.001 043	1.6940	417.36	2088.7	2506.1	417.46	2258.0	2675.5	1.3026	6.0568	7.3594
0.125	105.99	0.001 048	1.3749	444.19	2069.3	2513.5	444.32	2241.0	2685.4	1.3740	5.9104	7.2844
0.150	111.37	0.001 053	1.1593	466.94	2052.7	2519.7	467.11	2226.5	2693.6	1.4336	5.7897	7.2233
0.175	116.06	0.001 057	1.0036	486.80	2038.1	2524.9	486.99	2213.6	2700.6	1.4849	5.6868	7.1717
0.200	120.23	0.001 061	0.8857	504.49	2025.0	2529.5	504.70	2201.9	2706.7	1.5301	5.5970	7.1271
0.225	**124.00**	**0.001 064**	**0.7933**	**520.47**	**2013.1**	**2533.6**	**520.72**	**2191.3**	**2712.1**	**1.5706**	**5.5173**	**7.0878**

0.250	127.44	0.001 067	0.7187	535.10	2002.1	2537.2	535.37	2181.5	2716.9	1.6072	5.4455	7.0527
0.275	130.60	0.001 070	0.6573	548.59	1991.9	2540.5	548.89	2172.4	2721.3	1.6408	5.3801	7.0209
0.300	133.55	0.001 073	0.6058	561.15	1982.4	2543.6	561.47	2163.8	2725.3	1.6718	5.3201	6.9919
0.325	136.30	0.001 076	0.5620	572.90	1973.5	2546.4	573.25	2155.8	2729.0	1.7006	5.2646	6.9652
0.350	138.88	0.001 079	0.5243	583.95	1965.0	2548.9	584.33	2148.1	2732.4	1.7275	5.2130	6.9405
0.375	141.32	0.001 081	0.4914	594.40	1956.9	2551.3	594.81	2140.8	2735.6	1.7528	5.1647	6.9175
0.40	143.63	0.001 084	0.4625	604.31	1949.3	2553.6	604.74	2133.8	2738.6	1.7766	5.1193	6.8959
0.45	147.93	0.001 088	0.4140	622.77	1934.9	2557.6	623.25	2120.7	2743.9	1.8207	5.0359	6.8565
0.50	151.86	0.001 093	0.3749	639.68	1921.6	2561.2	640.23	2108.5	2748.7	1.8607	4.9606	6.8213
0.55	155.48	0.001 097	0.3427	655.32	1909.2	2564.5	655.93	2097.0	2753.0	1.8973	4.8920	6.7893
0.60	158.85	0.001 101	0.3157	669.90	1897.5	2567.4	670.56	2086.3	2756.8	1.9312	4.8288	6.7600
0.65	162.01	0.001 104	0.2927	683.56	1886.5	2570.1	684.28	2076.0	2760.3	1.9627	4.7703	6.7331
0.70	164.97	0.001 108	0.2729	696.44	1876.1	2572.5	697.22	2066.3	2763.5	1.9922	4.7158	6.7080
0.75	167.78	0.001 112	0.2556	708.64	1866.1	2574.7	709.47	2057.0	2766.4	2.0200	4.6647	6.6847
0.80	170.43	0.001 115	0.2404	720.22	1856.6	2576.8	721.11	2048.0	2769.1	2.0462	4.6166	6.6628
0.85	172.96	0.001 118	0.2270	731.27	1847.4	2578.7	732.22	2039.4	2771.6	2.0710	4.5711	6.6421
0.90	175.38	0.001 121	0.2150	741.83	1838.6	2580.5	742.83	2031.1	2773.9	2.0946	4.5280	6.6226
0.95	177.69	0.001 124	0.2042	751.95	1830.2	2582.1	753.02	2023.1	2776.1	2.1172	4.4869	6.6041
1.00	179.91	0.001 127	0.194 44	761.68	1822.0	2583.6	762.81	2015.3	2778.1	2.1387	4.4478	6.5865
1.10	184.09	0.001 133	0.177 53	780.09	1806.3	2586.4	781.34	2000.4	2781.7	2.1792	4.3744	6.5536
1.20	187.99	0.001 139	0.163 33	797.29	1791.5	2588.8	798.65	1986.2	2784.8	2.2166	4.3067	6.5233
1.30	191.64	0.001 144	0.151 25	813.44	1777.5	2591.0	814.93	1972.7	2787.6	2.2515	4.2438	6.4953
1.40	195.07	0.001 149	0.140 84	828.70	1764.1	2592.8	830.30	1959.7	2790.0	2.2842	4.1850	6.4693
1.50	198.32	0.001 154	0.131 77	843.16	1751.3	2594.5	844.89	1947.3	2792.2	2.3150	4.1298	6.4448
1.75	205.76	0.001 166	0.113 49	876.46	1721.4	2597.8	878.50	1917.9	2796.4	2.3851	4.0044	6.3896
2.00	212.42	0.001 177	0.099 63	906.44	1693.8	2600.3	908.79	1890.7	2799.5	2.4474	3.8935	6.3409
2.25	218.45	0.001 187	0.088 75	933.83	1668.2	2602.0	936.49	1865.2	2801.7	2.5035	3.7937	6.2972
2.5	223.99	0.001 197	0.079 98	959.11	1644.0	2603.1	962.11	1841.0	2803.1	2.5547	3.7028	6.2575
3.0	233.90	0.001 217	0.066 68	1004.78	1599.3	2604.1	1008.42	1795.7	2804.2	2.6457	3.5412	6.1869

APPENDIX B2-2 (Continued)

Saturated Steam (pressure table)

Press. MPa p	Temp. °C T	Specific Volume Sat. Liquid v_f	Specific Volume Sat. Vapor v_g	Internal Energy Sat. Liquid u_f	Internal Energy Evap. u_{fg}	Internal Energy Sat. Vapor u_g	Enthalpy Sat. Liquid h_f	Enthalpy Evap. h_{fg}	Enthalpy Sat. Vapor h_g	Entropy Sat. Liquid s_f	Entropy Evap. s_{fg}	Entropy Sat. Vapor s_g
3.5	242.60	0.001 235	0.057 07	1045.43	1558.3	2603.7	1049.75	1753.7	2803.4	2.7253	3.4000	6.1253
4	250.40	0.001 252	0.049 78	1082.31	1520.0	2602.3	1087.31	1714.1	2801.4	2.7964	3.2737	6.0701
5	263.99	0.001 286	0.039 44	1147.81	1449.3	2597.1	1154.23	1640.1	2794.3	2.9202	3.0532	5.9734
6	275.64	0.001 319	0.032 44	1205.44	1384.3	2589.7	1213.35	1571.0	2784.3	3.0267	2.8625	5.8892
7	285.88	0.001 351	0.027 37	1257.55	1323.0	2580.5	1267.00	1505.1	2772.1	3.1211	2.6922	5.8133
8	295.06	0.001 384	0.023 52	1305.57	1264.2	2569.8	1316.64	1441.3	2758.0	3.2068	2.5364	5.7432
9	303.40	0.001 418	0.020 48	1350.51	1207.3	2557.8	1363.26	1378.9	2742.1	3.2858	2.3915	5.6772
10	311.06	0.001 452	0.018 026	1393.04	1151.4	2544.4	1407.56	1317.1	2724.7	3.3596	2.2544	5.6141
11	318.15	0.001 489	0.015 987	1433.7	1096.0	2529.8	1450.1	1255.5	2705.6	3.4295	2.1233	5.5527
12	324.75	0.001 527	0.014 263	1473.0	1040.7	2513.7	1491.3	1193.6	2684.9	3.4962	1.9962	5.4924
13	330.93	0.001 567	0.012 780	1511.1	985.0	2496.1	1531.5	1130.7	2662.2	3.5606	1.8718	5.4323
14	336.75	0.001 611	0.011 485	1548.6	928.2	2476.8	1571.1	1066.5	2637.6	3.6232	1.7485	5.3717
15	342.24	0.001 658	0.010 337	1585.6	869.8	2455.5	1610.5	1000.0	2610.5	3.6848	1.6249	5.3098
16	347.44	0.001 711	0.009 306	1622.7	809.0	2431.7	1650.1	930.6	2580.6	3.7461	1.4994	5.2455
17	352.37	0.001 770	0.008 364	1660.2	744.8	2405.0	1690.3	856.9	2547.2	3.8079	1.3698	5.1777
18	357.06	0.001 840	0.007 489	1698.9	675.4	2374.3	1732.0	777.1	2509.1	3.8715	1.2329	5.1044
19	361.54	0.001 924	0.006 657	1739.9	598.1	2338.1	1776.5	688.0	2464.5	3.9388	1.0839	5.0228
20	365.81	0.002 036	0.005 834	1785.6	507.5	2293.0	1826.3	583.4	2409.7	4.0139	.9130	4.9269
21	369.89	0.002 207	0.004 952	1842.1	388.5	2230.6	1888.4	446.2	2334.6	4.1075	.6938	4.8013
22	373.80	0.002 742	0.003 568	1961.9	125.2	2087.1	2022.2	143.4	2165.6	4.3110	.2216	4.5327
22.09	374.14	0.003 155	0.003 155	2029.6	0	2029.6	2099.3	0	2099.3	4.4298	0	4.4298

APPENDIX B3 SUPERHEATED STEAM

(T in °C, v in m³/kg, u and h in kJ/kg, s in kJ/kgK)

Superheated Steam

T	p = .010 MPa (45.81)				p = .050 MPa (81.33)				p = .10 MPa (99.63)			
	v	u	h	s	v	u	h	s	v	u	h	s
Sat.	14.674	2437.9	2584.7	8.1502	3.240	2483.9	2645.9	7.5939	1.6940	2506.1	2675.5	7.3594
50	14.869	2443.9	2592.6	8.1749								
100	17.196	2515.5	2687.5	8.4479	3.418	2511.6	2682.5	7.6947	1.6958	2506.7	2676.2	7.3614
150	19.512	2587.9	2783.0	8.6882	3.889	2585.6	2780.1	7.9401	1.9364	2582.8	2776.4	7.6134
200	21.825	2661.3	2879.5	8.9038	4.356	2659.9	2877.7	8.1580	2.172	2658.1	2875.3	7.8343
250	24.136	2736.0	2977.3	9.1002	4.820	2735.0	2976.0	8.3556	2.406	2733.7	2974.3	8.0333
300	26.445	2812.1	3076.5	9.2813	5.284	2811.3	3075.5	8.5373	2.639	2810.4	3074.3	8.2158
400	31.063	2968.9	3279.6	9.6077	6.209	2968.5	3278.9	8.8642	3.103	2967.9	3278.2	8.5435
500	35.679	3132.3	3489.1	9.8978	7.134	3132.0	3488.7	9.1546	3.565	3131.6	3488.1	8.8342
600	40.295	3302.5	3705.4	10.1608	8.057	3302.2	3705.1	9.4178	4.028	3301.9	3704.7	9.0976
700	44.911	3479.6	3928.7	10.4028	8.981	3479.4	3928.5	9.6599	4.490	3479.2	3928.2	9.3398
800	49.526	3663.8	4159.0	10.6281	9.904	3663.6	4158.9	9.8852	4.952	3663.5	4158.6	9.5652
900	54.141	3855.0	4396.4	10.8396	10.828	3854.9	4396.3	10.0967	5.414	3854.8	4396.1	9.7767
1000	58.757	4053.0	4640.6	11.0393	11.751	4052.9	4640.5	10.2964	5.875	4052.8	4640.3	9.9764
1100	63.372	4257.5	4891.2	11.2287	12.674	4257.4	4891.1	10.4859	6.337	4257.3	4891.0	10.1659
1200	67.987	4467.9	5147.8	11.4091	13.597	4467.8	5147.7	10.6662	6.799	4467.7	5147.6	10.3463
1300	72.602	4683.7	5409.7	11.5811	14.521	4683.6	5409.6	10.8382	7.260	4683.5	5409.5	10.5183

T	P = .20 MPa (120.23)				P = .30 MPa (133.55)				P = .40 MPa (143.63)			
	v	u	h	s	v	u	h	s	v	u	h	s
Sat.	.8857	2529.5	2706.7	7.1272	.6058	2543.6	2725.3	6.9919	.4625	2553.6	2738.6	6.8959
150	.9596	2576.9	2768.8	7.2795	.6339	2570.8	2761.0	7.0778	.4708	2564.5	2752.8	6.9299
200	1.0803	2654.4	2870.5	7.5066	.7163	2650.7	2865.6	7.3115	.5342	2646.8	2860.5	7.1706
250	1.1988	2731.2	2971.0	7.7086	.7964	2728.7	2967.6	7.5166	.5951	2726.1	2964.2	7.3789
300	1.3162	2808.6	3071.8	7.8926	.8753	2806.7	3069.3	7.7022	.6548	2804.8	3066.8	7.5662
400	1.5493	2966.7	3276.6	8.2218	1.0315	2965.6	3275.0	8.0330	.7726	2964.4	3273.4	7.8985

APPENDIX B3 (Continued)

Superheated Steam

T	v	u	h	s	v	u	h	s	v	u	h	s
	p = .20 MPa (120.23)				p = .30 MPa (133.55)				p = .40 MPa (143.63)			
500	1.7814	3130.8	3487.1	8.5133	1.1867	3130.0	3486.0	8.3251	.8893	3129.2	3484.9	8.1913
600	2.013	3301.4	3704.0	8.7770	1.3414	3300.8	3703.2	8.5892	1.0055	3300.2	3702.4	8.4558
700	2.244	3478.8	3927.6	9.0194	1.4957	3478.4	3927.1	8.8319	1.1215	3477.9	3926.5	8.6987
800	2.475	3663.1	4158.2	9.2449	1.6499	3662.9	4157.8	9.0576	1.2372	3662.4	4157.3	8.9244
900	2.706	3854.5	4395.8	9.4566	1.8041	3854.2	4395.4	9.2692	1.3529	3853.9	4395.1	9.1362
1000	2.937	4052.5	4640.0	9.6563	1.9581	4052.3	4639.7	9.4690	1.4685	4052.0	4639.4	9.3360
1100	3.168	4257.0	4890.7	9.8458	2.1121	4256.8	4890.4	9.6585	1.5840	4256.5	4890.2	9.5256
1200	3.399	4467.5	5147.3	10.0262	2.2661	4467.2	5147.1	9.8389	1.6996	4467.0	5146.8	9.7060
1300	3.630	4683.2	5409.3	10.1982	2.4201	4683.0	5409.0	10.0110	1.8151	4682.8	5408.8	9.8780
	p = .50 MPa (151.86)				p = .60 MPa (158.85)				p = .80 MPa (170.43)			
Sat.	.3749	2561.2	2748.7	6.8213	.3157	2567.4	2756.8	6.7600	.2404	2576.8	2769.1	6.6628
200	.4249	2642.9	2855.4	7.0592	.3520	2638.9	2850.1	6.9665	.2608	2630.6	2839.3	6.8158
250	.4744	2723.5	2960.7	7.2709	.3938	2720.9	2957.2	7.1816	.2931	2715.5	2950.0	7.0384
300	.5226	2802.9	3064.2	7.4599	.4344	2801.0	3061.6	7.3724	.3241	2797.2	3056.5	7.2328
350	.5701	2882.6	3167.7	7.6329	.4742	2881.2	3165.7	7.5464	.3544	2878.2	3161.7	7.4089
400	.6173	2963.2	3271.9	7.7938	.5137	2962.1	3270.3	7.7079	.3843	2959.7	3267.1	7.5716
500	.7109	3128.4	3483.9	8.0873	.5920	3127.6	3482.8	8.0021	.4433	3126.0	3480.6	7.8673
600	.8041	3299.6	3701.7	8.3522	.6697	3299.1	3700.9	8.2674	.5018	3297.9	3699.4	8.1333
700	.8969	3477.5	3925.9	8.5952	.7472	3477.0	3925.3	8.5107	.5601	3476.2	3924.2	8.3770
800	.9896	3662.1	4156.9	8.8211	.8245	3661.8	4156.5	8.7367	.6181	3661.1	4155.6	8.6033
900	1.0822	3853.6	4394.7	9.0329	.9017	3853.4	4394.4	8.9486	.6761	3852.8	4393.7	8.8153
1000	1.1747	4051.8	4639.1	9.2328	.9788	4051.5	4638.8	9.1485	.7340	4051.0	4638.2	9.0153

T	v	h	s	v	h	s	v	h	s
1100	1.2672	4256.3	4889.9	9.4224	1.0559	4256.1	4889.6	9.3381
1200	1.3596	4466.8	5146.6	9.6029	1.1330	4466.5	5146.3	9.5185
1300	1.4521	4682.5	5408.6	9.7749	1.2101	4682.3	5408.3	9.6906

T	v	h	s
1100	.7919	4255.6	4889.1	9.2050
1200	.8497	4466.1	5145.9	9.3855
1300	.9076	4681.8	5407.9	9.5575

p = 1.00 MPa (179.91)

T	v	h	s
Sat.	.194 44	2583.6	2778.1	6.5865
200	.2060	2621.9	2827.9	6.6940
250	.2327	2709.9	2942.6	6.9247
300	.2579	2793.2	3051.2	7.1229
350	.2825	2875.2	3157.7	7.3011
400	.3066	2957.3	3263.9	7.4651
500	.3541	3124.4	3478.5	7.7622
600	.4011	3296.8	3697.9	8.0290
700	.4478	3475.3	3923.1	8.2731
800	.4943	3660.4	4154.7	8.4996
900	.5407	3852.2	4392.9	8.7118
1000	.5871	4050.5	4637.6	8.9119
1100	.6335	4255.1	4888.6	9.1017
1200	.6798	4465.6	5145.4	9.2822
1300	.7261	4681.3	5407.4	9.4543

p = 1.20 MPa (187.99)

T	v	h	s
Sat.	.163 33	2588.8	2784.8	6.5233
200	.169 30	2612.8	2815.9	6.5898
250	.192 34	2704.2	2935.0	6.8294
300	.2138	2789.2	3045.8	7.0317
350	.2345	2872.2	3153.6	7.2121
400	.2548	2954.9	3260.7	7.3774
500	.2946	3122.8	3476.3	7.6759
600	.3339	3295.6	3696.3	7.9435
700	.3729	3474.4	3922.0	8.1881
800	.4118	3659.7	4153.8	8.4148
900	.4505	3851.6	4392.2	8.6272
1000	.4892	4050.4	4637.0	8.8274
1100	.5278	4254.6	4888.0	9.0172
1200	.5665	4465.1	5144.9	9.1977
1300	.6051	4680.9	5407.0	9.3698

p = 1.40 MPa (195.07)

T	v	h	s
Sat.	.140 84	2592.8	2790.0	6.4693
200	.143 02	2603.1	2803.3	6.4975
250	.163 50	2698.3	2927.2	6.7467
300	.182 28	2785.2	3040.4	6.9534
350	.2003	2869.2	3149.5	7.1360
400	.2178	2952.5	3257.5	7.3026
500	.2521	3121.1	3474.1	7.6027
600	.2860	3294.4	3694.8	7.8710
700	.3195	3473.6	3920.8	8.1160
800	.3528	3659.0	4153.0	8.3431
900	.3861	3851.1	4391.5	8.5556
1000	.4192	4049.5	4636.4	8.7559
1100	.4524	4254.1	4887.5	8.9457
1200	.4855	4464.7	5144.4	9.1262
1300	.5186	4680.4	5406.5	9.2984

p = 1.60 MPa (201.41)

T	v	h	s
Sat.	.123 80	2596.0	2794.0	6.4218
225	.132 87	2644.7	2857.3	6.5518
250	.141 84	2692.3	2919.2	6.6732
300	.158 62	2781.1	3034.8	6.8844
350	.174 56	2866.1	3145.4	7.0694
400	.190 05	2950.1	3254.2	7.2374
500	.2203	3119.5	3472.0	7.5390
600	.2500	3293.3	3693.2	7.8080
700	.2794	3472.7	3919.7	8.0535

P = 1.80 MPa (207.15)

T	v	h	s
Sat.	.110 42	2598.4	2797.1	6.3794
225	.116 73	2636.6	2846.7	6.4808
250	.124 97	2686.0	2911.0	6.6066
300	.140 21	2776.9	3029.2	6.8226
350	.154 57	2863.0	3141.2	7.0100
400	.168 47	2947.7	3250.9	7.1794
500	.195 50	3117.9	3469.8	7.4825
600	.2220	3292.1	3691.7	7.7523
700	.2482	3471.8	3918.5	7.9983

P = 2.00 MPa (212.42)

T	v	h	s
Sat.	.099 63	2600.3	2799.5	6.3409
225	.103 77	2628.3	2835.8	6.4147
250	.111 44	2679.6	2902.5	6.5453
300	.125 47	2772.6	3023.5	6.7664
350	.138 57	2859.8	3137.0	6.9563
400	.151 20	2945.2	3247.6	7.1271
500	.175 68	3116.2	3467.6	7.4317
600	.199 60	3290.9	3690.1	7.7024
700	.2232	3470.9	3917.4	7.9487

Superheated Steam

T	v	u	h	s	v	u	h	s	v	u	h	s
	$p = 1.60$ MPa (201.41)				$p = 1.80$ MPa (207.15)				$p = 2.00$ MPa (212.42)			
800	.3086	3658.3	4152.1	8.2808	.2742	3657.6	4151.2	8.2258	.2467	3657.0	4150.3	8.1765
900	.3377	3850.5	4390.8	8.4935	.3001	3849.9	4390.1	8.4386	.2700	3849.3	4389.4	8.3895
1000	.3668	4049.0	4635.8	8.6938	.3260	4048.5	4635.2	8.6391	.2933	4048.0	4634.6	8.5901
1100	.3958	4253.7	4887.0	8.8837	.3518	4253.2	4886.4	8.8290	.3166	4252.7	4885.9	8.7800
1200	.4248	4464.2	5143.9	9.0643	.3776	4463.7	5143.4	9.0096	.3398	4463.3	5142.9	8.9607
1300	.4538	4679.9	5406.0	9.2364	.4034	4679.5	5405.6	9.1818	.3631	4679.0	5405.1	9.1329
	$p = 2.50$ MPa (223.99)				$p = 3.00$ MPa (233.90)				$P = 3.50$ MPa (242.60)			
Sat.	.079 98	2603.1	2803.1	6.2575	.066 68	2604.1	2804.2	6.1869	.057 07	2603.7	2803.4	6.1253
225	.080 27	2605.6	2806.3	6.2639								
250	.087 00	2662.6	2880.1	6.4085	.070 58	2644.0	2855.8	6.2872	.058 72	2623.7	2829.2	6.1749
300	.098 90	2761.6	3008.8	6.6438	.081 14	2750.1	2993.5	6.5390	.068 42	2738.0	2977.5	6.4461
350	.109 76	2851.9	3126.3	6.8403	.090 53	2843.7	3115.3	6.7428	.076 78	2835.3	3104.0	6.6579
400	.120 10	2939.1	3239.3	7.0148	.099 36	2932.8	3230.9	6.9212	.084 53	2926.4	3222.3	6.8405
450	.130 14	3025.5	3350.8	7.1746	.107 87	3020.4	3344.0	7.0834	.091 96	3015.3	3337.2	7.0052
500	.139 98	3112.1	3462.1	7.3234	.116 19	3108.0	3456.5	7.2338	.099 18	3103.0	3450.9	7.1572
600	.159 30	3288.0	3686.3	7.5960	.132 43	3285.0	3682.3	7.5085	.113 24	3282.1	3678.4	7.4339
700	.178 32	3468.7	3914.5	7.8435	.148 38	3466.5	3911.7	7.7571	.126 99	3464.3	3908.8	7.6837
800	.197 16	3655.3	4148.2	8.0720	.164 14	3653.5	4145.9	7.9862	.140 56	3651.8	4143.7	7.9134
900	.215 90	3847.9	4387.6	8.2853	.179 80	3846.5	4385.9	8.1999	.154 02	3845.0	4384.1	8.1276
1000	.2346	4046.7	4633.1	8.4861	.195 41	4045.4	4631.6	8.4009	.167 43	4044.1	4630.1	8.3288
1100	.2532	4251.5	4884.6	8.6762	.210 98	4250.3	4883.3	8.5912	.180 80	4249.2	4881.9	8.5192
1200	.2718	4462.1	5141.7	8.8569	.226 52	4460.9	5140.5	8.7720	.194 15	4459.8	5139.3	8.7000
1300	.2905	4677.8	5404.0	9.0291	.242 06	4676.6	5402.8	8.9442	.207 49	4675.5	5401.7	8.8723

T	p = 4.0 MPa (250.40)			p = 4.5 MPa (257.49)			p = 5.0 MPa (263.99)		
Sat.	.049 78	2801.4	6.0701	.044 06	2798.3	6.0198	.039 44	2794.3	5.9734
275	.054 57	2886.2	6.2285	.047 30	2863.2	6.1401	.041 41	2838.3	6.0544
300	.058 84	2960.7	6.3615	.051 35	2943.1	6.2828	.045 32	2924.5	6.2084
350	.066 45	3092.5	6.5821	.058 40	3080.6	6.5131	.051 94	3068.4	6.4493
400	.073 41	3213.6	6.7690	.064 75	3204.7	6.7047	.057 81	3195.7	6.6459
450	.080 02	3330.3	6.9363	.070 74	3323.3	6.8746	.063 30	3316.2	6.8186
500	.086 43	3445.3	7.0901	.076 51	3439.6	7.0301	.068 57	3433.8	6.9759
600	.098 85	3674.4	7.3688	.087 65	3670.6	7.3110	.078 69	3666.5	7.2589
700	.110 95	3905.9	7.6198	.098 47	3903.0	7.5631	.088 49	3900.1	7.5122
800	.122 87	4141.5	7.8502	.109 11	4139.3	7.7942	.098 11	4137.1	7.7440
900	.134 69	4382.3	8.0647	.119 65	4380.6	8.0091	.107 62	4378.8	7.9593
1000	.146 45	4628.7	8.2662	.130 13	4627.2	8.2108	.117 07	4625.7	8.1612
1100	.158 17	4880.6	8.4567	.140 56	4879.3	8.4015	.126 48	4878.0	8.3520
1200	.169 87	5138.1	8.6376	.150 98	5136.9	8.5825	.135 87	5135.7	8.5331
1300	.181 56	5400.5	8.8100	.161 39	5399.4	8.7549	.145 26	5398.2	8.7055

T	P = 6.0 MPa (275.64)			P = 7.0 MPa (285.88)			P = 8.0 MPa (295.06)		
Sat.	.032 44	2784.3	5.8892	.027 37	2772.1	5.8133	.023 52	2758.0	5.7432
300	.036 16	2884.2	6.0674	.029 47	2838.4	5.9305	.024 26	2785.0	5.7906
350	.042 23	3043.0	6.3335	.035 24	3016.0	6.2283	.029 95	2987.3	6.1301
400	.047 39	3177.2	6.5408	.039 93	3158.1	6.4478	.034 32	3138.3	6.3634
450	.052 14	3301.8	6.7193	.044 16	3287.1	6.6327	.038 17	3272.0	6.5551
500	.056 65	3422.2	6.8803	.048 14	3410.3	6.7975	.041 75	3398.3	6.7240
550	.061 01	3540.6	7.0288	.051 95	3530.9	6.9486	.045 16	3521.0	6.8778
600	.065 25	3658.4	7.1677	.055 65	3650.3	7.0894	.048 45	3642.0	7.0206
700	.073 52	3894.2	7.4234	.062 83	3888.3	7.3476	.054 81	3882.4	7.2812
800	.081 60	4132.7	7.6566	.069 81	4128.2	7.5822	.060 97	4123.8	7.5173
900	.089 58	4375.3	7.8727	.076 69	4371.8	7.7991	.067 02	4368.3	7.7351
1000	.097 49	4622.7	8.0751	.083 50	4619.8	8.0020	.073 01	4616.9	7.9384
1100	.105 36	4875.4	8.2661	.090 27	4872.8	8.1933	.078 96	4870.3	8.1300

APPENDIX B3 (Continued)

Superheated Steam

T	p = 6.0 MPa (275.64)				p = 7.0 MPa (285.88)				p = 8.0 MPa (295.06)			
	v	u	h	s	v	u	h	s	v	u	h	s
1200	.113 21	4454.0	5133.3	8.4474	.097 03	4451.7	5130.9	8.3747	.084 89	4449.5	5128.5	8.3115
1300	.121 06	4669.6	5396.0	8.6199	.103 77	4667.3	5393.7	8.5473	.090 80	4665.0	5391.5	8.4842

T	p = 9.0 MPa (303.40)				p = 10.0 MPa (311.06)				p = 12.5 MPa (327.89)			
	v	u	h	s	v	u	h	s	v	u	h	s
Sat.	.020 48	2557.8	2742.1	5.6772	.018 026	2544.4	2724.7	5.6141	.013 495	2505.1	2673.8	5.4624
325	.023 27	2646.6	2856.0	5.8712	.019 861	2610.4	2809.1	5.7568				
350	.025 80	2724.4	2956.6	6.0361	.022 42	2699.2	2923.4	5.9443	.016 126	2624.6	2826.2	5.7118
400	.029 93	2848.4	3117.8	6.2854	.026 41	2832.4	3096.5	6.2120	.020 00	2789.3	3039.3	6.0417
450	.033 50	2955.2	3256.6	6.4844	.029 75	2943.4	3240.9	6.4190	.022 99	2912.5	3199.8	6.2719
500	.036 77	3055.2	3386.1	6.6576	.032 79	3045.8	3373.7	6.5966	.025 60	3021.7	3341.8	6.4618
550	.039 87	3152.2	3511.0	6.8142	.035 64	3144.6	3500.9	6.7561	.028 01	3125.0	3475.2	6.6290
600	.042 85	3248.1	3633.7	6.9589	.038 37	3241.7	3625.3	6.9029	.030 29	3225.4	3604.0	6.7810
650	.045 74	3343.6	3755.3	7.0943	.041 01	3338.2	3748.2	7.0398	.032 48	3324.4	3730.4	6.9218
700	.048 57	3439.3	3876.5	7.2221	.043 58	3434.7	3870.5	7.1687	.034 60	3422.9	3855.3	7.0536
800	.054 09	3632.5	4119.3	7.4596	.048 59	3628.9	4114.8	7.4077	.038 69	3620.0	4103.6	7.2965
900	.059 50	3829.2	4364.8	7.6783	.053 49	3826.3	4361.2	7.6272	.042 67	3819.1	4352.5	7.5182
1000	.064 85	4030.3	4614.0	7.8821	.058 32	4027.8	4611.0	7.8315	.046 58	4021.6	4603.8	7.7237
1100	.070 16	4236.3	4867.7	8.0740	.063 12	4234.0	4865.1	8.0237	.050 45	4228.2	4858.8	7.9165
1200	.075 44	4447.2	5126.2	8.2556	.067 89	4444.9	5123.8	8.2055	.054 30	4439.3	5118.0	8.0987
1300	.080 72	4662.7	5389.2	8.4284	.072 65	4460.5	5387.0	8.3783	.058 13	4654.8	5381.4	8.2717

T (°C)	p = 15.0 MPa (342.24)				p = 17.5 MPa (354.75)				p = 20.0 MPa (365.81)			
Sat.	.010 337	2455.5	2610.5	5.3098	.007 920	2390.2	2528.8	5.1419	.005 834	2293.0	2409.7	4.9269
350	.011 470	2520.4	2692.4	5.4421								
400	.015 649	2740.7	2975.5	5.8811	.012 447	2685.0	2902.9	5.7213	.009 942	2619.3	2818.1	5.5540
450	.018 445	2879.5	3156.2	6.1404	.015 174	2844.2	3109.7	6.0184	.012 695	2806.2	3060.1	5.9017
500	.020 80	2996.6	3308.6	6.3443	.017 358	2970.3	3274.1	6.2383	.014 768	2942.9	3238.2	6.1401
550	.022 93	3104.7	3448.6	6.5199	.019 288	3083.9	3421.4	6.4230	.016 555	3062.4	3393.5	6.3348
600	.024 91	3208.6	3582.3	6.6776	.021 06	3191.5	3560.1	6.5866	.018 178	3174.0	3537.6	6.5048
650	.026 80	3310.3	3712.3	6.8224	.022 74	3296.0	3693.9	6.7357	.019 693	3281.4	3675.3	6.6582
700	.028 61	3410.9	3840.1	6.9572	.024 34	3398.7	3824.6	6.8736	.021 13	3386.4	3809.0	6.7993
800	.032 10	3610.9	4092.4	7.2040	.027 38	3601.8	4081.1	7.1244	.023 85	3592.7	4069.7	7.0544
900	.035 46	3811.9	4343.8	7.4279	.030 31	3804.7	4335.1	7.3507	.026 45	3797.5	4326.4	7.2830
1000	.038 75	4015.4	4596.6	7.6348	.033 16	4009.3	4589.5	7.5589	.028 97	4003.1	4582.5	7.4925
1100	.042 00	4222.6	4852.6	7.8283	.035 97	4216.9	4846.4	7.7531	.031 45	4211.3	4840.2	7.6874
1200	.045 23	4433.8	5112.3	8.0108	.038 76	4428.3	5106.6	7.9360	.033 91	4422.8	5101.0	7.8707
1300	.048 45	4649.1	5376.0	8.1840	.041 54	4643.5	5370.5	8.1093	.036 36	4638.0	5365.1	8.0442

T (°C)	p = 25.0 MPa				p = 30.0 MPa				p = 35.0 MPa			
375	.001 973 1	1798.7	1848.0	4.0320	.001 789 2	1737.8	1791.5	3.9305	.001 700 3	1702.9	1762.4	3.8722
400	.006 004	2430.1	2580.2	5.1418	.002 790	2067.4	2151.1	4.4728	.002 100	1914.1	1987.6	4.2126
425	.007 881	2609.2	2806.3	5.4723	.005 303	2455.1	2614.2	5.1504	.003 428	2253.4	2373.4	4.7747
450	.009 162	2720.7	2949.7	5.6744	.006 735	2619.3	2821.4	5.4424	.004 961	2498.7	2672.4	5.1962
500	.011 123	2884.3	3162.4	5.9592	.008 678	2820.7	3081.1	5.7905	.006 927	2751.9	2994.4	5.6282
550	.012 724	3017.5	3335.6	6.1765	.010 168	2970.3	3275.4	6.0342	.008 345	2921.0	3213.0	5.9026
600	.014 137	3137.9	3491.4	6.3602	.011 446	3100.5	3443.9	6.2331	.009 527	3062.0	3395.5	6.1179
650	.015 433	3251.6	3637.4	6.5229	.012 596	3221.0	3598.9	6.4058	.010 575	3189.8	3559.9	6.3010
700	.016 646	3361.3	3777.5	6.6707	.013 661	3335.8	3745.6	6.5606	.011 533	3309.8	3713.5	6.4631
800	.018 912	3574.3	4047.1	6.9345	.015 623	3555.5	4024.2	6.8332	.013 278	3536.7	4001.5	6.7450
900	.021 045	3783.0	4309.1	7.1680	.017 448	3768.5	4291.9	7.0718	.014 883	3754.0	4274.9	6.9886
1000	.023 10	3990.9	4568.5	7.3802	.019 196	3978.8	4554.7	7.2867	.016 410	3966.7	4541.1	7.2064
1100	.025 12	4200.2	4828.2	7.5765	.020 903	4189.2	4816.3	7.4845	.017 895	4178.3	4804.6	7.4057

APPENDIX B3 (Continued)

Superheated Steam

T	v	u	h	s	v	u	h	s	v	u	h	s
	$p = 25.0$ MPa				$p = 30.0$ MPa				$p = 35.0$ MPa			
1200	.027 11	4412.0	5089.9	7.7605	.022 589	4401.3	5079.0	7.6692	.019 360	4390.7	5068.3	7.5910
1300	.029 10	4626.9	5354.4	7.9342	.024 266	4616.0	5344.0	7.8432	.020 815	4605.1	5333.6	7.7653
	$p = 40.0$ MPa				$p = 50.0$ MPa				$p = 60.0$ MPa			
375	.001 640 7	1677.1	1742.8	3.8290	.001 559 4	1638.6	1716.6	3.7639	.001 502 8	1609.4	1699.5	3.7141
400	.001 907 7	1854.6	1930.9	4.1135	.001 730 9	1788.1	1874.6	4.0031	.001 633 5	1745.4	1843.4	3.9318
425	.002 532	2096.9	2198.1	4.5029	.002 007	1959.7	2060.0	4.2734	.001 816 5	1892.7	2001.7	4.1626
450	.003 693	2365.1	2512.8	4.9459	.002 486	2159.6	2284.0	4.5884	.002 085	2053.9	2179.0	4.4121
500	.005 622	2678.4	2903.3	5.4700	.003 892	2525.5	2720.1	5.1726	.002 956	2390.6	2567.9	4.9321
550	.006 984	2869.7	3149.1	5.7785	.005 118	2763.6	3019.5	5.5485	.003 956	2658.8	2896.2	5.3441
600	.008 094	3022.6	3346.4	6.0114	.006 112	2942.0	3247.6	5.8178	.004 834	2861.1	3151.2	5.6452
650	.009 063	3158.0	3520.6	6.2054	.006 966	3093.5	3441.8	6.0342	.005 595	3028.8	3364.5	5.8829
700	.009 941	3283.6	3681.2	6.3750	.007 727	3230.5	3616.8	6.2189	.006 272	3177.2	3553.5	6.0824
800	.011 523	3517.8	3978.7	6.6662	.009 076	3479.8	3933.6	6.5290	.007 459	3441.5	3889.1	6.4109
900	.012 962	3739.4	4257.9	6.9150	.010 283	3710.3	4224.4	6.7882	.008 508	3681.0	4191.5	6.6805
1000	.014 324	3954.6	4527.6	7.1356	.011 411	3930.5	4501.1	7.0146	.009 480	3906.4	4475.2	6.9127
1100	.015 642	4167.4	4793.1	7.3364	.012 496	4145.7	4770.5	7.2184	.010 409	4124.1	4748.6	7.1195
1200	.016 940	4380.1	5057.7	7.5224	.013 561	4359.1	5037.2	7.4058	.011 317	4338.2	5017.2	7.3083
1300	.018 229	4594.3	5323.5	7.6969	.014 616	4572.8	5303.6	7.5808	.012 215	4551.4	5284.3	7.4837

APPENDIX B4 PROPERTIES OF WATER IN THE SATURATED SOLID-VAPOR REGION

(v_i and v_g in m³/kg; u_i, u_{ig}, and u_g in kJ/kg; h_i, h_{ig}, and h_g in kJ/kg; and s_i, s_{ig}, and s_g in kJ/kgK)
(G. J. Van Wylen and R. E. Sonntag, *Fundamentals of Classical Thermodynamics, SI Version*, 2nd Edition. New York: John Wiley & Son, Inc. © 1976. Reprinted by permission.)

Temp. °C T	Specific Volume			Internal Energy			Enthalpy			Entropy		
	Press. kPa P	Sat. Solid $v_i \times 10^3$	Sat. Vapor v_g	Sat. Solid u_i	Subl. u_{ig}	Sat. Vapor u_g	Sat. Solid h_i	Subl. h_{ig}	Sat. Vapor h_g	Sat. Solid s_i	Subl. s_{ig}	Sat. Vapor s_g
.01	.6113	1.0908	206.1	-333.40	2708.7	2375.3	-333.40	2834.8	2501.4	-1.221	10.378	9.156
0	.6108	1.0908	206.3	-333.43	2708.8	2375.3	-333.43	2834.8	2501.3	-1.221	10.378	9.157
-2	.5176	1.0904	241.7	-337.62	2710.2	2372.6	-337.62	2835.3	2497.7	-1.237	10.456	9.219
-4	.4375	1.0901	283.8	-341.78	2711.6	2369.8	-341.78	2835.7	2494.0	-1.253	10.536	9.283
-6	.3689	1.0898	334.2	-345.91	2712.9	2367.0	-345.91	2836.2	2490.3	-1.268	10.616	9.348
-8	.3102	1.0894	394.4	-350.02	2714.2	2364.2	-350.02	2836.6	2486.6	-1.284	10.698	9.414
-10	.2602	1.0891	466.7	-354.09	2715.5	2361.4	-354.09	2837.0	2482.9	-1.299	10.781	9.481
-12	.2176	1.0888	553.7	-358.14	2716.8	2358.7	-358.14	2837.3	2479.2	-1.315	10.865	9.550
-14	.1815	1.0884	658.8	-362.15	2718.0	2355.9	-362.15	2837.6	2475.5	-1.331	10.950	9.619
-16	.1510	1.0881	786.0	-366.14	2719.2	2353.1	-366.14	2837.9	2471.8	-1.346	11.036	9.690
-18	.1252	1.0878	940.5	-370.10	2720.4	2350.3	-370.10	2838.2	2468.1	-1.362	11.123	9.762
-20	.1035	1.0874	1128.6	-374.03	2721.6	2347.5	-374.03	2838.4	2464.3	-1.377	11.212	9.835
-22	.0853	1.0871	1358.4	-377.93	2722.7	2344.7	-377.93	2838.6	2460.6	-1.393	11.302	9.909
-24	.0701	1.0868	1640.1	-381.80	2723.7	2342.0	-381.80	2838.7	2456.9	-1.408	11.394	9.985
-26	.0574	1.0864	1986.4	-385.64	2724.8	2339.2	-385.64	2838.9	2453.2	-1.424	11.486	10.062
-28	.0469	1.0861	2413.7	-389.45	2725.8	2336.4	-389.45	2839.0	2449.5	-1.439	11.580	10.141
-30	.0381	1.0858	2943	-393.23	2726.8	2333.6	-393.23	2839.1	2445.8	-1.455	11.676	10.221
-32	.0309	1.0854	3600	-396.98	2727.8	2330.8	-396.98	2839.1	2442.1	-1.471	11.773	10.303
-34	.0250	1.0851	4419	-400.71	2728.7	2328.0	-400.71	2839.1	2438.4	-1.486	11.872	10.386
-36	.0201	1.0848	5444	-404.40	2729.6	2325.2	-404.40	2839.1	2434.7	-1.501	11.972	10.470
-38	.0161	1.0844	6731	-408.06	2730.5	2322.4	-408.06	2839.0	2430.9	-1.517	12.073	10.556
-40	.0129	1.0841	8354	-411.70	2731.3	2319.6	-411.70	2838.9	2427.2	-1.532	12.176	10.644

APPENDIX C1 PROPERTIES OF SATURATED FREON-12 (DICHLORODIFLUOROMETHANE)[a]

(Copyright 1955 and 1956, E. I. du Pont de Nemours & Company, Inc. Adapted from English units. Used with permission.)

Properties of Saturated Freon-12 (Dichlorodifluoromethane)[a]

Temp. °C	Abs. Press. MPa p	Specific Volume m³/kg			Enthalpy kJ/kg			Entropy kJ/kg K		
		Sat. Liquid v_f	Evap. v_{fg}	Sat. Vapor v_g	Sat. Liquid h_f	Evap. h_{fg}	Sat. Vapor h_g	Sat. Liquid s_f	Evap. s_{fg}	Sat. Vapor s_g
−90	0.0028	0.000 608	4.414 937	4.415 545	−43.243	189.618	146.375	−0.2084	1.0352	0.8268
−85	0.0042	0.000 612	3.036 704	3.037 316	−38.968	187.608	148.640	−0.1854	0.9970	0.8116
−80	0.0062	0.000 617	2.137 728	2.138 345	−34.688	185.612	150.924	−0.1630	0.9609	0.7979
−75	0.0088	0.000 622	1.537 030	1.537 651	−30.401	183.625	153.224	−0.1411	0.9266	0.7855
−70	0.0123	0.000 627	1.126 654	1.127 280	−26.103	181.640	155.536	−0.1197	0.8940	0.7744
−65	0.0168	0.000 632	0.840 534	0.841 166	−21.793	179.651	157.857	−0.0987	0.8630	0.7643
−60	0.0226	0.000 637	0.637 274	0.637 910	−17.469	177.653	160.184	−0.0782	0.8334	0.7552
−55	0.0300	0.000 642	0.490 358	0.491 000	−13.129	175.641	162.512	−0.0581	0.8051	0.7470
−50	0.0391	0.000 648	0.382 457	0.383 105	−8.772	173.611	164.840	−0.0384	0.7779	0.7396
−45	0.0504	0.000 654	0.302 029	0.302 682	−4.396	171.558	167.163	−0.0190	0.7519	0.7329
−40	0.0642	0.000 659	0.241 251	0.241 910	−0.000	169.479	169.479	−0.0000	0.7269	0.7269
−35	0.0807	0.000 666	0.194 732	0.195 398	4.416	167.368	171.784	0.0187	0.7027	0.7214
−30	0.1004	0.000 672	0.158 703	0.159 375	8.854	165.222	174.076	0.0371	0.6795	0.7165
−25	0.1237	0.000 679	0.130 487	0.131 166	13.315	163.037	176.352	0.0552	0.6570	0.7121
−20	0.1509	0.000 685	0.108 162	0.108 847	17.800	160.810	178.610	0.0730	0.6352	0.7082
−15	0.1826	0.000 693	0.090 326	0.091 018	22.312	158.534	180.846	0.0906	0.6141	0.7046
−10	0.2191	0.000 700	0.075 946	0.076 646	26.851	156.207	183.058	0.1079	0.5936	0.7014
−5	0.2610	0.000 708	0.064 255	0.064 963	31.420	153.823	185.243	0.1250	0.5736	0.6986
0	0.3086	0.000 716	0.054 673	0.055 389	36.022	151.376	187.397	0.1418	0.5542	0.6960
5	0.3626	0.000 724	0.046 761	0.047 485	40.659	148.859	189.518	0.1585	0.5351	0.6937

APPENDIX C1 (Continued).

10	0.4233	0.000 733	0.040 180	0.040 914	45.337	146.265	191.602	0.1750	0.5165	0.6916
15	0.4914	0.000 743	0.034 671	0.035 413	50.058	143.586	193.644	0.1914	0.4983	0.6897
20	0.5673	0.000 752	0.030 028	0.030 780	54.828	140.812	195.641	0.2076	0.4803	0.6879
25	0.6516	0.000 763	0.026 091	0.026 854	59.653	137.933	197.586	0.2237	0.4626	0.6863
30	0.7449	0.000 774	0.022 734	0.023 508	64.539	134.936	199.475	0.2397	0.4451	0.6848
35	0.8477	0.000 786	0.019 855	0.020 641	69.494	131.805	201.299	0.2557	0.4277	0.6834
40	0.9607	0.000 798	0.017 373	0.018 171	74.527	128.525	203.051	0.2716	0.4104	0.6820
45	1.0843	0.000 811	0.015 220	0.016 032	79.647	125.074	204.722	0.2875	0.3931	0.6806
50	1.2193	0.000 826	0.013 344	0.014 170	84.868	121.430	206.298	0.3034	0.3758	0.6792
55	1.3663	0.000 841	0.011 701	0.012 542	90.201	117.565	207.766	0.3194	0.3582	0.6777
60	1.5259	0.000 858	0.010 253	0.011 111	95.665	113.443	209.109	0.3355	0.3405	0.6760
65	1.6988	0.000 877	0.008 971	0.009 847	101.279	109.024	210.303	0.3518	0.3224	0.6742
70	1.8858	0.000 897	0.007 828	0.008 725	107.067	104.255	211.321	0.3683	0.3038	0.6721
75	2.0874	0.000 920	0.006 802	0.007 723	113.058	99.068	212.126	0.3851	0.2845	0.6697
80	2.3046	0.000 946	0.005 875	0.006 821	119.291	93.373	212.665	0.4023	0.2644	0.6667
85	2.5380	0.000 976	0.005 029	0.006 005	125.818	87.047	212.865	0.4201	0.2430	0.6631
90	2.7885	0.001 012	0.004 246	0.005 258	132.708	79.907	212.614	0.4385	0.2200	0.6585
95	3.0569	0.001 056	0.003 508	0.004 563	140.068	71.658	211.726	0.4579	0.1946	0.6526
100	3.3440	0.001 113	0.002 790	0.003 903	148.076	61.768	209.843	0.4788	0.1655	0.6444
105	3.6509	0.001 197	0.002 045	0.003 242	157.085	49.014	206.099	0.5023	0.1296	0.6319
110	3.9784	0.001 364	0.001 098	0.002 462	168.059	28.425	196.484	0.5322	0.0742	0.6064
112	4.1155	0.001 792	0.000 005	0.001 797	174.920	0.151	175.071	0.5651	0.0004	0.5655

Properties of Superheated Freon-12

Temp. °C	0.05 MPa v m³/kg	h kJ/kg	s kJ/kg K	0.10 MPa v m³/kg	h kJ/kg	s kJ/kg K	0.15 MPa v m³/kg	h kJ/kg	s kJ/kg K
−20.0	0.341 857	181.042	0.7912	0.167 701	179.861	0.7401			
−10.0	0.356 227	186.757	0.8133	0.175 222	185.707	0.7628	0.114 716	184.619	0.7318
0.0	0.370 508	192.567	0.8350	0.182 647	191.628	0.7849	0.119 866	190.660	0.7543
10.0	0.384 716	198.471	0.8562	0.189 994	197.628	0.8064	0.124 932	196.762	0.7763
20.0	0.398 863	204.469	0.8770	0.197 277	203.707	0.8275	0.129 930	202.927	0.7977
30.0	0.412 959	210.557	0.8974	0.204 506	209.866	0.8482	0.134 873	209.160	0.8186
40.0	0.427 012	216.733	0.9175	0.211 691	216.104	0.8684	0.139 768	215.463	0.8390
50.0	0.441 030	222.997	0.9372	0.218 839	222.421	0.8883	0.144 625	221.835	0.8591
60.0	0.455 017	229.344	0.9565	0.225 955	228.815	0.9078	0.149 450	228.277	0.8787
70.0	0.468 978	235.774	0.9755	0.233 044	235.285	0.9269	0.154 247	234.789	0.8980
80.0	0.482 917	242.282	0.9942	0.240 111	241.829	0.9457	0.159 020	241.371	0.9169
90.0	0.496 838	248.868	1.0126	0.247 159	248.446	0.9642	0.163 774	248.020	0.9354

Temp. °C	0.20 MPa v m³/kg	h kJ/kg	s kJ/kg K	0.25 MPa v m³/kg	h kJ/kg	s kJ/kg K	0.30 MPa v m³/kg	h kJ/kg	s kJ/kg K
0.0	0.088 608	189.669	0.7320	0.069 752	188.644	0.7139	0.057 150	187.583	0.6984
10.0	0.092 550	195.878	0.7543	0.073 024	194.969	0.7366	0.059 984	194.034	0.7216
20.0	0.096 418	202.135	0.7760	0.076 218	201.322	0.7587	0.062 734	200.490	0.7440
30.0	0.100 228	208.446	0.7972	0.079 350	207.715	0.7801	0.065 418	206.969	0.7658
40.0	0.103 989	214.814	0.8178	0.082 431	214.153	0.8010	0.068 049	213.480	0.7869
50.0	0.107 710	221.243	0.8381	0.085 470	220.642	0.8214	0.070 635	220.030	0.8075
60.0	0.111 397	227.735	0.8578	0.088 474	227.185	0.8413	0.073 185	226.627	0.8276
70.0	0.115 055	234.291	0.8772	0.091 449	233.785	0.8608	0.075 705	233.273	0.8473
80.0	0.118 690	240.910	0.8962	0.094 398	240.443	0.8800	0.078 200	239.971	0.8665
90.0	0.122 304	247.593	0.9149	0.097 327	247.160	0.8987	0.080 673	246.723	0.8853
100.0	0.125 901	254.339	0.9332	0.100 238	253.936	0.9171	0.083 127	253.530	0.9038
110.0	0.129 483	261.147	0.9512	0.103 134	260.770	0.9352	0.085 566	260.391	0.9220

Temp (°C)	0.40 MPa			0.50 MPa			0.60 MPa		
20.0	0.045 836	198.762	0.7199	0.035 646	196.935	0.6999			
30.0	0.047 971	205.428	0.7423	0.037 464	203.814	0.7230	0.030 422	202.116	0.7063
40.0	0.050 046	212.095	0.7639	0.039 214	210.656	0.7452	0.031 966	209.154	0.7291
50.0	0.052 072	218.779	0.7849	0.040 911	217.484	0.7667	0.033 450	216.141	0.7511
60.0	0.054 059	225.488	0.8054	0.042 565	224.315	0.7875	0.034 887	223.104	0.7723
70.0	0.056 014	232.230	0.8253	0.044 184	231.161	0.8077	0.036 285	230.062	0.7929
80.0	0.057 941	239.012	0.8448	0.045 774	238.031	0.8275	0.037 653	237.027	0.8129
90.0	0.059 846	245.837	0.8638	0.047 340	244.932	0.8467	0.038 995	244.009	0.8324
100.0	0.061 731	252.707	0.8825	0.048 886	251.869	0.8656	0.040 316	251.016	0.8514
110.0	0.063 600	259.624	0.9008	0.050 415	258.845	0.8840	0.041 619	258.053	0.8700
120.0	0.065 455	266.590	0.9187	0.051 929	265.862	0.9021	0.042 907	265.124	0.8882
130.0	0.067 298	273.605	0.9364	0.053 430	272.923	0.9198	0.044 181	272.231	0.9061

Temp (°C)	0.70 MPa			0.80 MPa			0.90 MPa		
40.0	0.026 761	207.580	0.7148	0.022 830	205.924	0.7016	0.019 744	204.170	0.6982
50.0	0.028 100	214.745	0.7373	0.024 068	213.290	0.7248	0.020 912	211.765	0.7131
60.0	0.029 387	221.854	0.7590	0.025 247	220.558	0.7469	0.022 012	219.212	0.7358
70.0	0.030 632	228.931	0.7799	0.026 380	227.766	0.7682	0.023 062	226.564	0.7575
80.0	0.031 843	235.997	0.8002	0.027 477	234.941	0.7888	0.024 072	233.856	0.7785
90.0	0.033 027	243.066	0.8199	0.028 545	242.101	0.8088	0.025 051	241.113	0.7987
100.0	0.034 189	250.146	0.8392	0.029 588	249.260	0.8283	0.026 005	248.355	0.8184
110.0	0.035 332	257.247	0.8579	0.030 612	256.428	0.8472	0.026 937	255.593	0.8376
120.0	0.036 458	264.374	0.8763	0.031 619	263.613	0.8657	0.027 851	262.839	0.8562
130.0	0.037 572	271.531	0.8943	0.032 612	270.820	0.8838	0.028 751	270.100	0.8745
140.0	0.038 673	278.720	0.9119	0.033 592	278.055	0.9016	0.029 639	277.381	0.8923
150.0	0.039 764	285.946	0.9292	0.034 563	285.320	0.9189	0.030 515	284.687	0.9098

Temp. °C	1.00 MPa			1.20 MPa			1.40 MPa		
	v m³/kg	h kJ/kg	s kJ/kg K	v m³/kg	h kJ/kg	s kJ/kg K	v m³/kg	h kJ/kg	s kJ/kg K
50.0	0.018 366	210.162	0.7021	0.014 483	206.661	0.6812	0.012 579	211.457	0.6876
60.0	0.019 410	217.810	0.7254	0.015 463	214.805	0.7060	0.013 448	219.822	0.7123
70.0	0.020 397	225.319	0.7476	0.016 368	222.687	0.7293	0.014 247	227.891	0.7355
80.0	0.021 341	232.739	0.7689	0.017 221	230.398	0.7514	0.014 997	235.766	0.7575
90.0	0.022 251	240.101	0.7895	0.018 032	237.995	0.7727	0.015 710	243.512	0.7785
100.0	0.023 133	247.430	0.8094	0.018 812	245.518	0.7931	0.016 393	251.170	0.7988
110.0	0.023 993	254.743	0.8287	0.019 567	252.993	0.8129	0.017 053	258.770	0.8183
120.0	0.024 835	262.053	0.8475	0.020 301	260.441	0.8320	0.017 695	266.334	0.8373
130.0	0.025 661	269.369	0.8659	0.021 018	267.875	0.8507	0.018 321	273.877	0.8558
140.0	0.026 474	276.699	0.8839	0.021 721	275.307	0.8689	0.018 934	281.411	0.8738
150.0	0.027 275	284.047	0.9015	0.022 412	282.745	0.8867	0.019 535	288.946	0.8914
160.0	0.028 068	291.419	0.9187	0.023 093	290.195	0.9041			

Temp. °C	1.60 MPa			1.80 MPa			2.00 MPa		
	v m³/kg	h kJ/kg	s kJ/kg K	v m³/kg	h kJ/kg	s kJ/kg K	v m³/kg	h kJ/kg	s kJ/kg K
70.0	0.011 208	216.650	0.6959	0.009 406	213.049	0.6794	0.008 704	218.859	0.6909
80.0	0.011 984	225.177	0.7204	0.010 187	222.198	0.7057	0.009 406	228.056	0.7166
90.0	0.012 698	233.390	0.7433	0.010 884	230.835	0.7298	0.010 035	236.760	0.7402
100.0	0.013 366	241.397	0.7651	0.011 526	239.155	0.7524	0.010 615	245.154	0.7624
110.0	0.014 000	249.264	0.7859	0.012 126	247.264	0.7739	0.011 159	253.341	0.7835
120.0	0.014 608	257.035	0.8059	0.012 697	255.228	0.7944	0.011 676	261.384	0.8037
130.0	0.015 195	264.742	0.8253	0.013 244	263.094	0.8141	0.012 172	269.327	0.8232
140.0	0.015 765	272.406	0.8440	0.013 772	270.891	0.8332	0.012 651	277.201	0.8420
150.0	0.016 320	280.044	0.8623	0.014 284	278.642	0.8518	0.013 116	285.027	0.8603
160.0	0.016 864	287.669	0.8801	0.014 784	286.364	0.8698	0.013 570	292.822	0.8781
170.0	0.017 398	295.290	0.8975	0.015 272	294.069	0.8874	0.014 013	300.598	0.8955
180.0	0.017 923	302.914	0.9145	0.015 752	301.767	0.9046			

Temp	2.50 MPa			3.00 MPa			3.50 MPa		
90.0	0.006 595	219.562	0.6823						
100.0	0.007 264	229.852	0.7103	0.005 231	220.529	0.6770			
110.0	0.007 837	239.271	0.7352	0.005 886	232.068	0.7075	0.004 324	222.121	0.6750
120.0	0.008 351	248.192	0.7582	0.006 419	242.208	0.7336	0.004 959	234.875	0.7078
130.0	0.008 827	256.794	0.7798	0.006 887	251.632	0.7573	0.005 456	245.661	0.7349
140.0	0.009 273	265.180	0.8003	0.007 313	260.620	0.7793	0.005 884	255.524	0.7591
150.0	0.009 697	273.414	0.8200	0.007 709	269.319	0.8001	0.006 270	264.846	0.7814
160.0	0.010 104	281.540	0.8390	0.008 083	277.817	0.8200	0.006 626	273.817	0.8023
170.0	0.010 497	289.589	0.8574	0.008 439	286.171	0.8391	0.006 961	282.545	0.8222
180.0	0.010 879	297.583	0.8752	0.008 782	294.422	0.8575	0.007 279	291.100	0.8413
190.0	0.011 250	305.540	0.8926	0.009 114	302.597	0.8753	0.007 584	299.528	0.8597
200.0	0.011 614	313.472	0.9095	0.009 436	310.718	0.8927	0.007 878	307.864	0.8775

Temp	4.00 MPa		
120.0	0.003 736	224.863	0.6771
130.0	0.004 325	238.443	0.7111
140.0	0.004 781	249.703	0.7386
150.0	0.005 172	259.904	0.7630
160.0	0.005 522	269.492	0.7854
170.0	0.005 845	278.684	0.8063
180.0	0.006 147	287.602	0.8262
190.0	0.006 434	296.326	0.8453
200.0	0.006 708	304.906	0.8636
210.0	0.006 972	313.380	0.8813
220.0	0.007 228	321.774	0.8985
230.0	0.007 477	330.108	0.9152

APPENDIX D1 PROPERTIES OF SATURATED AMMONIA[a]

(Adapted from "Tables of Thermodynamic Properties of Ammonia," *National Bureau of Standards Circular No. 142*, 1923. Used with permission.)

Properties of Saturated Ammonia[a]

Temp. °C	Abs. Press. kPa p	Specific Volume m³/kg			Enthalpy kJ/kg			Entropy kJ/kg K		
		Sat. Liquid v_f	Evap. v_{fg}	Sat. Vapor v_g	Sat. Liquid h_f	Evap. h_{fg}	Sat. Vapor h_g	Sat. Liquid s_f	Evap. s_{fg}	Sat. Vapor s_g
−50	40.88	0.001 424	2.6239	2.6254	−44.3	1416.7	1372.4	−0.1942	6.3502	6.1561
−48	45.96	0.001 429	2.3518	2.3533	−35.5	1411.3	1375.8	−0.1547	6.2696	6.1149
−46	51.55	0.001 434	2.1126	2.1140	−26.6	1405.8	1379.2	−0.1156	6.1902	6.0746
−44	57.69	0.001 439	1.9018	1.9032	−17.8	1400.3	1382.5	−0.0768	6.1120	6.0352
−42	64.42	0.001 444	1.7155	1.7170	−8.9	1394.7	1385.8	−0.0382	6.0349	5.9967
−40	71.77	0.001 449	1.5506	1.5521	0.0	1389.0	1389.0	0.0000	5.9589	5.9589
−38	79.80	0.001 454	1.4043	1.4058	8.9	1383.3	1392.2	0.0380	5.8840	5.9220
−36	88.54	0.001 460	1.2742	1.2757	17.8	1377.6	1395.4	0.0757	5.8101	5.8858
−34	98.05	0.001 465	1.1582	1.1597	26.8	1371.8	1398.5	0.1132	5.7372	5.8504
−32	108.37	0.001 470	1.0547	1.0562	35.7	1365.9	1401.6	0.1504	5.6652	5.8156
−30	119.55	0.001 476	0.9621	0.9635	44.7	1360.0	1404.6	0.1873	5.5942	5.7815
−28	131.64	0.001 481	0.8790	0.8805	53.6	1354.0	1407.6	0.2240	5.5241	5.7481
−26	144.70	0.001 487	0.8044	0.8059	62.6	1347.9	1410.5	0.2605	5.4548	5.7153
−24	158.78	0.001 492	0.7373	0.7388	71.6	1341.8	1413.4	0.2967	5.3864	5.6831
−22	173.93	0.001 498	0.6768	0.6783	80.7	1335.6	1416.2	0.3327	5.3188	5.6515
−20	190.22	0.001 504	0.6222	0.6237	89.7	1329.3	1419.0	0.3684	5.2520	5.6205
−18	207.71	0.001 510	0.5728	0.5743	98.8	1322.9	1421.7	0.4040	5.1860	5.5900
−16	226.45	0.001 515	0.5280	0.5296	107.8	1316.5	1424.4	0.4393	5.1207	5.5600
−14	246.51	0.001 521	0.4874	0.4889	116.9	1310.0	1427.0	0.4744	5.0561	5.5305
−12	267.95	0.001 528	0.4505	0.4520	126.0	1303.5	1429.5	0.5093	4.9922	5.5015

−10	290.85	0.001 534	0.4169	0.4185	135.2	1296.8	1432.0	0.5440	4.9290	5.4730
−8	315.25	0.001 540	0.3863	0.3878	144.3	1290.1	1434.4	0.5785	4.8664	5.4449
−6	341.25	0.001 546	0.3583	0.3599	153.5	1283.3	1436.8	0.6128	4.8045	5.4173
−4	368.90	0.001 553	0.3328	0.3343	162.7	1276.4	1439.1	0.6469	4.7432	5.3901
−2	398.27	0.001 559	0.3094	0.3109	171.9	1269.4	1441.3	0.6808	4.6825	5.3633
0	429.44	0.001 566	0.2879	0.2895	181.1	1262.4	1443.5	0.7145	4.6223	5.3369
2	462.49	0.001 573	0.2683	0.2698	190.4	1255.2	1445.6	0.7481	4.5627	5.3108
4	497.49	0.001 580	0.2502	0.2517	199.6	1248.0	1447.6	0.7815	4.5037	5.2852
6	534.51	0.001 587	0.2335	0.2351	208.9	1240.6	1449.6	0.8148	4.4451	5.2599
8	573.64	0.001 594	0.2182	0.2198	218.3	1233.2	1451.5	0.8479	4.3871	5.2350
10	614.95	0.001 601	0.2040	0.2056	227.6	1225.7	1453.3	0.8808	4.3295	5.2104
12	658.52	0.001 608	0.1910	0.1926	237.0	1218.1	1455.1	0.9136	4.2725	5.1861
14	704.44	0.001 616	0.1789	0.1805	246.4	1210.4	1456.8	0.9463	4.2159	5.1621
16	752.79	0.001 623	0.1677	0.1693	255.9	1202.6	1458.5	0.9788	4.1597	5.1385
18	803.66	0.001 631	0.1574	0.1590	265.4	1194.7	1460.0	1.0112	4.1039	5.1151
20	857.12	0.001 639	0.1477	0.1494	274.9	1186.7	1461.5	1.0434	4.0486	5.0920
22	913.27	0.001 647	0.1388	0.1405	284.4	1178.5	1462.9	1.0755	3.9937	5.0692
24	972.19	0.001 655	0.1305	0.1322	294.0	1170.3	1464.3	1.1075	3.9392	5.0467
26	1033.97	0.001 663	0.1228	0.1245	303.6	1162.0	1465.6	1.1394	3.8850	5.0244
28	1098.71	0.001 671	0.1156	0.1173	313.2	1153.6	1466.8	1.1711	3.8312	5.0023
30	1166.49	0.001 680	0.1089	0.1106	322.9	1145.0	1467.9	1.2028	3.7777	4.9805
32	1237.41	0.001 689	0.1027	0.1044	332.6	1136.4	1469.0	1.2343	3.7246	4.9589
34	1311.55	0.001 698	0.0969	0.0986	342.3	1127.6	1469.9	1.2656	3.6718	4.9374
36	1389.03	0.001 707	0.0914	0.0931	352.1	1118.7	1470.8	1.2969	3.6192	4.9161
38	1469.92	0.001 716	0.0863	0.0880	361.9	1109.7	1471.5	1.3281	3.5669	4.8950
40	1554.33	0.001 726	0.0815	0.0833	371.7	1100.5	1472.2	1.3591	3.5148	4.8740
42	1642.35	0.001 735	0.0771	0.0788	381.6	1091.2	1472.8	1.3901	3.4630	4.8530
44	1734.09	0.001 745	0.0728	0.0746	391.5	1081.7	1473.2	1.4209	3.4112	4.8322
46	1829.65	0.001 756	0.0689	0.0707	401.5	1072.0	1473.5	1.4518	3.3595	4.8113
48	1929.13	0.001 766	0.0652	0.0669	411.5	1062.2	1473.7	1.4826	3.3079	4.7905
50	2032.62	0.001 777	0.0617	0.0635	421.7	1052.0	1473.7	1.5135	3.2561	4.7696

[a] Adapted from National Bureau of Standards Circular No. 142. *Tables of Thermodynamic Properties of Ammonia.*

APPENDIX D2 PROPERTIES OF SUPERHEATED AMMONIA

Properties of Superheated Ammonia

Abs. Press. kPa (Sat. Temp.) °C		-20	-10	0	10	20	30	40	50	60	70	80	100
50 (-46.54)	v	2.4474	2.5481	2.6482	2.7479	2.8473	2.9464	3.0453	3.1441	3.2427	3.3413	3.4397	
	h	1435.8	1457.0	1478.1	1499.2	1520.4	1541.7	1563.0	1584.5	1606.1	1627.8	1649.7	
	s	6.3256	6.4077	6.4865	6.5625	6.6360	6.7073	6.7766	6.8441	6.9099	6.9743	7.0372	
75 (-39.18)	v	1.6233	1.6915	1.7591	1.8263	1.8932	1.9597	2.0261	2.0923	2.1584	2.2244	2.2903	
	h	1433.0	1454.7	1476.1	1497.5	1518.9	1540.3	1561.8	1583.4	1605.1	1626.9	1648.9	
	s	6.1190	6.2028	6.2828	6.3597	6.4339	6.5058	6.5756	6.6434	6.7096	6.7742	6.8373	
100 (-33.61)	v	1.2110	1.2631	1.3145	1.3654	1.4160	1.4664	1.5165	1.5664	1.6163	1.6659	1.7155	1.8145
	h	1430.1	1452.2	1474.1	1495.7	1517.3	1538.9	1560.5	1582.2	1604.1	1626.0	1648.0	1692.6
	s	5.9695	6.0552	6.1366	6.2144	6.2894	6.3618	6.4321	6.5003	6.5668	6.6316	6.6950	6.8177
125 (-29.08)	v	0.9635	1.0059	1.0476	1.0889	1.1297	1.1703	1.2107	1.2509	1.2909	1.3309	1.3707	1.4501
	h	1427.2	1449.8	1472.0	1493.9	1515.7	1537.5	1559.3	1581.1	1603.0	1625.0	1647.2	1691.8
	s	5.8512	5.9389	6.0217	6.1006	6.1763	6.2494	6.3201	6.3887	6.4555	6.5206	6.5842	6.7072
150 (-25.23)	v	0.7984	0.8344	0.8697	0.9045	0.9388	0.9729	1.0068	1.0405	1.0740	1.1074	1.1408	1.2072
	h	1424.1	1447.3	1469.8	1492.1	1514.1	1536.1	1558.0	1580.0	1602.0	1624.1	1646.3	1691.1
	s	5.7526	5.8424	5.9266	6.0066	6.0831	6.1568	6.2280	6.2970	6.3641	6.4295	6.4933	6.6167
200 (-18.86)	v		0.6199	0.6471	0.6738	0.7001	0.7261	0.7519	0.7774	0.8029	0.8282	0.8533	0.9035
	h		1442.0	1465.5	1488.4	1510.9	1533.2	1555.5	1577.7	1599.9	1622.2	1644.6	1689.6
	s		5.6863	5.7737	5.8559	5.9342	6.0091	6.0813	6.1512	6.2189	6.2849	6.3491	6.4732
250 (-13.67)	v		0.4910	0.5135	0.5354	0.5568	0.5780	0.5989	0.6196	0.6401	0.6605	0.6809	0.7212
	h		1436.6	1461.0	1484.5	1507.6	1530.3	1552.9	1575.4	1597.8	1620.3	1642.8	1688.2
	s		5.5609	5.6517	5.7365	5.8165	5.8928	5.9661	6.0368	6.1052	6.1717	6.2365	6.3613
300 (-9.23)	v			0.4243	0.4430	0.4613	0.4792	0.4968	0.5143	0.5316	0.5488	0.5658	0.5997
	h			1456.3	1480.6	1504.2	1527.4	1550.3	1573.0	1595.7	1618.4	1641.1	1686.7
	s			5.5493	5.6366	5.7186	5.7963	5.8707	5.9423	6.0114	6.0785	6.1437	6.2693
350 (-5.35)	v			0.3605	0.3770	0.3929	0.4086	0.4239	0.4391	0.4541	0.4689	0.4837	0.5129
	h			1451.5	1476.5	1500.7	1524.4	1547.6	1570.7	1593.6	1616.5	1639.3	1685.2
	s			5.4600	5.5502	5.6342	5.7135	5.7890	5.8615	5.9314	5.9990	6.0647	6.1910
400 (-1.89)	v			0.3125	0.3274	0.3417	0.3556	0.3692	0.3826	0.3959	0.4090	0.4220	0.4478
	h			1446.5	1472.4	1497.2	1521.3	1544.9	1568.3	1591.5	1614.5	1637.6	1683.7
	s			5.3803	5.4735	5.5597	5.6405	5.7173	5.7907	5.8613	5.9296	5.9957	6.1228

Temperature, °C

Superheated steam table (values listed as v / h / s for each pressure; the figure in parentheses under each pressure is the saturation temperature). Column headings are temperatures.

p (sat.)	prop.	20	30	40	50	60	70	80	100	120	140	160	180
450 (1.26)	v			0.2752	0.2887	0.3017	0.3143	0.3266	0.3387	0.3506	0.3624	0.3740	0.3971
	h			1441.3	1468.1	1493.6	1518.2	1542.2	1565.9	1589.3	1612.6	1635.8	1682.2
	s			5.3078	5.4042	5.4926	5.5752	5.6532	5.7275	5.7989	5.8678	5.9345	6.0623
500 (4.14)	v	0.2698	0.2813	0.2926	0.3036	0.3144	0.3251	0.3357	0.3565	0.3771	0.3975		
	h	1489.9	1515.0	1539.5	1563.4	1587.1	1610.6	1634.0	1680.7	1727.5	1774.7		
	s	5.4314	5.5157	5.5950	5.6704	5.7425	5.8120	5.8793	6.0079	6.1301	6.2472		
600 (9.29)	v	0.2217	0.2317	0.2414	0.2508	0.2600	0.2691	0.2781	0.2957	0.3130	0.3302		
	h	1482.4	1508.6	1533.8	1558.5	1582.7	1606.6	1630.4	1677.7	1724.9	1772.4		
	s	5.3222	5.4102	5.4923	5.5697	5.6436	5.7144	5.7826	5.9129	6.0363	6.1541		
700 (13.81)	v	0.1874	0.1963	0.2048	0.2131	0.2212	0.2291	0.2369	0.2522	0.2672	0.2821		
	h	1474.5	1501.9	1528.1	1553.4	1578.2	1602.6	1626.8	1674.6	1722.4	1770.2		
	s	5.2259	5.3179	5.4029	5.4826	5.5582	5.6303	5.6997	5.8316	5.9562	6.0749		
800 (17.86)	v	0.1615	0.1696	0.1773	0.1848	0.1920	0.1991	0.2060	0.2196	0.2329	0.2459	0.2589	
	h	1466.3	1495.0	1522.2	1548.3	1573.7	1598.6	1623.1	1671.6	1719.8	1768.0	1816.4	
	s	5.1387	5.2351	5.3232	5.4053	5.4827	5.5562	5.6268	5.7603	5.8861	6.0057	6.1202	
900 (21.54)	v		0.1488	0.1559	0.1627	0.1693	0.1757	0.1820	0.1942	0.2061	0.2178	0.2294	
	h		1488.0	1516.2	1543.0	1569.1	1594.4	1619.4	1668.5	1717.1	1765.7	1814.4	
	s		5.1593	5.2508	5.3354	5.4147	5.4897	5.5614	5.6968	5.8237	5.9442	6.0594	
1000 (24.91)	v		0.1321	0.1388	0.1450	0.1511	0.1570	0.1627	0.1739	0.1847	0.1954	0.2058	0.2162
	h			1480.6	1537.7	1564.4	1590.3	1615.6	1665.4	1714.5	1763.4	1812.4	1861.7
	s			5.0889	5.2713	5.3525	5.4292	5.5021	5.6392	5.7674	5.8888	6.0047	6.1159
1200 (30.96)	v			0.1129	0.1185	0.1238	0.1289	0.1338	0.1434	0.1526	0.1616	0.1705	0.1792
	h				1497.1	1554.7	1581.7	1608.0	1659.2	1709.2	1758.9	1808.5	1858.2
	s				5.0629	5.2416	5.3215	5.3970	5.5379	5.6687	5.7919	5.9091	6.0214
1400 (36.28)	v			0.0944	0.0995	0.1042	0.1088	0.1132	0.1216	0.1297	0.1376	0.1452	0.1528
	h			1483.4	1515.1	1544.7	1573.0	1600.2	1652.8	1703.9	1754.3	1804.5	1854.7
	s			4.9534	5.0543	5.1419	5.2232	5.3053	5.4501	5.5836	5.7087	5.8273	5.9406
1600 (41.05)	v				0.0851	0.0895	0.0937	0.0977	0.1053	0.1125	0.1195	0.1263	0.1330
	h				1502.9	1534.0	1564.0	1592.3	1646.4	1698.5	1749.7	1800.5	1851.2
	s				4.9584	5.0530	5.1452	5.2273	5.3722	5.5104	5.6355	5.7555	5.8699
1800 (45.39)	v				0.0739	0.0781	0.0820	0.0856	0.0926	0.0992	0.1055	0.1116	0.1177
	h				1490.0	1523.5	1554.6	1584.1	1639.8	1693.1	1745.1	1796.5	1847.7
	s				4.8693	4.9715	5.0635	5.1482	5.3018	5.4409	5.5699	5.6914	5.8069
2000 (49.38)	v				0.0648	0.0688	0.0725	0.0760	0.0824	0.0885	0.0943	0.0999	0.1054
	h				1476.1	1512.0	1544.9	1575.6	1633.2	1687.6	1740.4	1792.4	1844.1
	s				4.7834	4.8930	4.9902	5.0786	5.2371	5.3793	5.5104	5.6333	5.7499

APPENDIX E PROPERTIES OF SATURATED MERCURY[a]

(Adapted from "Thermodynamic Properties of Mercury Vapor," Lucian A. Sheldon, *ASME*, 49A, 30, 1949. Used with permission.)

Properties of Saturated Mercury[a]

Press., MPa	Temp., °C	Enthalpy, kJ/kg			Entropy, kJ/kg K			Specific Volume Sat. Vapor, m³/kg
		Sat. Liquid	Evap.	Sat. Vapor	Sat. Liquid	Evap.	Sat. Vapor	
0.000 06	109.2	15.13	297.20	312.33	0.0466	0.7774	0.8240	259.6
0.000 07	112.3	15.55	297.14	312.69	0.0477	0.7709	0.8186	224.3
0.000 08	115.0	15.93	297.09	313.02	0.0487	0.7654	0.8141	197.7
0.000 09	117.5	16.27	297.04	313.31	0.0496	0.7604	0.8100	176.8
0.000 10	119.7	16.58	297.00	313.58	0.0503	0.7560	0.8063	160.1
0.0002	134.9	18.67	296.71	315.38	0.0556	0.7271	0.7827	83.18
0.0004	151.5	20.93	296.40	317.33	0.0610	0.6981	0.7591	43.29
0.0006	161.8	22.33	296.21	318.54	0.0643	0.6811	0.7454	29.57
0.0008	169.4	23.37	296.06	319.43	0.0666	0.6690	0.7356	22.57
0.0010	175.5	24.21	295.95	320.16	0.0685	0.6596	0.7281	18.31
0.002	195.6	26.94	295.57	322.51	0.0744	0.6305	0.7049	9.570
0.004	217.7	29.92	295.15	325.07	0.0806	0.6013	0.6819	5.013
0.006	231.6	31.81	294.89	326.70	0.0843	0.5842	0.6685	3.438
0.008	242.0	33.21	294.70	327.91	0.0870	0.5721	0.6591	2.632
0.010	250.3	34.33	294.54	328.87	0.0892	0.5627	0.6519	2.140
0.02	278.1	38.05	294.02	332.07	0.0961	0.5334	0.6295	1.128
0.04	309.1	42.21	293.43	335.64	0.1034	0.5039	0.6073	0.5942
0.06	329.0	44.85	293.06	337.91	0.1078	0.4869	0.5947	0.4113
0.08	343.9	46.84	292.78	339.62	0.1110	0.4745	0.5855	0.3163
0.1	356.1	48.45	292.55	341.00	0.1136	0.4649	0.5785	0.2581
0.2	397.1	53.87	291.77	345.64	0.1218	0.4353	0.5571	0.1377
0.3	423.8	57.38	291.27	348.65	0.1268	0.4179	0.5447	0.095 51
0.4	444.1	60.03	290.89	350.92	0.1305	0.4056	0.5361	0.073 78
0.5	460.7	62.20	290.58	352.78	0.1334	0.3960	0.5294	0.060 44
0.6	474.9	64.06	290.31	354.37	0.1359	0.3881	0.5240	0.051 37
0.7	487.3	65.66	290.08	355.74	0.1380	0.3815	0.5195	0.044 79
0.8	498.4	67.11	289.87	356.98	0.1398	0.3757	0.5155	0.039 78
0.9	508.5	68.42	289.68	358.10	0.1415	0.3706	0.5121	0.035 84
1.0	517.8	69.61	289.50	359.11	0.1429	0.3660	0.5089	0.032 66
1.2	534.4	71.75	289.19	360.94	0.1455	0.3581	0.5036	0.027 81
1.4	549.0	73.63	288.92	362.55	0.1478	0.3514	0.4992	0.024 29
1.6	562.0	75.37	288.67	364.04	0.1498	0.3456	0.4954	0.021 61
1.8	574.0	76.83	288.45	365.28	0.1515	0.3405	0.4920	0.019 49
2.0	584.9	78.23	288.24	366.47	0.1531	0.3359	0.4890	0.017 78
2.2	595.1	79.54	288.05	367.59	0.1546	0.3318	0.4864	0.016 37
2.4	604.6	80.75	287.87	368.62	0.1559	0.3280	0.4839	0.015 18
2.6	613.5	81.89	287.70	369.59	0.1571	0.3245	0.4816	0.014 16
2.8	622.0	82.96	287.54	370.50	0.1583	0.3212	0.4795	0.013 29
3.0	630.0	83.97	287.39	371.36	0.1594	0.3182	0.4776	0.012 52
3.5	648.5	86.33	287.04	373.37	0.1619	0.3115	0.4734	0.010 96
4.0	665.1	88.43	286.73	375.16	0.1641	0.3056	0.4697	0.009 78

Press., MPa	Temp., °C	Enthalpy, kJ/kg			Entropy, kJ/kg K			Specific Volume Sat. Vapor, m³/kg
		Sat. Liquid	Evap.	Sat. Vapor	Sat. Liquid	Evap.	Sat. Vapor	
4.5	680.3	90.35	286.44	376.79	0.1660	0.3004	0.4664	0.008 85
5.0	694.4	92.11	286.18	378.29	0.1678	0.2958	0.4636	0.008 09
5.5	707.4	93.76	285.93	379.69	0.1694	0.2916	0.4610	0.007 46
6.0	719.7	95.30	285.70	381.00	0.1709	0.2878	0.4587	0.006 93
6.5	731.3	96.75	285.48	382.23	0.1723	0.2842	0.4565	0.006 48
7.0	742.3	98.12	285.28	383.40	0.1736	0.2809	0.4545	0.006 09·
7.5	752.7	99.42	285.08	384.50	0.1748	0.2779	0.4527	0.005 75

[a] Adapted from *Thermodynamic Properties of Mercury Vapor,* by Lucian A. Sheldon. ASME, 49A, 30 (1949).

APPENDIX F1 PROPERTIES OF SATURATED NITROGEN[a]

(Adapted from "Thermodynamic Properties of Nitrogen Including Liquid and Vapor Phases from 63K to 2000K with Pressures to 10,000 Bar," R. T. Jacobsen and R. B. Stewart, *Journal of Phys. and Chem. Ref. Data, 2,* 757–922, 1973. Used with permission.)

Properties of Saturated Nitrogen[a]

Temp. K	Press. MPa p	Specific Volume m³/kg Sat. Liquid v_f	Evap. v_{fg}	Sat. Vapor v_g	Enthalpy kJ/kg Sat. Liquid h_f	Evap. h_{fg}	Sat. Vapor h_g	Entropy kJ/kg·K Sat. Liquid s_f	Evap. s_{fg}	Sat. Vapor s_g
63.143	0.01253	0.001 152	1.480 060	1.481 212	−150.348	215.188	64.840	2.4310	3.4076	5.8386
65	0.01742	0.001 162	1.093 173	1.094 335	−146.691	213.291	66.600	2.4845	3.2849	5.7694
70	0.03858	0.001 189	0.525 785	0.526 974	−136.569	207.727	71.158	2.6345	2.9703	5.6048
75	0.07612	0.001 221	0.280 970	0.282 191	−126.287	201.662	75.375	2.7755	2.6915	5.4670
77.347	0.101325	0.001 237	0.215 504	0.216 741	−121.433	198.645	77.212	2.8390	2.5706	5.4096
80	0.1370	0.001 256	0.162 794	0.164 050	−115.926	195.089	79.163	2.9083	2.4409	5.3492
85	0.2291	0.001 296	0.100 434	0.101 730	−105.461	187.892	82.431	3.0339	2.2122	5.2461
90	0.3608	0.001 340	0.064 950	0.066 290	−94.817	179.894	85.077	3.1535	2.0001	5.1536
95	0.5411	0.001 392	0.043 504	0.044 896	−83.895	170.877	86.982	3.2688	1.7995	5.0683
100	0.7790	0.001 452	0.029 861	0.031 313	−72.571	160.562	87.991	3.3816	1.6060	4.9876
105	1.0843	0.001 524	0.020 745	0.022 269	−60.691	148.573	87.882	3.4930	1.4150	4.9080
110	1.4673	0.001 613	0.014 402	0.016 015	−48.027	134.319	86.292	3.6054	1.2209	4.8263
115	1.9395	0.001 797	0.009 696	0.011 493	−34.157	116.701	82.544	3.7214	1.0145	4.7359
120	2.5135	0.001 904	0.006 130	0.008 034	−18.017	93.092	75.075	3.8450	0.7803	4.6253
125	3.2079	0.002 323	0.002 568	0.004 891	+6.202	50.114	56.316	4.0356	0.3989	4.4345
126.1	3.4000	0.003 184	0.000 000	0.003 184	+30.791	0.000	30.791	4.2269	0.0000	4.2269

[a]Adapted from "Thermodynamic Properties of Nitrogen Including Liquid and Vapor Phases from 63 K to 2000 K with Pressures to 10,000 Bar," R. T. Jacobsen and R. R. Stewart, *Jour. of Phys. and Chem. Ref. Data,* **2:** 757–922 (1973).

APPENDIX F2 PROPERTIES OF SUPERHEATED NITROGEN

Properties of Superheated Nitrogen

Temp. K	0.1 MPa			0.2 MPa			0.5 MPa		
	v m³/kg	h kJ/kg	s kJ/kg K	v m³/kg	h kJ/kg	s kJ/kg K	v m³/kg	h kJ/kg	s kJ/kg K
100	0.290 978	101.965	5.6944	0.142 475	100.209	5.4767	0.055 520	94.345	5.1706
125	0.367 217	128.505	5.9313	0.181 711	127.371	5.7194	0.073 422	123.824	5.4343
150	0.442 619	154.779	6.1228	0.220 014	153.962	5.9132	0.090 150	151.470	5.6361
175	0.517 576	180.935	6.2841	0.257 890	180.314	6.0760	0.106 394	178.434	5.8025
200	0.592 288	207.029	6.4234	0.295 531	206.537	6.2160	0.122 394	205.063	5.9447
225	0.666 552	233.085	6.5460	0.332 841	232.690	6.3388	0.138 173	231.459	6.0690
250	0.741 375	259.122	6.6561	0.370 418	258.796	6.4491	0.154 006	257.828	6.1801
275	0.815 563	285.144	6.7550	0.407 619	284.876	6.5485	0.169 642	284.076	6.2800
300	0.890 205	311.158	6.8457	0.445 047	310.937	6.6393	0.185 346	310.273	6.3715

Temp. K	1.0 MPa			2.0 MPa			4.0 MPa		
	v m³/kg	h kJ/kg	s kJ/kg K	v m³/kg	h kJ/kg	s kJ/kg K	v m³/kg	h kJ/kg	s kJ/kg K
125	0.033 065	117.422	5.1872	0.014 021	101.489	4.8878	0.008 234	115.716	4.8384
150	0.041 884	147.176	5.4042	0.019 546	137.916	5.1547	0.011 186	154.851	5.0804
175	0.050 125	175.255	5.5779	0.024 155	168.709	5.3449	0.013 648	187.521	5.2553
200	0.058 096	202.596	5.7237	0.028 436	197.609	5.4992	0.015 894	217.757	5.3976
225	0.065 875	229.526	5.8502	0.035 697	225.578	5.6309	0.018 060	246.793	5.5202
250	0.073 634	256.220	5.9632	0.036 557	253.032	5.7469	0.020 133	275.056	5.6277
275	0.081 260	282.720	6.0639	0.040 485	280.132	5.8501	0.022 178	302.848	5.7248
300	0.088 899	309.173	6.1563	0.044 398	307.014	5.9436			

	6.0 MPa			8.0 MPa			10.0 MPa		
150	0.004 413	87.090	4.5667	0.002 917	61.903	4.3518	0.002 388	48.687	4.2287
175	0.006 913	140.183	4.8966	0.004 863	125.536	4.7470	0.003 750	112.489	4.6239
200	0.008 772	177.447	5.0961	0.006 390	167.680	4.9726	0.005 016	158.578	4.8709
225	0.010 396	210.139	5.2410	0.007 691	202.867	5.1384	0.006 104	196.079	5.0474
250	0.011 934	240.806	5.3796	0.008 903	235.141	5.2750	0.007 112	229.861	5.1900
275	0.013 383	270.222	5.4917	0.010 034	265.676	5.3910	0.008 046	261.450	5.3103
300	0.014 800	298.907	5.5916	0.011 133	295.219	5.4942	0.008 950	291.800	5.4163

	15.0 MPa			20.0 MPa		
150	0.001 956	36.922	4.0798	0.001 781	33.637	3.9956
175	0.002 603	92.284	4.4213	0.002 186	83.453	4.3029
200	0.003 369	140.886	4.6813	0.002 685	130.291	4.5535
225	0.004 106	182.034	4.8752	0.003 208	172.307	4.7511
250	0.004 808	218.710	5.0303	0.003 728	210.456	4.9127
275	0.005 461	252.465	5.1845	0.004 223	245.640	5.0467
300	0.006 091	284.523	5.2707	0.004 704	278.942	5.1629

APPENDIX G1 PROPERTIES OF SATURATED OXYGEN[a]

(Adapted from L. A. Weber, *Journal of Research of the National Bureau of Standards*, 74A, 93, 1970. Used with permission.)

Properties of Saturated Oxygen[a]

Temp. K	Press. MPa p	Specific Volume m³/kg Sat. Liquid v_f	Evap. v_{fg}	Sat. Vapor v_g	Enthalpy kJ/kg Sat. Liquid h_f	Evap. h_{fg}	Sat. Vapor h_g	Entropy kJ/kg K Sat. Liquid s_f	Evap. s_{fg}	Sat. Vapor s_g
54.3507	0.00015	0.000 765	92.9658	92.9666	−193.432	242.553	49.121	2.0938	4.4514	6.5452
60	0.00073	0.000 780	21.3461	21.3469	−184.029	238.265	54.236	2.2585	3.9686	6.2271
70	0.00623	0.000 808	2.9085	2.9093	−167.372	230.527	63.155	2.5151	3.2936	5.8087
80	0.03006	0.000 840	0.681 04	0.681 88	−150.646	222.289	71.643	2.7382	2.7779	5.5161
90	0.09943	0.000 876	0.226 49	0.227 36	−133.758	213.070	79.312	2.9364	2.3663	5.3027
100	0.25425	0.000 917	0.094 645	0.095 562	−116.557	202.291	85.734	3.1161	2.0222	5.1383
110	0.54339	0.000 966	0.045 855	0.046 821	−98.829	189.320	90.491	3.2823	1.7210	5.0033
120	1.0215	0.001 027	0.024 336	0.025 363	−80.219	173.310	93.091	3.4401	1.4445	4.8846
130	1.7478	0.001 108	0.013 488	0.014 596	−60.093	152.887	92.794	3.5948	1.1766	4.7714
140	2.7866	0.001 230	0.007 339	0.008 569	−37.045	125.051	88.006	3.7567	0.8935	4.6502
150	4.2190	0.001 480	0.003 180	0.004 660	−7.038	79.459	72.421	3.9498	0.5301	4.4799
154.576	5.0427	0.002 293	0.000 000	0.002 293	32.257	0.000	32.257	4.1977	0.0000	4.1977

[a] Adapted from L. A. Weber, *Journal of Research of the National Bureau of Standards*, **74A**: 93 (1970).

APPENDIX G2 PROPERTIES OF SUPERHEATED OXYGEN

Properties of Superheated Oxygen

Temp. K	0.10 MPa v m³/kg	h kJ/kg	s kJ/kg K	0.20 MPa v m³/kg	h kJ/kg	s kJ/kg K	0.50 MPa v m³/kg	h kJ/kg	s kJ/kg K
100	0.253 503	88.828	5.4016	0.123 394	86.864	5.2083			
125	0.320 717	112.214	5.6107	0.158 268	110.988	5.4241	0.060 674	107.093	5.1650
150	0.386 914	135.301	5.7787	0.192 016	134.440	5.5947	0.075 039	131.788	5.3448
175	0.452 645	158.255	5.9202	0.225 276	157.609	5.7376	0.088 842	155.643	5.4919
200	0.518 127	181.145	6.0427	0.258 282	180.638	5.8609	0.102 371	179.105	5.6175
225	0.583 465	204.007	6.1502	0.291 140	203.596	5.9688	0.115 746	202.359	5.7268
250	0.648 711	226.869	6.2468	0.323 906	226.529	6.0657	0.129 025	225.506	5.8246
275	0.713 895	249.769	6.3369	0.356 610	249.483	6.1560	0.142 242	248.621	5.9156
300	0.779 036	272.720	6.4140	0.389 271	272.475	6.2332	0.155 415	271.740	5.9932

Temp. K	1.00 MPa v m³/kg	h kJ/kg	s kJ/kg K	2.00 MPa v m³/kg	h kJ/kg	s kJ/kg K	4.00 MPa v m³/kg	h kJ/kg	s kJ/kg K
125	0.027 869	99.653	4.9431						
150	0.035 976	127.112	5.1433	0.016 270	116.476	4.9130	0.005 526	81.481	4.5475
175	0.043 341	152.269	5.2986	0.020 544	145.112	5.0899	0.009 029	128.618	4.8414
200	0.050 394	176.508	5.4283	0.024 395	171.150	5.2293	0.011 376	159.715	5.0080
225	0.057 282	200.280	5.5401	0.028 051	196.052	5.3464	0.013 444	187.333	5.1380
250	0.064 068	223.795	5.6394	0.031 597	220.348	5.4491	0.015 378	213.374	5.2480
275	0.070 790	247.185	5.7314	0.035 073	244.309	5.5433	0.017 233	238.560	5.3469
300	0.077 467	270.516	5.8098	0.038 502	268.076	5.6263	0.019 039	263.234	5.4300

	6.00 MPa		
175	0.005 051	107.496	4.6431
200	0.007 027	147.232	4.8565
225	0.008 589	178.304	5.0029
250	0.009 991	206.340	5.1214
275	0.011 306	232.848	5.2253
300	0.012 570	258.464	5.3116

	8.00 MPa		
175	0.003 002	79.513	4.4384
200	0.004 864	133.760	4.7308
225	0.006 181	169.069	4.8973
250	0.007 316	199.317	5.0251
275	0.008 360	227.219	5.1344
300	0.009 351	253.797	5.2240

	10.00 MPa		
175	0.002 020	52.661	4.2573
200	0.003 603	119.767	4.6189
225	0.004 757	159.686	4.8072
250	0.005 730	192.401	4.9455
275	0.006 606	221.685	5.0572
300	0.007 432	249.262	5.1533

	20.00 MPa		
175	0.001 343	24.551	4.0086
200	0.001 727	75.318	4.2798
225	0.002 236	122.595	4.5024
250	0.002 755	163.109	4.6739
275	0.003 241	198.021	4.8069
300	0.003 700	229.655	4.9174

APPENDIX H CONSTANT-PRESSURE SPECIFIC HEATS OF VARIOUS IDEAL GASES[a]
(Adapted from T. C. Scott and R. E. Sonntag, Univ. of Michigan, unpublished, 1971,
except C_2H_6, C_3H_{10}, from K. A. Kobe, *Petroleum Refiner*, 28, No. 2, 113, 1949.
Used with permission.)

Constant-Pressure Specific Heats of Various Ideal Gases[a]

$$\bar{c}_{po} = kJ/kg\text{-}mol \cdot K$$

$$\theta = T(\text{Kelvin})/100$$

Gas		Range K	Max. Error %
N_2	$\bar{C}_{po} = 39.060 - 512.79\theta^{-1.5} + 1072.7\theta^{-2} - 820.40\theta^{-3}$	300–3500	0.43
O_2	$\bar{C}_{po} = 37.432 + 0.020102\theta^{1.5} - 178.57\theta^{-1.5} + 236.88\theta^{-2}$	300–3500	0.30
H_2	$\bar{C}_{po} = 56.505 - 702.74\theta^{-0.75} + 1165.0\theta^{-1} - 560.70\theta^{-1.5}$	300–3500	0.60
CO	$\bar{C}_{po} = 69.145 - 0.70463\theta^{0.75} - 200.77\theta^{-0.5} + 176.76\theta^{-0.75}$	300–3500	0.42
OH	$\bar{C}_{po} = 81.546 - 59.350\theta^{0.25} + 17.329\theta^{0.75} - 4.2660\theta$	300–3500	0.43
NO	$\bar{C}_{po} = 59.283 - 1.7096\theta^{0.5} - 70.613\theta^{-0.5} + 74.889\theta^{-1.5}$	300–3500	0.34
H_2O	$\bar{C}_{po} = 143.05 - 183.54\theta^{0.25} + 82.751\theta^{0.5} - 3.6989\theta$	300–3500	0.43
CO_2	$\bar{C}_{po} = -3.7357 + 30.529\theta^{0.5} - 4.1034\theta + 0.024198\theta^2$	300–3500	0.19
NO_2	$\bar{C}_{po} = 46.045 + 216.10\theta^{-0.5} - 363.66\theta^{-0.75} + 232.550\theta^{-2}$	300–3500	0.26
CH_4	$\bar{C}_{po} = -672.87 + 439.74\theta^{0.25} - 24.875\theta^{0.75} + 323.88\theta^{-0.5}$	300–2000	0.15
C_2H_4	$\bar{C}_{po} = -95.395 + 123.15\theta^{0.5} - 35.641\theta^{0.75} + 182.77\theta^{-3}$	300–2000	0.07
C_2H_6	$\bar{C}_{po} = 6.895 + 17.26\theta - 0.6402\theta^2 + 0.00728\theta^3$	300–1500	0.83
C_3H_8	$\bar{C}_{po} = -4.042 + 30.46\theta - 1.571\theta^2 + 0.03171\theta^3$	300–1500	0.40
C_4H_{10}	$\bar{C}_{po} = 3.954 + 37.12\theta - 1.833\theta^2 + 0.03498\theta^3$	300–1500	0.54

[a] From T. C. Scott and R. E. Sonntag, Univ. of Michigan, unpublished (1971), except
C_2H_6, C_3H_8, C_4H_{10} from K. A. Kobe, Petroleum Refiner 28 No. 2, 113 (1949).

Enthalpy of Formation at 25°C, Ideal Gas Enthalpy and Absolute Entropy at 100 kPa Pressure

Temp. K	Nitrogen, Diatomic (N_2) (9/30/65) $(\overline{h}_f^\circ)_{298} = 0$ kJ/kg-mol $M = 28.013$		Nitrogen, Monatomic (N) (3/31/61) $(\overline{h}_f^\circ)_{298} = 472\ 646$ kJ/kg-mol $M = 14.007$	
	$(\overline{h}^\circ - \overline{h}_{298}^\circ)$ kJ/kmol	$\overline{s}^\circ$ kJ/kmol K	$(\overline{h}^\circ - \overline{h}_{298}^\circ)$ kJ/kmol	$\overline{s}^\circ$ kJ/kmol K
0	−8 669	0	−6 197	0
100	−5 770	159.813	−4 117	130.596
200	−2 858	179.988	−2 042	145.006
298	0	191.611	0	153.302
300	54	191.791	38	153.432
400	2 971	200.180	2 117	159.411
500	5 912	206.740	4 197	164.051
600	8 891	212.175	6 276	167.842
700	11 937	216.866	8 351	171.047
800	15 046	221.016	10 431	173.821
900	18 221	224.757	12 510	176.268
1000	21 460	228.167	14 590	178.461
1100	24 757	231.309	16 669	180.440
1200	28 108	234.225	18 749	182.247
1300	31 501	236.941	20 824	183.803
1400	34 936	239.484	22 903	185.452
1500	38 405	241.878	24 983	186.887
1600	41 903	244.137	27 062	188.230
1700	45 430	246.275	29 142	189.490
1800	48 982	248.304	31 217	190.678
1900	52 551	250.237	33 296	191.799
2000	56 141	252.078	35 376	192.866
2100	59 748	253.836	37 455	193.883
2200	63 371	255.522	39 535	194.850
2300	67 007	257.137	41 614	195.774
2400	70 651	258.689	43 698	196.661
2500	74 312	260.183	45 777	197.511
2600	77 973	261.622	47 861	198.326
2700	81 659	263.011	49 949	199.113
2800	85 345	264.350	52 036	199.875
2900	89 036	265.647	54 124	200.607
3000	92 738	266.902	56 220	201.318
3200	100 161	269.295	60 421	202.674
3400	107 608	271.555	64 647	203.954
3600	115 081	273.689	68 906	205.171
3800	122 570	275.714	73 199	206.330
4000	130 076	277.638	77 534	207.443
4200	137 603	279.475	81 923	208.514
4400	145 143	281.228	86 370	209.548
4600	152 699	282.910	90 881	210.552
4800	160 272	284.521	95 462	211.527
5000	167 858	286.069	100 115	212.477
5200	175 456	287.559	104 847	213.401
5400	183 071	288.994	109 663	214.309
5600	190 703	290.383	114 558	215.201
5800	198 347	291.726	119 537	216.075
6000	206 008	293.023	124 600	216.933

[a] The thermochemical data in Table A.11 are calculated from the JANAF Thermochemical Tables, Thermal Research Laboratory, The Dow Chemical Company, Midland, Michigan. The date each table was issued is indicated.

Enthalpy of Formation at 25°C, Ideal Gas Enthalpy and Absolute
Entropy at 100 kPa Pressure

Temp. K	Oxygen, Diatomic (O$_2$) (9/30/65) $(\overline{h}_f^\circ)_{298} = 0$ kJ/kg-mol $M = 31.999$		Oxygen, Monatomic (O) (6/30/62) $(\overline{h}_f^\circ)_{298} = 249\ 195$ kJ/kg-mol $M = 16.00$	
	$(\overline{h}^\circ - \overline{h}_{298}^\circ)$ kJ/kmol	$\overline{s}^\circ$ kJ/kmol K	$(\overline{h}^\circ - \overline{h}_{298}^\circ)$ kJ/kmol	$\overline{s}^\circ$ kJ/kmol K
0	−8 682	0	−6 728	0
100	−5 778	173.306	−4 519	135.947
200	−2 866	193.486	−2 188	152.156
298	0	205.142	0	161.060
300	54	205.322	42	161.198
400	3 029	213.874	2 209	167.432
500	6 088	220.698	4 343	172.202
600	9 247	226.455	6 460	176.063
700	12 502	231.272	8 569	179.314
800	15 841	235.924	10 669	182.118
900	19 246	239.936	12 770	184.590
1000	22 707	243.585	14 862	186.795
1100	26 217	246.928	16 949	188.787
1200	29 765	250.016	19 041	190.603
1300	33 351	252.886	21 125	192.272
1400	36 966	255.564	23 213	193.820
1500	40 610	258.078	25 296	195.260
1600	44 279	260.446	27 380	196.603
1700	47 970	262.685	29 464	197.866
1800	51 689	264.810	31 547	199.059
1900	55 434	266.835	33 631	200.184
2000	59 199	268.764	35 715	201.251
2100	62 986	270.613	37 798	202.268
2200	66 802	272.387	39 882	203.238
2300	70 634	274.090	41 961	204.163
2400	74 492	275.735	44 045	205.050
2500	78 375	277.316	46 133	205.899
2600	82 274	278.848	48 216	206.720
2700	86 199	280.329	50 304	207.506
2800	90 144	281.764	52 392	208.268
2900	94 111	283.157	54 484	209.000
3000	98 098	284.508	56 576	209.711
3200	106 127	287.098	60 768	211.063
3400	114 232	289.554	64 973	212.339
3600	122 399	291.889	69 191	213.544
3800	130 629	294.115	73 425	214.686
4000	138 913	296.236	77 676	215.778
4200	147 248	298.270	81 948	216.820
4400	155 628	300.219	86 236	217.816
4600	164 046	302.094	90 546	218.774
4800	172 502	303.893	94 876	219.694
5000	180 987	305.621	99 224	220.585
5200	189 502	307.290	103 596	221.439
5400	198 037	308.901	107 985	222.267
5600	206 593	310.458	112 395	223.071
5800	215 166	311.964	116 821	223.849
6000	223 756	313.420	121 269	224.602

Enthalpy of Formation at 25°C, Ideal Gas Enthalpy and Absolute
Entropy at 100 kPa Pressure

| | Hydrogen, Diatomic (H_2) (3/31/61) | | Hydrogen, Monatomic (H) (9/30/65) | |
| | $(\bar{h}_f^\circ)_{298} = 0$ kJ/kg-mol $M = 2.016$ | | $(\bar{h}_f^\circ)_{298} = 217\ 986$ kJ/kg-mol $M = 1.008$ | |
Temp. K	$(\bar{h}^\circ - \bar{h}_{298}^\circ)$ kJ/kmol	$\bar{s}^\circ$ kJ/kmol K	$(\bar{h}^\circ - \bar{h}_{298}^\circ)$ kJ/kmol	$\bar{s}^\circ$ kJ/kmol K
0	−8 468	0	−6 197	0
100	−5 293	102.145	−4 117	92.011
200	−2 770	119.437	−2 042	106.417
298	0	130.684	0	114.718
300	54	130.864	38	114.847
400	2 958	139.215	2 117	120.826
500	5 883	145.738	4 197	125.466
600	8 812	151.077	6 276	129.257
700	11 749	155.608	8 351	132.458
800	14 703	159.549	10 431	135.236
900	17 682	163.060	12 510	137.684
1000	20 686	166.223	14 590	139.872
1100	23 723	169.118	16 669	141.855
1200	26 794	171.792	18 749	143.662
1300	29 907	174.281	20 824	145.328
1400	33 062	176.620	22 903	146.867
1500	36 267	178.833	24 983	148.303
1600	39 522	180.929	27 062	149.641
1700	42 815	182.929	29 142	150.905
1800	46 150	184.833	31 217	152.093
1900	49 522	186.657	33 296	153.215
2000	52 932	188.406	35 376	154.281
2100	56 379	190.088	37 455	155.294
2200	59 860	191.707	39 535	156.265
2300	63 371	193.268	41 610	157.185
2400	66 915	194.778	43 689	158.072
2500	70 492	196.234	45 769	158.922
2600	74 090	197.649	47 848	159.737
2700	77 718	199.017	49 928	160.520
2800	81 370	200.343	52 007	161.277
2900	85 044	201.636	54 082	162.005
3000	88 743	202.887	56 162	162.708
3200	96 199	205.293	60 321	164.051
3400	103 738	207.577	64 475	165.311
3600	111 361	209.757	68 634	166.499
3800	119 064	211.841	72 793	167.624
4000	126 846	213.837	76 948	168.691
4200	134 700	215.753	81 107	169.704
4400	142 624	217.594	85 266	170.670
4600	150 620	219.372	89 420	171.595
4800	158 682	221.087	93 579	172.482
5000	166 808	222.744	97 734	173.327
5200	174 996	224.351	101 893	174.143
5400	183 247	225.907	106 052	174.930
5600	191 556	227.418	110 207	175.683
5800	199 924	228.886	114 365	176.415
6000	208 346	230.313	118 524	177.118

Enthalpy of Formation at 25°C, Ideal Gas Enthalpy and Absolute
Entropy at 100 kPa Pressure

Temp. K	Water (H_2O) (3/31/61)		Hydroxyl (OH) (3/31/66)	
	$(\bar{h}_f^\circ)_{298} = -241\,827$ kJ/kg-mol $M = 18.015$		$(\bar{h}_f^\circ)_{298} = 39\,463$ kJ/kg-mol $M = 17.007$	
	$(\bar{h}^\circ - \bar{h}_{298}^\circ)$ kJ/kmol	$\bar{s}^\circ$ kJ/kmol K	$(\bar{h}^\circ - \bar{h}_{298}^\circ)$ kJ/kmol	$\bar{s}^\circ$ kJ/kmol K
0	−9 904	0	−9 171	0
100	−6 615	152.390	−6 138	149.587
200	−3 280	175.486	−2 975	171.591
298	0	188.833	0	183.703
300	63	189.038	54	183.892
400	3 452	198.783	3 033	192.465
500	6 920	206.523	5 991	199.063
600	10 498	213.037	8 941	204.443
700	14 184	218.719	11 903	209.004
800	17 991	223.803	14 878	212.979
900	21 924	228.430	17 887	216.523
1000	25 978	232.706	20 933	219.732
1100	30 167	236.694	24 025	222.677
1200	34 476	240.443	27 158	225.405
1300	38 903	243.986	30 342	227.949
1400	43 447	247.350	33 568	230.342
1500	48 095	250.560	36 840	232.598
1600	52 844	253.622	40 150	234.736
1700	57 685	256.559	43 501	236.769
1800	62 609	259.371	46 890	238.702
1900	67 613	262.078	50 308	240.551
2000	72 689	264.681	53 760	242.325
2100	77 831	267.191	57 241	244.020
2200	83 036	269.609	60 752	245.652
2300	88 295	271.948	64 283	247.225
2400	93 604	274.207	67 839	248.739
2500	98 964	276.396	71 417	250.200
2600	104 370	278.517	75 015	251.610
2700	109 813	280.571	78 634	252.974
2800	115 294	282.563	82 266	254.296
2900	120 813	284.500	85 918	255.576
3000	126 361	286.383	89 584	256.819
3200	137 553	289.994	96 960	259.199
3400	148 854	293.416	104 387	261.450
3600	160 247	296.676	111 859	263.588
3800	171 724	299.776	119 378	265.618
4000	183 280	302.742	126 934	267.559
4200	194 903	305.575	134 528	269.408
4400	206 585	308.295	142 156	271.182
4600	218 325	310.901	149 816	272.885
4800	230 120	313.412	157 502	274.521
5000	241 957	315.830	165 222	276.099
5200	253 839	318.160	172 967	277.617
5400	265 768	320.407	180 736	279.082
5600	277 738	322.587	188 531	280.500
5800	289 746	324.692	196 351	281.873
6000	301 796	326.733	204 192	283.203

Enthalpy of Formation at 25°C, Ideal Gas Enthalpy and Absolute
Entropy at 100 kPa Pressure

| | Carbon Dioxide (CO$_2$) (9/30/65) | | Carbon Monoxide (CO) (9/30/65) | |
| | $(\bar{h}_f^\circ)_{298} = -393\ 522$ kJ/ kg-mol $M = 44.01$ | | $(\bar{h}_f^\circ)_{298} = -110\ 529$ kJ/ kg-mol $M = 28.01$ | |
Temp. K	$(\bar{h}^\circ - \bar{h}^\circ_{298})$ kJ/kmol	$\bar{s}^\circ$ kJ/kmol K	$(\bar{h}^\circ - \bar{h}^\circ_{298})$ kJ/kmol	$\bar{s}^\circ$ kJ/kmol K
0	−9 364	0	−8 669	0
100	−6 456	179.109	−5 770	165.850
200	−3 414	199.975	−2 858	186.025
298	0	213.795	0	197.653
300	67	214.025	54	197.833
400	4 008	225.334	2 975	206.234
500	8 314	234.924	5 929	212.828
600	12 916	243.309	8 941	218.313
700	17 761	250.773	12 021	223.062
800	22 815	257.517	15 175	227.271
900	28 041	263.668	18 397	231.066
1000	33 405	269.325	21 686	234.531
1100	38 894	274.555	25 033	237.719
1200	44 484	279.417	28 426	240.673
1300	50 158	**283.956**	31 865	243.426
1400	55 907	**288.216**	35 338	245.999
1500	61 714	**292.224**	38 848	248.421
1600	67 580	**296.010**	42 384	250.702
1700	73 492	**299.592**	45 940	252.861
1800	79 442	**302.993**	49 522	254.907
1900	85 429	**306.232**	53 124	256.852
2000	91 450	**309.320**	56 739	258.710
2100	97 500	312.269	60 375	260.480
2200	103 575	315.098	64 019	262.174
2300	109 671	317.805	67 676	263.802
2400	115 788	320.411	71 346	265.362
2500	121 926	322.918	75 023	266.865
2600	128 085	325.332	78 714	268.312
2700	134 256	327.658	82 408	269.705
2800	140 444	329.909	86 115	271.053
2900	146 645	332.085	89 826	272.358
3000	152 862	334.193	93 542	273.618
3200	165 331	338.218	100 998	276.023
3400	177 849	342.013	108 479	278.291
3600	190 405	345.599	115 976	280.433
3800	202 999	349.005	123 495	282.467
4000	215 635	352.243	131 026	284.396
4200	228 304	355.335	138 578	286.241
4400	241 003	358.289	146 147	287.998
4600	253 734	361.122	153 724	289.684
4800	266 500	363.837	161 322	291.299
5000	279 295	366.448	168 929	292.851
5200	292 123	368.963	176 548	294.349
5400	304 984	371.389	184 184	295.789
5600	317 884	373.736	191 832	297.178
5800	**330 821**	376.004	199 489	298.521
6000	**343 791**	378.205	207 162	299.822

Enthalpy of Formation at 25°C, Ideal Gas Enthalpy and Absolute
Entropy at 100 kPa Pressure

Temp. K	Nitric Oxide (NO) (6/30/63) $(\bar{h}_f^\circ)_{298} = 90\ 592$ kJ/ kg-mol $M = 30.006$		Nitrogen Dioxide (NO₂) (6/30/63) $(\bar{h}_f^\circ)_{298} = 33\ 723$ kJ/ kg-mol $M = 46.005$	
	$(\bar{h}^\circ - \bar{h}^\circ_{298})$ kJ/kmol	$\bar{s}^\circ$ kJ/kmol K	$(\bar{h}^\circ - \bar{h}^\circ_{298})$ kJ/kmol	$\bar{s}^\circ$ kJ/kmol K
0	−9 192	0	−10 196	0
100	−6 071	177.034	−6 870	202.431
200	−2 950	198.753	−3 502	225.732
298	0	210.761	0	239.953
300	54	210.950	67	240.183
400	3 042	219.535	3 950	251.321
500	6 058	226.267	8 150	260.685
600	9 146	231.890	12 640	268.865
700	12 309	236.765	17 368	276.149
800	15 548	241.091	22 288	282.714
900	18 857	244.991	27 359	288.684
1000	22 230	248.543	32 552	294.153
1100	25 652	251.806	37 836	299.190
1200	29 121	254.823	43 196	303.855
1300	32 627	257.626	48 618	308.194
1400	36 166	260.250	54 095	312.253
1500	39 731	262.710	59 609	316.056
1600	43 321	265.028	65 157	319.637
1700	46 932	267.216	70 739	323.022
1800	50 559	269.287	76 345	326.223
1900	54 204	271.258	81 969	329.265
2000	57 861	273.136	87 613	332.160
2100	61 530	274.927	93 274	334.921
2200	65 216	276.638	98 947	337.562
2300	68 906	278.279	104 633	340.089
2400	72 609	279.856	110 332	342.515
2500	76 320	281.370	116 039	344.846
2600	80 036	282.827	121 754	347.089
2700	83 764	284.232	127 478	349.248
2800	87 492	285.592	133 206	351.331
2900	91 232	286.902	138 942	353.344
3000	94 977	288.174	144 683	355.289
3200	102 479	290.592	156 180	359.000
3400	110 002	292.876	167 695	362.490
3600	117 545	295.031	179 222	365.783
3800	125 102	297.073	190 761	368.904
4000	132 675	299.014	202 309	371.866
4200	140 260	300.864	213 865	374.682
4400	147 863	302.634	225 430	377.372
4600	155 473	304.324	237 003	379.946
4800	163 101	305.947	248 580	382.410
5000	170 736	307.508	260 161	384.774
5200	178 381	309.006	271 751	387.046
5400	186 042	310.449	283 340	389.234
5600	193 707	311.847	294 934	391.343
5800	201 388	313.194	306 532	393.376
6000	209 074	314.495	318 130	395.343

APPENDIX 12 PROPERTIES OF AIR AT LOW PRESSURE[a]

(G. J. Van Wylen and R. E. Sonntag, *Fundamentals of Classical Thermodynamics, SI Version*, 2nd Edition. New York: John Wiley & Sons, Inc. © 1976. Reprinted by permission.)

Properties of Air at Low Pressure[a]

T, K	h, kJ/kg	p_r	u, kJ/kg	v_r	$s°$, kJ/kg K
100	99.76	0.029 90	71.06	2230	1.4143
110	109.77	0.041 71	78.20	1758.4	1.5098
120	119.79	0.056 52	85.34	1415.7	1.5971
130	129.81	0.074 74	92.51	1159.8	1.6773
140	139.84	0.096 81	99.67	964.2	1.7515
150	149.86	0.123 18	106.81	812.0	1.8206
160	159.87	0.154 31	113.95	691.4	1.8853
170	169.89	0.190 68	121.11	594.5	1.9461
180	179.92	0.232 79	128.28	515.6	2.0033
190	189.94	0.281 14	135.40	450.6	2.0575
200	199.96	0.3363	142.56	396.6	2.1088
210	209.97	0.3987	149.70	351.2	2.1577
220	219.99	0.4690	156.84	312.8	2.2043
230	230.01	0.5477	163.98	280.0	2.2489
240	240.03	0.6355	171.15	251.8	2.2915
250	250.05	0.7329	178.29	227.45	2.3325
260	260.09	0.8405	185.45	206.26	2.3717
270	270.12	0.9590	192.59	187.74	2.4096
280	280.14	1.0889	199.78	171.45	2.4461
290	290.17	1.2311	206.92	157.07	2.4813
300	300.19	1.3860	214.09	144.32	2.5153
310	310.24	1.5546	221.27	132.96	2.5483
320	320.29	1.7375	228.45	122.81	2.5802
330	330.34	1.9352	235.65	113.70	2.6111
340	340.43	2.149	242.86	105.51	2.6412
350	350.48	2.379	250.05	98.11	2.6704
360	360.58	2.626	257.23	91.40	2.6987
370	370.67	2.892	264.47	85.31	2.7264
380	380.77	3.176	271.72	79.77	2.7534
390	390.88	3.481	278.96	74.71	2.7796
400	400.98	3.806	286.19	70.07	2.8052
410	411.12	4.153	293.45	65.83	2.8302
420	421.26	4.522	300.73	61.93	2.8547
430	431.43	4.915	308.03	58.34	2.8786
440	441.61	5.332	315.34	55.02	2.9020
450	451.83	5.775	322.66	51.96	2.9249
460	462.01	6.245	329.99	49.11	2.9473
470	472.25	6.742	337.34	46.48	2.9693
480	482.48	7.268	344.74	44.04	2.9909
490	492.74	7.824	352.11	41.76	3.0120
500	503.02	8.411	359.53	39.64	3.0328
510	513.32	9.031	366.97	37.65	3.0532
520	523.63	9.684	374.39	35.80	3.0733

$T,$ K	$h,$ kJ/kg	p_r	$u,$ kJ/kg	v_r	$s°,$ kJ/kg K
530	533.98	10.372	381.88	34.07	3.0930
540	544.35	11.097	389.40	32.45	3.1124
550	554.75	11.858	396.89	30.92	3.1314
560	565.17	12.659	404.44	29.50	3.1502
570	575.57	13.500	411.98	28.15	3.1686
580	586.04	14.382	419.56	26.89	3.1868
590	596.53	15.309	427.17	25.70	3.2047
600	607.02	16.278	434.80	24.58	3.2223
610	617.53	17.297	442.43	23.51	3.2397
620	628.07	18.360	450.13	22.52	3.2569
630	638.65	19.475	457.83	21.57	3.2738
640	649.21	20.64	465.55	20.674	3.2905
650	659.84	21.86	473.32	19.828	3.3069
660	670.47	23.13	481.06	19.026	3.3232
670	681.15	24.46	488.88	18.266	3.3392
680	691.82	25.85	496.65	17.543	3.3551
690	702.52	27.29	504.51	16.857	3.3707
700	713.27	28.80	512.37	16.205	3.3861
710	724.01	30.38	520.26	15.585	3.4014
720	734.20	31.92	527.72	15.027	3.4156
730	745.62	33.72	536.12	14.434	3.4314
740	756.44	35.50	544.05	13.900	3.4461
750	767.30	37.35	552.05	13.391	3.4607
760	778.21	39.27	560.08	12.905	3.4751
770	789.10	41.27	568.10	12.440	3.4894
780	800.03	43.35	576.15	11.998	3.5035
790	810.98	45.51	584.22	11.575	3.5174
800	821.94	47.75	592.34	11.172	3.5312
810	832.96	50.08	600.46	10.785	3.5449
820	843.97	52.49	608.62	10.416	3.5584
830	855.01	55.00	616.79	10.062	3.5718
840	866.09	57.60	624.97	9.724	3.5850
850	877.16	60.29	633.21	9.400	3.5981
860	888.28	63.09	641.44	9.090	3.6111
870	899.42	65.98	649.70	8.792	3.6240
880	910.56	68.98	658.00	8.507	3.6367
890	921.75	72.08	666.31	8.233	3.6493
900	932.94	75.29	674.63	7.971	3.6619
910	944.15	78.61	682.98	7.718	3.6743
920	955.38	82.05	691.33	7.476	3.6865
930	966.64	85.60	699.73	7.244	3.6987
940	977.92	89.28	708.13	7.020	3.7108
950	989.22	93.08	716.57	6.805	3.7227
960	1000.53	97.00	725.01	6.599	3.7346
970	1011.88	101.06	733.48	6.400	3.7463
980	1023.25	105.24	741.99	6.209	3.7580
990	1034.63	109.57	750.48	6.025	3.7695

T, K	h, kJ/kg	p_r	u, kJ/kg	v_r	$s°,$ kJ/kg K
1000	1046.03	114.03	759.02	5.847	3.7810
1020	1068.89	123.12	775.67	5.521	3.8030
1040	1091.85	133.34	793.35	5.201	3.8259
1060	1114.85	143.91	810.61	4.911	3.8478
1080	1137.93	155.15	827.94	4.641	3.8694
1100	1161.07	167.07	845.34	4.390	3.8906
1120	1184.28	179.71	862.85	4.156	3.9116
1140	1207.54	193.07	880.37	3.937	3.9322
1160	1230.90	207.24	897.98	3.732	3.9525
1180	1254.34	222.2	915.68	3.541	3.9725
1200	1277.79	238.0	933.40	3.362	3.9922
1220	1301.33	254.7	951.19	3.194	4.0117
1240	1324.89	272.3	969.01	3.037	4.0308
1260	1348.55	290.8	986.92	2.889	4.0497
1280	1372.25	310.4	1004.88	2.750	4.0684
1300	1395.97	330.9	1022.88	2.619	4.0868
1320	1419.77	352.5	1040.93	2.497	4.1049
1340	1443.61	375.3	1059.03	2.381	4.1229
1360	1467.50	399.1	1077.17	2.272	4.1406
1380	1491.43	424.2	1095.36	2.169	4.1580
1400	1515.41	450.5	1113.62	2.072	4.1753
1420	1539.44	478.0	1131.90	1.9808	4.1923
1440	1563.49	506.9	1150.23	1.8942	4.2092
1460	1587.61	537.1	1168.61	1.8124	4.2258
1480	1611.80	568.8	1187.03	1.7350	4.2422
1500	1635.99	601.9	1205.47	1.6617	4.2585

[a] Adapted from Table 1 in *Gas Tables*, by Joseph H. Keenan and Joseph Kaye. Copyright 1948, by Joseph H. Keenan and Joseph Kaye. Published by John Wiley & Sons, Inc., New York.

(*Steam Tables*, Keenan et al., © by John Wiley & Sons, Inc.
Reprinted by permission of John Wiley & Sons, Inc.)

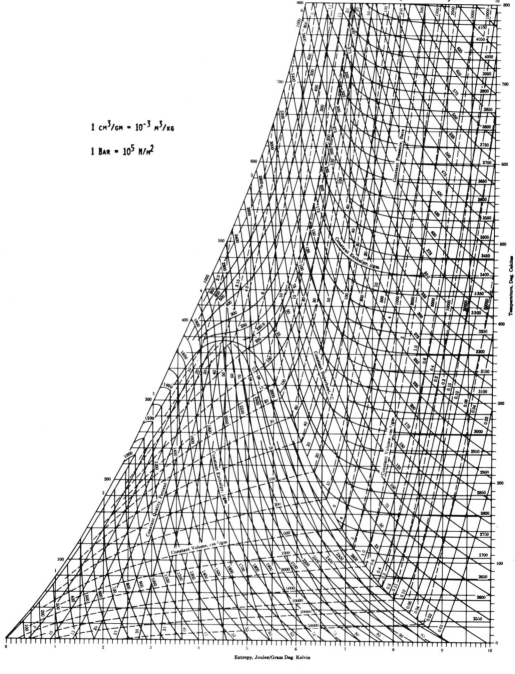

APPENDIX J1(b)   MOLLIER DIAGRAM FOR STEAM

(*Steam and Air Tables in SI Units*, T. F. Irvine, Jr. and J. P. Hartnett, © 1976, Hemisphere Publishing Corporation. Used with permission.)

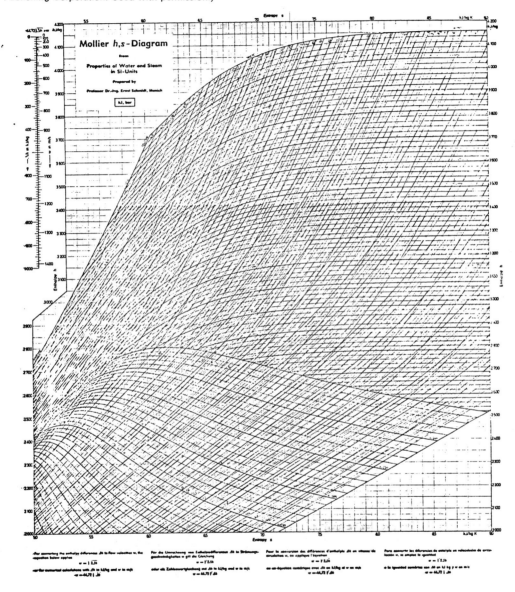

APPENDIX J2 THE PRESSURE-ENTHALPY DIAGRAM FOR FREON-12
(Thermodynamic Tables in SI (metric) Units, R. W. Haywood & J. H. Matthewman,
1968, University Press, Cambridge. Used with permission.)

PRESSURE-ENTHALPY DIAGRAM
FOR
DICHLORODIFLUOROMETHANE (CCl₂F₂)
(REFRIGERANT-12)

ABSOLUTE PRESSURE, p, MPa

SPECIFIC ENTHALPY, h, kJ/kg

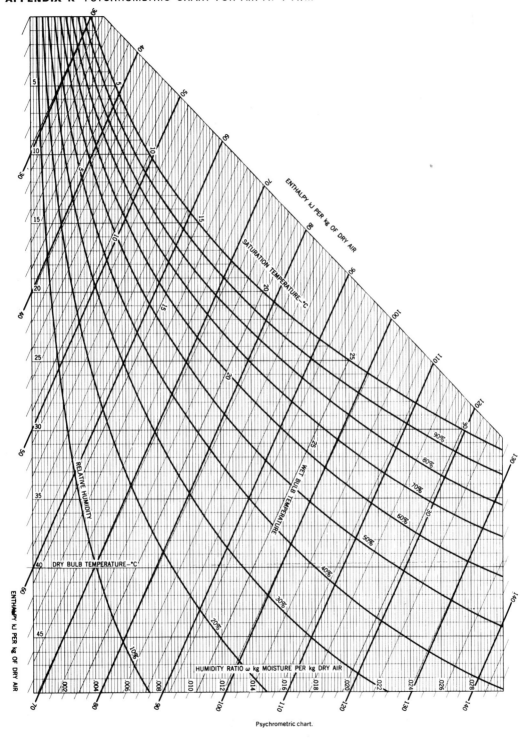

Psychrometric chart.

APPENDIX L CRITICAL CONSTANTS OF SEVERAL SUBSTANCES[a]
(Reprinted with permission from *Chemical Review*, 52, 117–236.
Copyright 1953, American Chemical Society.)

Critical Constants of Several Substances[a]

Substance	Formula	Molar Mass	Temp. K	Pressure MPa	Volume m³/kg mole
Ammonia	NH_3	17.03	405.5	11.28	.0724
Argon	Ar	39.948	151	4.86	.0749
Bromine	Br_2	159.808	584	10.34	.1355
Carbon Dioxide	CO_2	44.01	304.2	7.39	.0943
Carbon Monoxide	CO	28.011	133	3.50	.0930
Chlorine	Cl_2	70.906	417	7.71	.1242
Deuterium (Normal)	D_2	4.00	38.4	1.66	—
Helium	He	4.003	5.3	0.23	.0578
Helium³	He	3.00	3.3	0.12	—
Hydrogen (Normal)	H_2	2.016	33.3	1.30	.0649
Krypton	Kr	83.80	209.4	5.50	.0924
Neon	Ne	20.183	44.5	2.73	.0417
Nitrogen	N_2	28.013	126.2	3.39	.0899
Nitrous Oxide	N_2O	44.013	309.7	7.27	.0961
Oxygen	O_2	31.999	154.8	5.08	.0780
Sulfur Dioxide	SO_2	64.063	430.7	7.88	.1217
Water	H_2O	18.015	647.3	22.09	.0568
Xenon	Xe	131.30	289.8	5.88	.1186
Benzene	C_6H_6	78.115	562	4.92	.2603
n-Butane	C_4H_{10}	58.124	425.2	3.80	.2547
Carbon Tetrachloride	CCl_4	153.82	556.4	4.56	.2759
Chloroform	$CHCl_3$	119.38	536.6	5.47	.2403
Dichlorodifluoromethane	CCl_2F_2	120.91	384.7	4.01	.2179
Dichlorofluoromethane	$CHCl_2F$	102.92	451.7	5.17	.1973
Ethane	C_2H_6	30.070	305.5	4.88	.1480
Ethyl Alcohol	C_2H_5OH	46.07	516	6.38	.1673
Ethylene	C_2H_4	28.054	282.4	5.12	.1242
n-Hexane	C_6H_{14}	86.178	507.9	3.03	.3677
Methane	CH_4	16.043	191.1	4.64	.0993
Methyl Alcohol	CH_3OH	32.042	513.2	7.95	.1180
Methyl Chloride	CH_3Cl	50.488	416.3	6.68	.1430
Propane	C_3H_8	44.097	370	4.26	.1998
Propene	C_3H_6	42.081	365	4.62	.1810
Propyne	C_3H_4	40.065	401	5.35	—
Trichlorofluoromethane	CCl_3F	137.37	471.2	4.38	.2478

[a] K. A. Kobe and R. E. Lynn, Jr., *Chem. Rev.*, **52**: 117–236 (1953).

APPENDIX M LOGARITHMS TO THE BASE e OF THE EQUILIBRIUM CONSTANT K

Logarithms to the Base e of the Equilibrium Constant K

For the reaction $\nu_A A + \nu_B B \rightleftharpoons \nu_C C + \nu_D D$ the equilibrium constant K is defined as

$$K_p = \frac{a_C^{\nu_C} a_D^{\nu_D}}{a_A^{\nu_A} a_B^{\nu_B}}$$

Base on thermodynamic data given in the JANAF Thermochemical Tables, Thermal Research Laboratory, The Dow Chemical Company, Midland, Michigan.

Temp. K	$H_2 \rightleftharpoons 2H$	$O_2 \rightleftharpoons 2O$	$N_2 \rightleftharpoons 2N$	$H_2O \rightleftharpoons H_2 + \frac{1}{2}O_2$	$H_2O \rightleftharpoons \frac{1}{2}H_2 + OH$	$CO_2 \rightleftharpoons CO + \frac{1}{2}O_2$	$\frac{1}{2}N_2 + \frac{1}{2}O_2 \rightleftharpoons NO$
298	−164.005	−186.975	−367.480	−92.208	−106.208	−103.762	−35.052
500	−92.827	−105.630	−213.372	−52.691	−60.281	−57.616	−20.295
1000	−39.803	−45.150	−99.127	−23.163	−26.034	−23.529	−9.388
1200	−30.874	−35.005	−80.011	−18.182	−20.283	−17.871	−7.569
1400	−24.463	−27.742	−66.329	−14.609	−16.099	−13.842	−6.270
1600	−19.637	−22.285	−56.055	−11.921	−13.066	−10.830	−5.294
1800	−15.866	−18.030	−48.051	−9.826	−10.657	−8.497	−4.536
2000	−12.840	−14.622	−41.645	−8.145	−8.728	−6.635	−3.931
2200	−10.353	−11.827	−36.391	−6.768	−7.148	−5.120	−3.433
2400	−8.276	−9.497	−32.011	−5.619	−5.832	−3.860	−3.019
2600	−6.517	−7.521	−28.304	−4.648	−4.719	−2.801	−2.671
2800	−5.002	−5.826	−25.117	−3.812	−3.763	−1.894	−2.372
3000	−3.685	−4.357	−22.359	−3.086	−2.937	−1.111	−2.114
3200	−2.534	−3.072	−19.937	−2.451	−2.212	−0.429	−1.888
3400	−1.516	−1.935	−17.800	−1.891	−1.576	0.169	−1.690
3600	−0.609	−0.926	−15.898	−1.392	−1.088	0.701	−1.513
3800	0.202	−0.019	−14.199	−0.945	−0.501	1.176	−1.356
4000	0.934	0.796	−12.660	−0.542	−0.044	1.599	−1.216
4500	2.486	2.513	−9.414	0.312	0.920	2.490	−0.921
5000	3.725	3.895	−6.807	0.996	1.689	3.197	−0.686
5500	4.743	5.023	−4.666	1.560	2.318	3.771	−0.497
6000	5.590	5.963	−2.865	2.032	2.843	4.245	−0.341

Length	1 in = 0.08333 ft
	1 mile = 1609.26 m
	1 cm = 0.03281 ft
	1 m = 3.281 ft
Mass	1 lb_m = 0.4535 kg
	1 kg = 2.205 lb_m
	1 slug = 32.1736 lb_m
Force	1 lb_f = 4.448 N
	1 N = 0.2248 lb_f
Energy	1 Btu = 778.16 ft lb_f
	1 Btu = 1055.07 J
	1 Btu = 0.2520 Kcal
	1 J = 0.9478×10^{-3} Btu
	1 kw hr = 3413 Btu
Power	1 Btu/hr = 0.293 W
	1 hp = 2545 Btu/hr
	1 hp = 745.7 W
	1 Kcal = 1.1626 Watt-hr
	1 W = 3.413 Btu/hr
Pressure	1 atm = 2116 psf
	1 ft water = 62.43 psf
	1 in Hg = 70.77 psf
	1 atm = 1.01325×10^5 Pa
	1 lb_f/in^2 = 6.895 kPa
Density	1 lb_m/in^3 = 1728 lb_m/ft^3
	1 lb_m/in^3 = 2.77×10^4 kg/m^3
	1 kg/m^3 = 0.06243 lb_m/ft^3
Temperature	$T(°R) = T(°F) + 459.69$
	$T(K) = T(°C) + 273.16$
	$T(°F) = (9/5) T(°C) + 32$
	1 K = 1.8°R
Volume	1 gal (U.S.) = 0.1337 ft^3
	1 ft^3 = 28.32 liters = 0.02832 m^3
Heat flux	1 Btu/hr-ft^2 = 3.1537 W/m^2
	1 W/m^2 = 0.317 Btu/hr-ft^2
Specific Heat	1 Btu/lb_m°F = 4180 J/kg°C
	1 J/kg°C = 2.392×10^{-4} Btu/lb_m°F

Index

Index

A

Absorber, 407
Absorption refrigeration, 348, 407–9
Adiabatic bulk modulus, 434
Adiabatic compressibility, 434
Adiabatic flame temperature, 492
Adiabatic saturation, 450
Air conditioning, 457–59
Air-fuel ratio, 188, 476
Air-standard cycle, 286
Amagat-Leduc Law, 442
Atomic weight, 21
Avagadro's number, 21
Availability, 277
Availability, rate of, 278
Available energy, 262
Available portion of heat, 262
Available work, 261–69
Average velocity, 170
Azeotropes, 464

B

Barometer, 11
Beattie-Bridgeman equation, 158

Berthelot equation, 157
Binary vapor cycle, 387
Bore, 287
Bottom dead center, 287
Boundary, 27
Boyle's law, 32
Boyle temperature, 159
Brake work output, 304
Brayton cycle:
    deviations, 328
    intercooling and reheat, 334
    optimum pressure ratio, 325–27
    regenerative, 327
Bridgeman, 418

C

Caloric, 2
Calorimeter, separating, 125
Calorimeter, throttling, 125
Carnot heat engine, 225
Carnot refrigerator, 225
Carrier, 407
Celsius scale, 17
Centrigrade scale, 17
Charles law, 32
Chemical compound, 117

Chemical equilibrium, 504–7
Chemical potential, 459
Clapeyron-Clausius equation, 425
Clapeyron equation, 424
Clausius inequality, 230–33
Clausius statement, 223
Clearance factor, 288
Clearance volume, 287
Closed feed-water heater, 373
Coefficient of isobaric expansion, 149–51
Coefficient of isothermal compressibility,
      149–51
Coefficient of performance, 205
Coefficient of performance for heat
      pump, 205, 238
Coefficient of performance for refrigera-
      tor, 205, 238
Combined first and second law, 247
Combustion chamber, 178, 188
Compressibility chart, 152
Compressibility factor, 151
Compression ratio, 109–288
Compressor, 178
Compressor work, 290–91
Concentrations, 464
Condition line, 382
Conservation of mass, 168
Conservation of mass, vector equation,
      210–11
Constant pressure process, 33, 34
Constant temperature process, 31, 32, 34
Constant volume process, 33, 34
Contour integration, 67
Control surface, 165
Control volume, 165
Control volume, energy equation for,
      174–78
Cooling tower, 469
COP of a Carnot heat pump, 390
COP of a Carnot refrigeration cycle, 389
COP of a heat pump, 390
COP of an absorption system, 408
COP of a refrigeration cycle, 389, 391
COP of a reversed Brayton cycle, 402
Cutoff ratio, 309
Cutoff volume, 309
Cycle:
      Atkinson, 322–23
      Brayton, 196–99, 323–39
      Carnot, 102–5, 236–38
      diesel, 309–14

      dual, 317–18
      Ericsson, 322, 334
      Otto, 107-10, 297–303
      Rankine, 350–54
      Stirling, 318–22
Cycle, Carnot, 236–38
      COP, Carnot heat pump, 238
      COP, Carnot refrigerator, 238
Cyclic process, 34

D

Degree of saturation, 447
Density, 5
Determination of γ, 91
Dew point temperature, 447
Diesel engine, thermal efficiency, 313
Dieterici equation, 157
Differentials, exact, 65–66
Differentials, inexact, 65–66
Diffuser, 178
Dulong and Petit, 86

E

Economizer, 352
Efficiency, combined, 101
Efficiency, thermal, 101
Efficiency of a boiler, 381
Endothermic, 471
Energy, 1, 277
Energy, high grade, 219
Energy, internal, 73–81
Energy, kinetic, 73
Energy, low grade, 219
Energy, potential, 73
Energy, specific, 73
Energy of a system, 73
Enthalpy, 84
Enthalpy of formation, 483
Enthalpy of reaction, 485
Entropy, 233–35
Entropy, standard state, 251
Entropy change for control volume,
      257–60
Entropy change for ideal gas, 249
Entropy of a pure substance, 254

Equation of state, 20
Equilibrium, 502–4
Equilibrium, static, 9
Equilibrium constant, 506
Exothermic, 471
Explosion pressure, 493
Extent of reactions, 510–11

F

Feed pump, 178
First law, 74
First law for cyclic process, 80
First TdS equation, 421
Flow work, 171–73
Fluid, 123
Force, body, 9
Force, surface, 9
Four-stroke cycle, 302
Free expansion, 58
Free piston engine, 323
Fuel cell, 188

G

Gas constant, 20–21
Gas tables, 255–57
Gas turbine plant, 199–200
Gay Lussac's Law, 33
Generator, 408
Gibbs-Dalton Law, 440–41
Gibbs equation, 421
Gibbs function, 422
Gibbs function of formation, 499
Gibbs-Helmholtz equation, 508
Gram atomic weight, 21
Gravimetric proportions, 474

H

Heat, 62
Heat, sign convention, 63
Heat engines, 100
Heat exchanger, 178
Heat pump, 203–5

Heating value, 188
Heating value, higher, 485
Heating value, lower, 485
Heat of combustion, 79
Heat of evaporation, 126
Heat of sublimation, 145
Helmoltz function, 422
Henry's law, 464
Humidity ratio, 446

I

Ice point, 17
Ideal gas law, 20
Incinerator, 188
Indicated work output, 304
Indicator diagram, 303
Inner dead center, 287
Intercooling, 295
Internal combustion engine, 188
Internal energy of formation, 482
Internal energy of reaction, 485
International Practical Temperature
    Scale, 19
Inversion curve, 194–95
Irreversibility, 271
Irreversibility, external, 223
Irreversibility, internal, 223
Isentropic efficiency of a feed pump, 379
Isentropic efficiency of a turbine, 382
Isothermal bulk modulus, 434
Isothermal compression efficiency, 292

J

Jet propulsion, 339
Joule, 2, 73, 88
Joule-Thomson coefficient, 195, 434

K

Kelvin-Planck statement, 224
Kelvin scale, 19
Kelvin temperature scale, 227

L

Latent heat of fusion, 143
Latent heat of vaporization, 126
Law of corresponding states, 152
Linde cycle, 406
Liquefaction of gases, 405
Liquid, compressed, 119
Liquid, ideal, 131
Liquid, incompressible, 131
Liquid, saturated, 119
Liquid, subcooled, 119
Liquid, supercooled, 149
Liquid, superheated, 149
Load intensity, 6
Local equilibrium, 167
Lost work, 59, 245

M

Mach number, 215
Macroscopic, 4
Manometer, 12
Mass fraction, 439
Maxwell relations, 421–23
Mean effective pressure, 304
Mechanical efficiency of a pump, 379
Mechanical efficiency of a turbine, 383
Metastable state, 149
Microscopic, 4
Mixture molar mass, 441
Mixtures of ideal gases, 439
Moist air, 446
Moisture, 125
Molar mass, 21
Molecular weight, 21
Mole fraction, 439
Mollier diagram, 349–50
Multi-component system, 117

N

Nernst-Simon theorem, 495
Nozzle, 178
Nucleation site, 149
Number of moles, 439

O

One-dimensional flow, 170
Open feed-water heater, 372
Optimum intermediate pressure, 297
Otto Cycle:
    actual, 300
    air standard, 297
    thermal efficiency, 299
Outer dead center, 287
Outward normal, 210

P

Path function, 67, 111
Perfect gas equation, 20
Performance, 100
Perpetual motion machine, 225
Phase, liquid, 118–24
Phase, solid, 143
Phase, solid and liquid, 143
Phase, vapor, 118–24
Phase diagram, 145
Piston displacement, 287
Point function, 111
Power, 48
Pressure, 6–7
Pressure, absolute, 12
Pressure, atmospheric, 10
Pressure, critical, 123
Pressure, partial, 439
Pressure, ratio, 198
Pressure, reduced, 151
Pressure, relative, 255
Pressure, total, 439
Pressure field, 9
Pressure gauge, 12, 14
Pressure ratio, 198
Pressure regulator, 193
Prime mover, 286
Process, 29
Process, adiabatic, 63, 89
Process, cyclic, 80, 100
Process, irreversible, 219–23
Process, isenthalpic, 195
Process, isentropic, 235
Process, reversible, 219–23
Property, extensive, 5

Property, intensive, 5
Psychrometric chart, 453
p-v-T surface for ideal gas, 23

Q

Quality, 124–25
Quasi-equilibrium, 31
Quasi-static, 29

R

Rankine cycle, 350–54
Rankine cycle, actual, 379–84
Rankine cycle, regenerative, 369
Rankine cycle with extraction, 372–75
Rankine cycle with reheat, 364
Rankine cycle with superheat, 361
Raoult's law, 462
Reciprocating compressor, 288
Reciprocating device, 287
Redlich-Kwong, 157
Refrigerants, 392
Refrigerator, 100, 205–7
Refrigerator, multistage, 397
Refrigerator, vapor compression, 390
Regenerator, 319–20
Regenerator efficiency, 328
Relative humidity, 447
Reservoir, 63
Reversed Brayton cycle, 400–3
Reversible work, 270
Rumford, 2

S

Saturated liquid line, 143
Saturated moist air, 446
Saturated solid line, 143
Saturated state, 143
Saturated vapor line, 122
Saturation dome, 124
Second law, 223
Second Tds equation, 421

Single-component system, 117
Sink, 32
Slug, 3
Solar collector, 178
Solutions, 461
Source, 32
Specific heat, effect of pressure, 88–89
Specific heat, effect of temperature,
    88–89
Specific heat at constant pressure, 85
Specific heat at constant volume, 85
Specific heat at zero pressure, 86–87
Stage of a turbine, 381
Standard dry air, 446
State postulate, 417
State variables, 20
Steam generator, 188
Steam point, 17
Steam tables, 126
Stoichiometric equation, 474
Stroke, 287, 302
Stroke, compression, 302
Stroke, exhaust, 302
Stroke, expansion, 302
Stroke, power, 302
Stroke, suction, 302
Sublimation, 145
Supercritical, 123
Superheater, 188
Surroundings, 28
System, 27
System, closed, 27
System, isolated, 63, 245
System, open, 28

T

Temperature, 16
Temperature, absolute, 19
Temperature, reduced, 151
Temperature, saturation, 119
Theoretical air, 475
Thermal equilibrium, 16
Thermal mass, 318
Thermocouple, 37
Thermodynamic equilibrium, 29
Thermodynamic relations, 417–34
Thermometer, constant volume, 17
Thermometer, mercury-in-glass, 17

Third law of thermodynamics, 495
Throttle condition, 381
Throttle valve, 192–93
Throttling, 194–95
Time rate, 79
Ton of refrigeration, 392
Top dead center, 287
Torr, 12
Torricelli, 11
Triple line, 145
Triple point, 145
Turbine, 178, 181
Two-stroke cycle, 302

U

Ultimate analysis, 472
Unattainability statement, 495
Unavailable energy, 263
Units, CGS, 3
Units, English, 3
Units, SI, 3–4
Universal gas constant, 21
Universe, 28
Unsteady flow, 207

V

Vacuum, 12
van der Waals equation, 157
Van't-Hoff equation, 508
Vapor, saturated, 120
Vapor, superheated, 121
Vapor, wet, 125
Vapor compression refrigeration, 348

Venturimeter, 212
Virial coefficients, 159
Virial equations, 158
Volume, 5
Volume, molal, 22
Volume, reduced, 151
Volume, relative, 255
Volumetric efficiency, 289

W

Waste heat, 101
Wet-bulb thermometer, 451
Work, 40
Work, adiabatic, 93
Work, compressible system, 40
Work, constant pressure, 51
Work, constant temperature, 50
Work, constant volume, 51
Work, elastic, 43
Work, electrical, 47
Work, gravitational, 42
Work, negative, 42
Work, nonuseful, 57
Work, polytropic, 54
Work, positive, 42
Work, specific, 49
Work, spring, 45
Work, surface tension, 47
Work, thermodynamic, 40–41
Work, useful, 57

Z

Zeroth law, 16